Eukaryotic Genes
Their Structure, Activity and Regulation

Eukaryotic Genes
Their Structure, Activity and Regulation

Edited by

N. Maclean, BSc, PhD
Reader in Biology, Department of Biology, University of Southampton

S. P. Gregory, BSc, PhD
Research Assistant, Department of Biochemistry, University College, London

R. A. Flavell, PhD
President, Biogen, Cambridge, Massachusetts

Butterworths
London Boston Durban Singapore Sydney Toronto Wellington

First published 1983

© **Butterworth & Co. (Publishers) Ltd, 1983**

British Library Cataloguing in Publication Data

Eukaryotic genes
    1. Genetics   2. Eukaryotic cells
    I. Maclean, N.   II. Gregory, S. P.
    III. Flavell, R. A.
    574.87'322   QH521

    ISBN 0-408-10824-X

Library of Congress Cataloging in Publication Data

Main entry under title:

Eukaryotic genes
    Bibliography: p.
    Includes index
    1. Cytogenetics. 2. Eukaryotic cells
    I. Maclean, Norman, 1932–   II. Gregory, S. P.
    (Stephen P.) III. Flavell, R. A.
    [DNLM: 1. Genes. 2. Cells. QH 603.E8 E87]
    QH430.E88   1983   574.87'322   83-7533

    ISBN 0-408-10824-X

Photoset by Butterworths Litho Preparation Department
Printed in Great Britain at the University Press, Cambridge

Preface

Those with a particular interest in molecular genetics are surely privileged to be living at this time. The past few years have proved to be a period of immense revolution in genetics, with recent technology in gene manipulation making it possible to study genetics from the bottom upwards. In the past, genetical research was pursued largely by detecting characters expressed in the phenotype and, through observation on the pattern of their inheritance, drawing conclusions about the structure and function of the genes which constituted the underlying genotype. The current revolution has suddenly made it possible to dissect out genes from animals and plants, establish them in cloning vectors, and determine their base sequences directly. From this data, problems of gene regulation and expression can be attacked from the gene upwards rather than from the phenotype downwards. The next decade will undoubtedly see a great expansion in the applied aspects of gene technology, but the new approach will also continue to make important contributions to the basic sciences of genetics and biology.

This book presents the cream of the new information in a relatively condensed format. We have chosen the genes about which most is now known, devoting a chapter to each gene, gene family, or, in some cases, to an entire gene system. To ensure that this information is seen in context, we have also devoted early sections of the book to consideration of chromatin and chromosome structure and to some of the cellular techniques which have proved most useful in exploiting the products of the gene manipulation technology.

We would like to take this opportunity to thank our team of contributors for giving of so much time and trouble in assembling the data and writing the chapters. Of necessity those best fitted by experience to write a chapter in a book of this sort, are also those whose time is highly committed at the laboratory bench. We are also indebted to many of our colleagues for helping us to choose authors of the highest calibre and in sharing in the refereeing and editing of the individual chapters as they came in. To the team at Butterworths, for guidance, determination and patience, we also owe our best thanks.

Norman Maclean
Stephen Gregory
Richard A. Flavell

Contributors

J. M. Adams
The Walter and Eliza Hall Institute of Medical Research, P.O. Royal Melbourne Hospital, Victoria, Australia

H. N. Arst, Jr
Department of Genetics, The University, Newcastle upon Tyne, UK

E. K. F. Bautz
Institut für Molekulare Genetik, Universität Heidelberg, West Germany

A. P. Bird
MRC Mammalian Genome Unit, Edinburgh, UK

P. Borst
Section for Medical Enzymology and Molecular Biology, Laboratory of Biochemistry, University of Amsterdam, Jan Swammerdam Institute, Amsterdam, The Netherlands

M. E. Buckingham
Department of Molecular Biology, Pasteur Institute, Paris, France

S. G. Clarkson
Department of Microbiology, University of Geneva Medical School, Geneva, Switzerland

S. Cory
The Walter and Eliza Hall Institute of Medical Research, P.O. Royal Melbourne Hospital, Victoria, Australia

C. M. Croce
The Wistar Institute of Anatomy and Biology, Philadelphia, Pennsylvania, USA

M. A. DiBerardino
Department of Physiology and Biochemistry, The Medical College of Pennsylvania, Pennsylvania, USA

L. D. Etkin
Department of Zoology, University of Tennessee, Knoxville, Tennessee, USA

A. Flavell
Imperial Cancer Research Fund, Mill Hill Laboratories, London, UK

R. A. Flavell
Biogen Research Corporation, Cambridge, Massachusetts, USA

D. M. Glover
Cancer Research Campaign, Eukaryotic Molecular Genetics Research Group,
Department of Biochemistry, Imperial College of Science and Technology,
London, UK

R. B. Goldberg
Department of Biology, University of California, Los Angeles, California, USA

S. P. Gregory
Department of Biochemistry, University College, London, UK

L. A. Grivell
Section for Medical Enzymology and Molecular Biology, Laboratory of
Biochemistry, University of Amsterdam, Jan Swammerdam Institute, Amsterdam,
The Netherlands

F. G. Grosveld
National Institute for Medical Research, Mill Hill, London, UK

R. Kabisch
Institut für Molekulare Genetik, Universität Heidelberg, West Germany

L. Kedes
Howard Hughes Medical Institute Laboratory and Department of Medicine,
Stanford Medical School and Veterans Administration Medical Center, Palo Alto,
California, USA

L. J. Kleinsmith
Division of Biological Sciences, University of Michigan, Ann Arbor, Michigan,
USA

J. J. Lucas
Department of Microbiology, Health Sciences Center, State University of New
York at Stony Brook, New York, USA

N. Maclean
Department of Biology, University of Southampton, UK

R. Maxson
Howard Hughes Medical Institute Laboratory and Department of Medicine,
Stanford Medical School and Veterans Administration Medical Center, Palo Alto,
California, USA

J. R. Miller
Cancer Research Campaign, Eukaryotic Molecular Genetics Research Group,
Department of Biochemistry, Imperial College of Science and Technology,
London, UK

A. J. Minty
Department of Molecular Biology, Pasteur Institute, Paris, France

T. Mohun
Howard Hughes Medical Institute Laboratory and Department of Medicine,
Stanford Medical School and Veterans Administration Medical Center, Palo Alto,
California, USA

M. J. Pocklington
Department of Biology, University of Southampton, UK

G. E. Rogers
Department of Biochemistry, University of Adelaide, South Australia

G. U. Ryffel
Institut für Genetik, Kernforschungszentrum Karlsruhe, West Germany

A. E. Sippel
Institut für Genetik der Universität zu Köln, Köln, West Germany

G. Stein
Department of Biochemistry and Molecular Biology, University of Florida Medical
School, Gainesville, Florida, USA

J. Stein
Department of Immunology and Medical Microbiology, University of Florida
Medical School, Gainesville, Florida, USA

H. F. Tabak
Section for Medical Enzymology and Molecular Biology, Laboratory of
Biochemistry, University of Amsterdam, Jan Swammerdam Institute, Amsterdam,
The Netherlands

J. O. Thomas
Department of Biochemistry, University of Cambridge, UK

M. T. Vlad
Department of Biological Sciences, University of Warwick, UK

W. Wahli
Institut de Biologie Animale, Lausanne, Switzerland

Contents

Chapter 1

Introduction

Since 1953, when Crick, Watson and Wilkins provided insight into the double helical structure of DNA, the steady snowfall of new knowledge about genes and gene structure has turned into a blizzard, a blizzard that all involved in genetics must endeavour to live with and work in. This book sets out to make life amidst the rapid accumulation of new knowledge easier, to mark a road through the snow. To do so is ambitious and beyond the skills of a single individual. What we have therefore sought to do has been to compile a book, with chapters written by workers who can talk authoritatively in their field, thus providing up to date information in a format which reflects the comparative importance of different levels of information in terms of gene sequences, chromatin and chromosomes.

How can the present blizzard of new information be accounted for? Essentially it results from the high efficiency but comparative simplicity of the genetic code, consisting as it does of triplet codons of only four different bases and coding for little more than twenty different bits of translated information. This coding simplicity, together with the complementary nature of the sequence on the two sister strands of the DNA helix, implies that if ways could be found to locate gene sequences within chromatin and to determine base sequences within the isolated gene, then it would be possible to begin a study of genetics from precisely the opposite end to that which had occupied man's attentions over the centuries. Instead of studying phenotype, familial inheritance, or at best, the expression of genes in their constituent proteins, it has rather suddenly become possible to read out the genetic code itself and attempt to unscramble what is significant in the message so revealed. While it must be acknowledged that interpreting the facts about inheritance from a knowledge of DNA sequence is not by any means straightforward, and that many aspects of genetics will only be effectively understood by knowing the whole story from gene sequence to phenotypic character, yet it cannot be denied that the new approach has introduced a major revolution in the study of genetics and that our understanding of the processes of inheritance has made a considerable advance.

New technology has provided new insight

It is worthwhile to consider what the new techniques have been which have proved so valuable in genetics. Some of the technology is discussed in greater detail in the Introduction to individual sections of this book, and to a lesser extent in some of

1

the chapters themselves. But here it is appropriate to consider the range of techniques which have found application at many different levels in the genetic framework.

(a) at the level of the chromosome

Although the chromosomal location of the genes had been assumed since the turn of the century and known with certainty for a few decades, chromosomes have proved difficult to study. In most cells they were only visible during mitosis and many looked superficially similar. Even the chromosomal location of large blocks of genes could not be readily ascertained as is revealed by the early ambiguity about the chromosomes involved in human mongolism. The great step forward came with chromosome banding. Following on from the application of fluorescence techniques to the staining of chromosomes, it became apparent that particular cytochemical methods involving strong alkali or formamide would also specifically stain the centromeric chromatin. As such C banding became commonplace; it also became clear that somewhat similar techniques, when followed by Giemsa staining, would reveal a banded pattern over the entire chromsome, called the G band pattern. What was particularly useful was that different chromosomes displayed different G band patterns, so permitting for the first time unequivocal identification of particular chromosomes.

Ascribing a gene to a particular chromosome or even a unique locus on a chromosome remains difficult. Such assignment of genes has traditionally depended on linkage studies and measurement of cross-over frequencies where possible, but there are now staining techniques which, in at least some circumstances, can reveal the location of specific genes e.g. the use of silver stain to localize active ribosomal RNA sequences. Genes which are repetitious and occur in large blocks such as the transfer RNA genes can be localized by autoradiography following *in situ* hybridization with radioactive cDNA. Some work has been carried out on the location of unique structural gene sequences by autoradiography but the weak signal obtained renders the interpretation of such experiments with complementary cDNA extremely difficult. A quite different technology which can permit unequivocal assignment of a particular structural gene to an individual chromosome is that of cell fusion and the formation of bispecific hybrid cells. In particular by the exploitation of Sendai virus-mediated cell fusion between mouse and human cells, followed by the culture of the hybrids, often with the gradual loss of most of the human chromosome set, cell lines have been obtained which contain an 'all-mouse' genome with the exception of a single human chromosome. Banding of the chromosomes reveals the identity of the single human chromosome while a prolonged assay for specific protein expression, say of the thymidine kinase gene, permits unequivocal assignment of a particular structural gene to an individual chromosome.

Useful as these advances in whole chromosome technology have been, they are certainly less startling than the developments in the fields of chromatin and actual gene sequence, and these latter areas are chiefly what this volume seeks to consider.

(b) at the level of chromatin

Early theoretical attempts to understand the physical relationship between histone and DNA in chromatin largely assumed that DNA formed an inner core and that the histone molecules, applied to the surface of the double helix, could be regarded as the outer sheath. The true structure of chromatin was initially indicated by the simple but crucial experiments of Hewish and Burgoyne. These Australian workers digested rat liver chromatin with an endogenous calcium activated nuclease enzyme and produced the now familiar 200 base pair repeat ladder on electrophoresis of the DNA fragments from the digest. This result clearly implied that the DNA was in some way differentially protected from digestion, except for certain exposed sites located some 200 base pairs apart. When this observation was put alongside the results of chromatin fine structure analysis in the electron microscope, the notion of a histone bead as a repeating structure in chromatin soon emerged. Nuclear magnetic resonance (n.m.r.) studies soon established that, contrary to the earliest assumptions about histone/DNA conformation, the DNA was wound on the outside of the histone octamer. So was conceived what is now accepted as an accurate picture of the subunit of most chromatin, the nucleosome. Some of the earlier chapters in this book provide an up to date view of the nucleosome and how its structure is implicated with such other aspects of genetics as gene regulation and the arrangement of chromatin in chromosomes.

It is now clear that nucleosomes are the invariable chromatin subunit in all transcriptionally inactive chromatin (except sperm) and probably in most active chromatin as well. Arrangement of nucelosomes in the higher order structure of chromatin is less clear but already exploitation of techniques such as nuclear magnetic resonance have provided persuasive evidence that nucleosomes are arranged as helical solenoids, perhaps each solenoid representing a loop or domain of DNA attached to a backbone or scaffold in the intact chromosome.

(c) at the level of the DNA

It is at the level of the DNA itself that the greatest strides have been made in recent years, thus permitting the study of genetics to advance both from the level of the phenotype inwards to the gene and from the DNA sequence outwards to the expressed character. As outlined in greater detail in the introduction to Section V, a number of techniques have combined to make possible the sudden accumulation of information about gene sequences and their significance.

One area of rapid progress has been the development of methods for determining the sequences of bases along the length of a specific DNA molecule and the gradual speeding up of the process time necessary to operate these methods. For example, in the technique known as direct sequencing, a template DNA for which the sequence is being sought is conjugated with a short primer sequence. Radioactive nucleotides are then provided along with a repair DNA polymerase and the reaction stopped at different sites in parallel experiments by the inclusion of a dideoxynucleoside. The mixture of molecules from each reaction is separated on an acrylamide gel and autoradiography is used to locate the end labelled molecules in the gel ladder.

An alternative method of DNA sequencing by chemical degradation, originally described by Maxam and Gilbert, is also widely used. Both of these approaches, together with many detailed refinements, have permitted the rapid determination of DNA sequences thousands of bases in length.

Rapid sequencing has gone hand in hand with the techniques of gene location following shotgun experiments and production of numerous copies of a particular sequence by cloning. Nucleic acid hybridization experiments devised some years before gene cloning became an established method have also provided a powerful way of determining the complexity and sequence frequency of both cellular RNA and DNA and has revealed such phenomena as the interspersion of unique and repetitious sequences in eukaryotic genomes and, in combination with cloning and complementary DNA production, also revealed the astonishing phenomenon of intron sequences within eukaryotic coding sequences.

The aspect of molecular genetics which has proved most resistant to easy understanding despite the widespread use of computerized scanning of homologous sequences, is that of gene regulation. Information about CAT and TATA boxes and other less defined consensus sequences has rapidly accumulated, but the steps which ensure that a particular gene sequence is transcribed in one cell and not in another remain largely obscure.

The layout of the book

The presentation of the information and discussion in this volume follows the logic of beginning with the more cellular and chromosomal aspects before proceeding to consider some eukaryotic gene systems which are now fairly well understood, at least in terms of structure. Section I, entitled *Gene Regulation*, is devoted mainly to a consideration of chromatin and the way in which DNA and protein modification may possibly be involved in the regulation of gene activity. Here will be found the main consideration of histones, acidic proteins associated with chromatin such as high mobility group (HMG) proteins, ubiquitin and polymerases, and the ways in which the DNA and protein of chromatin form the nucleosome, the higher order arrangement of nucleosomes and ultimately the chromosomes.

Consideration of chromosome structure in Section I then leads one naturally to Section II and to a discussion of the form and function of extranuclear genes in mitochondria and chloroplasts, and to the particular contribution made to our knowledge of chromosome organization and function by the lampbrush chromosomes of vertebrate oocytes and the giant polytene chromosomes of Dipteran insects. Both of these chromosomes have a distinct claim to fame in that both are transcriptionally active whilst cytologically detectable. A detailed presentation of chromosome gross morphology and most aspects of chromosome genetics have been omitted from the book since much of this information is not essential background to the main thrust of the book, which is the structure, activity and regulation of eukaryotic genes.

Section III provides opportunity for discussion of particular techniques not covered in Section IV and involving experimental approaches at a cellular rather than a purely molecular level. Many of these techniques have proved of great potential value and elegance when combined with the methodology of gene cloning and sequencing. Thus it has become possible to assay for individual transcripts with cloned cDNA probes, and also to introduce into cells, in tissue culture, oocytes and in embryos, novel gene sequences which have been isolated from genome libraries and multiplied up by gene cloning.

Section IV forms the major part of the book and represents an in-depth consideration of a number of specific eukaryotic gene systems. Obviously the scope for inclusion of various genes in this section is far greater than space allows us. Inevitably, the chapters found here reflect, to a certain extent, the editors' personal preferences. However, in compiling this section we have attempted to present a broad spectrum of those eukaryotic genes that have revealed most about their structure, activity and regulation. The conclusions drawn from these studies provide a generalized framework of gene action in eukaryotic nuclei into which many other genetic systems will fit. The discussion of mobile genetic elements is included in this section for although they may not be true expressing systems in themselves, transposable elements can have profound effects on the activity of regions of DNA with which they are associated.

As a consequence of the speed and direction of scientific research in recent years, the genes discussed in Section IV are almost exclusively derived from higher animals. To redress the balance and promote the other side of the coin, the book ends with Section V which comprises brief discourses on the molecular genetics of fungi and higher plants. These serve to emphasize the fact that the organization and expression of genetic material in eukaryotes, whether plant, animal or fungi, are subject to common rules in operation. It is the regulations, which give rise to the observed phenotypic differences, that remain largely unknown.

Gene regulation

The chapters in this first section all share the common theme of gene regulation. References to the possible regulatory mechanisms operating at the level of specific gene expression are to be found throughout this volume, especially in Section IV. The chapters presented here, however, are more concerned with those features of eukaryotic chromatin that influence the ability of DNA in general to partake in its functions, particularly with respect to its transcription. The primary characteristics which determine whether a particular DNA sequence is capable of being transcribed or not are likely to be ubiquitous within eukaryotic cells. We know, for instance, that in all interphase nuclei there are regions of chromatin that are transcriptionally active, intermingled with regions where DNA transcription is highly repressed. Usually the latter regions (often referred to as heterochromatin) predominate. The topics addressed by the chapters in this section thus concern the state of the genetic material in eukaryotic nuclei and possible features of this form that may be associated with a division into expressed or non-expressed structures.

In the complex, continually changing environment of the nucleus it is comforting to find some degree of constancy. This attribute resides with the histones, the major protein components of the nucleus. There are five major species of histones and together these occur in an approximate 1:1 ratio with nuclear DNA. This value is more or less independent of tissue or species of origin which suggests that histones play a basic role in the packaging of DNA into chromatin. Our present understanding of the form of this histone–DNA interaction, both in terms of the primary unit of chromatin organization (the nucleosome) and the higher order structures these may adopt, is covered in Chapter 2. The debate as to whether nucleosomes are present in regions of DNA that are being actively transcribed continues, although in at least some instances the evidence for their presence is highly suggestive. This raises the question of whether these nucleosomes are arranged in a specific sequence-dependent (phase) relationship. Nucleosomes in general do not exhibit any DNA sequence specificity, but there are some situations where an ordered distribution of nucleosomes has been observed along the length of a specific gene. This applies to a single tissue type in which the relevant gene may be either active or inactive. If such phasing is shown to be a real phenomenon then it remains to be determined whether it is a primary or secondary effect. Specific binding of other proteins to DNA may force nucleosomes into adopting a phase relationship.

The nature and function of the other proteins in the nucleus, loosely termed the non-histone proteins, form the basis of Chapter 3. It is within this fraction of the nucleus that specific gene regulatory molecules are likely to reside. Different cell types or cells in differing states of growth, development or differentiation show qualitative and quantitative differences in their non-histone protein complements. Some of these proteins exhibit sequence-dependent DNA binding interactions. Others show preferences for single stranded regions in DNA. In addition, there is a group of relatively abundant non-histone proteins, the HMG proteins, which exhibit few differences between cell types. These presumably function in organizing the overall structure of chromatin, possibly by interacting with nucleosomes to generate active or inactive chromatin domains. The activity of a number of these non-histone proteins is dependent on their phosphorylation state, so that protein kinases are also likely to play an integral part in the regulation of genetic expression.

Post-synthetic modification of DNA may also be important and this is discussed in Chapter 4. In this case the modification involves the methylation of cytosine residues, principally at CG base pairs, to yield 5-methylcytosine. From prokaryotic precedents, 5-methylcytosine is known to be important in modulating DNA–protein interactions. It is attractive to speculate that it will also have a modulatory role in eukaryotic DNA. The presence of methylated and unmethylated domains in chromatin and the general positive correlation between undermethylation and transcription of a particular gene sequence are suggestive of a functional role. However, distinguishing cause from effect can prove difficult; as in many aspects of molecular genetics, the available data are conflicting. Absence of methylation is generally associated with the capacity for expression but is not confined to actively transcribed sequences. In this respect it seems to be correlated with the 'open' conformation of chromatin, as defined by sensitivity to DNase I digestion, although here again there are exceptions. The ultimate attraction of this process of cytosine methylation is that it provides a pattern of heritable, but reversible, changes in DNA which could provide an explanation for the maintenance of the committed state of differentiation through subsequent cell generations.

Chapter 2

Chromatin structure and superstructure

J. O. Thomas

Introduction

Essentially all the DNA in a eukaryotic nucleus is complexed with histones in a basic 10 nm diameter chromatin fibre, which may be further folded into a 25–30 nm diameter fibre. The basic fibre (or nucleosome filament) is a linear array of connected nucleosomes which abut each other. Each nucleosome contains about 200 base pairs (bp) of DNA (166–241 bp depending upon source) associated with an octameric protein core comprising pairs of each of four types of histone – the lysine-rich histones H2A and H2B, and the arginine-rich histones H3 and H4 which have been particularly well conserved during evolution. One molecule of the fifth histone, H1, is associated with the outside of each nucleosome, binding partly to the linker DNA between nucleosome cores[50, 51, 114].

There have been several recent reviews of the nucleosome[53] and of chromatin more generally[41, 77], so no attempt is made here to be comprehensive. Rather, some aspects of chromatin structure that are currently of particular interest will be discussed. These include the detailed structure of the nucleosome and nucleosome filament, and the question of whether the nucleosome and its placement has any functional significance over and above its structural significance as a packing unit. Specifically, is there 'phasing' of nucleosomes and DNA sequences? The nature of the next level of folding of the nucleosome filament will be discussed; the stability of such higher order structure could in principle determine whether a particular region of chromatin is rendered accessible for transcription, or masked and inaccessible. Finally our current view of the structure of transcriptionally active chromatin, recently reviewed elsewhere[41, 76, 124], will be summarized.

Structure of the nucleosome

An outline

The nucleosome filament shows periodic differential susceptibility to several endonucleases e.g. endogenous nucleases, micrococcal nuclease, DNase II. Evidently as the DNA leaves the surface of one histone octamer and passes to the next it becomes more sensitive to these enzymes than the DNA protected by histones in the core, and nucleosomes or oligonucleosomes are released as a result

9

of double strand cuts in these regions of 'linker DNA'. Mononucleosomes initially contain a full repeat length of DNA (e.g. 200 bp in rat liver). Continued digestion of these particles by micrococcal nuclease, acting as an exonuclease, reveals barriers to further digestion, (presumably) imposed by histone–DNA interactions. The first such impediment arises when the DNA is trimmed from its full repeat length (e.g. 200 bp) to 166 bp, which occurs without any change in histone composition of the particles. A more substantial barrier arises when the DNA is trimmed to 146 bp (originally assigned as 140 bp), giving a particle depleted of H1, which is a metastable intermediate in digestion and protected from further attack by the octameric histone core. The particles isolated at the two stages of digestion have been termed chromatosomes (166 bp) and nucleosome core particles (146 bp). The digestion of the nucleosome by micrococcal nuclease can therefore be summarized as follows:

Nucleosome	$\rightarrow$	Chromatosome	$\rightarrow$	Nucleosome core particle
200 bp*		166 bp		146 bp
+		+		+
octamer		octamer		octamer
+		+		
H1		H1		

All three particles sediment at about 11–12S. Analysis of their DNA sizes reveals that the broad band observed for the intact nucleosome (width ~60 bp) is sharpened during digestion so that the nucleosome core particle DNA is much less heterogeneous (146±6 bp). This relative homogeneity in DNA content is an important factor in achieving the crystallization of core particles (see below). An external location of the DNA in the nucleosome is indicated both by the accessibility of the DNA along its whole length to DNase I[83], and by neutron scattering studies[37, 86]. The nicking by DNase I at intervals of about 10 nucleotides along each strand is taken to reflect the periodicity of B-form DNA wound around a histone core (see also next section).

X-ray diffraction by single crystals, combined with electron microscopy of crystals, has shown the 146 bp core particle to be a slightly wedge-shaped disc, 11 nm in diameter and 5.5 nm high, containing about 1.75 turns of DNA with about 80 bp per turn[29]. The simplest assumption, in the absence of evidence to the contrary, is that the DNA duplex is smoothly bent around the histone core in a regular superhelix which is left-handed[32, 71]. The presence of a dyad axis in the nucleosome core particle has been demonstrated unequivocally by a higher resolution X-ray diffraction study[28], and by neutron diffraction by single crystals[11].

Extension of the 1.75 turns of DNA in the 146 bp nucleosome core particle by 10 bp at each end would give a particle containing two complete turns of DNA. This corresponds to the 166 bp chromatosome[103] isolated as the first metastable intermediate in the digestion of nucleosomes with micrococcal nuclease. Loss of H1 when chromatosomes are trimmed to core particles suggests that H1 binds to the 10 bp extensions at the ends of the core particle. The presence of H1 gives the nucleosome filament a zigzag appearance in electron micrographs, in contrast with a beads-on-a-string appearance when H1 is absent, and prevents the unfolding of nucleosomes at very low ionic strength (<0.2 mmol/ℓ), at least on the microscope

*or whatever the repeat length of the chromatin e.g. 166–241 bp

grid[112]. The conclusion drawn from all these observations is that H1 seals two complete turns of DNA around the nucleosome and fixes the entry and exit points close together.

H1 has three distinct regions of amino acid sequence[120] which appear to correspond to three 'domains' of structure, namely a central globular region of about 80 residues, which is relatively conserved in amino acid sequence from species to species, flanked by two very basic regions which appear to be flexible and not tightly folded, at least in solution[36]. The globular portion, isolated after tryptic removal of the flanking regions, seems to be sufficient to restore to H1-depleted chromatin a pause in the digestion at 166 bp[3], suggesting that this domain of H1 binds to the two terminal 10 bp regions in the chromatosome, closing two superhelical turns of DNA. For condensation of chromatin the basic flanking regions are required, and these probably interact with the linker DNA, in a manner as yet unclear (see also section on p. 23).

Both the nucleosome core particle and the core histone octamer (see later p. 12) possess a dyad axis of symmetry, raising the possibility that there may be *two* binding sites for H1. Although the number of H1 molecules per nucleosome has been in some dispute, recent measurements have shown that nuclei from several sources contain only sufficient H1 for one H1 per nucleosome on average (reference 8 and references cited therein).

The linking number paradox

The paradox is that although the X-ray analysis (see above) shows that there are two superhelical turns of DNA around the nucleosome, the change in linking number of closed circular DNA extracted from SV40 minichromosomes is only − 1.25 per nucleosome[32]. The change in linking number (where the linking number is essentially the number of times one strand of the DNA duplex crosses another) arises from some combination of changes in the twist and the writhe of the DNA (see reference 22 for a definition of these terms). The number of physical superhelical turns of DNA around the nucleosome (essentially, the writhe) will therefore equal the change in linking number only if the helical periodicity of the DNA (i.e. the twist) is the same for DNA free in solution and DNA associated with histones in the nucleosome. It was pointed out[29] that the paradox might be resolved if the helical periodicity of the DNA was reduced from 10.5 bp per turn in solution to 10.0 bp on the nucleosome, the latter being the value suggested by the DNase I digestion pattern (see above). Other proposals have also been made for resolution of the paradox[106, 131].

DNA in solution has subsequently been found, by two independent methods of measurement[94, 122], to have an average helical periodicity of about 10.6 bp per turn. It would therefore seem that the linking number paradox disappears. However, redetermination of the distance between DNase I cutting sites in chromatin gave a value of 10.4 bp on average[91], rather than the earlier value of 10.0 bp. Thus the periodicity of cutting by DNase I on the nucleosome is essentially the same as the helical periodicity of DNA in solution, and the linking number paradox remains! However, it has been argued[48, 91] that the nuclease cutting periodicity of 10.4 bp on the nucleosome is compatible with a helical repeat of 10.0 bp since access of the relatively large enzyme (DNase I) to the DNA duplex will be restricted by the other superhelical turn of DNA, so that the enzyme does not cut at the otherwise maximally exposed phosphodiester bonds. This explanation would remove the linking number paradox, since the helical periodicity of the DNA (the twist) changes, on binding to the histone octamer, from 10.6 bp to 10.0 bp per turn. X-ray diffraction patterns from nucleosome core particle crystals do indeed show strong intensities at 0.34 and 1.2 nm characteristic of B-form DNA with 10.0 bp per turn.

The detailed arrangement of histones and DNA in the nucleosome

A perspective on current X-ray and neutron diffraction analysis of the core particle

Complete solution of the structure of the nucleosome core particle which has a molecular weight of about 200 000 daltons is a formidable task. The current X-ray diffraction analysis, with crystals diffracting to 0.5 nm in which there is one core particle per asymmetric unit[28], will permit the histones and DNA to be distinguished if isomorphous heavy atom derivatives can be found. At least two such derivatives are required to solve the phase problem. The approach of choice to ensure that crystals of a derivative are isomorphous with those of the unmodified material is to introduce heavy atoms by soaking the crystals in a solution of a suitable reagent (rather than crystallizing material modified in solution). Since the thiol group(s) of H3 appear to be inaccessible in the native state, the other obvious targets for attachment of heavy atoms are the ε-amino groups of lysine residues, if ways can be found to restrict reaction to a limited number of sites, and to avoid disruption of possibly crucial electrostatic interactions of lysine ε-amino groups and phosphates by introduction of bulky groups.

Neutron diffraction studies with single crystals to a resolution of 2.5 nm have been carried out[11] in parallel with the X-ray crystallography[28]. The neutron analysis, with contrast matching of the very different scattering densities of, alternately, protein and DNA by suspension of the crystals in the appropriate D_2O/H_2O mixtures, enables the DNA and protein to be looked at separately. The data corresponding to projections along the three crystal axes are in complete agreement with the results from electron microscopy and X-ray diffraction[28, 29] and with a model in which about 1.8 turns of a DNA superhelix of pitch 2.75 nm and radius 4.2 nm are wound around a protein core. This core has dimensions about 5 × 5 × 6 nm and its projection along each of the three axes of the crystal is very similar to that of an independent three-dimensional reconstruction of the histone octamer described below. If ways can be found to solve the phase problem (analogous to that encountered in X-ray crystallography), future neutron analysis will also aim for a three-dimensional map of the nucleosome core particle.

Image reconstruction analysis of the histone octamer and a model for the nucleosome core particle

The histone octamer is released intact from the DNA in 2 mol/ℓ NaCl at neutral pH or above[115]. During attempts to crystallize the octamer for structural studies, ordered aggregates were generated which proved to be sufficiently regular to permit a complete image reconstruction analysis (for a review of the technique, see reference 23) resulting in a model for the octamer[49]. The octamer is wedge-shaped (*Figure 2.1a*), as is the nucleosome core particle, and has a dyad axis of symmetry. The wedge is 8 nm in diameter in its broadest aspect, and 5.5 nm high at right

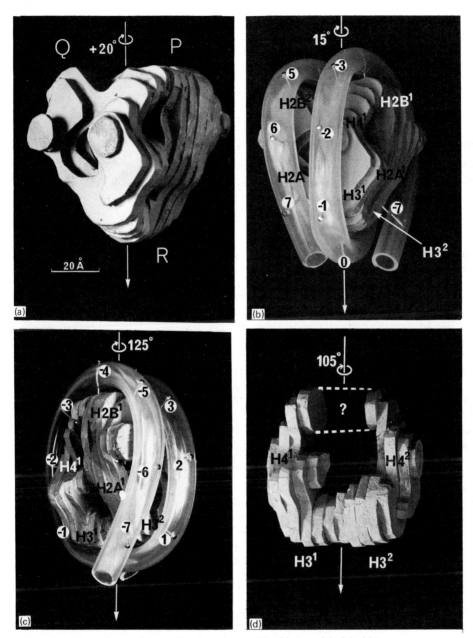

Figure 2.1. The structure of the nucleosome core particle. (a) A model of the histone octamer obtained by image reconstruction analysis. The arrow indicates the dyad axis. P,Q and R are the three ridges (see text). (b) and (c) Two views of the histone octamer structure with two turns of a DNA superhelix wound on it. (The DNA diameter is actually slightly larger than indicated.) Distances along the DNA are indicated by numbers −7 to +7, taking the dyad axis as origin, to mark the 14 repeats of the double helix contained in the 146 bp of DNA in the nucleosome core particle, which correspond to sites of DNase I attack. (d) The (H3)$_2$(H4)$_4$ tetramer dissected out of the octamer model. The region marked ? is a no-man's land that cannot be assigned unambiguously to one or other of the H4 and H2B molecules bordering it. The periphery of the tetramer constitutes about one turn, or somewhat less, of a flat left-handed helix. The views in (a)–(d) are related by rotation about the dyad axis as indicated. Adapted from Klug *et al*[49].

angles, as expected for the histone core of the nucleosome, given that the diameter of the DNA is 2 nm, and that it is wound outside the core. The path that the DNA would occupy is suggested by the presence of the three ridges P, Q and R which, together with the ridges on the sides of the octamer, can be envisaged as forming part of a left-handed helical ramp on which to wind the DNA. Such a path is shown in *Figure 2.1b*, in which the two-fold axes of symmetry in the protein and the DNA coincide. This orientation of histone and DNA is shown to be correct by the neutron diffraction analysis of crystals of nucleosome core particles discussed above, with density matching of, alternately, the protein and DNA. (This is shown in *Figures 6* and *3* of reference 11). Thus the two-fold symmetry of the DNA superhelix allows it to interact along a common two-fold axis with its wedge-shaped histone core which is the major determinant of nucleosome architecture.

The location of the individual histones cannot be distinguished in the structure of the octamer, which is to a resolution of 2.2 nm. However, knowledge of the shape of the octamer allowed the arrangement of histones within it to be deduced by consideration of histone–histone and histone–DNA crosslinking results. The latter identify contacts of particular histone types along the DNA relative to its 5′-radiolabelled end, and essentially give a linear map of histone contacts along the DNA[79]. Neither of the two chemical crosslinking approaches can distinguish the two copies of each histone present, so without further information they alone do not lead to a unique model. It is worth noting that in both approaches crosslinking is to the lysine sidechains in the histones (except in cases of tyrosine crosslinks) so most of the information is likely to be about the basic N-terminal regions ('arms' or 'tails'; see p. 15) which are the most lysine-rich regions. The same is true of a direct visualization approach outlined briefly[108] in which nucleosome cores are reconstituted with one (or two) of the four types of histone chemically modified on ε-amino groups of lysine to provide sites of attachment for platinum atoms, and then visualized directly in the scanning transmission electron microscope.

How is the number of possible arrangements of histones limited by a knowledge of the shape of the octamer? It was reasoned, for example, that the ridge R through which the dyad axis passes (*Figure 2.1a*) is too narrow to accommodate more than one thickness of a histone molecule. Since the two H3 molecules are known to lie within a S–S bond distance of each other[15] they must be on the dyad; and from DNA–histone crosslinking results it is known that H3 makes contacts with the end and the middle of 146 bp core particle DNA. All these considerations taken together place the two H3 molecules along the ridge R. Further reasoning will not be given here (see reference 49), but the assignment of histones in the structure is shown in *Figure 2.1b*. The linear order of Mirzabekov *et al.*[79] may now be made more specific by the addition of superscripts 1 and 2 to distinguish the two copies of each histone (*Figure 2.2*).

In the chromatosome, H1 would bind to the two ends of the DNA, and indeed this is consistent with the major sites of crosslinking in a DNA–histone crosslinking study of 175 bp untrimmed chromatosomes[10]. H1 would also be in close proximity to the $(H3)_2$ dimer in the core, thus accounting for the chemical crosslinking of H1 to H3 in nuclei and chromatin[33]. H1 has an extended shape, at least in solution, and this may account for its crosslinking to H2A in mononucleosomes (for references see reference 49). An extended shape has also been invoked to account for additional crosslinking of H1 at intervals along almost the whole length of chromatosome DNA[10], although the possibility that the basic N- and C-terminal regions of H1 are not tightly bound in the chromatosome under the reaction

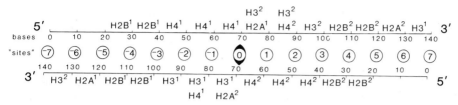

Figure 2.2. Assignment of DNA–histone crosslinks to individual histone molecules. The diagram is adapted from Figure 4 of Mirzabekov *et al.*[79] by addition of superscripts to distinguish between the two copies of each histone type, based on the models in *Figure 2.1*. From Klug *et al.*[49]

conditions, and are therefore free to make transient non-specific contacts with the DNA, should not be overlooked.

Dissection of the octamer: the roles of the arginine-rich and lysine-rich histones

Dissection of the octamer model according to the known dissociation pattern for the octamer in solution, namely into a tetramer $(H3)_2(H4)_2$ and two H2A–H2B dimers, reveals the roles of the two classes of histone (H3, H4 and H2A, H2B) in the organization of the nucleosome. As proposed by Kornberg[50] the tetramer[54] has a critical role. As shown in *Figure 2.1d*, it has the form of a 'dislocated horseshoe' and seems ideally suited to organize a central turn of DNA. Two H2A–H2B dimers each stabilize a further half-turn of DNA above and below the plane of the tetramer (*Figure 2.1b*).

This arrangement of the arginine-rich, and lysine-rich, histones in the nucleosome explains many of the results of nucleosome reconstitution experiments with various combinations of histones and DNA, as well as many of the properties of H3,H4–DNA complexes (see references 49 and 77 for references). In particular it explains why an $(H3)_2(H4)_2$ tetramer can prefold the DNA, at least *in vitro* and possibly *in vivo*, to receive H2A–H2B dimers to complete a nucleosome, whereas the converse is not true. It also explains the observation that a single tetramer protects about 70–80 bp from digestion by micrococcal nuclease in complexes of H3 and H4 with DNA in the absence of H2A and H2B[107], although 146 bp can be folded, albeit loosely, by only one tetramer[118].

The picture of nucleosome assembly that emerges from this analysis is therefore as follows. The $(H3)_2(H4)_2$ tetramer performs a central role in organizing the DNA into the 'central turn' of a primary particle. H2A–H2B dimers then add to this intermediate, stabilizing it by binding firmly to two remaining half-turns of DNA (previously bound loosely in the H3,H4–DNA complex), thereby completing the 166 bp nucleosome core. H1 then binds to the core to seal the two turns, and also provides a binding site for an adjacent length of linker DNA (in one, or both, directions along the DNA from the core), thus completing the nucleosome. Such a physical pathway of assembly would be consistent with the temporal order of deposition of histones on to newly replicated DNA[102, 130]. Nucleosome assembly has recently been reviewed[59].

The role of the basic N-terminal regions of the core histones

The N-terminal third or so of each of the core histone molecules contains a high proportion of the total lysine and arginine residues, and these basic N-terminal regions are likely to be involved in electrostatic interactions with DNA phosphates. That they perform an important structural role in chromatin is suggested by their relatively high amino acid sequence conservation, particularly in H3 and H4, and less so in H2A and H2B[120], and by the site-specific enzymic modifications, again most notably in H3 and H4, that occur in different functional states of chromatin. For example, transcriptionally active regions of chromatin are 'hyperacetylated', and contain increased proportions of the tetra-acetylated forms of H4 and H3 (reviewed in reference 76); cycles of acetylation and deacetylation might be essential for assembly of newly synthesized histones into chromatin and for attainment of the final correct fit[69]. It is therefore of some interest to pinpoint the location of the N-terminal regions in chromatin.

An early observation was that trypsin cleaves about 20–30 amino acid residues from the N-terminus of the four core histones in chromatin. This observation, together with nuclear magnetic resonance (n.m.r.) studies, led to the view that the basic N-terminal regions are flexible and unstructured in free histones and histone complexes (thus becoming known as 'tails' or 'arms' or 'fingers') and that they interact with DNA in the core particle, by encompassing it, thus remaining accessible to attack by trypsin (reviewed in references 51 and 114).

In attempts to assess the role of the N-terminal regions of the core histones in core particle structure, two approaches have been taken. In one, the N-terminal regions are removed with trypsin and the effect on structure tested in various ways; in the other, N-terminal regions are acetylated, usually *in vivo* in tissue culture cells, by butyrate inhibition of the deacetylase. However, although in both cases positive charge is removed from the N-terminal regions, the two situations cannot be equated. Trypsin removes 30–40% of the total positive charge in a histone, including both lysine and arginine residues; on the other hand in a (maximally) tetra-acetylated H4 only about 15% of the positive charge is removed, all from lysine. Caution is thus needed in extrapolating from the roles of tails deduced from studies of trypsinized histones, to modulation of the interaction of tails with DNA by acetylation *in vivo*.

Several observations suggest that the basic N-terminal regions are not essential for folded particles. Trypsinization lowers the sedimentation coefficient of core particles from 11S to 9.2S[65], consistent with about 25% loss of mass but maintenance of a folded structure, and DNase I still cuts at intervals of about 10 nucleotides along the whole length of the DNA. Similar particles can be reconstituted from DNA and a truncated octamer isolated from trypsinized core particles[126]. However the tails do seem to contribute to the stability of the core particle, since removal leads to a decrease in thermal stability, and to an increase in the rate of DNase I digestion[65] – in particular to increased frequency of cutting 20–35 and 60–80 bases from the 5'-ends of the core particle DNA[125]. In the model for the nucleosome core particle (*Figure 2.1b*), these sites are in contact with H2B, and the $(H3)_2$ dimer respectively, suggesting that the 'tails' of these histones do interact with the DNA. However n.m.r. spectra of core particles have been interpreted in terms of bound H3 and H4 N-terminal regions and unbound H2A and H2B regions[18]. The effect of acetylation on core particle structure and stability seems to be rather minimal (see reference 76).

Since some or all of the N-terminal basic regions are not essential in the 146 bp core particle structure, and yet they are relatively well conserved in amino acid sequence between species, and are the sites of specific enzymic modifications which seem to have little effect on the core particle, it is natural to seek a possible role for these regions in the nucleosome filament or in the higher order structure of chromatin. As regards the former possibility the extended arginine-rich N-terminal region of sea urchin sperm H2B has been correlated with the increased length of linker DNA in that tissue[137]. Evidence for the involvement of the basic N-terminal regions in higher order structure is scant, although the possibility that acetylation of these regions might lead to relaxation of such structure to allow gene expression is an attractive one. The effect of deleting the tails has recently been tested[4]. H1 was reconstituted on to H1-depleted chromatin lacking core histone tails (at two molecules, rather than the native one mole of H1 per nucleosome) and the ability to form higher order structures assessed. The conclusion was that the tails are required for folding, but that they act by electrostatic shielding of the DNA charge, since extraneous basic polypeptides could substitute for them. This apparent non-specificity is puzzling in view of the expected specificity in interactions of the tails (see above), but it may reflect only the relative insensitivity of the sedimentation and electron microscopy assays. Indeed electric dichroism, which was also used, suggested that the folded state achieved without core histone tails and with extraneous polypeptides was not identical with the native folded structure.

Nucleosome spacing in chromatin

When nuclei from different sources were digested with micrococcal nuclease it became apparent that the nucleosome repeat length differed in different species, tissues, and cell types within a tissue. The range of reported repeat lengths is from about 165 bp in yeast and cortical neurones to about 240 bp in sea urchin sperm (for reviews see references 19, 51, 114). In addition the repeat length may change during development[12, 27, 123]. The cause of these changes and their structural and functional consequences remain unclear.

Whatever the repeat length, core particles containing about 146 bp of DNA are generated on extensive digestion with micrococcal nuclease, so the variation is in the length of linker DNA. Assuming that in each case the two-turn chromatosome is the minimum common particle (see section *Structure of the Nucleosome*, p. 9), the linker DNA, which connects chromatosomes, varies from essentially zero in yeast and cortical neurons up to about 80 bp in sea urchin sperm. It is not clear what determines the length of linker DNA, although H1 and H2B have both been suggested[80, 137]. Another suggestion is that variations in linker length among different cell types could result from differences in histone to DNA ratio at the time of synthesis and assembly of chromatin[90].

The bands obtained on gel electrophoresis of the DNA after micrococcal nuclease digestion of chromatin are relatively broad. This could be due either to variation in the length of linker DNA within a cell type or to variation in the site of cleavage by micrococcal nuclease along the linker DNA, the former possibility being favoured by a statistical analysis[68]. That the linker length does indeed vary *within* a cell type has been demonstrated directly. Nucleosome dimers obtained by micrococcal nuclease digestion of rat liver chromatin were trimmed down to the position corresponding to 166 bp in chromatosomes, by digestion with exonuclease III and S1 nuclease, and dinucleosomes containing their full complement of H1 were selected for analysis (to avoid any contribution from sliding in H1-depleted dimers)[90]. The distribution of DNA sizes in the trimmed dimer was broad, indicating a heterogeneity in linker length. The range was from about zero to 80 bp, the mean value of 43 ± 3 bp being roughly as expected for nucleosomes from rat liver, given that the bulk repeat length in nuclei is about 200 bp. These experiments show convincingly that linker lengths vary within a cell type and do not reveal any obvious quantization or clustering of linker lengths on a large scale. However, although in many cases the nucleosome repeat length of specific genes is the same as the bulk repeat (e.g. references 9, 13, 16, 127), 5S RNA genes in *Xenopus* erythrocytes appear to have a repeat length shorter than that of the bulk chromatin[34, 40].

Several observations do, however, suggest that linkers may be quantized in integral multiples of about 10 bp, although in some cases the interpretation must for the moment be tentative. DNase I makes single strand cuts at about 10 base intervals over a DNA length longer than a single repeat length[67], (although this pattern was most pronounced in yeast, with zero-length linker, of the three sources examined and might therefore arise from regions of the chromatin with cores in contact). Further, extensive digestion of dinucleosomes with exonuclease III appears to proceed in steps of 10 bases[95] (but in this case there is a possibility of loss of H1 and sliding together of cores during the extensive digestion). Finally there is some evidence that micrococcal nuclease cuts with a 10 bp periodicity between 146 bp and the full DNA repeat length in chromatin[2, 113].

Nucleosome phasing?

The notion of specific 'phasing'[52] requires the specific placement of histone octamers with respect to DNA sequences in all cells of the same type. This would be an attractive way of allowing selective exposure of particular DNA sequences (e.g. promoters) by their location in the more accessible linker region between the cores of adjacent nucleosomes. Alteration of the phase would then ensure exposure of different sequences in different tissues; histones, through their specific placement, might, therefore, after all, directly influence gene expression. Clearly, to allow for a different phase in different cells or tissues of the same organism (indicated by different average nucleosome DNA repeat lengths), the phase would have to be set by something other than the genomic DNA sequence, e.g. the sequence-specific binding of some other protein. Over the past two or three years there has been intensive effort to establish whether phasing exists, and the experiments and their interpretation have generated considerable debate (for reviews see references 52, 136).

Early experiments with SV40 minichromosomes suggested a random location of nucleosomes with respect to DNA sequences[21, 87], although in polyoma virus nucleosomes appeared to occupy a number of possible alternative defined locations[88]. The first test for phase in cellular DNA, which was for single-copy DNA sequences in rat liver, indicated no bulk phasing[89]. It was subsequently deduced that there was no specific phasing of nucleosomes on 5S rRNA genes in rat liver, based on the representation of 5S gene sequences in nucleosome core particles[5]. When core particle DNA was hybridized to radioactive RNA and unprotected 5S RNA removed by digestion with ribonuclease, the protected RNA was not of the unique length expected from specific phasing, but was of all lengths up to 146 bases, much as expected for a random arrangement of nucleosomes on DNA sequence.

No case of bulk phasing has been reported. There have, however, been many recent reports that the arrangement of nucleosomes with respect to *particular* DNA sequences is not random, but that nucleosomes occupy preferred, and in some cases even unique, positions. The first report, for tRNA genes in chicken embryos[127], was followed by similar reports for other genes[13, 35, 63, 70, 98, 133, 134] and for satellite DNA in rat[42]. An early claim for precise phasing of nucleosomes in the α-satellite DNA of African green monkey cells[81] does not appear to be valid because of the sequence specificity of micrococcal nuclease[31] (see also below).

In many of the experiments that have been interpreted in terms of phasing, the 'indirect end labelling' approach has been used[83, 132]. Chromatin is digested very lightly with micrococcal nuclease, and the DNA is then extracted and digested to completion with a restriction enzyme that cuts within the sequence of interest at one or very few sites. Southern blots of these digests are then hybridized with a restriction fragment probe for the sequence of interest that abuts the restriction site in the original DNA. A defined set of bands (rather than a smear) is taken to indicate precise locations of nucleosomes with respect to the micrococcal nuclease cleavage site. However, with the recent demonstration of the marked sequence specificity of micrococcal nuclease[25, 38] the interpretation of some of these experiments, which depend critically on the assumption that micrococcal nuclease cleaves DNA essentially randomly with respect to DNA sequence, is thrown into some doubt. Control digests of free DNA have to be carried out, and since DNA and chromatin may be digested at very different rates some effort must be made to examine digestion patterns at various times before it can reasonably be concluded that the digestion pattern of chromatin is truly different from that of DNA. If the pattern is the same as that of naked DNA, the simplest explanation is that the arrangement of nucleosomes is indeed random, as observed in some cases (e.g. references 9, 13).

In one case that does suggest, with reasonable certainty, non-random positioning of nucleosomes with respect to DNA sequence, that of the tRNA genes in chicken embryos[127], the indirect end labelling approach was not used. By selecting tetranucleosomes from a fairly extensive micrococcal nuclease digest as starting material to be enriched for tRNA genes, the effect of sequence specificity of the enzyme was diminished and the nucleosomal nature of the starting material established. Three tRNA genes were found to occur in a small number of alternative positions on the DNA sequences, spaced with the nucleosome repeat periodicity of bulk chromatin, and with the start of the tRNA structural gene about 20 bp within the nucleosome. However this phased association of nucleosomes did not extend further than about five nucleosomes. In a more recent study[128] the same workers described a reconstitution system for site-directed nucleosome assembly on a cloned tRNA gene isolated from tetranucleosome DNA, and concluded that transcription of the gene depended on nucleosome position.

These studies provide evidence for a non-random location of nucleosomes and the interpretation is that this plays a functional role[127, 128]. An alternative interpretation of the data is that a non-histone protein involved in gene function imposes a degree of local order on the arrangement of nucleosomes and that this order serves no particular purpose[52]. The essence of the argument here is that the sequence-specific binding of a (regulatory) protein, or even the sequence-specific exclusion of nucleosomes[55], would form a boundary for a region within which nucleosomes are deposited at random. A statistical model for deposition of histones on to such a region of DNA predicts a pattern of nucleosome location which has the observed periodicity and which falls off 5–10 nucleosomes from the boundary[110], in keeping with the observations for chicken tRNA genes mentioned above. In the case of the reiterated tRNA gene clusters of *Xenopus*, if a sequence-specific protein bound once every 3.18 kb cluster (containing about 16 nucleosomes) no nucleosome would be more than eight nucleosomes away from such a boundary[13].

To summarize, the picture that emerges is one in which there is certainly no widespread positioning of nucleosomes on particular DNA sequences. There may be short range order in specific regions, arising from the effect of sequence-specific binding proteins or even occasional sequence specificity of the histones themselves[20]. Possible functional roles for such short range order include effects on initiation of transcription, and on packaging of non-transcribed DNA[13], but it may also be that the order observed is a secondary effect that serves no further purpose at all.

Higher order structure of chromatin: folding of the nucleosome filament

Introduction: Continuous or discontinuous folding?

The next level of folding of the nucleosome filament is taken to be the thick fibre of 25 nm diameter seen in electron micrographs of metaphase chromosomes and interphase nuclei in the absence of chelating agents (see, for example, references 96, 97). The nature of this higher level of folding has been the subject of some debate but there is general agreement that it requires H1. The first proposal, based on electron microscopy of condensed chromatin fragments, taken together with electron micrographs of thin sections of erythrocyte nuclei[24], was that the 25 nm fibre is a contact helix or 'solenoid' with six nucleosomes per turn, and a pitch of about 10 nm now known to correspond to the diameter of a disc-shaped nucleosome[30]. An alternative view, based partly on electron microscopy of chromatin released from nuclease-digested nuclei in 50–80 mmol/ℓ NaCl, and partly on the presence of peaks in sucrose gradients, was that the 25 nm fibre was discontinuously arranged in 'superbeads', clusters of six to twelve nucleosomes[39, 93, 109]. However, the case for such an organization has been weakened by reports that the apparent presence of superbeads and multiples thereof in sucrose gradients may, at least in some cases, be due to contamination of the chromatin preparation with ribonucleoprotein particles[82, 121]. The same seems likely to be true of similar globular structures termed 'nucleomers' and multinucleomers, stabilized by Mg^{2+} ions[47]. A single superbead of defined composition could in principle be interpreted as an element (in the disassembly) of an essentially helical structure, perhaps punctuated in some way, but in view of its apparent heterogeneity the possible structural and functional significance of the superbead seems suspect.

Physical studies of the folding of the nucleosome filament

A common approach has been to study reconstitution of the folded structure by addition of sodium chloride or magnesium ions to extended chromatin prepared at low ionic strength. Electron microscopy of samples fixed at a range of ionic strengths from 1 to 100 mmol/ℓ[7, 112] show gradual compaction of the zigzag arrangement of nucleosomes (see p. 10) to a fairly regular fibre of 25–30 nm diameter, interpreted as a solenoid, at ionic strength 60 mmol/ℓ and above. At about 40 mmol/ℓ the appearance of the fibre is reminiscent of superbeads (see *Figure 6c* of reference 7), and it therefore seems likely that 'superbeads' are 'detached' (non-interacting) turns of a solenoid structure.

Measurements of sedimentation coefficients confirm a gradual compaction with increasing ionic strength[7,14], and study of a range of chromatin fragment sizes suggests additional details of the higher order structure. In a series of short fragments, a change in behaviour at hexanucleosomes suggests that six nucleosomes are required to form a complete unit of higher order structure, and that this occurs at ionic strength 25 mmol/ℓ[7,14]. Discontinuities in behaviour at about this oligomer size were also found in another sedimentation study[85] and in measurements of electrical birefringence[74] and diffusion coefficient[73]. The absence of any further discontinuity in the sedimentation behaviour of increasingly larger fragments up to 50 nucleosomes suggests strongly that the folding process is continuous, as would be expected of helical condensation (for example, into a solenoid) and not discontinuous as would be expected of an array of superbeads[14]. Other studies also favour a solenoidal model with 6–7.5 nucleosomes per turn[62,78,111].

A second discontinuity in behaviour is observed for very large fragments (weight average size >50 nucleosomes for rat liver, or >60 nucleosomes for chicken erythrocytes), namely a jump in sedimentation coefficient at ionic strength 45 mmol/ℓ as the ionic strength is increased. This jump, which suggests a cooperative transition to a more compact structure, is consistent with the onset of stabilizing interactions between the turns of a solenoid at this ionic strength, individual turns apparently already being present at ionic strength 25 mmol/ℓ[7,14,116]; the structure can be 'locked' in its compact form, which survives low ionic strength by brief treatment with a lysine–lysine crosslinking reagent above ionic strength 45 mmol/ℓ, but not below[14]. The jump is observed only for long fragments, probably because such molecules are more susceptible to hydrodynamic shear than are short fragments, and require stronger interactions between turns to withstand disruption of the helix. In other words, if solenoids of over 50 nucleosomes (~8 turns) from rat liver chromatin are subjected to decreasing ionic strength, they are disrupted by shear at ionic strength 45 mmol/ℓ and below.

Since the jump in sedimentation coefficient appears to reflect the 'break point' of a solenoid as the ionic strength is decreased, it may prove to be a useful parameter for comparison of the stability of higher order structures of chromatins from different sources, e.g. with different repeat lengths and in different states of acetylation or phosphorylation. Indeed the critical nucleosome oligomer size for the onset of the jump is 10 nucleosomes (almost two turns of solenoid) longer in the case of chicken erythrocyte chromatin than it is for rat liver chromatin, consistent with a more stable higher order structure for the transcriptionally inactive chicken erythrocyte chromatin[7]. Since unravelling of the higher order structure is almost certainly a prerequisite for gene expression, an understanding of the factors that determine the stability of higher order structures is likely to be an important step in understanding the relation of chromatin structure to gene activity.

The arrangement of nucleosomes in the solenoid

It has been proposed on the basis of electron microscopy[112] that the disc-shaped chromatosomes are arranged in the solenoid with their faces projecting radially from the solenoid axis, and aligned roughly parallel to the axis, as shown in *Figure*

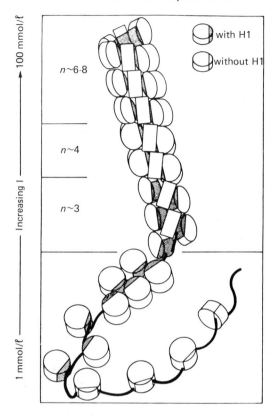

Figure 2.3. A model for chromatin condensation. An
idealized drawing of helical superstructures formed by
chromatin containing H1 with increasing ionic strength.
The open zigzag of nucleosomes (bottom left) closes up
to form helices with increasing numbers of nucleosomes
per turn (n). The solenoid formed at high ionic strength
probably has six nucleosomes per turn (see text). When
H1 is absent (bottom right), no zigzags or definite
higher order structures are found. From Thoma *et al.*[112]

2.3. Chromatosomes in neighbouring turns are in edge-to-edge contact, thus giving
the solenoid a pitch of 10 nm and probably accounting for the 10 nm reflection in
the X-ray pattern of folded chromatin fibres[105]. The results of electric dichroism
studies[62, 78, 135] and neutron scattering studies[111] are largely in accord with this
picture, as are earlier laser light-scattering measurements (see reference 112). The
dichroism measurements, carried out on Mg^{2+}-condensed solenoids[78] or on
solenoids locked in the compact state by first crosslinking with dimethyl
suberimidate at high ionic strength[61, 62], lead to slightly different values for the
angle by which the chromatosome faces are tilted from the solenoid axis, possibly
because of the difference in sample preparation (e.g. see reference 61).

Precise information on many of the details of the solenoid are still lacking, for
example on the location of linker DNA and of histone H1, and whether the central
hole, whose diameter is roughly that of a nucleosome, is 'empty' or not. However,
some suggestions have been made (e.g. references 78, 112, 129).

A role for histone H1 in higher order structure?

There is abundant evidence to suggest that H1 is essential for the formation of higher order structures (e.g. references 14, 112 and references therein). H1 molecules on adjacent nucleosomes are in close proximity, as shown by the formation of H1 polymers on crosslinking nuclei and chromatin with a variety of reagents, whereas no such polymers are observed for H1 free in solution (see reference 117 and references therein, and reference 43). A polymer of H1 molecules is an important feature of the model of Thoma, Koller and Klug[112] (*Figure 2.3*), and moreover it is suggested that it is aggregation of H1 into a helix that controls solenoid formation, the chromatosomes essentially following passively. Evidence for a conformational change in H1 at ionic strength ~20 mmol/ℓ[104], and for a change from a non-cooperative to a cooperative interaction with DNA at ionic strength 20–40 mmol/ℓ[92] lends support to this idea.

The formation of crosslinked H1 polymers for chromatin in the extended state, and for short oligonucleosomes, shows that H1 molecules are already in proximity or in contact *along* the nucleosome filament and not exclusively in the higher order structure[117]. However, there could well be additional contacts between H1 molecules in the folded state, and to determine whether this is so detailed dissection of the crosslinked polymers produced in various states of compaction of chromatin is needed. One reasonable possibility is that H1 molecules on adjacent nucleosomes are arranged head-to-tail (i.e. N-terminal basic region to C-terminal basic region) along the nucleosome filament, each H1 interacting with a nucleosome core through its central globular region and having its two basic regions extending along the linker in opposite directions towards the two adjacent nucleosomes. Increasing ionic strength might cause aggregation of the globular regions, possibly by hydrophobic bonding. In the condensed state new interactions (e.g. of the core histone tails, see p. 16) might occur, resulting in stabilization – but without further evidence there is little point in speculating. However, it seems reasonable to expect that if there are indeed H1–H1 contacts unique to the folded state, and important for its stability, higher order structures of different stability are likely to result from the presence of the different H1 subtypes produced, for example, at different developmental stages. In this context, the idea of 'lattices' of H1 subtypes, giving rise to regions of differential stability along a chromsome is an attractive one. However the distribution of H1 subtypes in the chromosome is not yet known.

Finally, a possibility that cannot be discounted is that H1 has no distinct role in higher order structure *per se*. Rather it may act only at the level of the nucleosome, stabilizing two complete turns of DNA, and neutralizing the charge on the linker, the solenoid being generated or stabilized through interactions between the core histones of different nucleosomes.

Exchange of H1 between chromatin fragments

The ability of H1 molecules to migrate between binding sites on different chromatin fragments at ionic strengths 30–40 mmol/ℓ and greater is a factor that should be borne in mind in any structural studies of H1 binding and distribution. When two chromatin fragments of different length, one radiolabelled *in vivo* with ^{14}C-lysine and the other unlabelled, are mixed at ionic strength 75 mmol/ℓ and then separated again, the 'hot' and 'cold' H1 (but none of the core histones) are found to have equilibrated between the two fragments[17]. Exchange can first be detected at an ionic strength as low as 30 mmol/ℓ. H1 on rat liver chromatin and H5 of chicken erythrocyte chromatin exchange under similar conditions[119], and since H5 may be regarded as an extreme H1 variant, more homologous H1 variants are also likely to exchange under these conditions *in vitro*.

At ionic strength 75 mmol/ℓ chromatin exists in the compact form. However there need be no contradiction between the phenomenon of rapid exchange of H1 between sites and the existence of H1-dependent higher order structures, since the overall structure is likely to be maintained provided most of the H1 binding sites are occupied most of the time.

Folding of the 25 nm fibre in metaphase chromosomes

The 25 nm fibre must undergo further coiling or folding in metaphase chromosomes, where the packing ratio for the DNA is estimated as about 10^4, compared with about 40 for a solenoid with six nucleosomes per turn. The difficulties in defining the nature of the further levels of folding are considerable. The metaphase chromosome is such a dense mass that conventional electron microscopy gives little information; substructure is visible only when the chromosome is swollen and thus already partially disrupted. Scanning electron micrographs have been interpreted in terms of further helical coiling of thick fibres in metaphase chromsomes[101], as have light and electron micrographs of fragmented metaphase chromosomes[6].

Continuing studies by Laemmli and his coworkers support the idea of loops of the 25–30 nm fibre, containing 50–100 kb of DNA, constrained by a protein scaffold during metaphase. The suggestion of a radial arrangement of loops attached to a central protein scaffold[57] receives some support from electron micrographs of transverse sections through swollen chromosomes[1, 75]. The radial loop model and the helical coil model for the compaction of the 25 nm fibre in the metaphase chromosome are compared in *Figure 7* of reference 75. The 'scaffold' in chromosomes prepared by improved methods has very recently been shown to consist of two major proteins (170 000 and 135 000 daltons) and to require Cu^{2+} or Ca^{2+} ions for stability[64]. A metalloprotein structure involving either Cu^{2+} or Ca^{2+} is also involved in the organization of the DNA in histone-depleted structures from interphase nuclei[60]. The looped nature of the DNA at interphase (for reviews see references 41, 114) has led to the attractive idea that chromatin 'domains' may be significant functional or structural units[41].

Active chromatin

The properties of transcriptionally active chromatin have recently been extensively reviewed[41, 76, 124] and no attempt will be made here to cover this ground again. However, to summarize, the fairly generally held view is that transcribed sequences are packaged in nucleosomes whose linkers are relatively susceptible to micrococcal nuclease, and that these nucleosomes contain hyperacetylated core histones and possibly undermethylated DNA, and are associated with particular non-histone proteins, namely the so-called high mobility group proteins, HMG14 and 17 (or, in trout testis, the homologous protein H6) which appear to be responsible for the DNase I sensitivity of DNA sequences that are capable of being transcribed in a particular cell type.

While many of these points seem well supported, some aspects may need to be re-evaluated in light of the following:

1. It is not clear whether HMG14 and 17 are exclusively located in active chromatin; all nucleosomes will in fact bind HMG14 and 17 in reconstitution experiments *in vitro*, up to a level of two per nucleosome[72, 99]. The HMG proteins and their possible role in chromatin are the subject of an entire recent volume[44] and are also discussed in Chapter 3 of this book.

2. Although nuclease digestion studies have suggested that genes presumed to be active in a particular tissue are packaged in nucleosomes, electron microscopy of active genes has given mixed results, showing beaded chromatin fibres in some cases and smooth fibres in others (reviewed in references 41, 76, 114). A difficulty with the electron microscopic studies lies in distinguishing nucleosomes from RNA polymerase molecules. A recent systematic study of active nucleolar chromatin from *Xenopus* oocytes provides an explanation for many of the earlier observations and suggests that ribosomal genes are free of nucleosomes and other structures except for RNA polymerase molecules[56].

3. Although some recent reports again suggest that genes which are assumed to be active yield a typical nucleosomal pattern of DNA fragments on digestion with micrococcal nuclease[13, 16, 66], it is becoming increasingly clear, with detailed analysis of micrococcal nuclease digestion patterns, that at least some actively transcribed sequences may not be packaged in a regular nucleosomal array, e.g. heat shock genes[46, 63, 134] and the ovalbumin gene[9] (see below).

4. There is an apparent absence of nucleosomes at the 5'-end flanking region of many actively transcribed genes. This region is often about an order of magnitude more sensitive to DNase I than active sequences generally[133] (for further references see references 9, 26, 41). In many cases the 5'-flanking region is also hypersensitive to micrococcal nuclease (e.g. reference 46). One possible reason for the hypersensitivity is suggested by electron microscopy which reveals a nucleosome-free region, several hundreds of base pairs long, upstream from transcribed sequences in some cases[58, 100]. Other proteins may well be bound at the nucleosome-free regions and it is obviously tempting to wonder whether, being located at the 5'-flanking region of a transcribed sequence, they will have a role in gene regulation. Bellard *et al.*[9] studied the chromatin structure of about 3000 bp of DNA at the 5'-end flanking region of the active ovalbumin gene, using the indirect end-labelling approach (see *Nucleosome Phasing?* on p. 18) with probes for ovalbumin sequences. They propose binding of proteins of some kind to explain disappearance of bands seen in naked DNA, and possibly acquisition of some single strand character of the DNA to account for additional cleavages.

Further evidence will be needed before any general statements can be made with certainty about the organization of transcribed sequences in nucleosomes, but the recent nuclease digestion results described above lend weight to earlier conclusions drawn by microscopists that nucleosomes might be absent from at least some types of transcription unit. Taking at face value the rather extensive literature on mononucleosomes enriched in transcribed sequences and containing HMG proteins, acetylated histones, etc., as outlined above, it seems unlikely that all transcriptionally active chromatin is always free of nucleosomes. However, stringent characterization of putative active nucleosomes is now called for in any future work. In this context a complete characterization of the 'A particle' produced by micrococcal nuclease digestion of *Physarum* nuclei, and enriched in transcribed sequences[45] will be of interest.

Conclusion

An attempt has been made in this chapter to summarize the present position in several areas of current interest relating to chromatin structure and the role of the histones. Limitations of space have meant that some other major areas in which histones are involved, e.g. replication, have not been discussed at all.

On the structural front in the next few years, we can hope to see a complete high-resolution analysis of the nucleosome core particle, to understand the relationship between one nucleosome and the next in higher order structure, and to clarify the way in which sequences are exposed for transcription in active chromatin. We may also hope to understand in more detail the structure of the metaphase chromosome, and the processes that underlie the reversible condensation and decondensation of chromosomes during the cell cycle.

References

1. Adolph, K. W. (1980). Isolation and structural organization of human mitotic chromosomes. *Chromosoma*, **76**, 23–33
2. Allan, J., Cowling, G. J., Harborne, N., Cattini, P., Craigie, R. and Gould, H. (1981). Regulation of the higher order structure of chromatin by histone H1 and H5. *Journal of Cell Biology*, **90**, 279–288
3. Allan, J., Hartman, P. G., Crane-Robinson, C. and Aviles, F. X. (1980). The structure of histone H1 and its location in chromatin. *Nature, London*, **288**, 675–679
4. Allan, J., Harborne, N., Rau, D. C. and Gould, H. (1982). Participation of core histone 'tails' in the stabilization of the chromatin solenoid. *Journal of Cell Biology*, **93**, 285–297
5. Baer, B. W. and Kornberg, R. D. (1979). Random location of nucleosomes on genes for 5S rRNA. *Journal of Biological Chemistry*, **254**, 9678–9681
6. Bak, A. L., Zeuthen, J. and Crick, F. H. C. (1977). Higher-order structure of human mitotic chromosomes. *Proceedings of the National Academy of Sciences, USA*, **74**, 1595–1599
7. Bates, D. L., Butler, P. J. G., Pearson, E. C. and Thomas, J. O. (1981). Stability of the higher-order structure of chicken erythrocyte chromatin in solution. *European Journal of Biochemistry*, **119**, 469–476
8. Bates, D. L. and Thomas, J. O. (1981). Histones H1 and H5: one or two molecules per nucleosome? *Nucleic Acids Research*, **9**, 5883–5894
9. Bellard, M., Dretzen, G., Bellard, F., Oudet, P. and Chambon, P. (1982). Disruption of the typical chromatin structure in a 2500 base pair region at the 5' end of the actively transcribed ovalbumin gene. *EMBO Journal*, **1**, 223–230.
10. Belyavsky, A. V., Bavykin, S. G., Goguadze, E. G. and Mirzabekov, A. D. (1980). Primary organization of nucleosomes containing all five histones and DNA 175 and 165 base pairs long. *Journal of Molecular Biology*, **139**, 519–536
11. Bentley, G. A., Finch, J. T. and Lewit-Bentley, A. (1981). Neutron diffraction studies on crystals of nucleosome cores using contrast variation. *Journal of Molecular Biology*, **145**, 771–784
12. Brown, I. (1978). Postnatal appearance of a short DNA repeat length in neurons of the cerebral cortex. *Biochemical and Biophysical Research Communications*, **84**, 285–292
13. Bryan, P. N., Hofstetter, H. and Birnstiel, M. (1981). Nucleosome arrangement on tRNA genes of *Xenopus laevis*. *Cell*, **27**, 459–466
14. Butler, P. J. G. and Thomas, J. O. (1980). Changes in chromatin folding in solution. *Journal of Molecular Biology*, **140**, 505–529
15. Camerini-Otero, R. D. and Felsenfeld, G. (1977). Supercoiling energy and nucleosome formation: the role of the arginine-rich histone kernel. *Nucleic Acids Research*, **4**, 1159–1181
16. Camerini-Otero, R. D. and Zasloff, M. A. (1980). Nucleosomal packaging of the thymidine kinase gene of herpes simplex virus transferred into mouse cells: an actively expressed single copy gene. *Proceedings of the National Academy of Sciences, USA*, **77**, 5079–5083
17. Caron, F. and Thomas, J. O. (1981) . Exchange of histone H1 between segments of chromatin. *Journal of Molecular Biology*, **146**, 513–537
18. Cary, P. D., Moss, T. and Bradbury, E. M. (1978). High-resolution proton magnetic resonance studies of chromatin core particles. *European Journal of Biochemistry*, **89**, 475–482
19. Chambon, P. (1977). Summary: the molecular biology of the eukaryotic genome is coming of age. *Cold Spring Harbor Symposium on Quantitative Biology*, **42**, 1209–1234

20. Chao, M. V., Gralla, J. and Martinson, H. G. (1979). DNA sequence directs placement of histone cores on restriction fragments during nucleosome formation. *Biochemistry*, **18**, 1068–1074
21. Crémisi, C., Pignatti, P. F. and Yaniv, M. (1976). Random location and absence of movement of the nucleosomes on SV40 nucleoprotein complex isolated from infected cells. *Biochemical and Biophysical Research Communications*, **73**, 548–554
22. Crick, F. H. C. (1976). Linking numbers and nucleosomes. *Proceedings of the National Academy of Sciences, USA*, **73**, 2639–2643
23. Crowther, R. A. and Klug, A. (1975). Structure analysis of macromolecular assemblies by image reconstruction from electron micrographs. *Annual Review of Biochemistry*, **44**, 161–182
24. Davies, H. G. and Haynes, M. E. (1976). Electron microscope observations on cell nuclei in various tissues of a teleost fish: the nucleolus associated monolayer of chromatin structural units. *Journal of Cell Science*, **21**, 315–327
25. Dingwall, C., Lomonossoff, G. P. and Laskey, R. A. (1981). High sequence specificity of micrococcal nuclease. *Nucleic Acids Research*, **9**, 2659–2673
26. Elgin, S. C. R. (1981). DNase I-hypersensitive sites of chromatin. *Cell*, **27**, 413–415
27. Ermini, M. and Kuenzle, C. C. (1978). The chromatin repeat length of cortical neurons shortens during early postnatal development. *FEBS Letters*, **90**, 167–172
28. Finch, J. T., Brown, R. S., Rhodes, D., Richmond, T., Rushton, B., Lutter, L. C. and Klug, A. (1981). X-ray diffraction study of a new crystal form of the nucleosome core showing higher resolution. *Journal of Molecular Biology*, **145**, 757–769
29. Finch, J. T., Lutter, L. C., Rhodes, D., Brown, R. S., Rushton, B., Levitt, M. and Klug, A. (1977). Structure of nucleosome core particles of chromatin. *Nature, London*, **269**, 29–36
30. Finch, J. T. and Klug, A. (1977). Solenoidal model for superstructure in chromatin. *Proceedings of the National Academy of Sciences, USA*, **73**, 1897–1901
31. Fittler, F. and Zachau, H. G. (1979). Subunit structure of α-satellite DNA containing chromatin from African green monkey cells. *Nucleic Acids Research*, **7**, 1–13
32. Germond, J. E., Hirt, B., Oudet, P., Gross-Bellard, M. and Chambon, P. (1975). Folding of the DNA double helix in chromatin-like structures from Simian Virus 40. *Proceedings of the National Academy of Sciences, USA*, **72**, 1843–1847
33. Glotov, B. O., Itkes, A. V., Nikolaev, L. G. and Severin, E. S. (1978). Evidence for the close proximity of histones H1 and H3 in chromatin of intact nuclei. *FEBS Letters*, **91**, 149–152
34. Gottesfeld, J. M. (1980). Organization of 5S genes in chromatin of *Xenopus laevis*. *Nucleic Acids Research*, **8**, 905–922
35. Gottesfeld, J. M. and Bloomer, L. S. (1980). Non-random alignment of nucleosomes on 5S RNA genes of *X. laevis*. *Cell*, **21**, 751–760
36. Hartman, P. G., Chapman, G. E., Moss, T. and Bradbury, E. M. (1977). Studies on the role and mode of operation of the very lysine-rich histone H1 in eukaryotic chromatin. *European Journal of Biochemistry*, **77**, 45–51
37. Hjelm, R. P., Kneale, G. G., Suau, P., Baldwin, J. P., Bradbury, E. M. and Ibel, K. (1977). Small angle neutron scattering studies of chromatin subunits in solution. *Cell*, **10**, 139–151
38. Hörz, W. and Altenburger, W. (1981). Sequence specific cleavage of DNA by micrococcal nuclease. *Nucleic Acids Research*, **9**, 2643–2658
39. Hozier, J., Renz, M. and Nehls, P. (1977). The chromosome fiber: evidence for an ordered superstructure of nucleosomes. *Chromosoma*, **62**, 301–317
40. Humphries, S. E., Young, D. and Carroll, D. (1979). Chromatin structure of the 5S ribonucleic acid genes of *Xenopus laevis*. *Biochemistry*, **18**, 3223–3231
41. Igo-Kemenes, T., Hörz, W. and Zachau, H. G. (1982). Chromatin. *Annual Review of Biochemistry*, **51**, 89–121
42. Igo-Kemenes, T., Omori, A. and Zachau, H. G. (1980). Non-random arrangement of nucleosomes in satellite 1 containing chromatin of rat liver. *Nucleic Acids Research*, **8**, 5377–5390
43. Itkes, A. V., Glotov, B. O., Nikolaev, L. G., Preem, S. R. and Severin, E. S. (1980). Repeating oligonucleosomal units. A new element of chromatin structure. *Nucleic Acids Research*, **8**, 507–527
44. Johns, E. W. (1982). *The HMG Chromosomal Proteins*. New York, Academic Press
45. Johnson, E. M., Campbell, G. R. and Allfrey, V. G. (1978). Different nucleosome structures on transcribing and non-transcribing ribosomal gene sequences. *Science*, **206**, 1192–1194
46. Keene, M. A. and Elgin, S. C. R. (1981). Micrococcal nuclease as a probe of DNA sequence organization and chromatin structure. *Cell*, **27**, 57–64
47. Kiryanov, G. I., Smirnova, T. A. and Polyakov, V. Yu. (1982). Nucleomeric organization of chromatin. *European Journal of Biochemistry*, **124**, 331–338
48. Klug, A. and Lutter, L. C. (1981). The helical periodicity of DNA in the nucleosome. *Nucleic Acids Research*, **9**, 4267–4283

49. Klug, A., Rhodes, D., Smith, J., Finch, J. T. and Thomas, J. O. (1980). A low resolution structure for the histone core of the nucleosome. *Nature, London*, **287**, 509–516
50. Kornberg, R. D. (1974). Chromatin structure: a repeating unit of histones and DNA. *Science*, **184**, 868–871
51. Kornberg, R. D. (1977). Structure of chromatin. *Annual Review of Biochemistry*, **46**, 931–954
52. Kornberg, R. D. (1981). The location of nucleosomes in chromatin: specific or statistical? *Nature, London*, **292**, 579–580
53. Kornberg, R. D. and Klug, A. (1981). The nucleosome. *Scientific American*, **244**, 52–64
54. Kornberg, R. D. and Thomas, J. O. (1974). Chromatin structure: oligomers of the histones. *Science*, **184**, 865–868
55. Kunkel, G. R. and Martinson, H. G. (1981). Nucleosomes will not form on double-stranded RNA or over poly (dA) poly (dT) tracks in recombinant DNA. *Nucleic Acids Research*, **9**, 6869–6888
56. Labhart, P. and Koller, Th. (1982). Structure of the active nucleolar chromatin of *Xenopus laevis* oocytes. *Cell*, **28**, 279–292
57. Laemmli, U. K., Cheng, S. M., Adolph, K. W., Paulson, J. R., Brown, J. A. and Baumbach, W. R. (1977). Metaphase chromosome structure: the role of non-histone proteins. *Xenopus laevis* ooaytes. *Cold Spring Harbor Symposium on Quantitative Biology*, **42**, 351–360
58. Lamb, M. M. and Daneholt, B. (1979). Chracterization of active transcription units in Balbiani rings of *Chironomus tentans*. *Cell*, **17**, 835–848
59. Laskey, R. A. and Earnshaw, W. C. (1980). Nucleosome assembly. *Nature, London*, **286**, 763–767
60. Lebkowski, J. S. and Laemmli, U. K. (1982). Evidence for two levels of DNA folding in histone-depleted HeLa interphase nuclei. *Journal of Molecular Biology*, **156**, 309–324; Non-histone proteins and long range organizations of HeLa interphase DNA. *Journal of Molecular Biology*, **156**, 325–344
61. Lee, K. S. and Crothers, D. M. (1982). Influence of ionic strength on the dichroism properties of polynucleosomal fibres. *Biopolymers*, **21**, 101–116
62. Lee, K. S., Mandelkern, M. and Crothers, D. M. (1981). Solution structural studies of chromatin fibers. *Biochemistry*, **20**, 1438–1445
63. Levy, A. and Noll, M. (1981). Chromatin fine structure of active and repressed genes. *Nature, London*, **289**, 198–203
64. Lewis, C. D. and Laemmli, U. K. (1982). Higher order metaphase chromosome structure: evidence for metalloprotein interactions. *Cell*, **29**, 171–181
65. Lilley, D. M. J. and Tatchell, K. (1977). Chromatin core particle unfolding induced by tryptic cleavage of histones. *Nucleic Acids Research*, **4**, 2039–2055
66. Lohr, D. E. (1981). Detailed analysis of the nucleosomal organization of transcribed DNA in yeast chromatin. *Biochemistry*, **20**, 5966–5972
67. Lohr, D., Tatchell, K. and Van Holde, K. E. (1977). On the occurrence of nucleosome phasing in chromatin. *Cell*, **12**, 829–836
68. Lohr, D. and Van Holde, K. E. (1979). Organization of spacer DNA in chromatin. *Proceedings of the National Academy of Sciences, USA*, **76**, 6326–6330
69. Louie, A. J. and Dixon, G. H. (1972). Synthesis, acetylation and phosphorylation of histone IV and its binding to DNA during spermatogenesis in trout. *Proceedings of the National Academy of Sciences, USA*, **69**, 1975–1979
70. Louis, C. and Schedl, P. (1980). Chromatin structure of the 5S RNA genes of *D. melanogaster*. *Cell*, **22**, 387–392
71. Lutter, L. C. (1978). Kinetic analysis of deoxyribonuclease 1 cleavages in the nucleosome core: evidence for a DNA superhelix. *Journal of Molecular Biology*, **124**, 391–420
72. Mardian, J. K. W., Paton, A. E., Bunick, G. J. and Olins, D. E. (1980). Nucleosome cores have two specific binding sites for non-histone chromosomal proteins HMG14 and HMG17. *Science*, **209**, 1534–1536
73. Marion, C., Bezot, P., Hesse-Bezot, C., Roux, B. and Bernengo, J.-C. (1981). Conformation of chromatin oligomers. *European Journal of Biochemistry*, **120**, 169–176
74. Marion, C. and Roux, B. (1978). Nucleosome arrangement in chromatin. *Nucleic Acids Research*, **5**, 4431–4449
75. Marsden, M. P. F. and Laemmli, U. K. (1979). Metaphase chromosome structure: evidence for a radial loop model. *Cell*, **17**, 849–858
76. Mathis, D., Oudet, P. and Chambon, P. (1980). Structure of transcribing chromatin. *Progress in Nucleic Acid Research and Molecular Biology*, **24**, 1–55
77. McGhee, J. D. and Felsenfeld, G. (1980). Nucleosome structure. *Annual Review of Biochemistry*, **49**, 1115–1156
78. McGhee, J. D., Rau, D. C., Charney, E. and Felsenfeld, G. (1980). Orientation of the nucleosome within the higher order structure of chromatin. *Cell*, **22**, 87–96

79. Mirzabekov, A. D., Shick, V. V., Belyavsky, A. V. and Bavykin, S. G. (1978). Primary organization of nucleosome core particles of chromatin: sequence of histone arrangement along DNA. *Proceedings of the National Academy of Sciences, USA*, **75**, 4184–4188
80. Morris, N. R. (1976). A comparison of the structure of chicken erythrocyte and chicken liver chromatin. *Cell*, **9**, 627–632
81. Musich, P. R., Maio, J. J. and Brown, F. L. (1977). Subunit structure of chromatin and the organization of eukaryotic highly repetitive DNA: indication of a phase relation between restriction sites and chromatin subunits in African green monkey and calf nuclei. *Journal of Molecular Biology*, **117**, 657–677
82. Muyldermans, S., Lasters, I., Wyns, L. and Hamers, R. (1980). Upon the observations of superbeads in chromatin. *Nucleic Acids Research*, **8**, 2165–2172
83. Nedospasov, S. A. and Georgiev, G. (1980) Non-random cleavage of SV40 DNA in the compact mini chromosome and free in solution by micrococcal nuclease. *Biochemical and Biophysical Research Communications*, **92**, 532–539
84. Noll, M. (1974). Subunit structure of chromatin. *Nucleic Acids Research*, **1**, 1573–1578
85. Osipova, T. N., Pospelov, V. A., Svetlikova, S. B. and Vorob'ev, V. I. (1980). The role of histone H1 in compaction of nucleosomes. *European Journal of Biochemistry*, **113**, 183–188
86. Pardon, J. F., Worcester, D. L., Wooley, J. C., Tatchell, K., Van Holde, K. E. and Richards, B. M. (1975). Low-angle neutron scattering from chromatin subunit particles. *Nucleic Acids Research*, **2**, 2163–2176
87. Polisky, B. and McCarthy, B., (1975). Location of histones on Simian Virus 40 DNA. *Proceedings of the the National Academy of Sciences, USA*, **72**, 2895–2899
88. Ponder, B. A. J. and Crawford, L. V. (1977). The arrangement of nucleosomes in nucleoprotein complexes from polyoma virus and SV40. *Cell*, **11**, 35–49
89. Prunell, A. and Kornberg, R. D. (1977). Relation of nucleosomes to DNA sequences. *Cold Spring Harbor Symposium on Quantitative Biology*, **42**, 103–108
90. Prunell, A. and Kornberg, R. D. (1982). Variable center to center distance of nucleosomes in chromatin. *Journal of Molecular Biology*, **154**, 515–523
91. Prunell, A., Kornberg, R. D., Lutter, L. C., Klug, A., Levitt, M. and Crick, F. H. C. (1979). Periodicity of deoxyribonuclease 1 digestion of chromatin. *Science*, **204**, 855–858
92. Renz, M. and Day, L. A. (1976) Transition from non-cooperative to cooperative and selective binding of histone H1 to DNA. *Biochemistry*, **15**, 3220–3228
93. Renz, M., Nehls, P. and Hozier, J. (1977). Involvement of histone H1 in the organization of the chromosome fiber. *Proceedings of the National Academy of Sciences, USA*, **74**, 1879–1883
94. Rhodes, D. and Klug, A. (1980). Helical periodicity of DNA determined by enzyme digestion. *Nature, London*, **286**, 573–578
95. Riley, D. and Weintraub, H. (1978). Nucleosomal DNA is digested to repeats of 10 bases by exonuclease III. *Cell*, **13**, 281–293
96. Ris, H. and Korenberg, J. (1979). Chromosome structure and levels of chromosome organization. In *Cell Biology*, Vol. 2, (Prescott, D. M. and Goldstein, L. Eds.), pp. 267–361. New York, Academic Press
97. Ris, H. and Kubai, D. F. (1970). Chromosome structure. *Annual Review of Genetics*, **4**, 263–294
98. Samal, B. and Worcel, A. (1981). Chromatin structure of the histone genes of *D. melanogaster*. *Cell*, **23**, 401–409
99. Sandeen, G., Wood, W. I. and Felsenfeld, G. (1980). The interaction of high mobility proteins HMG14 and 17 with nucleosomes. *Nucleic Acids Research*, **8**, 3757–3778
100. Saragosti, S., Moyne, G. and Yaniv, M. (1980). Absence of nucleosomes in a fraction of SV40 chromatin between the origin of replication and the region coding for the late leader RNA. *Cell*, **20**, 65–73
101. Sedat, J. and Manuelidis, L. (1977). A direct approach to the structure of eukaryotic chromosomes. *Cold Spring Harbor Symposium on Quantitative Biology*, **42**, 331–350
102. Senshu, T., Fukuoa, M. and Ohashi, M. (1978). Preferential association of newly synthetized H3 and H4 histones with newly replicated DNA. *Journal of Biochemistry (Japan)*, **84**, 985–988
103. Simpson, R. T. (1978). Structure of the chromatosome, a chromatin particle containing 160 base pairs of DNA and all the histones. *Biochemistry*, **17**, 5524–5531
104. Smerdon, M. J. and Isenberg, I. (1976). Conformational changes in sub-fractions of calf thymus histone H1. *Biochemistry*, **15**, 4233–4242
105. Sperling, L. and Klug, A. (1977). X-ray studies on 'native' chromatin. *Journal of Molecular Biology*, **112**, 253–263
106. Stein, A. (1980). DNA wrapping in nucleosomes. The linking number problem re-examined. *Nucleic Acids Research*, **8**, 4803–4820
107. Stockley, P. G. and Thomas, J. O. (1979). A nucleosome-like particle containing an octamer of the arginine rich histones H3 and H4. *FEBS Letters*, **99**, 129–135

108. Stoeckert, C., Beer, M., Wiggins, J. W. and Hjelm, R. P. (1981). Position of histone pairs; H3, H4 and H2A, H2B within the nucleosome. In *Proceedings of the 39th Annual EMSA Meeting* , (G. W. Bailey, Ed.), pp. 444–445. Baton Rouge, Louisiana, Claitor's Publishing Division

109. Strätling, W., Müller, U. and Zentgraf, H. (1978). The higher order repeat structure of chromatin is built up of globular particles containing eight nucleosomes. *Experimental Cell Research*, **117**, 301–311

110. Stryer, L. and Kornberg, R. D. unpublished results, cited in reference 52

111. Suau, P, Bradbury, E. M. and Baldwin, J. P., (1979). Higher-order structures of chromatin in solution. *European Journal of Biochemistry*, **97**, 593–602

112. Thoma, F., Koller, Th. and Klug, A. (1979). Involvement of histone H1 in the organization of the nucleosome and of salt-dependent superstructures of chromatin. *Journal of Cell Biology*, **83**, 403–427

113. Thomas, J. O. unpublished results

114. Thomas, J. O. (1978). In *International Review of Biochemistry, Biochemistry of Nucleic Acids, II* (Clark, B. F. C., Ed.) Vol. 17, pp. 181–232. Baltimore, University Park Press

115. Thomas, J. O. and Butler, P. J. G. (1977). Characterization of the octamer of histones free in solution. *Journal of Molecular Biology*, **116**, 769–781

116. Thomas, J. O. and Butler, P. J. G. (1980). Size-dependence of a stable higher-order structure of chromatin. *Journal of Molecular Biology*, **144**, 89–93

117. Thomas, J. O. and Khabaza, A. J. A. (1980). Cross-linking of histone H1 in chromatin. *European Journal of Biochemistry*, **112**, 501–511

118. Thomas, J. O. and Oudet, P. (1979). Complexes of the arginine-rich histone tetramer $(H3)_2(H4)_2$ with negatively supercoiled DNA: electron microscopy and chemical cross-linking, *Nucleic Acids Research*, **7**, 611–623

119. Thomas, J. O. and Rees, C. (1983) Exchange of histones H1 and H5 between chromatin fragments. A preference of H5 for higher-order structure. *European Journal of Biochemistry* (in press)

120. Von Holt, C., Strickland, W. N., Brandt, W. F. and Strickland, M. S. (1979). More histone structures. *FEBS Letters*, **100**, 201–218

121. Walker, B. W., Lothstein, L., Baker, C. L. and LeSturgeon, W. M. (1980). The release of 40S hnRNP particles by brief digestion of HeLa nuclei with micrococcal nuclease. *Nucleic Acids Research*, **8**, 3639–3657

122. Wang, J. (1979). Helical repeat of DNA in solution. *Proceedings of the National Academy of Sciences, USA*, **76**, 200–203

123. Weintraub, H. (1978). Nucleosome repeat length increases during erythropoiesis in chick. *Nucleic Acids Research*, **5**, 1179–1188

124. Weisbrod, S. (1982). Active chromatin. *Nature, London*, **297**, 289–295

125. Whitlock, J. P. Jr. and Simpson, R. T. (1977). Localization of the sites along nucleosomal DNA which interact with NH_2-terminal histone regions. *Journal of Biological Chemistry*, **252**, 6516–6520

126. Whitlock, J. P. Jr. and Stein, A. (1978). Folding of DNA by histones which lack their NH_2-terminal regions. *Journal of Biological Chemistry*, **253**, 3857–3861

127. Wittig, B. and Wittig, S. (1979). A phase relationship associates tRNA structural gene sequences with nucleosome cores. *Cell*, **18**, 1173–1183

128. Wittig, S. and Wittig, B. (1982). Function of the tRNA gene promoter depends on nucleosome position. *Nature*, **297**, 31–38

129. Worcel, A. and Benyajati, C. (1977). Higher order coiling of DNA in chromatin. *Cell*, **12**, 83–100

130. Worcel, A., Han, S. and Wong, M. L. (1978). Assembly of newly replicated chromatin. *Cell*, **15**, 969–977

131. Worcel, A., Strogatz, S. and Riley, D. (1981). Structure of chromatin and the linking number of DNA. *Proceedings of the National Academy of Sciences, USA*, **78**, 1461–1465

132. Wu, C. (1980). The 5'ends of *Drosophila* heat shock genes in chromatin are hypersensitive to DNase I. *Nature, London*, **286**, 854–860

133. Wu, C., Bingham, P. M., Livak, K. J., Holmgren, R. and Elgin, S. C. R. (1979). Chromatin structure of specific genes. I. Evidence for higher order domains of defined DNA sequence. *Cell*, **16**, 797–806

134. Wu, C., Wong, Y.-C. and Elgin, S. C. R. (1979). Chromatin structure of specific genes. II. Disruption of chromatin structure during gene activity. *Cell*, **16**, 807–814

135. Yabuki, H., Dattagupta, N. and Crothers, D. M. (1982). Orientation of nucleosomes in the thirty-nanometer chromatin fiber. *Biochemistry*, **21**, 5015–5020

136. Zachau, H. G. and Igo-Kemenes, T. (1981). Face to phase with nucleosomes. *Cell*, **24**, 597–598

137. Zalenskaya, I. A., Pospelov, V. A., Zalensky, A. O. and Vorob'ev, V. I. (1981). Nucleosomal structure of sea urchin and starfish sperm chromatin. Histone 2B is possibly involved in determining the length of linker DNA. *Nucleic Acids Research*, **9**, 473–487

Chapter 3

Non-histone proteins and gene regulation

G. Stein, J. Stein and L. J. Kleinsmith

Introduction

It has been known for a long time that the chromosomal DNA of eukaryotes is associated with a large quantity of histone and non-histone proteins. Though the role of histones in packaging DNA into nucleosomes has been well established in recent years (see Chapter 2 of this volume), the functional significance of the non-histone proteins is not nearly so clear. Our limited understanding of this latter group of proteins is at least in part due to the heterogeneity of this fraction, the relative insolubility of many of its components, and the limited availability of *in vitro* assays for chromosomal protein function which accurately reflect the *in vivo* situation. In spite of these limitations, however, it has become gradually apparent that amongst the non-histone chromosomal proteins are macromolecules involved in chromatin structure, in catalysing nuclear reactions, and perhaps in determining whether or not specific genetic sequences are available for transcription. In this chapter, we will summarize the properties of this diverse class of proteins which have led to the above generalizations, placing an emphasis on the relationship between genomic structure and the regulation of gene expression.

General properties of non-histone proteins

Among the general characteristics which a group of molecules would be expected to exhibit if they were involved in regulating gene expression are heterogeneity, tissue and species specificity, variations which correlate with changes in gene activity, and sequence-specific interactions with DNA. In the following sections on the properties of non-histone proteins, it will become apparent that these molecules not only meet the above criteria, but also exhibit several other attributes which further reinforce the notion that they play a role in the regulation of chromatin structure and function.

Heterogeneity and tissue/species specificity of non-histone chromosomal proteins

With present levels of resolution attainable by two-dimensional gel electrophoresis, the non-histone proteins in chromatin have been shown to contain at least several hundred distinct species[16, 70]. This level of complexity rivals that of the cytoplasmic proteins, and contrasts markedly with the comparative lack of heterogeneity of the histones.

The question of the tissue specificity of non-histone proteins has been extensively investigated by both electrophoretic and immunological methods. Both approaches demonstrate the existence of tissue specificity in the non-histone protein fraction, though it is also clear that most tissues share with each other a common set of major non-histone proteins. Differences in extraction techniques and the limited resolving power of one-dimensional electrophoresis have led to some disagreement as to how extensive a variation exists in non-histone proteins between tissues. However, the introduction of high resolution two-dimensional electrophoretic systems has led to the conclusion that as many as half the non-histone chromosomal proteins of Friend and HeLa cells differ from each other[71]. This degree of divergence is less than that observed for the cytoplasmic proteins (three quarters of which differ), indicating that many of the non-histone chromosomal proteins may be under relatively strict evolutionary conservation. The tissue- and/or species-specific chromosomal proteins presumably include those involved in the control of cell-specific gene expression, whereas the remainder of the proteins may be conserved because they have to interact with DNA or other molecules in general processes such as DNA replication, transcription, and the maintenance of chromosomal structure. One subclass of non-histone proteins which has little tissue or species specificity is the high mobility group (HMG) proteins, a group of molecules whose role in chromatin structure will be discussed later.

Quantitative variations in non-histone chromosomal protein content

Although histones are present in chromatin from most tissues at about a 1:1 (w/w) ratio to DNA, the proportion of non-histone proteins in chromatin from different tissues varies considerably. Most tissues have a non-histone protein to DNA ratio around 1.0^9, but this ratio varies from as low as 0.1 in certain types of sperm to as high as 9 in the slime mould. The quantity of non-histone proteins present in chromatin generally varies with the biological state of the cell, with increased non-histone protein content correlating with increased levels of RNA synthesis, and vice versa.

Non-histone chromosomal proteins of 'active' and 'inactive' chromatin

Quantitative variations in the non-histone chromosomal protein content of active and inactive tissues suggests that chromatin template activity is influenced by the non-histone chromosomal proteins. This hypothesis is further supported by the finding that when chromatin is fractionated into 'active' and 'inactive' components, the non-histone protein to DNA ratio is generally greater in the active fraction. Electrophoretic analysis has revealed both qualitative and quantitative differences between the non-histone proteins from the different chromatin fractions. Certain specific non-histone proteins with defined enzyme activities, such as nuclear protein kinase, have been found to be preferentially localized in the active fraction. These results indicate that some non-histone proteins may be involved in transcriptional activation and/or maintenance of a diffuse chromatin structure, whereas others may be involved in the inhibition of transcriptional activity or in the condensation of chromatin structure.

Changes in non-histone chromosomal proteins associated with modifications in gene activity

A regulatory role for components of the non-histone chromosomal protein fraction is suggested by the changes which occur in these macromolecules in association with modifications in gene expression. For example, alterations in non-histone proteins have been observed following the stimulation of resting cells to proliferate, during differentiation and development, after hormone and drug stimulation, and during malignant transformation. Some of the many experimental systems in which such observations have been made will be summarized below. For detailed references, the reader is referred to reference 87.

Changes in non-histone chromosomal proteins associated with cell proliferation

An increase in the rate of non-histone chromosomal protein synthesis has been observed in many different systems where non-dividing cells are stimulated to proliferate, e.g. salivary glands stimulated by isoproterenol, human fibroblasts stimulated by serum, lymphocytes stimulated by phytohaemagglutinin, concanavalin A, leucoagglutinin, or anti-immunoglobulin, liver regeneration following partial hepatectomy, and refeeding of starved *Physarum polycephalum* and *Tetrahymena*. Electrophoretic analysis has revealed that, in many cases, a change in the overall quantity of non-histone chromosomal proteins can be accounted for by an increase in the levels of a few specific proteins. The increase in the synthesis of specific non-histone chromosomal proteins which occurs when cells are stimulated to divide precedes increases in RNA and DNA synthesis, and so it is reasonable to postulate that these proteins are involved in the control of gene transcription at the onset of cell proliferation. However, the levels of some non-histone proteins decrease when cells are stimulated to divide; these may be involved in the maintenance of the quiescent state.

Changes in non-histone chromosomal proteins during development and differentiation

During the course of development and differentiation, some batteries of genes are activated, while others are repressed; thus, developing and differentiating systems provide good opportunities for studying the molecules involved in controlling selective gene expression. Quantitative and qualitative stage-specific changes in non-histone chromosomal proteins have been observed in sea urchin embryos, *Xenopus laevis* tadpoles, *Oncopeltus* (milkweed bug) embryos, and during the development of chick oviduct and embryonic red blood cells. Alterations in the complement of non-histone chromosomal proteins have also been observed during the differentiation of pollen from *Hippeastrum belladonna*, during the conversion of lymphoid spleen to an erythroid organ, and in the dimethyl sulphoxide-stimulated erythroid differentiation of Friend cells. In all the above instances, changes in the non-histone proteins parallel alterations in the transcriptional activity and structure of the chromatin involved. Further evidence for a link between the level of non-histone chromosomal proteins in a cell and the RNA synthetic capacity of that cell is provided by the reduction in amount and the loss of specific non-histone proteins observed during the condensation of chromatin and concomitant repression of RNA synthesis which occurs during spermatogenesis and the maturation of erythroid cells.

In contrast to the above findings, some groups have detected only minimal changes in the complement of non-histone chromosomal proteins during prostaglandin- and cAMP-induced differentiation of neuroblastoma cells, during normal or hydrocortisone-induced embryonic development of the neural retina, and during normal development of various brain tissues. Thus, gross changes in non-histone proteins do not necessarily accompany the process of cell differentiation.

Changes in non-histone chromosomal proteins associated with gene activation by hormones and drugs

Many hormones and drugs are now known to alter the expression of specific genes in their target tissues. In several of these cases, e.g. insulin, aldosterone, ecdysone, testosterone, cortisone and phenobarbitone, increases in the synthesis of non-histone proteins have been found to accompany the induction process. The changes which occur in the non-histone protein fraction are both hormone and target tissue specific. In the case of the insect hormone ecdysone, one can directly observe the accumulation of specific non-histone proteins in the puffed regions of the polytene chromosomes after hormone administration[32]. In several instances, controls have been carried out to demonstrate the specificity of hormone effects on non-histone proteins. For example, the changes in non-histone proteins which occur in prostate gland in response to testosterone are prevented when the anti-androgen, cyproterone acetate, is administered together with testosterone *in vivo*[59].

In addition to the finding that hormones induce alterations in the synthesis and accumulation of non-histone proteins, there is a large body of evidence which suggests that components of the non-histone protein fraction may play a direct role in the mechanism of steroid hormone binding to chromatin[83, 91].

Changes in non-histone chromosomal proteins associated with malignant transformation

Another finding which supports the hypothesis that the non-histone chromosomal proteins are involved in the control of gene expression is that they undergo quantitative and qualitative changes during malignant transformation, some such changes being evident within hours of treatment of normal tissues with carcinogens. The possible involvement of non-histone chromosomal proteins as mediators of aberrant gene expression in neoplastic cells has been discussed in a recent review[86], to which the reader is referred for detailed references. In brief, changes in the levels and/or rates of synthesis of non-histone chromosomal proteins have been observed in cancers of the breast, liver, colon, brain, prostate and lymphocytes. Differences between the non-histone proteins of normal and neoplastic tissues have been confirmed by immunological techniques including complement fixation, enzyme-linked immunoassay, and immunohistochemistry. A particularly interesting observation made in Busch's laboratory is that a protein which appears in lymphocytes only when they become leukaemic apparently corresponds to a protein present in hepatomas, but not in normal liver[101, 102]. This protein may be involved in some general aspects of carcinogenesis, such as the perturbation of replicative control.

Non-histone chromosomal proteins also seem to be altered during viral transformation of cells. It has been shown by immunological techniques that the non-histone proteins of SV40-transformed WI-38 human fibroblasts differ from those of untransformed cells. Gel electrophoretic and radioactive labeling studies have also demonstrated differences in levels and rates of synthesis of specific non-histone proteins in SV40-transformed and untransformed WI-38 cells and mouse embryo fibroblasts, as well as in Rous sarcoma virus infected chick embryo fibroblasts.

Phosphorylation of non-histone chromosomal proteins

One of the most unusual properties of the non-histone proteins is the extensive degree to which they are phosphorylated. The protein-bound phosphate content of this fraction as a whole is enough to account for several phosphorylated amino acids for every hundred amino acid residues present, giving the non-histone protein fraction the highest concentration of phosphorylated polypeptides of any protein fraction in the cell. All the properties discussed thus far as characterizing the non-histone protein fraction as a whole (i.e. heterogeneity, tissue/species specificity, association with active chromatin and changes in association with alterations in gene activity) apply to the phosphorylated non-histone proteins as well[44, 45, 63]. In regard to these properties, especially striking is the vast number of different experimental systems in which changes in non-histone protein phosphorylation have been observed in association with changing nuclear metabolism, chromatin structure, and gene activity (*Table 3.1*).

Such observations have stimulated interest in isolating and identifying individual phosphorylated non-histone proteins so that their specific roles in nuclear functions can be investigated. In the process, it has become gradually apparent that many different kinds of non-histone proteins are phosphorylated, and that the physiological roles of these various proteins may be quite diverse. Thus, among the non-histone nuclear proteins now known to be phosphorylated one can distinguish at least a dozen different categories: chromatin associated proteins which influence transcription, nucleolar proteins, the S-100 (brain-specific) protein, RNA polymerase and associated factors, DNA-binding proteins, poly(A) polymerase, histone deacetylase, glucocorticoid receptors, nucleosomal proteins (including HMG proteins), proteins associated with heterodisperse nuclear RNA, nuclear envelope proteins, and nuclear matrix proteins. A detailed discussion of the properties of each of these groups of phosphorylated proteins is beyond the scope of this chapter, especially since the subject has been recently reviewed in depth elsewhere[63]. In the present context, only a few general issues raised by this diverse group of phosphorylated proteins will be touched upon.

The most critical issue regards the significance of the fact that these proteins are all phosphorylated. The functional significance of protein phosphorylation has been most clearly established for cytoplasmic enzymes such as glycogen phosphorylase, phosphorylase kinase, and glycogen synthetase, where such modifications serve to either increase or decrease catalytic activity, depending on the enzyme in question[27, 49, 51, 78, 88]. By analogy, it has generally been assumed that phosphorylation serves to regulate the activities of other proteins known to be phosphorylated, though in most instances such a regulatory role is yet to be clearly demonstrated. In the case of the phosphorylated non-histone proteins listed above,

TABLE 3.1. Summary of experimental systems in which changes in non-histone protein phosphorylation have been observed (for detailed references, see reference 63)

Growth and Development	*Hormonal Stimulation*
Lymphocyte activation	Testosterone (prostate)
Avian erythrocytes	Antiandrogens (prostate)
Physarum polycephalum	Glucocorticoids (insect salivary glands)
HeLa cells	Glucocorticoids (liver)
Reuben hepatoma cells	Glucocorticoids (lymphosarcoma)
Yoshida sarcoma	Aldosterone (kidney)
Sea urchin embryos	Oestradiol (uterus)
Trout embryos	Oestradiol (mammary carcinoma)
Muscle differentiation	Oestradiol (oviduct)
Neuronal and glial cells	Oestradiol (brain)
Kidney regeneration	Prolactin (mammary gland)
Barley germination	Chorionic gonadotropin (ovary)
Chinese hamster cells (K_{12})	Calcitonin (bone cells)
WI-38 fibroblasts	Triiodothyronine (liver)
Chick embryo fibroblasts	Triiodothyronine (heart)
Hamster kidney cells ($BHK_{21}C_{13}$)	Thyrotropin (thyroid)
Testis and epididymis	Abscisic acid (*Lemna*)
Uterus	Cyclic AMP (rat liver)
ChO cells (heat shock)	Cyclic AMP (rat heart)
Liver (aging)	Cyclic AMP (lymphocytes)
Liver (regeneration)	Cyclic AMP (mammary carcinoma)
Liver (cell culture)	Cyclic AMP (adrenal medulla)
Landschutz tumor cells (amino acid starvation)	Cyclic AMP (salivary gland)
	Cyclic AMP (*Neurosopora*)
	Cyclic AMP (pineal gland)
Chemical Agents	Cyclic AMP (lymphocytes)
Isoproterenol (salivary gland)	Cyclic AMP (salivary gland)
Isoproterenol (pineal gland)	
Acetylcholine (lymphocytes)	*Malignant and Transformed Cells*
Norepinephrine (glioma)	Mammary carcinoma
Carbachol (lymphocytes)	Azo-dye carcinogenesis
Phenobarbitone (liver)	Dimethylbenzanthracene carcinogenesis
α-1,2,3,4,5,6-Hexachlorocyclohexane (liver)	Methylcholanthrene carcinogenesis
Morphine (brain)	SV-40 transformed fibroblasts
Hemin (liver)	Morris hepatomas
Vitamin D (kidney, liver)	Novikoff hepatoma
Prostaglandin (lymphocytes)	Friend erythroleukemia
2,4-Dichlorophenoxyacetic acid (soybeans)	Ehrlich ascites
Helanalin (Ehrlich ascites)	Walker tumor
Cytarabine (salivary gland)	Adenovirus transformed cells
Bleomycin (salivary gland)	Murine sarcoma virus transformed cells
Chlorambucil (Yoshida sarcoma)	Human leukemia
Melphalan (Yoshida sarcoma)	
Cyclophosphamide (Yoshida sarcoma)	
Eupahyssopin (Ehrlich ascites)	

phosphorylation has been shown to affect activation of gene transcription[14,15,47,50], non-histone protein binding to DNA[35, 50, 61, 67], poly(A) polymerase activity[75,76], RNA polymerase activity[48] and glucocorticoid binding activity[65]. For other phosphorylated non-histone proteins, such as S-100 histone deacetylase, nucleosomal phosphoproteins, hnRNA-associated phosphoproteins, and nuclear envelope and nuclear matrix phosphoproteins, no direct evidence concerning the functional significance of phosphorylation is available. Though there is no shortage of attractive ideas (e.g. phosphorylation of hnRNA-associated proteins may be involved in regulating RNA processing and/or transport, phosphorylation of histone deacetylase may regulate its catalytic activity, phosphorylation of nucleosomal proteins may influence chromatin structure, etc.), one should exercise extreme caution in such instances where there are no definitive data available.

The fact that the phosphorylation state of non-histone proteins influences their ability to stimulate RNA synthesis and bind to DNA *in vitro* raises the question of the possible role of non-histone protein phosphorylation in the control of specific gene expression. This question is difficult to answer unequivocally because of the severe limitations of the cell-free systems in which such investigations have been carried out. However, after discussing the remaining properties of the non-histone protein fraction, we will return to this general issue in more detail, discussing both the limitations of the current data implicating non-histone proteins in specific gene regulation and the prospects for obtaining more definitive information in the future.

DNA binding of non-histone chromosomal proteins

Selective gene expression implies the restriction of some genetic loci and the specific activation of others. If the non-histone proteins play a role in these processes, it is reasonable to postulate that the mechanism through which transcriptional control is achieved will depend, in part, on associations between these proteins and DNA. In prokaryotes there are several examples of DNA binding proteins which regulate RNA synthesis, for example the *lac* repressor-mediated inhibitor of transcription of the *lac* operon[2,26] and the cAMP receptor-mediated facilitation of initiation of *lac* and *gal* transcription[4,104].

Specific DNA binding non-histone proteins have been isolated by several groups of workers. In this section we will discuss some of the properties of these DNA binding proteins which may have a bearing on their postulated role in the control of gene expression.

Sequence-specific DNA binding non-histone chromosomal proteins

DNA binding non-histone proteins have been most commonly isolated by affinity chromatography on columns containing immobilized homologous or heterologous DNAs. Only a small proportion of the proteins in non-histone preparations bind specifically to homologous DNA[46,92] and many of these are phosphoproteins[43]. Considerable confusion exists as to the heterogeneity of these proteins, some groups observing a highly complex population of non-histone proteins[43,89], whereas others do not[68,92,94]. Similarly, most groups have found that non-histones bind selectively to homologous DNA, though others have found little species-specific binding and argue that their interaction with DNA has severely restricted the evolutionary diversity of these proteins. Such data must generally be considered with some caution because binding conditions vary from group to group, different non-histone protein preparations are used, in most cases considerable overlap of proteins in DNA binding fractions has been observed, and rechromatography of 'specific' DNA binding protein fractions has yielded contradictory results.

Specific binding of non-histone proteins to DNA probably occurs under a limited set of conditions, and optimization of specific binding *in vitro* requires the determination of a number of binding parameters. The nitrocellulose filter binding of nucleoprotein complexes has provided a rapid and flexible DNA binding assay.

Sevall and co-workers[39, 40, 81], for example, used tandem heterologous and homologous DNA affinity chromatography to fractionate a rat liver non-histone protein preparation, and then employed a filter binding assay to analyse the binding specificities of the protein fractions obtained. At saturation, the heterologous DNA binding proteins bound 60–70% of whole rat liver DNA whereas the homologous DNA binding proteins bound only 35% implying the existence of some sequence specificity. Competition experiments established that the homologous DNA binding protein fraction binds homologous DNA preferentially, with no such preference being observed for the heterologous DNA binding protein fraction. Analysis of the DNA retained by the filters showed that it is enriched four-fold in unique DNA sequences and depleted in repetitive sequences relative to total DNA.

Cloned DNA sequences have also been used to identify sequence-specific binding. Hsieh and Brutlag[36] isolated a protein from *Drosophila* embryos which specifically binds to a cloned repeating unit of satellite DNA when the plasmid and its insert are in superhelical conformation. Restriction endonuclease mapping of this DNA–protein complex revealed that the protein protects one specific restriction site on the inserted DNA fragment. Similarly, Weideli *et al.*[96] have isolated a DNA binding protein (DB-2) from *Drosophila* and bound it to random or restriction fragments of whole cell DNA. Bound DNA fragments were recovered from nitrocellulose filters, inserted into plasmids, and cloned in bacteria by standard recombinant DNA procedures. The cloned DNA fragments were reacted with the DNA binding protein again, and bound DNA fragments were cloned a second time. This process yielded a cloned DNA fragment which selectively binds to the DB-2 protein at high ionic strength ($2.0 \, mol/\ell \, NaCl$). Immunocytochemical localization of the DB-2 protein revealed that it is concentrated specifically on chromsome 3 (section 95a/b) in intact cells, supporting the postulate that the protein is associated with a specific region of the *Drosophila* genome *in vivo*.

One obvious disadvantage to experiments such as those described above, in which one monitors the binding of proteins to purified DNA sequences, is that *in vivo* the DNA is part of a chromatin complex which may behave quite differently from naked DNA. A model system which attempts to overcome this difficulty has been devised by Chao *et al.*[13] who used a 203 base pair DNA fragment containing the *E. coli lac* operator sequences to reconstitute nucleosomal structures with calf thymus histones. When such nucleosomes were bound to the *lac* repressor protein, sequence-specific recognition could be detected without the displacement of histones. Since the binding was quantitative and specific, the results suggest that the reconstitution protocol yields structures where the bound DNA strand faces the outer surface of the nucleo–histone complex. Chemical crosslinking of the histones inhibits binding of the repressor molecule to the DNA, indicating that conformational changes in the nucleosome are necessary for high-affinity binding. Binding of the *lac* repressor only occurs when the operator sequence is positioned in the region of the nucleosome stabilized by H2A–H2B interaction (but not H3–H4 interactions). This implies that sequence-specific binding in eukaryotic chromatin may be modulated by DNA interactions with histone proteins which define the accessibility of sequences to intermolecular interactions.

Histones may not be the only chromatin component which influences the binding of non-histone proteins to DNA. Dastugue and Crepin[19] have reported that the binding of non-histone proteins to isolated DNA exhibits less species specificity than the binding of the same proteins to intact chromatin. This observation suggests that the specific binding of non-histone proteins to DNA may be influenced by chromatin proteins other than histones, since the histones do not vary much from species to species. Although the physiological significance, if any, of the *in vitro* binding of non-histone proteins to chromatin is not certain, it is interesting to note that such binding is accompanied by a stimulation of transcription in homologous, but not heterologous, protein/chromatin mixtures[18].

An alternative approach to the identification of specific DNA:non-histone protein complexes formed *in vivo* has been provided by Bekhor and coworkers[6, 7, 23, 66]. Essentially, chromatin is extensively washed with 2.0 mol/ℓ NaCl to remove loosely bound proteins, and the protein-depleted DNA is then dialysed against 10 mmol/ℓ Tris-HCl. Low speed centrifugation yields a nucleoprotein complex containing some 1% of the total nuclear DNA complexed to a specific non-histone protein fraction. The DNA in this complex is enriched twenty-fold in transcribed sequences, and is depleted of non-transcribed sequences, inactive gene sequences, and highly repetitive DNA sequences. Since the proteins bind specific DNA sequences with such high affinity, rearrangement during isolation is an unlikely explanation for this nucleoprotein complex.

Proteins which bind to single stranded DNA

Binding of non-histone proteins to single stranded DNA has been studied ever since it was first observed that such proteins stimulate DNA synthesis in prokaryotes[3]. Eukaryotic single stranded DNA binding proteins have subsequently been identified in a wide variety of species and tissues. Some of these proteins have the characteristics of DNA unwinding or 'helix-destabilizing' proteins[20, 33, 90], some stimulate *in vitro* DNA synthesis by eukaryotic DNA polymerase[10, 67], some have nicking-closing activity[20] and some are potent nuclease inhibitors[64]. In one instance, phosphorylation of a DNA binding protein has been shown to both reduce its binding to single stranded (but not duplex) DNA and abolish its ability to stimulate DNA polymerase activity, suggesting a role for this helix-destabilizing protein in DNA replication[67]. Meiotic cells have also been reported to contain a helix-destabilizing protein whose properties are influenced by phosphorylation[35, 61], but this protein is present mainly between S-phase of the cell cycle and the termination of chromosome pairing. Its virtual absence during DNA replication and its prominence during the period of chromosome pairing and crossing over suggest a possible function in genetic recombination rather than DNA synthesis.

Functional significance of DNA binding proteins

While DNA binding undoubtedly represents a powerful tool for isolating subsets of chromosomal proteins, caution must be exercised in ascribing physiological significance to such binding interactions observed under artificial conditions. Attempts to study physiological alterations in DNA binding proteins occurring *in vivo* have led to correlations with cell growth rate and malignant transform-ation[11, 58, 102], but such observations do not make the functional significance of the individual proteins obvious. It is clear that because one is dealing with a heterogeneous group of molecules in the DNA binding protein fraction, a clear understanding of their physiological roles will ultimately require the isolation and characterization of individual protein species. One such group of DNA binding proteins which has been so purified and characterized is the HMG proteins discussed below.

The high mobility group (HMG) proteins

Other than specific enzymes (to be discussed in the next section), the HMG proteins are the most extensively characterized group of non-histone proteins. They are readily identified on the basis of their high solubility in mineral acids, their lack of tissue or species specificity, and their high concentration in the nucleus[28, 73, 80, 95]. Four main HMG proteins have been characterized in calf thymus, designated HMG1, HMG2, HMG14 and HMG17 in order of increasing electrophoretic mobility in acid urea; similar proteins have been identified in most other tissues and species examined. One possible exception is trout testes, which has only two major HMG proteins: the H6 protein which has 50% sequence homology with calf HMG17, and HMG-T, which has been compared with calf HMG1 and HMG2[56, 95]. The HMG proteins are all present in nucleosomes isolated from chromatin by limited digestion with micrococcal nuclease, but can be completely dissociated from chromatin by treatment with 0.35 mol/ℓ NaCl. Isolated HMG proteins will rebind to DNA at low ionic strength and although there is no apparent sequence specificity in this binding, HMG1 and HMG2 have a preferential affinity for single stranded DNA and the potential to unwind double stranded DNA[1, 38, 41, 82, 103].

Experiments in which transcriptionally active chromatin has been selectively solubilized by various types of nuclease treatment have led to the general conclusion that HMG proteins are selectively localized in the regions of active genes. This conclusion is not without controversy and apparent contradictions, however, probably because HMG protein release has not always been adequately quantified and it is sometimes difficult to compare HMG proteins from different organisms. Thus preferential release of HMG1 and HMG2 by selective digestion of active chromatin with DNase I has been reported for calf thymus, pig thymus, mouse brain, and duck erythrocytes[53, 72, 93], but such release has not been confirmed in studies on rabbit thymus and liver[29]. Likewise HMG-T, a trout testis HMG protein which is supposedly similar to HMG1 and HMG2, is not preferentially released by DNase I[57]. On the other hand preferential release of H6 has been reported in trout testis digests[55, 56], while HMG17, which is sometimes compared to H6, is not released in some laboratories[29, 52, 72], but is released in others[98]. All the HMG proteins have been found associated with 0.1 mol/ℓ NaCl-soluble mononucleosomes[30, 53, 54], which are reportedly enriched in transcribed DNA sequences.

The sub-nucleosomal localization of the HMG proteins is difficult to define with precision at present. In trout testis, H6 has been found on nucleosomal core particles in transcribed chromatin, whereas HMG-T appears to be associated with the internucleosomal linker regions immediately adjacent to those core nucleosomes containing protein H6[54]. Similarly, in other organisms HMG14 and HMG17 appear to be closely associated with core particles[30, 62], while HMG1 and HMG2 may be present on linker DNA. The polar nature of HMG14 and HMG17 is believed to permit their binding to both the histone octamer and the nucleosomal DNA in core particles, whereas HMG1 and HMG2 bind DNA, are apparently found as homopolymers *in vivo*[37] and can partake in hydrophobic interaction[17], all of which may point to their presence on linker DNA where they may either crosslink the chromatin filament by head-to-tail interactions with other HMG proteins or with histone H1. Alternatively, it has been proposed that they replace histone H1 on the linker DNA, and are thereby involved in the decondensation of transcribed chromatin.

The role of HMG14 and HMG17 in the structure of transcribed chromatin is less controversial. It has been observed that transcribed genomic sequences in chromatin lose their preferential sensitivity to DNase I in the absence of HMG14 and HMG17, but this sensitivity can be restored by reconstitution of HMG14 and HMG17 back to HMG-depleted chromatin[25, 31, 84, 98, 99, 100]. This observation is also true for crude mononucleosome preparations, indicating that the HMG proteins sensitize the nucleosomal DNA to DNase I in a linker DNA-independent fashion. Such a conclusion is further supported by the observation that in whole chromatin, HMG14 and HMG17 sensitize transcribed sequences in the absence of histone H1 (or H5). The reconstitution studies and the use of affinity chromatography with immobilized HMG14 and HMG17 have established that HMG14 and HMG17 selectively bind to nucleosomes containing actively transcribed sequences (about one out of twenty nucleosomes overall), conferring upon them sensitivity to DNase I digestion. Thus, HMG14 and HMG17 appear to bind to nucleosomes in a sequence-specific fashion, in spite of the fact that they bind naked DNA with no apparent sequence specificity.

The biochemical basis for the specific binding of HMG14 and HMG17 to active nucleosomes is not clear at present. Attempts to identify the factor(s) involved have included the use of nucleosomes washed with $0.6\,mol/\ell$ NaCl, as well as nucleosomes dissociated in high salt urea and then reconstituted, but in both cases the specificity of HMG binding is retained[99]. In addition, no individual type of non-histone protein is present on nucleosomes in sufficient amounts to imply a role in HMG–nucleosome recognition. In fact, the $0.6\,mol/\ell$ NaCl washed nucleosomes have no detectable non-histone proteins at all. Nucleosomes to which HMG14 and HMG17 bind selectively have been reported to be distinguished by an enrichment in acetylated histones and undermethylated DNA sequences[97, 99]. While the role of the undermethylated DNA sequences in HMG recognition is uncertain, the levels of histone acetylation appear to be insufficient in themselves to explain the molar specificity of HMG binding to active nucleosomes. Competition studies have shown that the histone H2A–H2B dimer, but not the H3–H4 dimer, will compete with HMG14 and HMG17 for binding sites on active nucleosomes, suggesting that the HMG proteins bind to a site which normally interacts with histones H2A and H2B, perhaps the H3–H4:DNA complex. HMG14 and HMG17 thus appear to be able to recognize a specific nucleosomal conformation which may be induced by a combination of histone and DNA post-synthetic modifications, implying that enzymatic non-histone proteins present in sub-stoichiometric amounts may be ultimately responsible for determining the structure of chromosomal sequences destined to be transcribed or repressed.

Enzymatic non-histone proteins

In addition to the structural and regulatory non-histone proteins described in the previous sections, a variety of different enzymes are present in the non-histone protein fraction (*Table 3.2*). The major components in this category are enzymes catalysing nucleic acid synthesis and post-synthetic modifications of nucleic acids

TABLE 3.2. Summary of enzymatic components of the non-histone chromosomal protein fraction (for detailed references see reference 87)

Enzyme category	Function
Nucleic acids as substrates	
DNA polymerases	Polymerization of deoxyribonucleotides into DNA
RNA polymerases	Polymerization of ribonucleotides into RNA
Nucleases	Processing or degradation of DNA and/or RNA
Nucleotide ligase	Joining DNA segments during DNA replication and repair
Nucleotide exotransferases	
DNA	Addition of nucleotides to the ends of nucleic acids
RNA (poly(A) polymerase)	
DNA methylase	Methylation of DNA
DNA-untwisting, unwinding, or relaxing enzymes (helix-destabilizing proteins)	Unwind DNA double helix and stabilize the resulting single stranded DNA
Chromosomal proteins as substrates	
Proteases	Processing or degradation of protein
Acetylases	Acetylation of proteins
Deacetylases	Removal of acetate groups from proteins
Protein kinases	Phosphorylation of proteins
(Histone)	
(Non-histone)	
Histone methylases	Methylation of histones
Poly(ADP-ribose) synthetase	Addition of ADP-ribose moieties to chromsomal proteins
Poly(ADP-ribose) glycohydrolase	Removal of ADP-ribose moieties from chromosomal proteins

and proteins. Some enzymes in this group, such as histone kinase, non-histone kinase, histone acetylase, and DNA methylase, may be involved in catalysing conformational changes in chromatin components associated with alterations in gene expression.

Direct evidence for non-histone proteins as regulatory molecules

Effects of non-histone proteins on gene transcription in chromatin

We have now seen that many of the properties of the non-histone protein fraction are consistent with the notion that some molecules in this fraction are involved in the control of gene transcription. Attempts to provide more direct evidence for this idea have involved experiments in which the effects of non-histone proteins on RNA synthesis catalysed by bacterial or homologous RNA polymerase have been studied directly *in vitro*. In the late 1960s and early 1970s many studies were carried out, and it was generally found that one can alter the rate of RNA synthesis by adding, removing, or changing the phosphorylation state of non-histone proteins. Nucleic acid hybridization techniques were employed to show that the type of RNA synthesized is specifically influenced by the tissue of origin of the non-histone proteins. As specific cDNA probes for defined RNA sequences became available they were applied to these systems, yielding data which suggested that specific non-histone proteins are involved in the regulation of globin, ovalbumin, histone, and ribosomal gene transcription. In the case of histone and ribosomal gene transcription, the phosphorylation state of the non-histone proteins was also shown to influence the efficiency of gene transcription. (For a detailed discussion and references for the above points, see reference 87.)

The ultimate goal of studies of the type described above has been to isolate and purify individual non-histone proteins and implicate them in the regulation of particular genes. Because of the difficulties associated with studying chromatin transcription *in vitro*, this goal has rarely, if ever, been achieved. The most progress in this direction has been made studying ribosomal RNA transcription, where a non-histone phosphoprotein of 139 000 daltons has been purified from the slime mould *Physarum polycephalum*[5,50]. This phosphoprotein has been shown to stimulate the RNA polymerase I catalysed synthesis of ribosomal RNA from a homologous nucleolar chromatin template, and to bind selectively to DNA restriction fragments containing the symmetry axis of the palindromic ribosomal DNA. Removal of phosphate groups from the phosphoprotein inhibits both its ability to bind to ribosomal DNA and its ability to stimulate ribosomal RNA synthesis, suggesting that the phosphorylation state of this phosphoprotein may be involved in regulating ribosomal gene transcription.

While data such as those described in the above two paragraphs have been interpreted as being consistent with a possible role of some non-histone proteins in regulating specific gene transcription, extreme caution must be exercised in interpreting results like those obtained with *in vitro* transcription systems. Because of the potential for numerous kinds of artifacts in such systems, the following criteria need to be met before one can justify the use of isolated chromatin as a bona fide model which accurately reflects *in vivo* transcription of mRNA sequences:

1. The probes used to identify specific RNA sequences synthesized *in vitro* should be homologous and should accurately reflect the nucleotide sequence of the primary gene transcript. Sufficient amounts of probe should be available to carry out hybridization analysis of gene transcripts in RNA as well as in DNA excess. It is also desirable to have probes which represent the completely processed polysomal mRNAs.
2. Conditions for RNA transcription, isolation of transcripts and hybridization analysis should be such that endogenous, chromatin-associated RNAs do not interfere with quantitation of gene transcripts.
3. Precautions must be taken to eliminate or account for RNA-dependent RNA synthesis when transcription is carried out with *E. coli* RNA polymerase. It is desirable to transcribe chromatin with the appropriate homologous RNA polymerase, in which case it would be necessary to demonstrate that only the enzyme responsible for transcription *in vivo* is operative in the *in vitro* system.
4. It should be established that initiation of RNA synthesis is taking place *in vitro*, not merely elongation of RNA molecules initiated *in vivo*.
5. It is necessary to establish that initiation and termination of transcription *in vitro* occur at the same sites as they do *in vivo*.
6. It should be shown that the degree of symmetric or asymmetric transcription of genes from chromatin is the same as that which takes place in intact cells.
7. In addition to establishing fidelity of gene transcription from chromatin *in vitro*, it is necessary to show that the availability *and* lack of availability of other genetic sequences for *in vitro* transcription from chromatin is the same as in intact cells.

8. Any stimulatory or inhibitory effects of chromosomal proteins on DNA transcription *in vitro* must be observed under conditions, particularly after reconstitution, where structural properties of the chromosomal protein–DNA complexes reflect those occurring *in vivo*.

Because the vast majority of experiments have not been carried out under conditions where all the above criteria are satisfied, questions have been raised concerning the validity of the conclusion that non-histone proteins are involved in specific gene regulation. It is clear that in spite of the large number of chromatin transcription studies in which non-histone proteins have been shown to affect gene transcription, more rigorously controlled systems consisting of well-defined, purified components will ultimately be needed to resolve the issue and to precisely pinpoint the role of specific non-histone proteins. A major step in this direction has recently been made possible by the development of recombinant DNA techniques for cloning specific genomic sequences for subsequent use in cell-free transcription studies.

Utilization of cloned genes for *in vitro* transcription studies

Because of the complexity of the eukaryotic genome, it is difficult to study factors involved in controlling transcription by simply incubating chromatin containing total cellular DNA with purified RNA polymerase. The transcription product of any particular gene represents but a tiny fraction of the total RNA synthesized from such a template, and the situation is further complicated by an inability to define the primary RNA transcript of any particular gene without precise information concerning the *in vivo* initiation and termination sites. The advent of nucleic acid cloning and sequencing techniques has provided a powerful new weapon for overcoming these obstacles, for with cloned genes one can precisely define the organization of a gene, including its transcriptional start and stop sites, and use cloned DNAs containing specific genes and their adjacent sequences as templates for *in vitro* transcription studies. By creating mutants in which specific regions have been deleted, one can even define with precision the particular DNA base sequences involved in regulating transcription.

One of the most striking examples of how this approach can be used to define the role of non-histone proteins in gene regulation has been provided by the elegant work of Brown's laboratory on 5S ribosomal RNA transcription in *Xenopus*[8, 79]. By creating deletion mutants in which varying length segments were removed from the 5' and 3' flanking regions of cloned 5S genes, they were able to show that a region in the centre of the gene (located approximately between nucleotides 50 and 80) is required for directing accurate initiation of transcription. Thus mutants deleted as far as 50 bases into the gene from the 5' end still initiate transcription at a precise point 50 bases upstream, producing a fused transcript consisting of both plasmid and 5S RNA sequences. Further studies revealed that when the control region located in the centre of the 5S gene is subcloned in the absence of any other 5S DNA sequences, it still directs initiation of transcription at a specific point 50 bases upstream in whatever surrounding DNA it is placed adjacent to.

Shortly after the above findings were reported, Roeder's laboratory isolated a protein from *Xenopus* ovaries which specifically stimulates transcription of cloned 5S genes[21]. This factor was found to specifically bind to a region within the 5S RNA gene located in the same position as the control region defined by deletion analysis.

Pelham and Brown[69] subsequently showed that this non-histone protein factor is identical to an abundant cytoplasmic protein known to be complexed to 5S RNA in immature *Xenopus* oocytes. Thus a negative feedback mechanism seems to be operative *in vivo* in which the synthesis of the non-histone transcription factor turns on 5S RNA synthesis, producing 5S RNA molecules which in turn bind to the transcription factor and thereby inhibit its ability to further activate 5S RNA transcription.

The above system is clearly the best understood example of how a specific non-histone protein can control the transcription of a particular eukaryotic gene. Unfortunately, the 5S gene is atypical in many ways: it is much smaller (120 bases) than typical structural genes, its transcription is catalysed by a different enzyme (RNA polymerase III), and its gene product is synthesized in massive amounts. Hence, the usefulness of the 5S gene as a general model for how non-histone proteins regulate gene transcription in eukaryotes is open to question. On the other hand, the experimental approaches developed in the process of unraveling the mechanisms involved in controlling transcription of this gene are certain to form the basis for future investigations on the more complex eukaryotic structural genes.

Concluding remarks

In this chapter we have attempted to review progress which has been made during the past several years toward elucidating the role of non-histone chromosomal proteins in determining the structure and functional properties of the genome. It is clear that at present, we are only at the threshold of understanding the nature of eukaryotic regulatory macromolecules and their mode of action. With regard to the regulation of gene expression, histones appear to act as non-specific repressors of DNA-dependent RNA synthesis. In contrast, amongst the complex and heterogeneous non-histone chromosomal proteins are components which appear to regulate the transcription of specific genetic sequences. Yet, many perplexing problems concerning the non-histone chromosomal proteins remain to be resolved.

The specific non-histone chromosomal proteins responsible for rendering particular genetic sequences transcribable generally remain to be identified. It is usually assumed that specific regulatory proteins comprise only a small percentage of the non-histone chromosomal proteins, but experimental evidence to support this assumption is lacking. As an alternative possibility, one could postulate a model for gene regulation in which multiple copies of regulatory proteins are associated with the genome, with only a limited number of these proteins existing in a 'functional interaction'. An important concept which should be considered is that a single protein may regulate several genes, particularly in situations where cellular events are functionally interrelated or coupled. For example, one may envisage several genes involved with genome replication being controlled by a single regulatory protein. This concept is consistent with our recent identification of cloned genomic human sequences whose mRNAs, like histone mRNAs, are present on the polysomes only during S phase and are selectively lost from the polysomes when DNA synthesis is inhibited[60, 74]. One possible interpretation of these results is that the expression of functionally related genes is being observed. A similar situation may exist with hormone-stimulated processes. It is not clear whether regulatory proteins should comprise a subset of the non-histone chromosomal proteins with common characteristics such as molecular weight, charge or structure. In this regard it will be interesting to establish whether similar types of proteins control 'single copy genes', such as globin genes, as opposed to 'reiterated genes', such as histone and ribosomal genes.

A basic question to be answered is whether activation of genes is generally brought about by newly synthesized non-histone chromosomal proteins or modifications of pre-existing genome-associated non-histone chromosomal proteins. Alternatively, proteins residing in the cytoplasm or in the nucleoplasm may be modified in such a manner that they become associated with the genome and thereby render genes transcribable. Johnson, Karn and Allfrey[42] have observed that activation of lymphocytes by mitogenic agents results in accumulation of pre-existing cytoplasmic proteins in the nucleus. A nucleoplasmic pool of non-histone chromosomal proteins has also been reported[85]. These latter two observations are consistent with the possibility that alterations in gene readout involve recruitment of proteins from the cytoplasm or nucleoplasm and their subsequent association with the genome. However, other studies suggest that protein synthesis is required for activation of transcription in human diploid fibroblasts following stimulation to proliferate[77]. We have also discussed earlier in this chapter the evidence that the presence or absence of phosphate groups on non-histone chromosomal proteins is important in rendering genes transcribable.

Elucidating the manner in which non-histone chromosomal proteins are associated with other genome components should significantly enhance our understanding of the mechanisms by which regulatory proteins interact with defined regions of the genome to render specific genes transcribable. While tenaciously bound as well as readily dissociable non-histone chromosomal proteins have been purported to be the subclass of the non-histone protein fraction containing the regulatory macromolecules, direct experimental evidence to distinguish between these two alternatives is at present limited. It is also presently unclear if non-histone chromosomal proteins interact directly with DNA or with histone–DNA complexes.

In addition to understanding the mechanism by which non-histone chromosomal proteins activate transcription of specific genes, the mechanism by which transcription is turned off must also be accounted for. One may envisage inactivation of a gene or set of genes via degradation of the activator protein or proteins. Proteases which utilize non-histone chromosomal proteins as substrates have been shown to be associated with chromatin. An alternative mechanism for inactivation of regulatory proteins may involve acetate and/or phosphate groups added or removed from non-histone chromosomal proteins. Deacetylases as well as phosphatases have been identified within the nucleus, lending credence to such speculation.

But perhaps most importantly, caution must be exercised in making unqualified generalizations regarding regulation of eukaryotic gene expression and eukaryotic gene regulators. Many of the genes thus far examined in a comprehensive manner, such as globin and ovalbumin genes, reflect a long term commitment of a cell to expression of a differentiated gene product. These genes contain intervening sequences and their transcriptions require a complex series of processing steps (cleavage, splicing and chemical modification) before they are translocated to the cytoplasm as templates for the synthesis of proteins. The types of transcriptional control operative in these situations and the nature of their regulators may be distinctively different from that associated with genes such as histone genes, or genes which code for metabolic enzymes, where expression is transient and acutely responsive to cellular requirements, e.g. DNA replication or substrate levels. From structural as well as functional standpoints, it may be instructive to bear in mind that at least some such genes do not contain intervening sequences and undergo a comparatively minimal amount of post-transcriptional processing. It would not be at all unrealistic to consider the possibility that such genes might be organized and regulated in a prokaryotic-type manner.

As fractionation and characterization of the non-histone chromosomal proteins progress, and additional information regarding the nature of sequences involved in control of specific genes becomes available, the functional properties of eukaryotic gene regulators should become more apparent. The recent dramatic advances in molecular cloning, nucleic acid sequencing, and monoclonal antibody techniques provide us with a powerful new set of weapons which should greatly facilitate future progress in this area.

References

1. Abercrombie, B. D., Kneale, G. G., Crane-Robinson, C., Bradbury, M., Goodwin, G. H., Walder, J. and Johns, E. W. (1978). Studies on the conformational properties of HMG17 and its interaction with DNA. *European Journal of Biochemistry*, **84**, 173
2. Adler, K., Beyreuther, K., Fanning, E., Geisler, N., Gronenborn, B., Klemm, A., Mueller-Hill, B., Pfahl, M. and Schmitz, A. (1972). How *lac* repressor binds to DNA. *Nature, London*, **237**, 322
3. Alberts, B. M. and Frey, L. (1970). T4 bacteriophage gene 32: a structural protein in the replication and recombination of DNA. *Nature, London*, **227**, 1313
4. Anderson, W. B., Schneider, A. B., Emmer, M., Perlman, P. L. and Pastan, I. (1971). Purification of and properties of the cyclic adenosine 3',5'-monophosphate-dependent gene transcription in *Escherichia coli*. *Journal of Biological Chemistry*, **246**, 5929
5. Atmar, V. J., Daniels, G. R., Kuehn, G. D. and Braun, R. (1980). Opposing kinetic effects of an acidic nucleolar phosphoprotein from *Physarum polycephalum* on homologous and heterologous transcription systems, *FEBS Letters*, **114**, 205
6. Bekhor, I. and Feldman, G. (1976). Assembly of DNA with histone and nonhistone proteins *in vitro*. *Biochemistry*, **15**, 4771
7. Bekhor, I. and Mirrell, C. J. (1979). Simple isolation of DNA hydrophobically complexed with presumed gene regulatory proteins (M3). *Biochemistry*, **18**, 609
8. Bogenhagen, D. F., Sakonju, S. and Brown, D. D. (1980). A control region in the center of the 5S RNA gene directs specific initiation of transcription: II. The 3' border of the region. *Cell*, **19**, 27
9. Busch, H., Ballal, N. R., Olson, M. O. J. and Yeoman, L. C. (1975). Chromatin and its nonhistone proteins. In *Methods in Cancer Research*, Vol XI, (Busch, H., Ed.), p. 43. New York, Academic Press
10. Carrara, G., Gattoni, S., Mercanst, B. and Tochini-Valentini, G. P. (1977). Purification of a DNA-binding protein from *Xenopus laevis* unfertilized eggs. *Nucleic Acids Research*, **4**, 2855
11. Catino, J. J., Yeoman, L. C., Mandel, M. and Busch, H. (1978). Characterization of DNA binding protein from rat liver chromatin which decreases during growth. *Biochemistry*, **17**, 983
12. Chao, M. N., Gralla, J. D. and Martinson, H. G. (1980). *Lac* operator nucleosomes. I. Repressor binds specifically to operator with the nucleosome core. *Biochemistry*, **19**, 3254
13. Chao, M. N., Gralla, J. D. and Martinson, H. G. (1980). *Lac* operator nucleosomes. 2. *Lac* nucleosomes can change conformation to strengthen binding by *lac* repressor. *Biochemistry*, **19**, 3260
14. Chuang, D. M., Hollenbeck, R. and Costa, E. (1976). Enhanced template activity in chromatin from adrenal medulla after phosphorylation of chromosomal proteins. *Science*, **193**, 60
15. Chuang, D. M., Hollenbeck, R. A. and Costa, E. (1977). Protein phosphorylation in nuclei of adrenal medulla incubated with cyclic adenosine 3',5'-monophosphate-dependent protein kinase. *Journal of Biological Chemistry*, **252**, 8365
16. Clark, B. F. C. (1981). Towards a total human map. *Nature, London*, **292**, 491
17. Conner, B. J. and Comings, D. E. (1981). Isolation of nonhistone chromosomal HMG protein from mouse liver nuclei by hydrophobic chromatography. *Journal of Biological Chemistry*, **258**, 3283

18. Crepin, M. and Dastugue, B. (1979). Regulation of transcription of DNA-bound nonhistone proteins: *European Journal of Biochemistry*, **99**, 499

19. Dastugue, B. and Crepin, M. (1979). Interaction of nonhistone proteins with DNA-bound nonhistone proteins. *European Journal of Biochemistry*, **99**, 491

20. Duguet, M. and de Recondo, A. M. (1978). A DNA unwinding protein isolated from regenerating rat liver. *Journal of Biological Chemistry*, **253**, 1660

21. Engelke, D. R., Ng, S. Y., Shastry, B. S. and Roeder, R. G. (1980). Specific interaction of purified transcription factor with an internal control region of 5S RNA genes. *Cell*, **19**, 717

22. Fujitani, H. and Holoubek, V. (1974). Non-histone nuclear proteins of rat brain. *Journal of Neurochemistry*, **23**, 1215

23. Gates, D. M. and Bekhor, I. (1979). DNA sequences selection by tightly bound nonhistone chromosomal proteins. *Nucleic Acids Research*, **6**, 1617

24. Gates, D. M. and Bekhor, I. (1966). DNA-binding activity of tightly bound nonhistone chromosomal proteins in chicken liver chromatin. *Nucleic Acids Research*, **6**, 3411

25. Gazit, B., Panet, A. and Cedar, H. (1980). Reconstitution of a DNase I structure on active genes. *Proceedings of the National Academy of Sciences, USA*, **77**, 1787

26. Gilbert, W. and Mueller-Hill, B. (1966). Isolation of the *lac* repressor. *Proceedings of the National Academy of Sciences, USA*, **56**, 1891

27. Glass, D. B. and Krebs, E. G. (1980). Protein phosphorylation catalyzed by cyclic AMP-dependent and cyclic GMP-dependent protein kinases. *Annual Review of Pharmacology and Toxicology*, **20**, 363

28. Goodwin, G. H. and Johns, E. W. (1973). Isolation and characterization of two calf thymus chromatin nonhistone proteins with high contents of acidic and basic amino acids. *European Journal of Biochemistry*, **40**, 215

29. Goodwin, G. H. and Johns, E. W. (1978). Are the HMG nonhistone chromosomal proteins associated with 'Active Chromatin'? *Biochimica et Biophysica Acta*, **519**, 279

30. Goodwin, G. H., Matthew, C. G. P., Wright, C. A., Venkov, C. D. and Johns, E. W. (1979). Analysis of the HMG proteins associated with salt soluble nucleosomes. *Nucleic Acids Research*, **7**, 1815

31. Groudine, M. and Wintraub, H. (1981). Activation of globin genes during chicken development. *Cell*, **24**, 393

32. Helmsing, P. and Berendes, H. (1971). Induced accumulation of nonhistone proteins in polytene nuclei of *Drosophila hydei*. *Journal of Cell Biology*, **50**, 893

33. Herrick, G. and Alberts, B. (1976). Nucleic acid helix-coil transitions mediated by helix-unwinding proteins from calf thymus. *Journal of Biological Chemistry*, **251**, 2133

34. Hotta, Y. and Stern, H. (1978). The effect of dephosphorylation on the properties of a helix-destabilizing protein from meiotic cells and its partial reversal *Physarum polycephalum*, *European Journal of Biochemistry*, **90**, 29

35. Hotta, Y. and Stern, H. (1979). The effect of dephosphorylation on the properties of a helix destabilizing protein from meiotic cells and its partial reversal by protein kinase. *European Journal of Biochemistry*, **95**, 31

36. Hsieh, T. S. and Brutlag, D. L. (1980). A protein that preferentially binds *Drosophila* DNA. *Proceedings of the National Academy of Sciences, USA*, **76**, 726

37. Itkes, A. V., Golotov, B. O., Nikolacv, L. G. and Severin, E. S. (1981). Clusters of nonhistone chromosomal protein HMG 1 molecules in intact chromatin. *FEBS Letters*, **118**, 63

38. Jackson, P. J., Fishback, J. L., Bidney, D. L. and Reeck, G. R. (1979). Preferential affinity of the high molecular weight HMG nonhistone chromosomal proteins for single stranded-DNA. *Journal of Biological Chemistry*, **254**, 5569

39. Jagodzinski, L. L., Chilton, J. C. and Sevall, J. S. (1978). DNA-binding nonhistone proteins: DNA site reassociation. *Nucleic Acids Research*, **5**, 1487

40. Jagodzinski, L. L., Castro, E., Sherrod, P., Lee, D. and Sevall, J. S. (1979). Reassociation kinetics of nonhistone-bound DNA-sites. *Journal of Biological Chemistry*, **254**, 3038

41. Javaherian, K. and Sadeghi, M. (1979). Nonhistone proteins HMG1 and HMG2 unwind the DNA double helix. *Nucleic Acids Research*, **6**, 3569

42. Johnson, E. M., Karn, J. and Allfrey, V. G. (1974). Early nuclear events in the induction of lymphocyte proliferation by mitogens. Effects of concanavalin A on the phosphorylation and distribution of non-histone chromatin proteins. *Journal of Biological Chemistry*, **249**, 4990

43. Kleinsmith, L. J. (1973). Specific binding of phosphorylated non-histone chromatin proteins to deoxyribonucleic acid. *Journal of Biological Chemistry*, **248**, 5448

44 Kleinsmith, L. J. (1974). Acidic nuclear proteins. In *Acidic Proteins of the Nucleus* (Cameron, I. L. and Jeterk J. R., Jr., Eds.) p. 103. New York, Academic Press

45. Kleinsmith, L. J. (1978). Phosphorylation of nonhistone proteins. In *The Cell Nucleus*, Vol. VI, (Busch, H., Ed.) p. 221. New York, Academic Press

46. Kleinsmith, L. J., Heidema, J. and Carroll, A. (1970). Specific binding of rat liver nuclear proteins to DNA. *Nature, London*, **226**, 1025

47. Kleinsmith, L. J., Stein, J. and Stein, G. (1976). Dephosphorylation of nonhistone proteins specifically alters the pattern of gene transcription in reconstituted chromatin. *Proceedings of the National Academy of Sciences, USA*, **73**, 1174

48. Kranias, E. G., Schweppe, J. S. and Jungmann, R. A. (1977). Phosphorylative and functional modifications of nucleoplasmic RNA polymerase II by homologous adenosine 3′,5′-monophosphate-dependent protein kinase from calf thymus and by heterologous phosphatase. *Journal of Biological Chemistry*, **252**, 6750

49. Krebs, E. G. and Beavo, J. A. (1979). Phosphorylation–dephosphorylation of enzymes. *Annual Review of Biochemistry*, **48**, 923

50. Kuehn, G. D., Affolter, H. U., Atmar, V. J., Seebeck, T., Gubler, U. and Braun, R. (1979). Polyamine-mediated phosphorylation of nucleolar protein from *Physarum polycephalum* that stimulates rRNA synthesis. *Proceedings of the National Academy of Sciences, USA*, **76**, 2541

51. Langan, T. A. (1973). Protein kinases and protein kinase substrates. *Advances in Cyclic Nucleotide Research*, **3**, 99

52. Levy, W. B. and Dixon, G. H. (1978). Partial purification of transcriptionally active nucleosomes from trout testes cells. *Nucleic Acids Research*, **5**, 4155

53. Levy, W. B. and Dixon, G. H. (1978). A study of the localization of the HMGs in chromatin. *Canadian Journal of Biochemistry*, **56**, 480

54. Levy, W. B., Connor, W. and Dixon, G. H. (1979). A subset of trout testis nucleosomes enriched in transcribed DNA sequences containing HMG proteins as major structural components. *Journal of Biological Chemistry*, **254**, 609

55. Levy, W. B., Kuehl, L. and Dixon, G. H. (1980). The release of HMG H6 and protamine gene sequences upon selective DNase degradation of trout testes chromatin. *Nucleic Acids Research*, **8**, 2859

56. Levy, W. B., Wong, N. C. W. and Dixon, G. H. (1977). Selective association of the trout-specific H6 protein with chromatin regions susceptible to DNase I and DNase II: possible location of HMG-T in the spacer region between core nucleosomes. *Proceedings of the National Academy of Sciences, USA*, **74**, 2810

57. Levy, W. B., Wong, N. C. W., Waton, D. C., Peters, E. H. and Dixon, G. H. (1977). Structure and function of the low-salt extractable chromosomal proteins: preferential association of trout testis proteins H6 and HMGT with chromatin regions selectively sensitive to nucleases. *Cold Spring Harbor Symposium on Quantitative Biology*, **42**, 793

58. Magim, B. E. and Dorsett, P. H. (1977). Changes in the synthesis of DNA-binding proteins during the onset of transformation in NRK cells transformed by Ts mutant of Rous Sarcoma virus. *Journal of Virology*, **22**, 469

59. Mainwaring, W. I. P., Rennie, P. S. and Keen, J. (1976). The androgenic regulation of prostate proteins with a high affinity for deoxyribonucleic acid. Evidence for a prostate deoxyribonucleic acid-unwinding protein. *Biochemical Journal*, **156**, 253

60. Marashi, F., Decker, L., Rickles, R., Sierra, F., Stein, J. and Stein, G. S. (1982). Histone proteins in HeLa S₃ cells are synthesized in a cell cycle stage specific manner. *Science* (in press)

61. Mather, J. and Hotta, Y. (1977). A phosphorylable DNA-binding protein associated with a lipoprotein fraction from rat spermatocyte nuclei. *Experimental Cell Research*, **109**, 181

62. Mathew, C. G. P., Goodwin, G. H. and Johns, E. W. (1979). Studies on the association of the HMG nonhistone proteins with isolated nucleosomes. *Nucleic Acids Research*, **6**, 167

63. Mitchell, S. J. and Kleinsmith, L. J. (1982). Nuclear protein kinases. In *Nonhistone Proteins* (L. S. Hnilica, Ed.). CRC Press (in press.)

64. Nass, K. and Frenkel, G. D. (1979). A DNA-binding protein from KB cells which inhibits deoxyribonuclease activity on single strand DNA. *Journal of Biological Chemistry*, **254**, 3407

65. Nielsen, C. J., Sando, J. J. and Pratt, W. B. (1977). Evidence that dephosphorylation inactivates glucocorticoid receptors. *Proceedings of the National Academy of Sciences, USA*, **74**, 1398

66. Norman, G. L. and Bekhor, I. (1981). Enrichment of selected active human gene sequences in the placental DNA fraction associated with tightly bound nonhistone chromosomal proteins. *Biochemistry*, **20**, 3568

67. Otto, B., Baynes, M. and Knippers, R. (1977). A single-strand-specific DNA-binding protein from mouse cells that stimulates DNA polymerase. Its modification by phosphorylation. *European Journal of Biochemistry*, **73**, 17

68. Patel, G. L. and Thomas, T. L. (1973). Some binding parameters of chromatin acidic proteins with high affinity for deoxyribonucleic acid. *Proceedings of the National Academy of Sciences, USA*, **70**, 2524

69. Pelham, H. R. B. and Brown, D. D. (1980). A specific transcription factor that can bind either the 5S RNA gene or 5S RNA. *Proceedings of the National Academy of Sciences, USA*, **77**, 4170

70. Peterson, J. L. and McConkey, E. H. (1976). Non-histone chromosomal proteins from HeLa cells. A survey by high resolution, two-dimensional electrophoresis. *Journal of Biological Chemistry*, **251**, 548

71. Peterson, J. L. and McConkey, E. H. (1976). Proteins of Friend leukemia cells. Comparison of hemoglobin-synthesizing and noninduced populations. *Journal of Biological Chemistry*, **251**, 555

72. Plumb, M. A. and MacGillivray, A. J. (1981). The distribution of HMG proteins in chromatin fractions produced by nuclease digestion of pig thymus nuclei. *Biochemical Society Transactions*, **9**, 143

73. Rabbani, A., Goodwin, G. H. and Johns, E. W. (1978). Studies on the tissue specificity of the HMG nonhistone chromosomal proteins from calf. *Biochemical Journal*, **173**, 497

74. Rickles, R., Marashi, F. F., Wells, J. and Stein, G. S. (1982). Analysis of histone gene expression during the cell cycle in HeLa cells using cloned human histone genes. *Proceedings of the National Academy of Sciences, USA*, (in press)

75. Rose, K. M. and Jacob, S. T. (1979). Phosphorylation of nuclear poly(A) polymerase. Comparison of liver and hepatoma enzymes. *Journal of Biological Chemistry*, **254**, 10256

76. Rose, K. M. and Jacob, S. T. (1980). Phosphorylation of nuclear poly (adenylic acid) polymerase by protein kinase: mechanism of enhanced poly (adenylic acid) synthesis. *Biochemistry*, **19**, 1472

77. Rovera, G., Farber, J. L. and Baserga, R. (1971). Gene activation in WI-38 fibroblasts stimulated to proliferate: requirement for protein synthesis. *Proceedings of the National Academy of Sciences, USA*, **68**, 1725

78. Rubin, C. S. and Rosen, O. M. (1975). Protein phosphorylation, *Annual Review of Biochemistry*, **44**, 831

79. Sakonju, S., Bogenhagen, D. F. and Brown, D. D. (1980). A control region in the center of the 5S RNA gene directs specific initiation of transcription: I. The 5' border of the region. *Cell*, **19**, 13

80. Sanders, C. (1977). A method for the fractionation of the HMG nonhistone chromosomal proteins. *Biochemical and Biophysical Research Communications*, **78**, 1034

81. Sevall, J. S., Cockburn, A., Savage, M. and Bonner, J. (1975). DNA–protein interactions of the rat liver nonhistone chromosomal protein. *Biochemistry*, **14**, 782

82. Shooter, K. V., Goodwin, G. H. and Johns, E. W. (1974). Interactions of a purified nonhistone chromosomal protein with DNA and histone. *European Journal of Biochemistry*, **47**, 263

83. Spelsberg, T. C. (1974). The role of nuclear acidic proteins in binding steroid hormones. In *Acidic Proteins of the Nucleus* (Cameron, I. L. and Jeter, Jr., J. R., Eds.), p.247. New York, Academic Press

84. Stadler, J., Larsen, A., Engel, J. D., Dolan, M., Groudine, M. and Weintraub, H. (1980). Tissue specific DNA cleavages in the globin chromatin domain introduced by DNase I. *Cell*, **20**, 451

85. Stein, G. S. and Thrall, C. L. (1973). Evidence for the presence of nonhistone chromosomal proteins in the nucleoplasm of HeLa S₃ cells. *FEBS Letters*, **32**, 41

86. Stein, G. S., Stein, J. L. and Thomson, J. A. (1978). Chromosomal proteins in transformed and neoplastic cells: a review. *Cancer Research*, **38**, 1181

87. Stein, G., Plumb, M., Stein, J. L. Phillips, I. R. and Shephard, E. A. (1982). Nonhistone proteins in genetic regulation. In *Nonhistone proteins* (L. S. Hnilica, Ed.). CRC Press (in press)

88. Taborsky, G. (1974). Phosphoproteins. *Advances in Protein Chemistry*, **28**, 1

89. Teng, C. S., Teng, C. T. and Allfrey, V. G. (1971). Studies of nuclear acidic proteins Evidence for their phosphorylation, tissue specificity, selective binding to deoxyribonucleic acid, and stimulatory effects on transcription. *Journal of Biological Chemistry*, **246**, 3597

90. Thomas, T. L. and Patel, G. L. (1976). DNA unwinding component of the nonhistone chromatin proteins. *Proceedings of the National Academy of Sciences, USA*, **73**, 4364

91. Thrall, C. L., Webster, R. A. and Spelsberg, T. C. (1982). Steroid receptor interaction with chromatin. In *The Cell Nucleus* (Busch, H., Ed.). New York, Academic Press (in press)

92. Van den Broek, H. W. J., Nooden, L. D., Sevall, J. S. and Bonner, J. (1973). Isolation purification, and fractionation of nonhistone chromosomal proteins. *Biochemistry*, **12**, 229

93. Vidali, G., Boffa, L. C. and Allfrey, V. G. (1977). Selective release of chromosomal proteins during limited DNase I digestion of avian erythrocytes chromatin. *Cell*, **12**, 409

94. Wakabayaski, K., Wang, S., Hord, G. and Hnilica, L. S. (1973). Tissue-specific nonhistone chromatin proteins with affinity for DNA. *FEBS Letters*, **32**, 46

95. Watson, D. C., Wong, N. C. and Dixon, G. H. (1979). The complete amino acid sequence of a trout testis nonhistone protein H6 localized in a subset of nucleosomes and its similarity to calf thymus nonhistone proteins HMG14 and HMG17. *European Journal of Biochemistry*, **95,** 193
96. Weideli, H., Brack, C. and Gehring, W. J. (1980). Characterization of *Drosophila* DNA-binding protein DB-2: demonstration of its sequence-specific interaction with DNA. *Proceedings of the National Academy of Sciences, USA*, **77,** 3773
97. Weintraub, H., Larsen, A. and Groudine, M. (1981). α-Globin-gene switching during the development of chicken embryos: Expression and chromatin structure. *Cell*, **24,** 333
98. Weisbrod, S. and Weintraub, H. (1979). Isolation of a subclass of nuclear proteins responsible for conferring a DNase I sensitive structure on globin chromatin. *Proceedings of the National Academy of Sciences, USA*, **76,** 630
99. Weisbrod, S. and Weintraub, H. (1981). Isolation of actively transcribed nucleosomes using immobilized HMG14 and HMG17 and an analysis of α-globin chromatin. *Cell*, **23,** 391
100. Weisbrod, S., Groudine, M. and Weintraub, H. (1980). Interaction of HMG14 and HMG17 with actively transcribed genes. *Cell*, **19,** 289
101. Yeoman, L. C., Taylor, C. W. Jordan, C. W. and Busch, H. (1975). Differences in chromatin proteins of growing and non-growing tissues. *Experimental Cell Research*, **91,** 207
102. Yeoman, L. C., Seeber, S., Taylor, C. W., Fernbach, D. J., Falletta, J. M., Jordan, J. J. and Busch, H. (1976). Differences in chromatin proteins of resting and growing human lymphocytes. *Experimental Cell Research*, **100,** 47
103. Yu, S. S., Li, H. J., Goodwin, G. H. and Johns, E. W. (1977). Interaction of nonhistone chromosomal proteins HMG1 and HMG2 with DNA. *European Journal of Biochemistry*, **78,** 497
104. Zubay, G., Schwartz, D. and Beckwith, J. (1970). Mechanism of activation of catabolite-sensitive genes: a positive control system. *Proceedings of the National Academy of Sciences, USA*, **66,** 104

Chapter 4

DNA modification

A. P. Bird

Introduction

Like many cellular macromolecules, DNA is subject to post-synthetic modification. In animals and plants the modification involves methylation of cytosine to give 5-methylcytosine (5mC) (*Figure 4.1*)[57, 68, 78], though in prokaryotes and unicellular eukaryotes other modifications are found[48]. This chapter is concerned with 5mC in animals, because it is here that recent progress has been most rapid[50]. Like most of molecular biology, the field has been transformed by the application of restriction endonucleases and cloning technology. In addition to this practical advance, the concept of restriction and modification as worked out in bacteria has inspired intriguing models for the function of 5mC in eukaryotes[28, 51, 53].

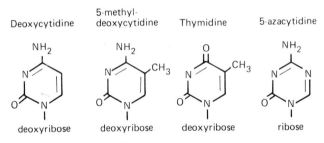

Figure 4.1. The structure of 5-methyl-2-deoxycytidine (5mC) compared with deoxycytidine, thymidine, and the drug 5-azacytidine. Note the structural similarity between 5mC and thymidine. The analogue 5-azacytidine has a nitrogen atom at the 5-position of the ring which prevents it from accepting a methyl group in DNA.

From the point of view of gene expression, interest in DNA methylation springs from two phenomena. First, it is known that the presence or absence of 5mC in bacteria can determine whether a protein (i.e. a restriction endonuclease) interacts productively or unproductively with its recognition sequence on DNA[42]. This raises the possibility that eukaryotic methylation is also involved in switching DNA–protein interactions. Secondly, once more by analogy with bacteria, the process of methylation has the properties of a replication system that is capable of perpetuating a pattern of methylation through many cell generations (see p. 55). The idea of a heritable, but reversible, signal on the DNA, whose presence or absence does not affect genetic coding, has obvious attractions.

Methylation is mainly at CG

5-Methylcytosine was first detected by base analysis of DNA from various sources[78]. Early attempts to sequence neighbouring bases showed that 5mC is usually followed by G on its 3' side[21, 25]. No specificity was observed in the 5' neighbour. This picture has been supported by more recent analyses of total cellular DNA[12] and directly demonstrated by sequencing studies. The latter experiments used the chemical hydrolysis method of nucleotide sequencing[41] which is able to distinguish between 5mC and unmodified C. Unfortunately this method is limited to genomic DNA fractions that can be purified by physical methods, since cloned sequences no longer retain their native methylation pattern. In spite of this limitation, several repetitive sequence fractions have been analysed (e.g. *Xenopus laevis* 5S genes[22, 43], mouse satellite DNA[30]), and in each case methylation is found exclusively at CG.

Although CG is by far the predominant methylated sequence, other 5mC-containing dinucleotides are observed. Pyrimidine tract analysis detects small amounts of mCC and mCT in animal DNA in addition to mCG[21, 25]. An example of the sequence mCC or mCmC has now been located near the fetal globin genes of man[67].

Mapping 5mC with restriction endonucleases

The most widespread method for analysing 5mC makes use of certain class II restriction endonucleases[5, 24]. Many endonucleases include CG in their cleavage

TABLE 4.1. Restriction endonucleases that are useful for studying DNA methylation in animals and plants

Endonuclease	Recognition sequence cleaved	Not cleaved
Hpa II/Hap II	CCGG	CmCGG
Msp I	CCGG, CmCGG	mCCGG
Hha I	GCGC	GmCGC
Sma I	CCCGGG	CCmCGGG
Xma I	CCCGGG, CCmCGGG	?
Xho I	CTCGAG	CTmCGAG
Ava I	CPyCGPuG	CPy.mCGPuG
Sal I	GTCGAC	GTmCGAC
Taq I	TCGA, TmCGA	TCGmA
Aos II	GPuCGPyC	GPu.mCGPyC

Note that all recognition sequences include the dinucleotide CG. References for cleavage sites and sensitivity to modification are collected in reference 36.

recognition sequence, and are unable to cleave if the C is methylated (*Table 4.1*). Cleavage within a particular sequence, therefore, indicates an unmethylated CG at that site. Failure to cleave, on the other hand, indicates either that sites contain mCG, or that there are no sites for the enzyme. This ambiguity necessitates a control digestion to determine how many times the restriction site occurs in the sequence of interest. The control can be achieved in two ways: either the cloned (i.e. unmethylated) sequence can be tested for the presence of restriction sites, or the genomic DNA can be digested with two enzymes that recognize the CG-containing site, but only one of which is blocked by CG methylation. The latter approach is simpler and more direct, and a suitable pair of enzymes, Hpa II and Msp I, is available (*Table 4.1*). Both enzymes cleave at CCGG, but Hpa II is blocked by methylation at the second C[40], while Msp I is not [14, 69]. Thus Msp I can be used to determined the number of CCGGs in a genomic sequence, while Hpa II shows how many of the sites are methylated. When combined with the blotting method[62], Hpa II and Msp I offer a powerful technique for detecting methylation of specific sequences in genomic DNA.

Replication of mCG

CGs occur in pairs on opposite strands of the DNA duplex (*Figure 4.2*). In theory, this symmetry provides the basis for replication of any pattern of methylated and unmethylated CG[28, 51]. Since DNA replication is semi-conservative, and since

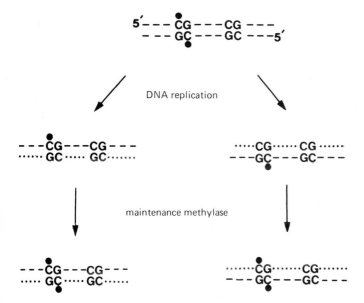

Figure 4.2. Replication of the pattern of CG methylation by a maintenance methylase following DNA replication. The short segment of parental DNA (top) contains a symmetrically methylated CG pair and an unmethylated CG pair. Following DNA replication a newly synthesized progeny strand (dotted line) is paired with each parental strand, but is not yet methylated. The maintenance methylase restores symmetry of the half-methylated sites, but is incapable of *de novo* methylation at the unmethylated CG pair. Thus a replica of the parental pattern of methylation is transmitted to each daughter cell.

methylation is a post-synthetic modification[13, 33, 55, 61], it follows that newly replicated DNA comprises a parental strand with an array of methylated or unmethylated CGs, while the progeny strand is not yet methylated. If we postulate an enzyme that only methylates progeny CGs that are opposite (i.e. base-paired with) mCG on the parental strand, the consequence is replication of the parental strand pattern onto the progeny strand (*Figure 4.2*).

There is now good evidence in favour of this model. Thus, in ribosomal DNA (rDNA) from *Xenopus laevis*, paired CGs are either symmetrically methylated or unmethylated, but not half-methylated[3]. This confirms the prediction of the replication model. Moreover, since most CGs in rDNA have a low but apparently random probability (about 1%) of existing in the unmethylated form, the majority of unmethylated CGs are not due to intrinsically unmethylatable sites. If CGs were methylated without reference to the parental strand, this would lead to more half-methylated than unmethylated sites. Since the reverse is found, the replication mechanism is strongly supported.

Direct evidence for replication has been obtained by introducing *in vitro* methylated DNA into cells. Two methods have been used: (a) transformation[76] of mouse cells lacking a functional thymidine kinase (*tk*) gene with a methylated *tk* gene or with an unmethylated *tk* gene plus methylated plasmid or bacteriophage DNA[47, 75]; (b) direct injection of methylated DNA into *Xenopus* eggs[27]. The DNA tested in each experiment was methylated with the Hpa II methylase[40]. Both approaches show that Hpa II sites can remain methylated after several cycles of DNA replication while CGs that are not part of Hpa II sites do not acquire methylation. Unmethylated Hpa II sites in control DNA also remain unmethylated. Thus the individual CG is a unit as regards inheritance of methylation.

Although CG methylation is retained, each experiment reveals that the methylation process is less than 100% efficient. Thus after 25 generations in mouse cells, only about 50% of the original methylated Hpa II sites in the chicken *tk* gene are still methylated[75]. It is not clear if this inefficiency is a normal feature of the methylation process, or if it is due to the fact that the test DNA is only methylated at Hpa II sites.

The enzyme that maintains the pattern of methylation between cell generations is referred to as the maintenance methylase[51]. Several DNA–methylase activities have been isolated from cultured cell nuclei, and they appear to have properties consistent with a maintenance function[1, 59].

De novo methylation

The demonstration that cells are able to replicate the pattern of CG methylation begs the question of how and when such patterns arise in the first place. Evidence on this subject is scanty, but two kinds of observation are relevant. The first is that when unmethylated sequences are introduced into culture cells by viral infection or by transformation, they occasionally acquire 5mC at multiple sites (see p. 58 for an example). We cannot estimate how often such *de novo* methylation occurs, as many cell generations elapse between the introduction of foreign DNA and the test for methylation, and consequently there is a chance for rare, but advantageous, events to be selected in the population. Co-transformation experiments with unmethylated (presumably selectively neutral) bacteriophage DNA suggest, however, that *de novo* methylation in cultured cells is a very rare event[47].

The second relevant observation concerns *de novo* methylation during development. Base composition analysis shows that bovine sperm DNA contains only 66% of the level of 5mC in somatic cells (Vanyushin *et al*, 1970)[68]. This is mainly due to greatly reduced levels of 5mC in the major bovine satellite DNA. In somatic tissues the satellite is methylated at 10% of C residues, but in sperm, less than 1% of satellite C is methylated[63]. It follows that sperm satellite DNA is methylated *de novo* during development. This evidence for developmental remethylation contrasts with results showing that a number of specific gene sequences are more heavily methylated in sperm than in somatic tissues (see p. 60). Further work is needed to see whether sequences are commonly 'remethylated' in each animal generation, or if sequences can also become methylated during evolution and remain so by virtue of a maintenance methylase operating at each germ cell division.

Variable levels of mCG among animals

Early studies of DNA base composition showed that levels of 5mC are highly variable between animal species[57]. At one extreme, 5mC was not reliably detected in insects (i.e. less than 0.2 moles % of bases), while in vertebrates 5mC accounted for 1–2% of bases. Levels of greater than 5% are common in plants. Recent application of a more sensitive technique to mosquito DNA shows levels close to the limits of detection at 0.03 moles%[1]. The range of methylation levels is more apparent when expressed as the percentage of all CGs that are methylated. Paucity of data limits the accuracy of the estimate for insects, but it is likely that in *Drosophila* less than 1% of CG is methylated, compared to typical levels of 50–90% in mammals[14, 60]. The wide range is visible when total DNA from a variety of animals is digested with Hpa II and Msp I (see *Figure 4.3*). Hpa II and Msp I generate indistinguishable fragments from insect DNA, whereas vertebrate DNA is poorly cleaved by Hpa I compared to Msp I[6, 49]. Intermediate between the extremes is a pattern of fractional methylation found in a range of invertebrates. Any model for the function of DNA methylation must explain this inter-species variability.

Domains of methylation

It is striking that in the animal species tested so far the level of CG methylation in genomic DNA is less than 100%. What is the spatial relationship between methylated and unmethylated CG? In the sea urchin the answer is that methylated CGs are clustered together in domains of DNA from 15 kb to over 50 kb in length[9]. Unmethylated CGs are clustered in domains of similar length. Thus the sea urchin genome can be thought of as divided into methylated and unmethylated 'compartments'. Similar compartments are observed in a wide range of other organisms, including several non-arthropod invertebrates[6], a slime mould[74] and a green plant[15].

Domains of methylation are not obvious in vertebrates. This may be because relatively few CGs are unmethylated, and because the dinucleotide CG is under-represented in vertebrate DNA (see p. 62). These properties would prevent unmethylated and methylated domains from separating clearly by size after

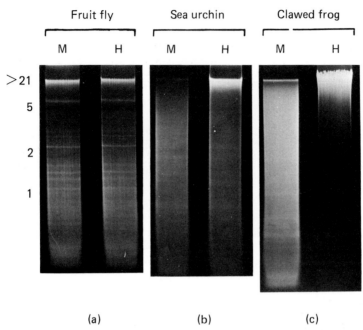

Figure 4.3. The wide range in levels of CG methylation in animals as detected by Hpa II(H) and Msp I(M). Both enzymes cleave the unmodified sequence CCGG, but Hpa II will not cleave CmCGG whereas Msp I does cleave this sequence. (a) *Drosophila melanogaster* (fruit fly) DNA gives indistinguishable fragment patterns indicating that most CCGG sites are unmethylated. (b) *Echinus esculentus* (sea urchin) DNA is partially resistant to Hpa II, suggesting that there are both methylated and unmethylated sites in the genome (see p. 59 and *Figure 4.4*). (c) *Xenopus laevis* (clawed frog) DNA appears largely undigested by Hpa II, indicating that most sites are methylated at CG. DNA from each animal was digested and fractionated on a 1.2% agarose gel. DNA was stained with ethidium bromide and photographed under short-wave ultraviolet light. Numbers refer to fragment size in kilobase pairs. Modified from reference 6 with permission.

digestion with methyl-sensitive restriction enzymes. It is clear, however, that vertebrate genomes do contain unmethylated domains in addition to the more abundant methylated sequences. Most ribosomal RNA genes, for example, are not detectably methylated in mammals, despite the presence of a high proportion of CGs in each repeating unit[7].

These results suggest that compartmentalization of genomes into methylated and unmethylated domains is a widespread, if not universal, phenomenon. An organization of this kind has implications for the mechanism of *de novo* methylation. It suggests that the unit of *de novo* methylation (as distinct from the unit of methyl replication) is not an individual CG, but a sequence domain within which all CGs become methylated at once[9]. This view is supported by the few cases of *de novo* methylation that have been observed so far. For example, a clone of mouse cells in which the viral *tk* gene is not expressed has been found to be methylated at many Hpa II sites in the *tk* gene[46]. In the original line, and several other revertant lines, the *tk* gene lacks methylation. Thus the methylation event appears to involve coordinate methylation of many CGs in a sequence block.

DNA methylation and transcription

An exciting possibility is that CG methylation is somehow involved in regulating gene expression. This section surveys some of the relevant evidence, mentioning its encouraging, as well at its less easily interpretable, aspects.

Transcription of genes in unmethylated domains

Cells containing integrated viral sequences may or may not express the viral genes. This variability provides the opportunity to look for correlations between methylation and transcription, by testing expressing and non-expressing cells for

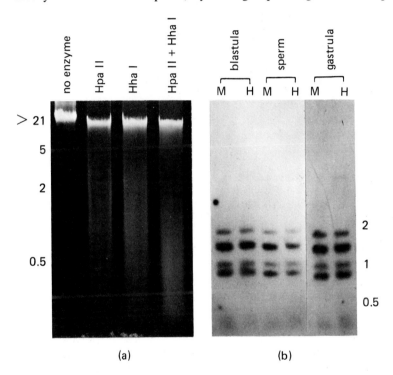

(a)   (b)

Figure 4.4. Domains of methylation in the sea urchin. (a) When sperm DNA is digested with Hpa II or Hha I a fraction of DNA remains apparently undigested, and a fraction is cleaved to fragments of about 1 kb. After digestion with Hpa II and Hha I together, the amount of resistant DNA is the same as with either enzyme alone, but the digested fraction is of lower average molecular weight. This, together with other experiments, demonstrates that sea urchin DNA comprises long stretches of heavily methylated DNA (the resistant fraction) and long stretches of unmethylated DNA (the digested fraction). (b) Absence of methylation of histone genes in expressing and non-expressing tissues. DNA from blastula (expressing), sperm (non-expressing) and gastrula (non-expressing) has been digested with Hpa II (H) and Msp I (M) and blotted from an agarose gel onto nitrocellulose. Histone coding sequences were identified by hybridizing the filter with a labelled clone of histone DNA followed by autoradiography. All band patterns are the same, indicating that CCGG is unmethylated in histone genes of the sea urchin, regardless of the level of histone gene transcription. Adapted from reference 9 with permission.

methylation of the viral genome. In several cases a correlation is found[16, 20, 64]. A particularly clear example involves cells that have been transformed with adenovirus-2[64]. In lines expressing the early viral genes, the region coding for early functions can be cleaved by mCG-sensitive restriction enzymes (i.e. it is undermethylated), while the unexpressed late region is resistant to these enzymes. In cells that express early and late genes, both gene sets are undermethylated. Thus in this and other viral cases, expressed sequences are part of unmethylated domains.

This is also true for many invertebrate genes. In the sea urchin, for example, the genes for 5S rRNA, 28S and 18S rRNA, and histones are all found in the unmethylated fraction of the genome[9]. Indeed, there is as yet no example of a methylated gene in the invertebrates. Correspondingly there is no evidence for a change in methylation accompanying expression. For example, the embryonic histone genes of sea urchins are unmethylated in sperm and gastrulae, where they are not expressed, but also in blastulae, where they are heavily transcribed (*Figure 4.4*[9]). In these and other cases (see below) absence of methylation is associated with the capacity for expression, but does not guarantee expression.

Expression of sequences in unmethylated domains is also apparent in the vertebrates. Unlike many vertebrate genes, the bulk of mouse rDNA lacks methylation, but a small fraction is heavily methylated[7]. In Balb/c mice, incubation of liver nuclei with DNase I degrades unmethylated rDNA more rapidly than methylated rDNA. This suggests that the unmethylated, but not the methylated, genes are in the active chromatin configuration[72].

Transcription of genes in methylated domains

The level of CG methylation in vertebrates is considerably higher than in invertebrates, and vertebrate genes are often found to be methylated. Nevertheless, the inverse correlation between capacity for transcription and methylation levels is still observed. Detailed study of several genes reveals that cells expressing the gene have lost methylation in its vicinity compared with most non-expressing tissues. For example, ovalbumin genes of the chicken oviduct lack methylation at specific intragenic and extragenic sites compared with the same sequence in non-expressing tissues[35, 39]. Similarly, β-globin genes have unmethylated sites in chicken erythrocyte DNA, but not in non-erythroid tissues[37, 39]. The same correlation is observed at the developmentally regulated β-like globin gene locus in man[67]. Expression of the fetal globins (γ^G and γ^A) is correlated with undermethylation in the vicinity of these genes, while the closely linked adult genes (δ and β) remain methylated. Conversely δ and β gene regions are undermethylated in adult erythroid cells, while the γ genes are not. In most non-erythroid tissues the locus remains heavily methylated, the highest levels being observed in sperm. The correlation with globin gene transcription is not absolute, however. Placental cells, HeLa cells and KB cells have the globin locus undermethylated, yet they do not express the globin genes. This suggests, in line with the results discussed above, that loss of cytosine methylation is not sufficient to ensure transcription, though it may be a necessary precondition.

Loss of methylation at specific extragenic sites is correlated with the onset of rRNA transcription in *Xenopus laevis* embryos[8]. Chromosomal rRNA genes are heavily methylated at all developmental stages, but a region of the non-transcribed spacer, upstream of the origin of transcription, is undermethylated in somatic tissues. Sperm rDNA, by contrast, is fully methylated in this region. The transition from sperm to somatic patterns occurs during the first day of development, near the time when rRNA synthesis is thought to commence. Localized loss of methylation is also indicated at the β-like globin gene locus of rabbits[58]. Three out of thirteen Hpa II sites are fully methylated in non-expressing tissues, but are incompletely methylated in tissues containing erythroid cells. Methylation at the other ten sites, some of which are within the coding regions, is poorly correlated with expression. This picture contrasts with a study of α-globin genes in the chicken, where the undermethylated region observed in expressing cells corresponds well with the transcription unit[73].

To summarize, several vertebrate genes are heavily methylated in sperm, somewhat less heavily methylated in somatic tissues that do not express the gene, and significantly undermethylated in expressing tissues. Undermethylation can also be observed in certain tissues that do not appear to express the gene. The pattern of undermethylation varies between genes, ranging from localized demethylation outside the transcribed region (e.g. *Xenopus laevis* rDNA and probably rabbit β-globin genes), to undermethylation throughout the transcribed region (chicken α-globin genes). Despite variation in detail the results establish a striking correlation between transcription and loss of CG methylation. The correlation encourages the view that removal of certain methyl groups is required before a gene in a methylated domain can be transcribed. The less interesting alternative is that transcription of a sequence leads to loss of CG methylation. Choice between these explanations awaits a rigorous test of the transcriptional capacity of a gene in various states of methylation.

DNA methylation and the active chromatin structure

Transcribed regions of the genome are in chromatin that is preferentially sensitive to the nuclease DNase I[72]. This is interpreted as a loosening of the chromatin structure accompanying transcription. In some systems the active chromatin structure persists in cells that have been, but are no longer, actively transcribing a particular gene. This suggests that adoption of the active structure is independent of transcription itself and may be a precondition for gene expression, though there is as yet no definite evidence for this. Are DNase I sensitivity and undermethylation correlated? In a developmental study of the α-globin genes in chicken, a striking temporal and spatial correlation is observed between loss of C methylation and DNase I sensitivity[73]. The boundaries of the undermethylated region in expressing cells correspond closely with the boundaries of the region hypersensitive to DNase I, and with the region coding for α-globin RNA. Furthermore, undermethylation and DNase I sensitivity appear to arise at the same time in development. From these results it is tempting to speculate that undermethylation and DNase I sensitivity are causally related processes.

Other data, however, are less compatible with this idea. For example, *Drosophila* has little detectable methylation, yet sequences become DNase I sensitive upon transcription, just as in vertebrates[77]. A specific case involves the ribosomal RNA genes in amphibian species hybrids. Hybrids between *Xenopus laevis* and *Xenopus borealis* preferentially express *X. laevis* rRNA genes[29]. Correspondingly, the active *X. laevis* genes are hypersensitive to DNase I compared with the *X. borealis* genes. Both parental rDNAs, however, are undermethylated in the non-transcribed spacer but heavily methylated in the transcribed regions[38]. Thus *X. borealis* rDNA is not in the active chromatin configuration, yet it is undermethylated in the spacer (see p. 60). This shows that undermethylation of these sites is not a direct trigger for the active chromatin structure. The results also show that undermethylation of the transcribed region of the *X. laevis* genes is not necessary for adoption of the active structure.

5-azacytidine

The cytidine analogue 5-azacytidine (*Figure 4.1*) can have dramatic effects on cell differentiation and gene expression. Permanent cell lines, for example, can be induced to differentiate into contractile muscle cells, adipocytes and chondrocytes following treatment with the drug[66]. Studies of specific genes show that 5-azacytidine treatment can activate genes that would otherwise remain quiescent. Examples of this include transcriptional activation of silent retroviral genes[26], a mouse metallothionein gene[17] and the human X-linked gene for hypoxanthine-phosphoribosyltransferase[44]. The latter gene was quiescent as a result of X-inactivation.

The structure of 5-azacytidine shows a nitrogen atom at the 5-position which would prevent methylation in DNA (*Figure 4.1*). It is therefore attractive to speculate that the gene-activating effects of 5-azacytidine are due to its interference in the process of DNA methylation. This is supported by the finding that methylation is inhibited by treatment with the drug[31]. Also, activation of retroviral and metallothionein genes following drug treatment is accompanied by a drop in the level of cytosine methylation at these sequences[17, 26].

Despite the appeal of these experiments some reservations about 5-azacytidine should be borne in mind. First, it is an inhibitor of many cellular processes (e.g. DNA synthesis, RNA synthesis, protein synthesis, tRNA methylation and RNA processing; see reference 31 for references). This means that its effects cannot automatically be attributed to inhibition of DNA methylation. Secondly, the inhibition of methylation does not seem to result entirely from incorporation of 5-azacytidine in place of cytidine, because methylation is 85% inhibited under conditions where only 5% of C is substituted[31]. Thus the effect on DNA methylation is, as yet, unexplained.

The 5S genes of *Xenopus* – an apparent exception to the correlation

The genes coding for 5S ribosomal RNA in *Xenopus* are among the best studied eukaryotic genes[10]. About 20 000 copies of the repeat unit are present in each cell, and their transcription is under dual control. Most of the genes are active only in oocytes (oocyte-type 5S genes), while a minority (somatic-type 5S genes) are active in somatic cells and oocytes[23, 71]. The basis for this dual control is not yet known, but transcription studies already show that 5S genes do not fit neatly with the correlation between DNA undermethylation and transcription. Direct sequencing of oocyte-type 5S genes from blood cells (where they are not expressed) indicates heavy methylation at CG[22, 43]. In spite of this, these methylated genes can be transcribed as efficiently as unmethylated genes when injected into frog oocytes either as purified DNA[11] or as somatic cell nuclei[34]. The simplest conclusion is that DNA methylation has no effect on transcription of 5S genes, though it remains possible that oocytes are unusual in being able to transcribe methylated sequences.

(a)

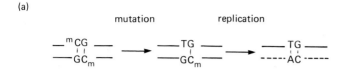

mutation   replication

(b)

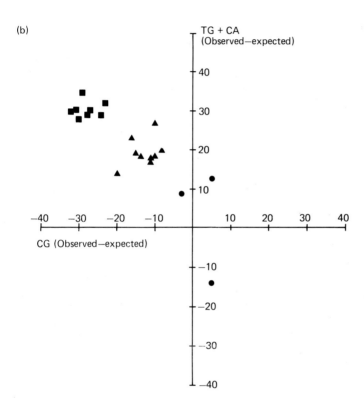

Figure 4.5. Evidence that 5-methylcytosine tends to mutate to thymine. (a) A scheme of mutation from mCG to TG and CA. Deamination of mC initially causes a mismatched T–G pair, which if not repaired, gives rise to a T–A pair after DNA replication. Two CG doublets are lost and one TG plus one CA are gained. (b) A correlation between CG deficiency and TG + CA excess in animals with different levels of methylation. Insects, in which there is little detectable methylation (see *Figure 4.3a*), have no CG deficiency and no reproducible TG + CA excess (circles). Non-arthropod inverterbrates, in which a fraction of the genome is methylated (see *Figure 4.3b*), have an intermediate deficiency of CG corresponding with a similar excess of TG + CA (triangles). Vertebrates, which are heavily methylated (see *Figure 4.3c*), have a marked deficiency of CG matched by a marked excess of TG + CA (squares). Doublet frequences are expressed as the difference between the observed frequency per 1000 dinucleotides and the expected frequency per 1000 dinucleotides. Sources are cited in reference 4. Taken from reference 4, with permission.

Methylation and mutation

The dinucleotide CG occurs at about 20% of its expected frequency in vertebrate DNA[32, 52, 65]. Since CG is also the predominant methylated sequence, it has been suggested that methylation and the CG deficiency are related phenomena[54]. In bacteria it is known that 5mC is a hotspot for mutation through deamination to T (see *Figures 4.1* and *4.5a*)[18]. Deamination of unmodified C, by contrast, gives U which, unlike T, is recognized as abnormal and exercised from the DNA. Three lines of evidence implicate a similar 5mC → T transition as the source of the CG deficiency in vertebrates (see *Figure 4.5a*). First, the extent of the CG deficiency in different animals is proportional to the extent of methylation of the genome[4]. Insects, for example, have undetectable CG methylation and have no significant CG deficiency. Secondly, the deficiency of CG is matched by a corresponding excess of TG and CA, as would be expected if mCG mutates to TG (*Figure 4.5b*)[4]. Animals with no CG deficiency have no TG + CA excess. Thirdly, there are indications that restriction sites that include CG are unusually polymorphic between individuals of the same species[39]. These results imply that cytosine methylation imposes a burden of increased mutability. Presumably the advantages conferred by the presence of 5mC outweigh this attendant disadvantage. An interesting suggestion that 5mC might change to T during development is not yet supported by experimental evidence[56].

DNA structure and methylation

It has been known for some time that relatively low levels of methylation can have measurable effects on physical properties of DNA, such as buoyant density and melting temperature[19]. Recent experiments show a much more dramatic effect on DNA conformation. The synthetic polymer poly (dG-dC) exists in the B-form in low salt solutions, but in high salt it adopts a 'left-handed' configuration known as Z-DNA[70]. When all Cs in the hexamer are methylated, however, Z-DNA forms at magnesium concentrations equivalent to those found in the cell[2]. Moreover, the Z-DNA conformation is disproportionately favoured when the sequence is only partially methylated. In view of indications that Z-DNA occurs *in vivo*[45], it will be of great interest to determine whether CG methylation plays a role in stabilizing or promoting this conformation.

References

1. Adams, R. L. P., McKay, E. L., Craig, L. M. and Burdon, R. H. (1979). Mouse DNA methylase: methylation of native DNA. *Biochimica et Biophysica Acta*, **561**, 345–357
2. Behe, M. and Felsenfeld, G. (1981). Effects of methylation on a synthetic polynucleotide: The B–Z transition of poly (dG–m^{5}dC)· poly (dG–m^{5}dC). *Proceedings of the National Academy of Sciences, USA*, **78**, 1619–1623
3. Bird, A. P. (1978). Use of restriction enzymes to study eukaryotic DNA methylation: II. The symmetry of methylated sites supports semi-conservative copying of the methylation pattern. *Journal of Molecular Biology*, **118**, 49–60
4. Bird, A. P. (1980). DNA methylation and the frequency of CpG in animal DNA. *Nucleic Acids Research*, **8**, 1499–1503

5. Bird, A. P. and Southern, E. (1978). Use of restriction enzymes to study eukaryotic DNA methylation: I. The methylation pattern in ribosomal DNA from *Xenopus laevis*. *Journal of Molecular Biology*, **118**, 27–47
6. Bird, A. P. and Taggart, M. H. (1980). Variable patterns of total DNA and rDNA methylation in animals. *Nucleic Acids Research*, **8**, 1485–1497
7. Bird, A. P., Taggart, M. H. and Gehring, C. (1981). Methylated and unmethylated ribosomal RNA genes in the mouse. *Journal of Molecular Biology*, **152**, 1–17
8. Bird, A. P., Taggart, M. H. and Macleod, D. (1981). Loss of rDNA methylation accompanies the onset of ribosomal gene activity in early development of *X. laevis*. *Cell*, **26**, 381–390
9. Bird, A. P., Taggart, M. H. and Smith, B. A. (1979). Methylated and unmethylated DNA compartments in the sea urchin genome. *Cell*, **17**, 889–901
10. Brown, D. D. (1981). Gene expression in eukaryotes. *Science*, **211**, 667–674
11. Brown, D. D. and Gurdon, J. (1977). High fidelity transcription of 5S DNA injected into *Xenopus* oocytes. *Proceedings of the National Academy of Sciences, USA*, **74**, 2064–2068
12. Browne, M. J. and Burdon, R. H. (1977). The sequence specificity of vertebrate DNA methylation. *Nucleic Acids Research*, **4**, 1025–1037
13. Burdon, R. H. and Adams, R. L. P. (1969). The *in vivo* methylation of DNA in mouse fibroblasts. *Biochimica et Biophysica Acta*, **174**, 322–329
14. Cedar, H., Solage, A., Glaser, G. and Razin, A. (1979). Direct detection of methylated cytosine in DNA by the use of the restriction enzyme Msp I. *Nucleic Acids Research*, **6**, 2125–2132
15. Clark, C. G. and Bird, A. P. (1982). Detection of total DNA and rDNA methylation in plants using restriction enzymes. Unpublished results
16. Cohen, J. C. (1980). Methylation of milk-borne and genetically transmitted mouse mammary tumor virus proviral DNA. *Cell*, **19**, 653–662
17. Compere, S. J. and Palmiter, R. D. (1981). DNA methylation controls the inducibility of the mouse metallothionein-I gene in lymphoid cells. *Cell*, **25**, 233–240
18. Coulondre, C., Miller, J. H., Farabough, P. J. and Gilbert, W. (1978). Molecular basis of base substitution hotspots in *Escherichia coli*. *Nature, London*, **27**, 775–780
19. Dawid, I. B., Brown, D. D. and Reeder, R. H. (1970). Composition and structure of chromosomal and amplified ribosomal DNAs of *Xenopus laevis*. *Journal of Molecular Biology*, **51**, 341–360
20. Desrosiers, R. C., Mulder, C. and Fleckenstein, B. (1979). Methylation of *Herpes saimiri* DNA in lymphoid tumor cell lines. *Proceedings of the National Academy of Sciences, USA*, **76**, 3839–3843
21. Doscocil, J. and Sorm, F. (1962). Distribution of 5-methylcytosine in pyrimidine sequences of deoxyribonucleic acids. *Biochimica et Biophysica Acta*, **55**, 953–959
22. Federoff, N. V. and Brown, D. D. (1978). The nucleotide sequence of oocyte 5S DNA in *Xenopus laevis*. 1. The AT-rich spacer. *Cell*, **13**, 701–716
23. Ford, P. J. and Southern, E. M. (1973). Different sequences for 5S RNA in kidney cells and ovaries of *Xenopus laevis*. *Nature, New Biology*, **241**, 7–10
24. Gautier, F., Bunemann, H. and Grotjahn, L. (1977). Analysis of calf thymus satellite DNA: evidence for specific methylation of cytosine in CG sequences. *European Journal of Biochemistry*, **80**, 175–183
25. Grippo, P., Iaccarino, M., Parisi, E. and Scarano, E. (1968). Methylation of DNA in developing sea urchin embryos. *Journal of Molecular Biology*, **36**, 195–208
26. Groudine, M., Eisenman, R. and Weintraub, H. (1981). Chromatin structure of endogenous retroviral genes and activation by an inhibitor of DNA methylation. *Nature, London*, **292**, 311–317
27. Harland, R. M. (1982). Inheritance of DNA methylation in microinjected eggs of *Xenopus laevis*. *Proceedings of the National Academy of Sciences, USA*, **79**, 2323–2327
28. Holliday, R. and Pugh, J. E. (1975). DNA modification mechanisms and gene activity during development. *Science*, **187**, 226–232
29. Honjo, T. and Reeder, R. H. (1973). Preferential transcription of *Xenopus laevis* ribosomal RNA in interspecies hybrids between *Xenopus laevis* and *Xenopus mulleri*. *Journal of Molecular Biology*, **80**, 217–228
30. Horz, W. and Altenburger, W. (1981). Nucleotide sequence of mouse satellite DNA. *Nucleic Acids Research*, **9**, 683–696
31. Jones, P. A. and Taylor, S. M. (1980). Cellular differentiation, cytidine analogs and DNA methylation. *Cell*, **20**, 85–93
32. Josse, J., Kaiser, A. A. and Kornberg, A. (1961). Enzymatic synthesis of deoxyribonucleic acid. VIII. Frequences of nearest neighbour base sequences in deoxyribonucleic acid. *Journal of Biological Chemistry*, **236**, 864–875
33. Kappler, J. W. (1970). The kinetics of DNA methylation in cultures of a mouse adrenal cell line. *Journal of Cell Physiology*, **75**, 21–32

34. Korn, L. J. and Gurdon, J. B. (1981). The reactivation of developmentally inert 5S genes in somatic nuclei injected into *Xenopus* oocytes. *Nature, London*, **289**, 461–465
35. Kuo, M. T., Mandel, J. L. and Chambon, P. (1979). DNA methylation: correlation with DNase I sensitivity of chicken ovalbumin and conalbumin chromatin. *Nucleic Acids Research*, **7**, 2105–2113
36. McClelland, M. (1981). The effects of sequence specific DNA methylation on restriction endonuclease cleavage. *Nucleic Acids Research*, **22**, 5859–5866
37. McGhee, J. D. and Ginder, G. D. (1979). Specific DNA methylation sites in the vicinity of the chicken β-globin genes. *Nature, London*, **280**, 419–420
38. Macleod, D. and Bird, A. (1982). DNase I sensitivity and methylation of active versus inactive rRNA genes in *Xenopus* species hybrids. *Cell* , **29**, 211–218
39. Mandel, J. L. and Chambon, P. (1979). DNA methylation: organ specific variations in the methylation pattern within and around ovalbumin and other chicken genes. *Nucleic Acids Research*, **7**, 2081–2103
40. Mann, M. B. and Smith, H. O. (1978). Specificity of Hpa II and Hae III DNA methylases. *Nucleic Acids Research*, **4**, 4211–4221
41. Maxam, A. M. and Gilbert, W. (1977). A new method for sequencing DNA. *Proceedings of the National Academy of Sciences, USA*, **74**, 560–564
42. Meselson, M., Yuan, R. and Heywood, J. (1972). Restriction and modification of DNA. *Annual Review of Biochemistry*, **41**, 447–466
43. Miller, J. R., Cartwright, E. M., Brownlee, G. G., Fedoroff, N. V. and Brown, D. D. (1978). The nucleotide sequence of oocyte 5S DNA in *Xenopus laevis*. II. The CG-rich region. *Cell*, **13**, 717–725
44. Mohandas, T., Sparkes, R. S. and Shapiro, L. J. (1981). Reactivation of an inactive human X-chromosome: evidence for X-inactivation by DNA methylation. *Science*, **211**, 393–396
45. Nordheim, A., Pardue, M. L., Lafer, E. M., Moller, A., Stollar, B. D. and Rich, A. (1981). Antibodies to left-handed Z-DNA bind to interband regions of *Drosophila* polytene chromosomes. *Nature, London*, **294**, 417–422
46. Pellicer, A., Robins, D., Wold, B., Sweet, R., Jackson, J., Lowy, I., Roberts, J., Sim, G. K., Silverstein, S. and Axel, R. (1980). Altering genotype and phenotype by DNA-mediated gene transfer. *Science*, **209**, 1414–1421
47. Pollack, Y., Stein, R., Razin, A. and Cedar, H. (1980). Methylation of foreign DNA sequences in eukaryotic cells. *Proceedings of the National Academy of Sciences, USA*, **77**, 6463–6467
48. Rae, P. M. M. and Steele, R. E. (1978). Modified bases in the DNAs of unicellular eukaryotes: an examination of distributions and possible roles, with emphasis on hydroxy-methyluracil in dinoflagellates. *Biosystems*, **10**, 37–53
49. Rae, P. M. M. and Steele, R. E. (1979). Absence of cytosine methylation at CCGG and GCGC sites in the rDNA coding regions and intervening sequences of *Drosophila* and the rDNA of other higher insects. *Nucleic Acids Research*, **6**, 2987–2995
50. Razin, A. and Riggs, A. D. (1980). DNA methylation and gene function. *Science*, **210**, 604–610
51. Riggs, A. D. (1975). X-Inactivation, differentiation and DNA methylation. *Cytogenetics and Cell Genetics*, **14**, 9–25
52. Russell, G. J., Walker, P. M. B., Elton, R. A. and Subak-Sharpe, J. H. (1976). Doublet frequency analysis of fractionated vertebrate nuclear DNA. *Journal of Molecular Biology*, **108**, 1–23
53. Sager, R. and Kitchin, R. (1975). Selective silencing of eukaryotic DNA. *Science*, **189**, 426–433
54. Salser, W. (1977). Globin mRNA sequences: analysis of base pairing and evolutionary implications. *Cold Spring Harbor Symposium on Quantitative Biology*, **42**, 985–1002
55. Scarano, E., Iaccarino, M., Grippo, P. and Winckelmans, D. (1965). On methylation of DNA during development of the sea urchin embryo. *Journal of Molecular Biology*, **14**, 603–607
56. Scarano, E., Iaccarino, M., Grippo, P. and Parisi, E. (1967). The heterogeneity of thymine methyl group origin in DNA pyrimidine isostichs of developing sea urchin embryos. *Proceedings of the National Academy of Sciences, USA*, **57**, 1394–1400
57. Shapiro, H. S. (1968). Distrubtion of purines and pyrimidines in deoxyribose-nucleic acids. In *Handbook of Biochemistry* (Sober, H. A., Ed.), pp. H39–H48. Cleveland, The Chemical Rubber Co.
58. Shen, C. J. and Maniatis, T. (1980). Tissue specific DNA methylation in a cluster of rabbit β-like globin genes. *Proceedings of the National Academy of Sciences, USA*, **77**, 6634–6638
59. Simon, D., Grunert, F., U.v. Acken, Doring, H. P. and Kroger, H. (1978). DNA-methylase from regenerating rat liver: purification and characterization. *Nucleic Acids Research*, **5**, 2153–2167
60. Singer, J., Roberts-Ems, J. and Riggs, A. D. (1979). Methylation of mouse liver DNA by means of restriction enzymes Hpa II and Msp I. *Science*, **203**, 1019–1021
61. Sneider, T. W. and Potter, V. R. (1969). Methylation of mammalian DNA: Studies on Novikoff Hepatoma cells in tissue culture. *Journal of Molecular Biology*, **42**, 271–284

62. Southern, E. M. (1975). Detection of specific sequences among DNA fragments separated by gel electrophoresis. *Journal of Molecular Biology*, **98**, 503–517

63. Sturm, K. S. and Taylor, J. H. (1981). Distribution of 5-methylcytosine in DNA of somatic and germline cells from bovine tissues. *Nucleic Acids Research*, **9**, 4537–4546

64. Sutter, D. and Doerfler, W. (1980). Methylation of integrated adenovirus type 12 DNA sequences in transformed cells is inversely correlated with viral gene expression. *Proceedings of the National Academy of Sciences, USA*, **77**, 253–256

65. Swartz, M. N., Trautner, T. A. and Kornberg, A. (1962). Enzymatic synthesis of deoxyribonucleic acid. XI. Further studies on nearest neighbor base sequences in deoxyribonucleic acids. *Journal of Biological Chemistry*, **237**, 1961–1967

66. Taylor, S. M. and Jones, P. A. (1979). Multiple new phenotypes induced in 10T½ and 3T3 cells treated with 5-azacytidine. *Cell*, **17**, 771–779

67. van der Ploeg, L. H. T. and Flavell, R. A. (1980). DNA methylation in the human γδβ-globin locus in erythroid and nonerythroid tissues. *Cell*, **19**, 947–958

68. Vanyushin, B. F., Tkacheva, S. G. and Belozersky, A. N. (1970). Rare bases in animal DNA. *Nature, London*, **225**, 948–951

69. Waalwijk, C. and Flavell, R. A. (1978). Msp I, an isoschizomer of Hpa II which cleaves both unmethylated and methylated Hpa II sites. *Nucleic Acids Research*, **5**, 3231–3236

70. Wang, A. H. J., Quigley, G. J., Kolpack, F. J., Crawford, J. L., van Boom, J. H., van der Marel, G. and Rich, A. Molecular structure of a left-handed double helical DNA fragment at atomic resolution. *Nature, London*, **282**, 680–686

71. Wegnez, M., Monier, R. and Denis, H. (1972). Sequence heterogeneity of 5S RNA in *Xenopus laevis*. *FEBS Letters*, **25**, 13–20

72. Weintraub, H. and Groudine, M. (1976). Chromosomal subunits in active genes have an altered conformation. *Science*, **193**, 848–856

73. Weintraub, H., Larsen, A. and Groudine, M. (1981). α-Globin gene switching during the development of chicken embryos: expression and chromosome structure. *Cell*, **24**, 333–344

74. Whittaker, P. A., McLachlan, A. and Hardman, N. (1981). Sequence organisation in nuclear DNA from *Physarum polycephalum*: methylation of repetitive sequences. *Nucleic Acids Research*, **9**, 801–814

75. Wigler, M., Levy, P. and Perucho, M. (1981). The somatic replication of DNA methylation. *Cell*, **24**, 33–40

76. Wigler, M., Sweet, R., Sim, G. K., Wold, B., Pellicer, A., Lacy, E., Maniatis, T., Silverstein, S. and Axel, R. (1979). Transformation of mammalian cells with genes from procaryotes and eucaryotes. *Cell*, **16**, 777–785

77. Wu, C., Wong, Y. C. and Elgin, S. C. R. (1979). The chromatin structure of specific genes: II. Disruption of chromatin structure during gene activity. *Cell*, **16**, 807–814

78. Wyatt, G. R. (1951). The purine and pyrimidine composition of deoxypentose nucleic acids. *Biochemical Journal*, **48**, 584–590

The organization of chromosomes

The title of this section should be interpreted in a strictly limited way, since the book does not attempt to deal seriously with the topic of the structure of mitotic or meiotic chromosomes. But since genes are located on chromosomes, some aspects of chromosome organization have a direct bearing on gene function. Most chromosomes are neither actively engaged in DNA replication nor DNA transcription and so have no pressing relevance to the structure and function of the gene. But two types of chromosomes found in nature are active in transcription, and so offer situations in which gene activity can be visualized in the light microscope. These two examples, the lampbrush chromosomes of vertebrate oocytes and the giant polytene chromosomes of some differentiated tissues of Dipteran flies, have made a very great contribution to genetics and particularly to our understanding of gene regulation and transcriptional activity. It is therefore appropriate to discuss these chromosomes in this volume and set in perspective what is to be learnt from them about gene arrangement and gene regulation.

Whether the genomes of mitochondria and chloroplasts ought properly to be called chromosomes is a matter for debate, but it has seemed to the editors appropriate to include this interesting aspect of cellular genetics in this section. Chapter 5 is therefore devoted to the genomes of mitochondria and chloroplasts and the characteristics of some of the genes found within them. The evolutionary origins of mitochondria and chloroplasts may or may not involve prokaryotic ancestral cells. But either way, these organelles possess genomes of considerable diversity, and with some features which are neither obviously prokaryotic nor eukaryotic. The diversity of these genetic systems is emphasized by the fact that mitochondrial genomes may be circular, linear or may exist as networks of catenated circles. Some genes of the mitochondria or chloroplasts of some organisms possess intron sequences and, in general, no copy of the mitochondrial or chloroplast genome has as yet been detected in the nuclear genome of any cell.

Chapter 6 discusses the lampbrush chromosome, a puzzling situation of widespread transcriptional activity throughout the genome. Many aspects of lampbrush chromosome transcription remain problematical. Although it is clear that most transcription represents the activity of RNA polymerase II, both middle repetitive and unique sequences (and even some highly repetitive sequences) are vigorously transcribed. Identification of 'unique' sequence genes represented in the localized lampbrush transcripts, although theoretically possible, has not as yet been unequivocally achieved. Only the moderately repetitious histone genes have been assigned to particular loci and loops and that only in *Triturus cristatus carnifex*. The function of the lampbrush chromosome also remains problematical and at present, fascinating as they are, these structures continue to pose more problems than they solve.

The presentation in Chapter 7 of present understanding of the polytene chromosome reveals that in this situation a more convincing story is beginning to emerge. Contrary to the weight of earlier evidence and opinion, it now appears likely that it is *interband* transcription which represents the routine activity of the few thousand 'housekeeping' genes and that the decondensation and transcriptional activity revealed in puffing of individual bands may well represent the specialized differentiated expression of 'luxury' protein genes. Such a conclusion is made easier by the upward revision from 5% to 26% of the proportion of the *Drosophila* genome which occurs in the interbands and not in the more condensed banded regions.

Localization of particular proteins within the polytene chromosomes is also discussed here and evidence from the use of fluorescently tagged monoclonal antibody is reviewed. Antibodies raised against RNA polymerase B (type II) and against the insect hormone ecdysone have proved especially useful and the use of antisera directed against such proteins as HMG proteins also holds promise.

In general the polytene chromosome seems to be easier to interpret than the lampbrush and, since it does represent interphase chromatin in action, any data about proportions of active and inactive gene loci have important implications for our understanding of eukaryotic gene regulation.

Chapter 5

Extranuclear genes

P. Borst, H. F. Tabak and L. A. Grivell

Introduction

There are three types of extranuclear genes known, viral genes, genes in mitochondria and genes in chloroplasts. This chapter will deal only with the organelle genes.

There are several reasons why organelle genes are of interest beyond their essential role in the synthesis of organelles Organelle genomes display principles of organization and control not encountered either in the eukaryotic nucleus, or in prokaryotes. For example, trypanosome mitochondria contain the smallest ribosomal RNAs (rRNAs) known in nature, the tRNAs of mammalian mitochondria are unusually small and low in modified bases; some yeast mitochondrial genes contain optional introns; and the processing of precusor RNAs in mitochondria follows highly complex pathways and, in yeast, yields interesting side products, like circular RNAs. Even though the number of genes in organelle DNA is low (see *Table 5.1*), these DNAs are usually present in multiple copies and may even form a major fraction of the total DNA. The contribution of the organelles to cellular DNA and RNA synthesis is correspondingly high. Some

TABLE 5.1. The contribution of organelle DNAs to cell DNA

Source of DNA	Organelles per cell	Organelle DNA (% of total)		Reference
		Amount	Complexity	
Mitochondrial DNA				
Mouse L-cell line	10^2	1	0.0005	23
Toad egg	10^7	99	0.0005	12
Yeast diploid	2–50[a]	15	0.6	7
Chloroplast DNA				
Chlamydomonas diploid	2	7–14	0.3	27
Tobacco leaves	10^2	10–50[b]	0.01	20

[a] Depends on growth conditions.
[b] Controversial.

knowledge of organelle genes and their expression could, therefore, be of use to every molecular biologist, and in particular to people working with plants, unicellular eukaryotes or animal embryos.

Genes in mitochondrial DNA – general characteristics[2, 6, 17, 22, 24]

Diversity is the hallmark of mitochondrial genetic systems and this is illustrated by the data summarized in *Tables 5.2* to *5.7* and *Figures 5.1* to *5.6*. Mitochondrial

TABLE 5.2. Size and structure of mitochondrial DNAs[6]

Organism	Structure	Size (kb)
Animals (from flatworm to man)	Circular	14–18[a]
Higher plants	Circular	100–150[b]
Fungi:		
Yeast (*Saccharomyces*)	Circular	73–78[a]
Kluyveromyces	Circular	33
Torulopsis	Circular	19
Protozoa:		
Malaria (*Plasmodium*)	Circular	27
Acanthamoeba	Circular	41
Paramecium	Linear	41
Tetrahymena	Linear	45–54[a]
Trypanosoma	Network[c]	6000
	Maxi-circle	21
	Mini-circle	1
Crithidia	Network[c]	30 000
	Maxi-circle	35
	Mini-circle	3
Algae:		
Chlamydomonas	Linear	15

[a] The molecular weights of the DNAs in different representatives of this group differ; in any one organism only a single type of molecule is found.
[b] In several plants like maize, corn, wheat, cucumber and tobacco, the mitochondrial DNA consists of several size classes of circles with a combined genetic information content exceeding 150 kb.
[c] This mitochondrial DNA is usually called kinetoplast DNA (see also *Figure 5.1*).

TABLE 5.3. Mitochondrial genetic codes

Codon	Usual assignment	Assignment in mitochondria of		
		Man	Neurospora	Yeast
UGA	Stop	Trp	Trp	Trp
(UGG	Trp	Trp	Trp	Trp)
CUA	Leu	Leu	Leu	Thr
(ACX	Thr	Thr	Thr	Thr)
AUA	Ile	Met	Ile	Ile
(AUG	Met	Met	Met	Met)
AGA	Arg	Stop	Arg	Arg
AGG	Arg	Stop	Arg	Arg
(CGN	Arg	Arg	Arg	Arg)

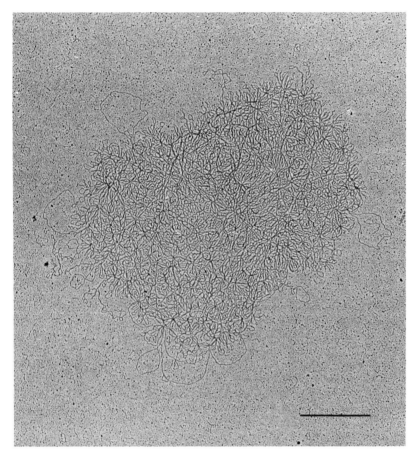

Figure 5.1. Electron micrograph of the kinetoplast DNA (= mitochondrial DNA) of *Trypanosoma brucei.* From reference 10. The bar is 1 µm.

genes may be present in circular DNA linear, DNA molecules or in large networks of catenated circles (*Table 5.2, Figure 5.1*). The genes may be tightly packed together and transcribed from one promoter as in human mitochondrial DNA (*Figures 5.2* and *5.5; Table 5.6*) or spread out and transcribed from multiple promotors as in yeast mitochondrial DNA (*Figures 5.3* and *5.5; Table 5.6*). The genes may resemble their bacterial counterparts or have changed drastically in evolution as the seryl-tRNA of man (*Figure 5.4*) or the rRNAs of trypanosomes (*Table 5.5*). The genes may or may not be riddled by introns (*Figures 5.2* and *5.3*). The gene order is very different in man and yeast (compare *Figures 5.2* and *5.3*) and even among different fungi, although the order is conserved over shorter evolutionary time spans, e.g. mouse–man. Even the genetic code differs in mitochondria of different organisms (*Table 5.3*).

Common features shared by all mitochondrial genetic systems include the location of the DNA in the matrix space and the presence in this DNA of a basic set of mitochondrial genes. This set includes genes for the mitochondrial rRNAs and tRNAs, for a few mitochondrial membrane proteins and for a series of minor proteins (specified by the unassigned reading frames (URFs)) of unknown function

74

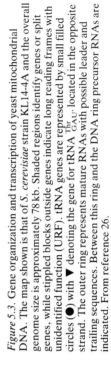

Figure 5.3. Gene organization and transcription of yeast mitochondrial DNA. The map shown is that of *S. cerevisiae* strain KL14-4A and the overall genome size is approximately 78 kb. Shaded regions identify genes or split genes, while stippled blocks outside genes indicate long reading frames with unidentified function (URF). tRNA genes are represented by small filled circles (●) with ▼ denoting the gene for tRNA $^{Thr}_{GAU}$ located in the opposite strand. The outer ring represents mature RNAs with possible leader and trailing sequences. Between this ring and the DNA ring precursor RNAs are indicated. From reference 26.

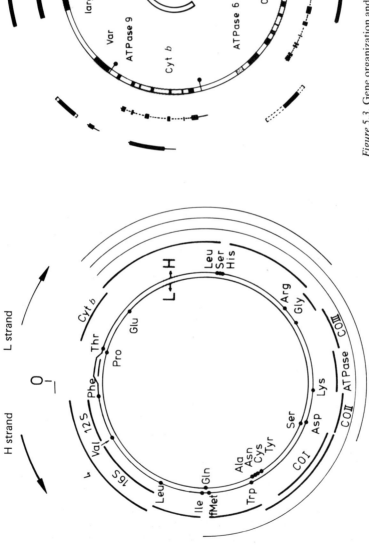

Figure 5.2. Transcription of human mitochondrial DNA. O is the origin of H strand replication with the D loop region on the right. Thin lines indicate L strand transcripts; thick lines H strand transcripts. (D loop, displacement loop.) L and H strand are the light and heavy DNA strand, respectively.) From reference 26.

TABLE 5.4. An inventory of mitochondrial genes in yeast and man

Mitochondrial component	Mitochondrial gene product in	
	Yeast[a] (a/α)	Man[b] (♀/♂)
Cytochrome c oxidase		
Subunit I	+	+
Subunit II	+	+
Subunit III	+	+
Ubiquinol-cytochrome c reductase		
Apocytochrome *b*	+	+
ATPase complex		
Subunit 6	+	+
Subunit 9	+	−
Large ribosomal subunit		
rRNA	+	+
Small ribosomal subunit		
rRNA	+	+
Ribosome-associated protein	+	−?
RNA processing enzymes		
Intron 2 COB maturase	+	−
Intron 4 COB maturase	+	−
tRNAs	about 25	22
URFs	≥8	8

[a] From reference 6.
[b] From reference 1.

(*Table 5.4*). In all organisms the bulk of the genes for mitochondrial proteins resides in the nucleus and these proteins are made on cell-sap ribosomes and imported. The set of genes present in mitochondrial DNA is not exactly the same in all mitochondria, however, as exemplified by the gene for subunit 9 of the mitochondrial ATPase complex (*Table 5.4*).

No systematic clustering of functionally-related genes has been observed in any mitochondrial DNA. The genes for subunits of cytochrome *c* oxidase, for instance, are scattered around the circles in *Figures 5.2* and *5.3* and even in mitochondrial DNAs where tRNA genes are not evenly spread around the circle, as in yeast (*Figure 5.3*) or *Neurospora*, the clustering is incomplete.

Mitochondrial rRNA and tRNA genes[26]

Table 5.5 shows that the rRNA genes in mitochondria are a diverse lot. In animal mitochondrial DNAs the genes for the two large subunits are separated only by a tRNA gene and the genes are transcribed in the direction 5′–small–large–3′ into a common precursor as in *Escherichia coli*. In trypanosomes the order of the two genes is reversed and they have shrunk to an unprecedented small size. In yeast mitochondrial DNA the two rRNA genes are more than 20 kb apart and are transcribed from separate promoters. The primary transcript of the gene for the large rRNA starts at the position of the 5′-end of the mature RNA but contains a 1.3 kb extension beyond the 3′-end. This extension includes at least one tRNA. The small (15S) rRNA is also made as precursor, but this contains 80 nucleotides extra at its 5′-end and no 3′ extension. The processing of the precursor RNAs from both

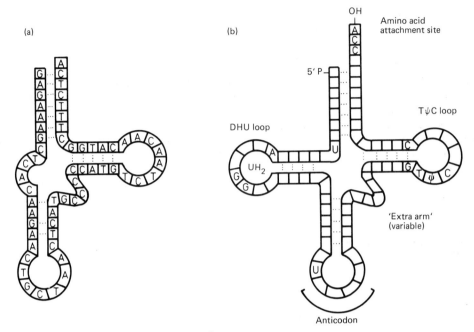

Figure 5.4. (a) Human mitochondrial tRNA^{Ser} compared with (b) the 'universal' structure of tRNAs elsewhere in nature.

ribosomal genes can occur in yeast petite mutants that cannot make mitochondrial proteins. The splicing and trimming enzymes required for processing are, therefore, imported from the cell sap.

The tRNA genes in animal mitochondrial DNA are scattered around the circle and distributed over both strands (*Figure 5.2*). Each strand of the DNA is probably transcribed completely into a large RNA (*Table 5.6*) which is then processed to gene-sized transcripts. The recognition by processing enzymes of the tRNA structures in these primary transcripts is considered to be essential for processing. In fungi most of the mitochondrial tRNAs are clustered (see *Figure 5.3*). In *Neurospora* the tRNA genes are surrounded by inverted repeats that are thought to play a role in the processing of a primary transcript containing these tRNAs. With yeast attempts to isolate long precursor RNAs containing several tRNAs have failed. Synthesis of most tRNAs does include a processing step, however, and this is known to require an RNA derived from a poorly-characterized region that lies

TABLE 5.5. The diversity mitochondrial rRNA genes

Organism	S value ribosomes	Size of rRNAs		Mole % G + C	Distance gene Large–small (bp)	Order of transcription	5S RNA present	Refs
		Large	Small					
Higher plants	78	3600	2100	44	?	?	+	11
Yeast								
(*Saccharomyces*)	73	3200	1600	23	>20000	No order	–	7
Tetrahymena	80	2600	1100	31	8000	No order	–	16
Man	60	1559	954	45	<100	S → L	–	11, 1
Trypanosomes								
(*Trypanosoma*)	?	1240	630	20	<100?	L → S	?	15

between the genes for tRNA^{Met} and 15S rRNA (see *Figure 5.3*). The nature and function of this RNA cofactor are not known.

Mitochondrial genes coding for proteins

The gene for apocytochrome *b* is found in all mitochondrial DNAs and the amino acid sequence of this protein has been conserved to a large extent throughout evolution, the homology between yeast and bovine apocytochrome *b* being about 50%. Nevertheless, the route from gene to protein is very different in yeast and man, as illustrated in *Figure 5.5*. In the yeast *Saccharomyces*, two versions of this gene are found – long COB genes with five introns and short COB genes, which only have the last two introns of the long version. Both versions are surrounded by AT-rich DNA segments of unknown function, like most yeast mitochondrial genes. Transcription of the apocytochrome *b* gene in yeast probably starts and ends at the

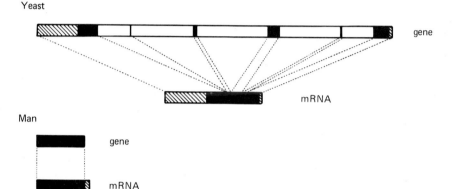

Figure 5.5. The gene and mRNA for apocytochrome *b*. The yeast gene is the long version of the gene found in strain KL14-4A. Coding sequences are shown as filled-in bars (■) and introns as open bars (□). The hatched regions are the non-translated sequences in the mRNA. From reference 9.

position where the termini of the mature messenger RNA (mRNA) are encoded. A complex splicing pathway removes the introns and results in a mRNA with a very long leader sequence of about 1000 nucleotides, a trailing 3′ sequence of about 100 bp and no poly(A) tail detectable by oligo(dT)-cellulose chromatography.

The apocytochrome *b* gene in human mitochondrial DNA is tightly wedged between tRNA genes and contains no introns. It is transcribed as part of a complete transcript of the H strand of the DNA. Processing of this transcript at the tRNA genes yields an ultra-short pre-mRNA, with neither leader nor trailing sequences. The sequence even lacks a proper stop codon and ends at a U. Polyadenylation converts this into a UAA ochre stop.

This example shows that very different strategies are used in different branches of the evolutionary tree to make a mitochondrial protein. There is no typical 'mitochondrial' way to control gene expression (*Table 5.6*).

Circles, (optional) introns and maturases[26]

Analysis of transcripts of the gene for subunit I of cytochrome *c* oxidase in yeast led to the discovery of three circular RNAs, all of which correspond to intron

TABLE 5.6. Mitochondrial transcription[26]

HeLa cell (human) mitochondria
    1. All transcripts probably start at a single promotor.
    2. Both DNA strands are completely transcribed (symmetric transcription).
    3. Punctuation with tRNAs may play key role in processing of precursor RNAs.
    4. Genes lack introns.
    5. mRNAs contain poly(A) tails

Yeast mitochondria
    1. At least five different promoters present as judged from the hybridization of transcripts carrying di- or triphosphate ends. These include the large rRNA and a precursor to the small rRNA.
    2. Asymmetric transcription.
    3. Introns are present in some genes.
    4. mRNAs lack standard poly(A) tails.

---

sequences. A fourth circle has since been found and this contains the sequence of the first intron of the long apocytochrome *b* gene. Although these circles behave as covalently closed, the joint may not be a normal 3'-5' phosphodiester bond, because reverse transcriptase is unable to pass the joint. An obvious possibility is that these circles arise as side products of splicing. It should be noted, however, that only some introns yield circles and that some of these circles are present at concentrations higher than the mature mRNA. The possibility that these circles have some function – as circular mRNAs or in regulatory interactions – cannot be dismissed.

    The introns in yeast mitochondrial genes have three interesting and unique features:

1. Some are optional, i.e. they are present in some yeast strains, but not in others (*Table 5.7*). No gross functional differences have been found between versions of the same gene with and without introns.
2. Some introns contain open reading frames and in two cases these have been shown to code for 'maturases', i.e. proteins involved in the splicing of mRNA precursors. *Figure 5.6* shows how the maturase encoded in intron 2 (I_2) of the apocytochrome *b* gene is made. Processing of the primary transcript starts with the excision of the I_1 sequence by an enzyme that is made outside the mitochondria and imported. This splice links the first exon, which encodes the N terminal half of apocytochrome *b* to an open reading frame in I_2 and this results

TABLE 5.7. Introns in mitochondrial genes in yeast[6]

Gene	Number of introns	
	Total	*Optional*
Large rRNA	1	1
Apocytochrome *b*	5[a]	3
Cytochrome *c* oxidase, subunit I	9[a]	5[a]
Small rRNA	0	0
tRNAs (25 genes analysed)	0	0
ATPase complex, subunits 6 and 9	0	0
Cytochrome *c* oxidase, subunits II and III	0	0

[a] Minimal estimate.

in the synthesis of a 42 kD fusion protein, the COB I_2 maturase. This maturase is required for the splicing of intron I_2. Because it helps to destroy the mRNA on which it is made, it regulates its own synthesis, a phenomenon called splicing homeostasis by Slonimski[24]. A second maturase is encoded in intron 4 (I_4) of the COB gene. This I_4 maturase is not only required for the excision of I_4, but also for the excision of one of the introns in the gene for subunit I of cytochrome c oxidase. What is the rationale of this complex arrangement? If the only advantage were in the increased gene length (and hence the increased

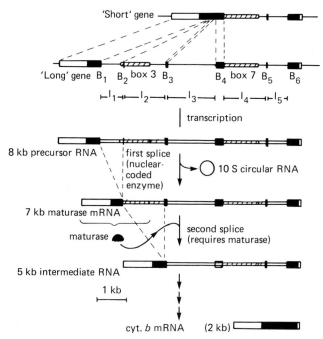

Figure 5.6. The structure of the 'long' and 'short' versions of the gene for apocytochrome *b* and of the mRNAs transcribed from the 'long' gene. Black boxes correspond to the exons B_1–B_6 which specify the amino acid sequence of apocytochrome *b*; open blocks designate the non-translated regions of the mRNA for cytochrome *b*; hatched blocks represent long open reading frames in the introns of the *b* gene. Whether the 8 kb precursor RNA is the primary transcript or not is uncertain. See text for further explanation. From reference 8.

opportunity for recombination), introduction of introns would suffice and the maturases would only result in unnecessary complications. It is, therefore, likely that the increased processing complexity of the precursor RNA itself provides an advantage, e.g. because it allows a finer control of apocytochrome *b* synthesis than is possible in intron-less genes. This hypothesis remains to be rigorously tested. The possibility that for some of these intronic URFs translation in itself may be required for splicing (for instance by stretching the RNA) without the resulting protein having function at all, cannot be excluded yet.

3. There are indications that some introns in yeast mitochondrial DNA have been inserted rather recently. One is that several introns show substantial sequence homology. Another is that the position of the intron in the large rRNA gene

differs from the position of the introns in nuclear or chloroplast rRNA genes. Mitochondrial introns may, therefore, have evolved from transposable genetic elements, even if they do not move around any more today or carry some of the sequence peculiarities that are typical of such elements[8].

The function of most URFs in yeast mitochondrial DNA and all URFs in human mitochondrial DNA is unknown. The finding that one of the human URFs has substantial sequence homology with an URF in the fungus *Aspergillus* shows that some URFs code for essential proteins conserved over long periods of evolutionary time. These URFs may specify RNA processing enzymes like the RNA maturases specified by the open reading frames in the introns in the yeast apocytochrome *b* gene.

Regulatory sequences in mitochondrial DNA[26]

Little is known about the sequences that determine the start and stop of mitochondrial RNA synthesis, the start of mitochondrial protein synthesis or the processing of RNA precursors. The only putative promoter sequence known is an 18 bp sequence present just in front of both rRNA genes in yeast. The sequences at the edge of introns differ from the consensus sequence found to border introns in nuclear DNAs and no common mitochondrial splice sequences can be defined (or may exist). Processing of long precursor RNAs in animal mitochondria apparently involves enzymes that recognize the tRNAs (or anti-tRNA sequences) used as interpunction between genes (*Figure 5.2*). The sequences involved in RNA trimming in yeast remain to be defined.

The structure, activity and organization of chloroplast genes[3, 14, 19, 25]

Chloroplast genes are present in circular DNA molecules that are similar in size throughout nature (*Table 5.8*). There are many more genes on chloroplast DNA than on the most complex mitochondrial DNA analysed thus far and only a fraction of these has been identified (*Table 5.9*). The tRNA and protein genes are scattered around the circle (*Figure 5.7*) without discernible functional clustering, e.g. the genes for subunits of the ATPase complex are found in two different locations.

Most chloroplast DNAs contain multiple copies of the rRNA genes. In most higher plants and in the alga *Chlamydomonas* there are two copies and these are present on large inverted repeats (*Figure 5.7*). Pea manages without this repeat, however, and has only one set of rRNA genes. In some *Euglena* strains there are three ribosomal transcription units in tandem. However, *Euglena* strains with one unit exist and look healthy[28]. It appears, therefore, that multiple rRNA genes are not a prerequisite for normal chloroplast function and that the selection pressure to maintain multiple copies is not strong.

The hallmark of chloroplast genetic systems is their close resemblance to that of *E. coli*. This was already clear from simple exchange experiments. Chloroplast ribosomes will work with *E. coli* enzymes to make protein; chloroplast ribosomal subunits will combine with *E. coli* subunits to make functional ribosomes; chloroplast genes can be used to programme an *E. coli* transcription translation system to make chloroplast proteins. Recently, this high degree of homology has been confirmed at the DNA sequence level. The overall organization of the chloroplast ribosomal transcription unit is the same as that of *E. coli* with the gene order 5'-16S-2 tRNAs-23S-4.5S-5S-3'[21]. Even the two tRNAs in this unit, tRNA[Ala]

TABLE 5.8. Size of selected plastid DNAs[19]

Higher plants	Size plastid DNA (μm)	Sum restriction fragments (kb)
Dicotyledons		
Phaseolus (bean)	40	—
Spinacia (spinach)	46	154
Pisum (pea)	39	121
Monocotyledons		
Tulipa	44	—
Zea mays	43	151
Spirodela (duckweed)	54	182
Archegoniatea		
Asplenium	44	
Sphaerocarpes	37	
Algae		
Euglena	44	132
Chlamydomonas	60	191
Acetabularia	Large?	

and tRNA^{Ile}, are the same as in *E. coli* and the unit is transcribed into a precursor that is processed in an analogous fashion as in *E. coli*. Maize 16S rRNA has a 74% sequence homology with *E. coli* 16S rRNA; the chloroplast tRNAs are typical prokaryotic tRNAs; prokaryotic transcription initiation and ribosome binding signals have been found at the correct position in front of chloroplast genes, and the control of highly regulated chloroplast genes appears to operate at the transcription initiation level.

Nevertheless, introns have also been observed in a few chloroplast genes. The gene for the 23S rRNA of *Chlamydomonas* chloroplasts contains a single intron. Since both copies of the gene contain the intron there is no doubt that this gene is functional and the chloroplast must, therefore, contain the enzymes to splice out the intron sequence from a precursor RNA. No introns have been found in any

TABLE 5.9. Genes in chloroplast DNA[14, 19]

Structural RNA genes
  35–40 tRNAs.
  1–3 sets of rRNA
Genes for ribosomal proteins
  At least one.
Genes for membrane proteins
  ATPase complex: 3 subunits (out of 5) of the CF$_1$ part. One subunit (out of 3) of the CF$_0$ part.
  'Photogene 32' (= gene for a 32 kD protein of the thylakoid membrane, which is made in response to light).
  Cytochrome *f*.
  Cytochrome *b*$_{599}$.
Genes for stromal proteins
  Large subunit of ribulose-bisphosphate carboxylase.
  Elongation factors G and T[a]

[a]Not in *Euglena*.

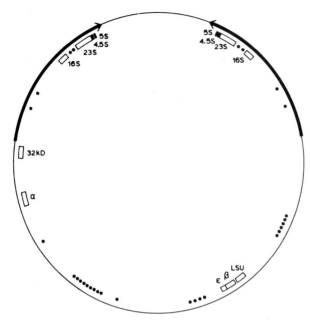

Figure 5.7. Gene organization in chloroplast DNA of higher plants. The two thick bars ending in an arrow are the long inverted repeats. 16S, 23S, 4.5S and 5S denote the rRNA genes. The black dots are tRNA genes. α, β and ϵ are genes for subunits of the chloroplast ATPase complex CF_1. LSU is the gene for the large subunit of ribulose-1,5-bisphosphate carboxylase and 32 kD is the gene for the light-induced protein of 32 kD. Redrawn from reference 19.

other chloroplast ribosomal gene, either in plants or in *Euglena*. Introns of 949 and 806 bp, respectively, are present, however, in the two tRNA genes in the ribosomal transcription unit of maize chloroplast DNA[21]. Several other tRNA genes are without introns. The presence of introns in the prokaryotic-like chloroplast DNA strongly supports the contention[4] that life started out with genes in pieces and that the introns were removed later from the genes of the prokaryotes present on earth today (at least from those analysed thus far). Why only some chloroplast genes have retained introns is not clear.

Defective organelle DNAs

Some unicellular eukaryotes can survive without functional mitochondria or chloroplasts and this may result in the emergence of mutants with defective organelle DNA or no organelle DNA at all. Three well characterized examples are known:

1. Yeast petite mutants. These are deletion mutants in which up to 99% of the genome may be deleted. An interesting aspect of this deletion is that the remaining segment is amplified to compensate for the loss. This results in long molecules (up to wild-type size), which contain tandem repetitions of a defined mitochondrial DNA segment. Virtually any mitochondrial DNA segment can be

'pre-cloned' in this way, although segments that contain a sequence related to the replication origin are preferred[7].

2. Trypanosomes. In the sub-genus *Trypanozoon* mitochondrial biogenesis is totally repressed in the vertebrate host and mutants with defective kinetoplast DNA (having lost part or all of their maxi-circle sequences, cf. *Table 5.2*) will arise[10]. An example is *Trypanosoma evansi*, a sub-species which is widely distributed over the world and in which the maxi-circle is absent.

3. Bleached algae. Algae may live on oxidative metabolism and forego photosynthesis. There are bleached variants of *Euglena*, for instance, in which a defective chloroplast DNA is found consisting mainly of ribosomal genes. Compensatory amplification does not seem to occur in this case and the amount of chloroplast DNA is low[18].

Other extranuclear DNAs

In addition to viral and organelle DNAs, many other examples of extranuclear DNAs can be found in the literature. Some of these–like the 2 µm circles of yeast or the heterogeneous small DNA circles found in many metazoa–probably function in the nucleus and their presence in the cytoplasm can be considered accident or artefact. Other reports on extranuclear DNAs have not been confirmed (kinetosome DNA, membrane DNA) or have been shown to be due to leakage of DNA from the nucleus (informational DNA).

Evolutionary origin and evolution of organelle DNAs[5, 13]

There are two hypotheses for the origin of organelle genes:

1. Eukaryotic nuclear DNA.
2. A domesticated prokaryotic endosymbiont.

An endosymbiont origin is now usually accepted for chloroplasts. The striking homology in organization and sequence of the ribosomal transcription unit in chloroplasts and in *E. coli* and the resemblance of chloroplast transcription and translation signals to their prokaryotic counterparts is not easily explained if the origin of chloroplast DNA were a eukaryotic one. On the basis of the sequence of 5S rRNAs Dayhoff and coworkers have even suggested that the colonization of eukaryotic cells by chloroplast progenitors occurred in several independent events throughout evolution and that chloroplasts in algae and higher plants are descendants from different prokaryotic ancestors.

The origin of mitochondrial DNA is more disputed, because present-day mitochondrial components do not have much homology with either prokaryotic or eukaryotic systems, with the exception of the plant mitochondrial rRNAs which have typical prokaryotic properties. It is thought that the lack of homology of most mitochondrial genetic systems with their putative bacterial ancestors is best explained by assuming a rapid evolution of mitochondrial genes. Evidence for this has come from sequence comparison of mitochondrial and nuclear genes in related eukaryotes. This indicates that mitochondrial genes evolve about ten times faster than comparable nuclear genes. Rapid gene evolution could also account for the enormous diversity in the mitochondrial DNAs analysed thus far and may have been made possible by the low number of genes in mitochondrial DNA and the very limited amount of regulation required in the expression of these genes because most eukaryotic cells need the same mitochondria all the time.

References

1. Anderson, S., Bankier, A. T., Barrell, B. G., De Bruijn, M. H. L., Coulson, A. R., Drouin, J., Eperon, I. C., Nierlich, D. P., Roe, B. A., Sanger, F., Schreier, P. H., Smith, A. J. H., Staden, R. and Young, I. G. (1981). Sequence and organization of the human mitochondrial genome. *Nature, London*, **290**, 457–465
2. Attardi, G. (1981). Organization and expression of the mammalian mitochondrial genome: A lesson in economy. *TIBS*, **6**, 86–90; 100–103
3. Bedbrook, J. R. and Kolodner, R. (1979). The structure of chloroplast DNA. *Annual Review of Plant Physiology*, **30**, 593–620
4. Blake, C. C. F. (1978). Do genes-in-pieces imply proteins-in-pieces? *Nature, London*, **273**, 267–268
5. Bogorad, L. (1982). Regulation of intracellular gene flow in the evolution of eukaryotic genomes. In *Origins of Chloroplasts* (Schiff, J. A. and Lyman, H., Eds.), pp. 277–295. Amsterdam, Elsevier
6. Borst, P. (1981). The biogenesis of mitochondria in yeast and other primitive eukaryotes. In *International Cell Biology 1980–1981* (Schweiger, H. G., Ed.), pp. 239–249. Berlin, Springer Verlag
7. Borst, P. and Grivell, L. A. (1978). The mitochondrial genome of yeast. *Cell*, **15**, 705–723
8. Borst, P. and Grivell, L. A. (1981a). One gene's intron is another gene's exon. *Nature, London*, **289**, 493–440
9. Borst, P. and Grivell, L. A. (1981b). Small is beautiful – portrait of a mitochondrial genome. *Nature, London*, **290**, 443–444
10. Borst, P. and Hoeijmakers, J. H. J. (1979). Kinetoplast DNA. *Plasmid*, **2**, 20–40
11. Buetow, D. E. and Wood, W. M. (1979). The mitochondrial translation system. *Subcellular Biochemistry*, **5**, 1–85
12. Dawid, I. B. (1965). Deoxyribonucleic acid in amphibian eggs. *Journal of Molecular Biology*, **12**, 581–599
13. Doolittle, W. F. (1980). Revolutionary concepts in evolutionary cell biology. *TIBS*, **5**, 146–149
14. Ellis, R. J. (1981). Chloroplast proteins: Synthesis, transport and assembly. *Annual Review of Plant Physiology*, **32**, 111–138
15. Eperon, I. C., Janssen, J. W. G., Hoeijmakers, J. H. J. and Borst, P. (1983). The major transcripts of the kinetoplast DNA of *Trypanosoma brucei* are very small ribosomal RNAs. *Nucleic Acids Research,* **11**, 105–125
16. Goldbach, R. W., Borst, P., Bollen-De Boer, J. E. and Van Bruggen, E. F. J. (1978). The organization of ribosomal RNA genes in the mitochondrial DNA of *Tetrahymena pyriformis* strain ST. *Biochimica et Biophysica Acta*, **521**, 169–186
17. Grivell, L. A. (1983). Mitochondrial DNA. *Scientific American* (in press)
18. Heizmann, P., Doly, J., Hussein, Y., Nicolas, P., Nigon, V. and Bernardi, G. (1981). The chloroplast genome of bleached mutants of *Euglena gracilis*. *Biochimica et Biophysica Acta*, **653**, 412–415
19. Herrmann, R. G. and Possingham, J. V. (1980). Plastid DNA–The plastome. In *Results and Problems in Cell Differentiation: Chloroplasts* (Reinert, J., Ed.), Volume 10, pp. 45–96. Berlin, Springer Verlag
20. Jope, C. A., Hirai, A. and Wildman, S. G. (1978). Evidence that the amount of chloroplast DNA exceeds that of nuclear DNA in mature leaves. *Journal of Cell Biology*, **79**, 631–636
21. Koch, W., Edwards, K. and Kössel, H. (1981). Sequencing of the 16S-23S spacer in a ribosomal RNA operon of *Zea mays* chloroplast DNA reveals two split tRNA genes. *Cell*, **25**, 203–213
22. Kroon, A. M. and Saccone, C. (Eds.) (1980). *Organization and Expression of the Mitochondrial Genome*. Amsterdam, North-Holland
23. Nass, M. M. K. (1976). Mitochondrial DNA. In *Handbook of Genetics* (King, R. C., Ed.), Volume 5, pp. 477–533. New York, Plenum Press
24. Slonimski, P. P., Borst, P. and Attardi, G. (Eds.) (1982). *Mitochondrial Genes*. Cold Spring Harbor Laboratory, Cold Spring Harbor, N.Y.
25. Stumpf, P. K. and Cann, E. E. (Eds.) (1981). *Biochemistry of Plants*. New York, Academic Press
26. Tabak, H. F., Grivell, L. A. and Borst, P. (1983). Transcription of mitochondrial DNA. In *CRC Critical Reviews in Biochemistry* (Fasman, G. D., Ed.). Boca Ratton, CRC Press (in press)
27. Wells, R. and Sager, R. (1971). Denaturation and the renaturation kinetics of chloroplast DNA from *Chlamydomonas reinhardi*. *Journal of Molecular Biology*, **58**, 611–622
28. Wurtz, B. A. and Buetow, D. E. (1981). Intraspecific variation in the structural organization and redundancy of chloroplast ribosomal DNA cistrons in *Euglena gracilis*. *Current Genetics*, **3**, 181–187

Chapter 6

Lampbrush chromosomes

M. T. Vlad

Distribution, historical background and method of preparation

Lampbrush chromosomes are transcriptionally active chromosomes present in the oocytes during the diplotene stage of the first meiotic division. They are almost universally found in animal oocytes, with the exception of some higher insects which have meroistic ovaries (i.e. the presence of nurse cells appears to obviate the need for chromosomes of the germinal vesicle to assume the lampbrush form). In organisms that have lampbrush chromosomes, all of them are in the lampbrush form, including the microchromosomes and supernumeraries. Lampbrush chromosomes are absent, or at the most scarcely developed in spermatocytes, with the notable exception of the Y chromosome in spermatocytes of *Drosophila*. They have not been found in meiotic cells (males or females) of higher plants. However lampbrushes are present in the 'primary nucleus' of the unicellular alga *Acetabularia*[53]. Hence they are not strictly confined to the Animal Kingdom. Most of the information concerning these structures has come from work in the urodele amphibia, namely newts, axolotls and salamanders. Although lampbrush chromosome maps are available for some anuran species[23, 42], the urodeles have larger lampbrush chromosomes which are more amenable to cytological studies.

Lampbrush chromosomes were first observed by Flemming[15] in histological sections through ovaries of the urodele *Siredon* in 1882. Ten years later Rückert[46] described them in isolated, fixed and stained oocyte nuclei from the shark *Pristiurus*. He noticed that the chromosomes at this stage are associated in pairs and that the lateral projections are loops, with both ends reflected back into the chromosome axis. Rückert[46] drew a comparison between the appearance of these chromosomes and the brushes then used for cleaning oil lamps. Hence, he named them lampbrush chromosomes. No further observations of real consequence were made until 1937, when Duryee[13] discovered that in manually isolated nuclei from amphibian oocytes, placed in a saline solution lacking calcium, the chromosomes maintained a 'life-like' condition and could be observed using light microscopy. Most of Duryee's speculations regarding these structures were erroneous due to the misinterpretation of optical artefacts, but the practical demonstration that these chromosomes could be readily isolated in fresh, unfixed condition was to prove valuable in later studies.

In 1954, Gall[17] developed a simple manual technique for dissecting out the chromosomes from the nucleus and examining them in phase contrast microscopy. Essentially, the oocyte is punctured with a needle and the nucleus is gently squeezed out through the hole. It is then picked up in a fine Pasteur pipette, transferred to a saline solution in a chamber constructed from a microscope slide with a central hole and a coverslip sealed across it with wax. In the chamber the nuclear envelope is removed and a coverslip placed over the saline. As the nuclear sap disperses the chromosomes spread apart and fall onto the bottom coverslip, where they can be examined from below, using a phase contrast microscope with its optical train inverted. With this method, Gall made it possible to observe, for the first time, morphologically intact and unfixed lampbrush chromosomes with the highest objective magnification obtainable in light microscopy. Later, Miller[36] made a significant advance in the preparation of lampbrush chromosomes for electron microscopy.

Most of the basic information concerning the morphology of lampbrush chromosomes is due to light microscopy studies made by Gall, Callan and MacGregor in newt oocytes. Although observations in other organisms have been limited there is nothing to suggest that their lampbrush chromosomes are significantly different from those in amphibia which are described in the next section.

Morphology of amphibian lampbrush chromosomes

Lampbrush chromosomes are diplotene bivalents, with the two homologous chromosomes associated at the chiasmata points. The chiasmata are confined to the chromosome axes and have never been observed to occur within the loops.

A lampbrush chromosome may be described as a row of dense granules containing deoxyribonucleoproteins (DNPs), called chromomeres, connected by a thin thread of the same material, the interchromomeric fibre (*Figure 6.1*). The chromomeres are ¼–2 μm in diameter and spaced approximately 2 μm apart along the chromosome axes. Each chromomere is associated with one or more pairs of loops.

It has been shown that the loops have a thin axis of DNA[8] surrounded by a matrix of ribonucleoproteins (RNPs)[32]. The loops are variable in length and extend laterally 5–50 μm from the chromosome axes. There are exceptionally long loops which may be 200 μm or more in length. Some loops have highly distinctive morphologies and textures of matrix. Such loops form obvious landmarks and together with other features are used to map lampbrush chromosomes. Maps of lampbrush chromosomes from several *Triturus* species and detailed descriptions of marker loops have been given by Callan and Lloyd[7].

The vast majority of loops have the same qualitative appearance showing an uncompacted, finely granular matrix (*Figures 6.2* and *6.3*). It has been estimated that there are a total of 5000 to 10000 loops over the entire karyotypes of *Triturus* and *Notophthalmus*[4, 19]. Sister loops arising from the same chromomere have the same structural characteristics, although occasionally they can differ in length. Many, though not all, loops have an assymetric appearance being thin at one end and becoming progressively thicker towards the other end (*Figure 6.2*). Some loops also show multiple tandem assymetries, going from thin to thick several times along their length (*Figure 6.3*).

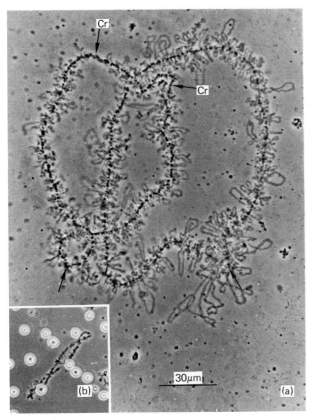

Figure 6.1. Phase contrast micrograph of two lampbrush
bivalents. (a) bivalent IX from *T. c. carnifex* which has a C value
of 27 pg DNA (*n* = 12). The two homologous chromosomes are
joined at the ends (arrows), cr = centromeric bars lacking lateral
loops; (b) a bivalent of middle length from *Xenopus laevis* which
has a C value of 3 pg DNA (*n* = 18). The two bivalents are shown
at the same magnification and illustrate clearly the relationship
between the C value and the size of lampbrush chromosomes,
particularly loop length.

The assymetry of the loop reflects the direction of RNA transcription. The RNA
synthesis starts at the thin end and proceeds towards the thick end of the loop.
Loops with multiple tandem assymetries are considered to contain several
transcription units (TUs), the number of TUs corresponding to the number of thin
to thick regions present within a loop (*Figure 6.3*). This interpretation derives from
electron microscopy studies of lampbrush chromosomes, prepared by the Miller
spreading technique. Lampbrush chromosomes are dispersed in distilled water,
adjusted to pH 8.5–9 with a minimal amount of borate buffer, attached to
carbon-coated grids, stained and viewed by electron microscopy. This procedure
results in the loss of most of the proteins associated with the chromosomes, leaving
only the DNA and the nascent RNA transcripts which are still associated with their
DNA template by the RNA polymerase. Such preparations show that lateral loops
consist of discrete and uninterrupted linear gradients of increasing length of RNA

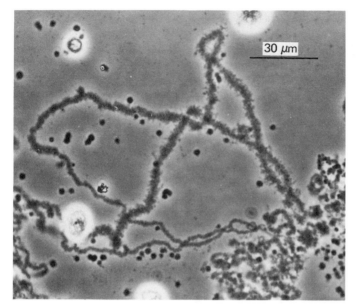

Figure 6.2. A pair of conspicuous loops on a lampbrush chromosome from *Plethodon vehiculum* containing one very long transcription unit and showing the thin and thick insertions. From M. Vlad, PhD Thesis, 1974.

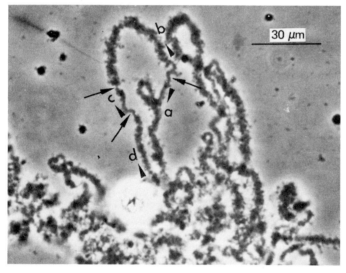

Figure 6.3. A loop from *Plethodon cinereus* containing four transcription units and showing multiple asymmetry (going from thin to thick) along its length. The starting points or thin ends are indicated by arrows. The direction of transcription for each transcription unit is indicated by arrow heads. The transcription unit b is of opposite polarity to a, c and d. From M. Vlad, PhD Thesis, 1974.

fibrils, originating from an initiation point on the DNA axes and extending to a termination point at a fixed distance away[37]. Between these points the polymerase molecules are closely packed and possibly transcribing at maximal rates (*Figure 6.4*). There may be several TUs of similar or different lengths, with short transcript-free spacers a few μm long. Neighbouring TUs may be of the same or opposite polarity[1, 47].

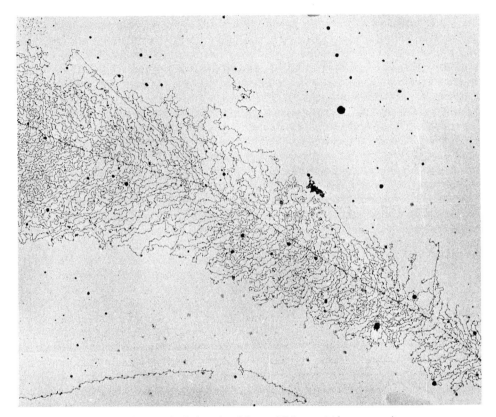

Figure 6.4. An electron micrograph of a lampbrush loop of *Triturus viridescens* near its thin insertion showing a gradient of increasing length of RNA fibrils attached to the loop DNA axis. Magnification ×38 000. From Miller and Hamkalo[38].

The centromeres of lampbrush chromosomes exhibit different features in various species, often being difficult to distinguish. In some species they appear as smooth round granules, whilst in others these granules are flanked by axial bars[4]. Finally, they may be simply bars a few μm in length, thicker or thinner than the chromosome axis. Invariably, centromeres lack visible lateral loops (*Figure 6.1*).

For the description of lampbrush chromosomes in *Drosophila* see Bostock and Sumner[3].

There is one interesting observation, the significance of which is not understood. Animals with large genomes have very long lampbrush loops and those with small genomes have much shorter loops (*Figure 6.1*).

Development of lampbrush chromosomes

The actual process of development of lampbrush chromosomes has not been systematically studied. It is obvious that they extend from relatively short and contracted chromosomes at the end of pachytene. It is not clear if there is a gradual transition from the pachytene type of chromosome organization to the lampbrush type of chromosome organization, as the chromosome becomes larger and progressively more finely 'chromomeric' in appearance, with loops unfolding from the chromomeres, as transcription starts at specific initiation points; or alternatively, there is an extensive decondensation of chromosomes at the end of the pachytene followed by the initiation of transcription at specific points, and condensation of chromatin that is not to be transcribed. Transcription and condensation could proceed together until the typical loop pair/chromomere pattern becomes distinguishable. Unfortunately, pre-lampbrush oocytes from most animals are too small to handle successfully in order to observe the process of development of lampbrush chromosomes with the light microscope.

Recently Hill and MacGregor[27] have studied the origin of transcription units in *Xenopus* oocytes by the Miller spreading technique described earlier. They have shown that dispersed chromatin from pachytene oocytes is entirely nucleosomal and untranscribed. As the oocytes enter diplotene, transcription begins at various sites with RNA polymerases sparsely distributed along transcriptional units. The level of polymerase initiation then increases and the TUs rapidly become covered in RNA transcripts all along their length. When this happens one would expect the TUs to look like lampbrush loops if it were possible to examine them with a phase contrast microscope. An interesting observation is that the lengths of TUs appear to be established very early in diplotene, and remain unchanged thereafter. Taken at its face value, this means that early diplotene is the stage when the 'decision' is taken as to which parts of the chromatin will form compact, chromomeric DNP and which will be transcribed, and form lateral loops. Miller's spreading technique causes a complete disorganization of chromosomal morphology and therefore it is not possible to distinguish between the two alternatives of development of lampbrush chromosomes, mentioned above.

In phase contrast microscopy it is apparent that the lampbrush chromosomes of small yolky oocytes are at the peak of their development in that the chromosomes and loops are at their longest and the chromomeres are very small. As the oocytes grow in size and approach metaphase the chromosome axes decrease in length by amalgamation of adjacent chromomeres and the lateral loops become shorter until eventually no trace of lampbrush structure remains.

Hormone treatment also affects development and regression of lateral loops and can induce other types of changes[4], but the significance of these changes is not clear.

The lampbrush phase lasts for a long time. In amphibians it is estimated at 1–6 months and there is intensive RNA synthesis throughout. In general, the larger the genome, the longer the lampbrush phase[30].

Structure and genetics of lampbrush chromosomes

Lampbrush chromosomes provide the only direct evidence for the uninemic nature of the eukaryotic chromosomes and the covalent integrity of the DNA molecule throughout the length of the chromosome. In Callan's own words: 'A DNA fibre runs continuously from one end of the chromosome out into each loop and between each chromomere until it reaches the other end'[7] (*Figure 6.5*). This statement has been confirmed by experiments involving measurements of strand thickness in electron micrographs of lampbrush chromosomes[36] and measurements of the kinetics of digestion of lampbrush chromosomes with deoxyribonuclease[20]. Any suggestion that metaphase chromosomes may be organized differently can be refuted by the observation that the linear order of landmarks is the same for lampbrush and metaphase chromosomes[22, 25].

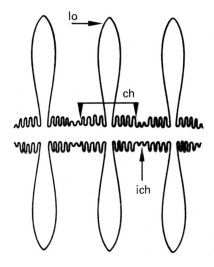

Figure 6.5. Diagramatic representation of the concept of continuity of DNA in chromosome axes and lateral loops; ch, chromomere; ich, interchromomeric region; lo, loop axis.

Since lampbrush chromosomes are bivalents it is possible to compare lateral loops and other structures on the two homologues. It is known that there are allelic alternatives for given loops in various *Triturus* subspecies, and the frequencies at which these alternatives appear, in wild populations, are as predicted by Hardy-Weinberg's law[6, 7]. Differences between homologous chromosomes in hybrids between subspecies of *Triturus*[4, 7] have shown that the chromomere/loop complexes behave like pairs of Mendelian alleles. Also the landmark structures of the karyotype are species-specific and constant from generation to generation. Such observations indicate that the loops are specific manifestations of the DNA sequences in the genome. These facts and the analogy with polytene chromosomes lead to the assumption that each chromomere and its loop represents a single gene. Doubt was cast on this assumption by a study of chromomere number in three related species of *Plethodon*[57]. One of these species, *P. cinereus*, has half of the C value of two others, *P. vehiculum* and *P. dunni*. All three species have nearly identical karyotypes but the chromosomes of *P. cinereus* are considerably shorter than their counterparts in *P. vehiculum* and *P. dunni* and still the size and the frequency of chromomeres along the chromosomes axes is much the same. Over the whole karyotype *P. vehiculum* and *P. dunni* have 60–70% more chromomeres than *P. cinereus*. If lampbrush chromosomes represented single genetic units, then one might expect to find roughly the same number in ll three species. This is certainly not the case. Unfortunately, virtually nothing is known of the genetics of urodeles, and it is not yet possible to make any correlations between chromosome structure and phenotype, as in *Drosophila*.

Data from molecular studies of genome organization point to the fact that a chromomere/loop complex is a mixture of sequences (repetitive and unique). Because of their constancy from generation to generation they constitute morphological units of chromosomes and as such they are useful for studies of chromosome organization.

Transcription on lampbrush chromosomes

The main feature of lampbrush chromosomes is their transcriptional activity, which makes them excellent material for the study of this process.

In 1962 Gall and Callan[21] showed by autoradiography that the lateral loops incorporate tritiated uridine. The overwhelming majority of the loops have a uniform rate of incorporation throughout their length. An exception occurs in the giant granular loop of *Triturus cristatus carnifex* in which there is a sequential incorporation of the label, starting at the thin end of the loop and progressing gradually towards the thick end, until by 14 days the whole loop is labelled. This pattern of sequential labelling together with observations on the recovery of this loop from RNA inhibition by actinomycin D[52], lent support to the suggestion that all lateral loops are continually extended from the chromomeres at their thinner insertions and retracted back into the chromomeres at their thicker insertions[7,21]. This hypothesis, however, turned out to be erroneous and the extraordinary behaviour of the giant granular loop is not yet understood.

Experiments involving treatment with various concentrations of α-amanitin and antibody raised against various types of RNA polymerases have shown that for the majority of the loops the RNA synthesis is due to RNA polymerase II[2,48].

Nascent RNA molecules on the lateral loops rapidly become associated with newly synthesized proteins to form the RNP matrix surrounding the loops' DNA axis[21]. The proteins associated with the lampbrush RNA transcripts cover a considerable size range and represent more than 95% of the chromosomal mass[35,50]. The RNP matrix is made up of strings of particles of 20–30 nm in diameter which aggregate in different ways in different loops[34,41]. Examples of various matrices are illustrated in *Figure 6.6.*. Antibodies prepared against a single nuclear polypeptide associated with lampbrush RNA transcripts bind to a relatively small number of loops (approximately 10) on several chromosomes[49]. When monoclonal antibodies to transcript RNA proteins become available, many more loop-specific polypeptides may be discovered. The conclusion from these studies is that both the organization of RNP granules and some of their protein components may be specific to given loops and probably explains their distinctive morphologies and textures of matrix seen in phase contrast microscopy.

The RNA synthesized on lampbrush loops consists of very long molecules and appears to be equivalent to heterogeneous nuclear RNA[11]. In well spread preparations of lampbrush chromosomes from *Pleurodeles*[1] and *Triturus*[51] the RNP fibrils extend to 3–30 μm at the terminal point of transcription. In *Notophthalmus* RNP fibrils measuring well over 10 μm in length have been observed at intermediate positions along the loop axes, and it is estimated that fibrils considerably longer, perhaps up to 100 μm or more in length, would be found at the thick insertion of some loops[37,38].

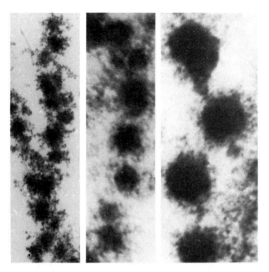

Figure 6.6. High voltage electron micrograph showing the arrangements and various morphological types of RNPs which constitute the matrix of lateral loops. Magnification ×25 000. From Malcolm and Sommerville[34].

Because of their large size and specific morphology, lampbrush chromosomes are excellent material for correlating molecular and cytological aspects of RNA transcription. In particular *in situ* hybridization of labelled DNA probes to nascent RNA transcripts on the lateral loops and autoradiography is a powerful technique for identification of loci at which specific sequences are transcribed[44]. An example of *in situ* hybridization and autoradiography of labelled DNA to RNA transcripts is shown in *Figure 6.7*. More recently labelled cRNA probes copied from single stranded DNA templates have been used in the same manner to identify RNA transcripts resulting from opposite strands of the DNA[12, 22]. Hybridization of labelled RNA or DNA probes to lampbrush chromosome preparations in which RNA transcripts have been removed by RNAase treatment, and the DNA has been denatured by alkaline solutions is used to identify sequences located in the chromosome axes.

In situ hybridization studies have demonstrated that there is widespread transcription of repetitive DNA sequences on the lateral loops. Middle repetitive DNA Cot fractions from *T. c. carnifex* were hybridized to lampbrush chromosome preparations from the same species. Autoradiographs of the preparations showed that a great many loops are distinctly and in some cases heavily labelled. Most of the heavily labelled loops were located on the cytologically unusual heteromorphic arms of the bivalent I[31]. Interpretation of such results is difficult because of the complexity of the DNA probe. The use of cloned DNA sequences overcomes this problem. We have selected five cloned repetitive DNA fragments from *T. c. carnifex* for hybridization to lampbrush chromosomes[58]. The repetition frequencies of the cloned sequences range from 3×10^2 to 6×10^4 copies per genome. The GC contents and interspersion patterns within the genome suggest that these sequences

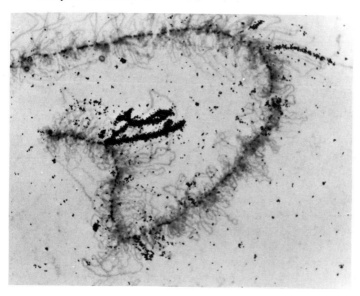

Figure 6.7. Autoradiograph of part of a lampbrush chromosome from *T. c. carnifex* following *in situ* hybridization of a cloned, repetitive DNA sequence to loop RNA transcripts. One pair of loops is heavily labelled throughout their length (centre). Another loop is labelled only part way along its length (upper right). For detailed explanation see text.

are typical for the major repetitive classes found in *T. c. carnifex* and represent members of individual repetitive families. All five sequences hybridize to nascent RNA transcripts on the lampbrush loops in a well defined pattern. One of these sequences is interesting in that it shows individual variation. Some females have a small number of loops (five pairs) hybridizing in the heteromorphic region of both arms of bivalent I, while other females have a large number of loops (12–20 pairs) hybridizing in the same region of bivalent I. These two distinct patterns of hybridization are constant for the individual female and may reflect a different distribution of some repetitive families between populations of the same species. The other four cloned sequences, some of them clustered and some highly interspersed, hybridize to only three pairs of loops on only one of the heteromorphic arms of bivalent I. In the group of clones we have analysed the least repeated sequences are transcribed at many more loci than the more highly repeated sequences. We do not yet know if these sequences are linked to coding sequences, or, indeed if they are themselves coding sequences[59].

The most surprising finding of such studies is that satellite DNA sequences are transcribed on the lampbrush loops. This has been demonstrated for two species. Varley *et al.*[55, 56] have shown that two cloned satellite DNAs (basic repeat 330 and 275 bp) hybridize to RNA transcripts of a large number of loops located again on the heteromorphic arms of bivalent I of *T. c. carnifex*. The pattern of hybridization is characteristic for each of the satellite sequences. Patterns and levels of labelling of individual loops suggested to the authors that in some loops RNA consists entirely of satellite transcripts and in others satellite sequences are probably part of a larger transcript. At this stage, the evidence points towards an unusual situation in which a substantial amount of repetitive DNA sequences have accumulated in the heteromorphic region of chromosome I. For further reading on the oddity of this chromosome in *T. c. carnifex* the reader is referred to the original papers[29, 33, 39]. When satellite sequences are hybridized to the chromomeric DNA it shows significant labelling. The pattern of labelling in DNA/DNA hybrids compares well with the pattern of loop labelling in the corresponding DNA/RNA transcript hybrids. This immediately suggests that there are more of the same sequence lying untranscribed in the chromosome axis.

An entirely different situation has been described for *Notophthalmus viridescens*[12]. In this case the transcription of a specific satellite sequence is strictly confined to loops located at or near the sphere loci of chromosomes II and VI. These loops may contain only one transcription unit or they may have multiple transcription units of different lengths and different polarities. By using single stranded probes for *in situ* hybridization, Diaz *et al.*[12] have demonstrated unequivocally that transcription can proceed in both directions within the length of a loop, and that opposite strands of the loop DNA axis are read in transcription units of opposite morphological polarities. The transcription of satellite sequences in *Notophthalmus* seems to be due to failure of transcription termination signals. In this species, tracts of various lengths of satellite sequences are interspersed with clusters of sequences coding for histone genes[54]. It is proposed that transcription initiates at the promoter site of a histone cluster, proceeds without interruption into the adjacent satellite sequence and stops at or near the next histone cluster. The length of the transcription units will vary according to the length of the satellite tract. The transcripts of the opposite strands of the DNA are explained by the fact that the coding sequences for H_1, H_3, H_2A and H_4 are in one strand and those coding for H_2B are in the other strand. Therefore both strands of the DNA must be transcribed in order to obtain messenger RNA for all five histone polypeptides. *In situ* hybridization of histone and satellite sequences to mitotic chromosomes established the location of both sequences on the same chromosomes and the same loci where the transcription is detected in lampbrush chromosomes[22]. Satellite sequences are also found in the centromeric regions of most chromosomes but satellite transcripts are only occasionally detected at the condensed centromeric bars in lampbrush chromosomes.

The transcription of histone genes has been studied in lampbrush chromosomes from *T. c. carnifex* and again the results are different from what has been found in *Notophthalmus*[43]. In this instance, cloned echinoderm histone DNA probes were utilized. After some initial problems involving the hybridization of a single repeat (CT/GA) present in the spacer between coding sequences it has been established that histone genes are transcribed on a small number of loops, on chromosome I, in the heteromorphic region[9]. Gall and his colleagues[22] have shown that the majority of histone sequences are also located at the sphere loci in *T. c. carnifex* (chromosomes V and VIII, DNA/DNA hybridization to mitotic chromosomes). Yet Callan *et al.*[9] have not detected any histone transcripts on the loops at the sphere loci. It is unlikely that these results are due to heterologous probes since the sequences of the echinoderm histone probe used by Callan are well known and they do cross react with histone sequences for *Notophthalmus* and *Triturus*[26, 54]. One possible explanation could be that the major histone gene clusters of sphere loci in *T. c. carnifex* are not transcribed during the lampbrush stage but minor clusters at other loci are active. This would be similar to the situation in the sea urchins which contain an oogenetic and a somatic set of histone genes. An analogous situation was described by Morgan *et al.*[40] for the 18S and 28S ribosomal sequences in *T. c.*

carnifex. The major clusters of ribosomal genes are located on chromosomes VI and IX[25]. When these sequences are hybridized to the RNA transcripts in lampbrush chromosome preparations, there is no hybridization at these loci, but a few loops at other loci do hybridize strongly, suggesting that isolated gene clusters are being transcribed.

Because of the amplified signal achieved by hybridization of DNA or RNA probes to the nascent RNA transcripts accumulated in the loop matrix, the lampbrush chromosomes are the only type of eukaryotic chromosomes (with the exception of polytenes) in which unique sequences can be located by *in situ* hybridization. Because of lack of cloned homologous probes we do not yet know where any unique coding sequences are located on lampbrush chromosomes, although many of them are known to be transcribed during oogenesis.

In conclusion to this section it should be emphasized that *in situ* hybridization studies have revealed some puzzling but interesting aspects of transcription in oogenesis.

The function of lampbrush chromosomes

Lampbrush chromosomes antedate the evolution of coelomates and since they represent an integral part of the meiotic cycle it is reasonable to suppose that they have an important role in oogenesis.

The classical hypothesis for lampbrush chromosome function is that they provide the stored oocyte RNAs that are needed for maintenance of the early embryonic development[14, 18]. Two observations have helped to sustain this hypothesis for many years. Firstly, mature oocytes contain large amounts of hnRNAs including polyadenylated messenger RNAs (pA$^+$ RNAs) which are synthesized during oogenesis; secondly, there is a reasonable correspondence between the proportion of genomic DNA that is represented in the lampbrush loops and the complexity of stored oocyte RNAs[11, 51]. A consequence of this hypothesis would be that RNAs would accumulate in oocytes throughout the length of lampbrush stage. However, this does not seem to be the case. Rosbash and his colleagues[24, 45] have found that in *Xenopus* oocytes pA$^+$ RNAs accumulate very rapidly before vitellogenesis and remain at a steady state level until the end of oogenesis. Moreover, pA$^+$ RNAs synthesized in previtellogenic oocytes are stable for the entire length of oogenesis[16]. The time at which pA$^+$ RNAs reach the steady state level corresponds to the early Dumont stage II of oogenesis. Lampbrush chromosomes are present during Dumont stage II and very probably during Dumont stage I[27], when the accumulation of pA$^+$ RNAs takes place. However, they are extended and fully active throughout vitellogenesis, a long time after the accumulation of pA$^+$RNAs has ceased.

The suggestion that RNA transcribed on lampbrush loops for the longest part of their existence does not contribute to the stored pA$^+$ RNA, together with the findings by *in situ* hybridization of transcription of unusual and probably untranslatable sequences (satellite and simple repeats) casts doubt on the importance of lampbrush chromosome transcription for early development.

There are other hypotheses that have been put forward to explain these remarkable structures. The 'Master/Slave hypothesis' introduced by Callan in 1967[5] was the first to propose a genetic function for the lampbrush loop. It was an immensely attractive theory but it has now been discredited. In 1975 Leon[28] proposed that lampbrush loops are regions where the chromosomal DNA is undergoing preprogramming for development such as exchanges or enzymatic modifications of regulatory molecules, including perhaps non-histone proteins. This hypothesis is entirely speculative and based on circumstantial evidence.

More recently Cavalier-Smith[10] proposed that the function of lampbrush chromosomes is to produce large amounts of hnRNAs in order to swell the germinal vesicle, so increasing the area of its nuclear envelope, and the number of nuclear pores available for export of the large amount of ribosomal material that is produced by the amplified nucleolar DNA. In principle this hypothesis fits better than any other that has been offered so far, particularly as it is discussed in the wide context of genome evolution. Nonetheless it seems significant that the oocytes from meroistic ovaries in some insects have large swollen nuclei, but no lampbrush chromosomes. A more general comment in relation to Cavalier-Smith's argument is that the total amount of transcripts made on lampbrush loops is very small compared with the massive amount of ribosomal transcripts produced by the amplified nucleolar DNA, and therefore is unlikely to be significant with respect to swelling of the nucleus except, perhaps, in the earliest lampbrush stages before the nucleolar DNA becomes fully active.

Whatever their function is, lampbrush chromosomes are of interest not only because of their intrinsic characteristics, but also because they have, and are still, providing information which is more widely applicable to chromosome structure, organization and RNA transcription.

For further reading on the organization and transcription in lampbrush chromosomes the reader is referred to a recent review by MacGregor[30].

References

1. Angelier, N. and Lacroix, J. C. (1975). Complexes de transcription d'origines nucléolaire et chromosomique d'ovocytes de *Pleurodeles waltlii* et *P. poireti* (Amphibiens, Urodèles). *Chromosoma*, **51**, 323–335
2. Bona, M., Scheer, V. and Bautz, E. K. F. (1981). Antibodies to RNA polymerase II (B) inhibit transcription in lampbrush chromosomes after microinjection into living Amphibian oocytes. *Journal of Molecular Biology*, **151**, 81–99
3. Bostock, C. J. and Sumner, A. T. (1978). *The Eukaryotic Chromosome*, pp.365–368. Amsterdam, Elsevier/North-Holland Biomedical Press
4. Callan, H. G. (1963). The nature of lampbrush chromosomes. *International Review of Cytology*, **15**, 1–34
5. Callan, H. G. (1967). The organization of genetic units in chromosomes. *Journal of Cell Science*, **2**, 1–7
6. Callan, H. G. and Lloyd, L. (1956). Visual demonstration of allelic differences within cell nuclei. *Nature, London*, **178**, 355–357
7. Callan, H. G. and Lloyd, L. (1960). Lampbrush chromosomes of crested newts *Triturus cristatus* (Laurenti). *Philosophical Transactions of the Royal Society*, **B243**, 135–219
8. Callan, H. G. and MacGregor, H. C. (1958). Action of deoxyribonuclease on lampbrush chromosomes. *Nature, London*, **181**, 1479–1480
9. Callan, H. G., Old, R. W. and Gross K. W. (1980). Problems exposed by the results of *in situ* hybridization to lampbrush chromosomes. *European Journal of Cell Biology*, **22**, 21
10. Cavalier-Smith, T. (1978). Nuclear volume control by nucleoskeletal DNA, selection for cell volume and cell growth rate, and the solution of the DNA C-value paradox. *Journal of Cell Science*, **34**, 247–278
11. Davidson, E. H. (1976). *Gene Activity in Early Development*, pp.367–378. New York and London, Academic Press

12. Diaz, M. O., Barsacchi-Pilone, G., Mahon, K. and Gall, J. C. (1981). Transcripts from both strands of a satellite DNA occur on lampbrush chromosome loops of the newt *Notophthalmus*. *Cell*, **24**, 649–659
13. Duryee, W. R. (1937). Isolation of nuclei and non-mitotic chromosome pairs from frog eggs. *Archiv für experimentelle Zellforschung*, **19**, 171–176
14. Duryee, W. R. (1950). Chromosomal physiology in relation to nuclear structure. *Annals of the New York Academy of Science*, **50**, 920–953
15. Flemming, W. (1882). *Zellsubstanz, Kern und Zelltheilung*. Liepzig, F. C. W. Vogel
16. Ford, P. J., Mathieson, T. and Rosbash, M. (1977). Very long-lived messenger RNA in ovaries of *Xenopus laevis*. *Developmental Biology*, **57**, 417–427
17. Gall, J. G. (1954). Lampbrush chromosomes from oocyte nuclei of the newt. *Journal of Morphology*, **94**, 283–351
18. Gall, J. G. (1955). Problems of structure and function of the amphibian oocyte nucleus. *Symposium of the Society for Experimental Biology. 9. Fibrous Protein and their Biological Significance*, pp. 358–370
19. Gall, J. G. (1963a). In *Cytodifferentiation and Macromolecular Synthesis* (Lock, M., Ed.), p.119. New York, Academic Press
20. Gall, J. G. (1963b). Kinetics of deoxyribonuclease action on chromosomes. *Nature, London*, **198**, 36–38
21. Gall, J. G. and Callan, H. G. (1962). ^{3}H-uridine incorporation in lampbrush chromosomes. *Proceedings of the National Academy of Sciences, USA*, **48**, 562–570
22. Gall, J. G., Stephenson, E. C., Eriba, H. P., Diaz, M. O. and Barsacchi-Pilone, G. (1981). Histone genes are located at the sphere loci of newt lampbrush chromosomes. *Chromosoma*, **84**, 159–171
23. Giorgi, F. and Galleni, L. (1972). The lampbrush chromosomes of *Rana esculenta* (Amphibia–Anura). *Caryologia*, **25**, 107–123
24. Golden, L., Shafer, U. and Rosbash, M. (1980). Accumulation of individual pA$^+$ RNA during oogenesis of *Xenopus laevis*. *Cell*, **22**, 835–844
25. Hennen, S., Mizuno, S. and MacGregor, H. C. (1975). *In situ* hybridization of ribosomal DNA labelled with ^{125}iodine to metaphase and lampbrush chromosomes from newts. *Chromosoma*, **50**, 349–369
26. Hilder, V. A., Livesey, N. R., Turner, P. C. and Vlad, M. (1981). Histone gene number in relation to C-value in amphibians. *Nucleic Acid Research*, **9**, 5737–5746
27. Hill, R. S. and MacGregor, H. C. (1980). The development of lampbrush chromosome-type transcription in the early diplotene oocyte of *Xenopus laevis*: An electron microscope analysis. *Journal of Cell Science*, **44**, 87–101
28. León, P. E. (1975). Function of lampbrush chromosomes: a hypothesis. *Journal of Theoretical Biology*, **55**, 481–490
29. MacGregor, H. C. (1979). *In situ* hybridization of highly repetitive DNA to chromosomes of *Triturus cristatus*. *Chromosoma*, **71**, 57–64
30. MacGregor, H. C. (1980). Recent developments in the study of lampbrush chromosomes. *Heredity*, **44**, 3–35
31. MacGregor, H. C. and Andrews, C. (1977). The arrangement and transcription of 'middle repetitive' DNA sequences on lampbrush chromosomes of *Triturus*. *Chromosoma*, **63**, 109–126
32. MacGregor, H. C. and Callan, H. G. (1962). The action of enzymes on lampbrush chromosomes. *Quarterly Journal of Microscopical Sciences*, **103**, 173–203
33. MacGregor, H. C. and Horner, H. (1980). Heteromorphism for chromosome I, a requirement for normal development in crested newt. *Chromosoma*, **76**, 111–122
34. Malcolm, D. B. and Sommerville, J. (1974). The structure of chromosome derived ribonucleoproteins in oocytes of *Triturus cristatus carnifex* (Laurenti). *Chromosoma*, **48**, 137–158
35. Malcolm, D. B. and Sommerville, J. (1977). The structure of nuclear ribonucleoproteins of amphibian oocytes. *Journal of Cell Science*, **24**, 143–165
36. Miller, O. L. Jr. (1964). Fine structure of lampbrush chromosomes. National Cancer Institute, Monograph, **18**, pp. 79–99
37. Miller, O. L. and Bakken, A. H. (1972). Morphological studies of transcription. *Acta Endocrinologica, Supplement*, **168**, 155–177
38. Miller, O. L. and Hamkalo, B. A. (1972). Visualization of RNA synthesis on chromosomes. *International Review of Cytology*, **33**, 1–25
39. Morgan, G. T. (1978). Absence of chiasmata from the heteromorphic region of chromosome I during spermatogenesis in *Triturus cristatus carnifex*. *Chromosoma*, **66**, 269–280
40. Morgan, G. T., MacGregor, H. C. and Coleman, A. (1980). Multiple ribosomal gene sites revealed by *in situ* hybridization of *Xenopus* rDNA to *Triturus* lampbrush chromosomes. *Chromosoma*, **80**, 309–330

41. Mott, M. R. and Callan, H. G. (1975). An electron microscope study of the lampbrush chromosomes of the newt *Triturus cristatus*. *Journal of Cell Science*, **17**, 241–261
42. Müller, W. P. (1974). The lampbrush chromosomes of *Xenopus laevis* (Daudin). *Chromosoma*, **47**, 283–296
43. Old, R. W., Callan, H. G. and Grass, K. W. (1977). Localization of histone gene transcripts in newt lampbrush chromosomes by *in situ* hybridization. *Journal of Cell Science*, **27**, 57–79
44. Pukkila, P. J. (1975). Identification of the lampbrush chromosome loops which transcribe 5S ribosomal RNA in *Notophthalmus (Triturus) viridescens*. *Chromosoma*, **53**, 71–89
45. Rosbash, M. and Ford, P. J. (1974). Polyadenylic acid-containing RNA in *Xenopus laevis* oocytes. *Journal of Molecular Biology*, **85**, 87–101
46. Rückert, J. (1892). Zur Entwickelungsgeschichte des Ovarioleies bei Selachiern. *Anatomischer Anzeiger*, **7**, 107–158
47. Scheer, V., Franke, W. W., Trendelenburg, M. F. and Spring, H. (1976). Classification of loops of lampbrush chromosomes according to the arrangement of transcription complexes. *Journal of Cell Science*, **22**, 503–520
48. Schultz, L. D., Kay, B. K. and Gall, J. G. (1981). *In vitro* RNA synthesis in oocytes nuclei of the newt *Notophthalmus*. *Chromosoma*, **82**, 171–187
49. Scott, S. E. M. and Sommerville, J. (1974). Location of nuclear proteins on the chromosomes of newt oocytes. *Nature, London*, **250**, 680–682
50. Sommerville, J. (1977). Gene activity in the lampbrush chromosomes of Amphibian oocytes. *International Review of Biochemistry*, **15**, 79–156
51. Sommerville, J. and Malcolm, D. B. (1976). Transcription of genetic information in Amphibian oocytes. *Chromosoma*, **55**, 183–208
52. Snow, M. H. L. and Callan, H. G. (1969). Evidence for a polarized movement of the lateral loops of newt lampbrush chromosomes during oogenesis. *Journal of Cell Science*, **5**, 1–25
53. Spring, H., Scheer, U., Franke, W. W. and Trendelenburg, M. F. (1975). Lampbrush-type chromosomes in the primary nucleus of the green alga *Acetabularia mediterranea*. *Chromosoma*, **50**, 25–43
54. Stephenson, E. C., Erba, H. P. and Gall, J. C. (1981). Histone gene clusters of the newt *Notophthalmus* are separated by long tracts of satellite DNA. *Cell*, **24**, 639–647
55. Varley, J. M. and MacGregor, H. C. (1980). Satellite DNA is transcribed in lampbrush chromosomes. *Nature, London*, **238**, 686–688
56. Varley, J. M., MacGregor, H. C., Nardi, J., Andrews, C. and Erba, H. P. (1980). Cytological evidence of transcription of highly repeated DNA sequences during the lampbrush stage of *Triturus cristatus carnifex*. *Chromosoma*, **80**, 289–307
57. Vlad, M. and MacGregor, H. C. (1975) Chromomere number and its genetic significance in lampbrush chromosomes. *Chromosoma*, **50**, 327–347
58. Vlad, M. and Hilder, V. A. (1980). *In situ* hybridization of cloned DNA sequences to lampbrush chromosomes from *Triturus cristatus carnifex*. *European Journal of Cell Biology*, **22**, 624
59. Vlad, M. and Hilder, V. A. (1983). A study of chromosomal organization of repetitive DNA sequences by '*in situ*' hybridization. *The Histochemical Journal*, **15**, (in press)

Chapter 7

Polytene chromosomes

E. K. F. Bautz and R. Kabisch

Introduction

During the last few years polytene chromosomes of Diptera have received particular attention by molecular biologists as they allow location of either genes or chromosomal proteins in genetically defined segments of chromosomes, the former by *in situ* hybridization[15], the latter by indirect immunofluorescence using antibodies directed against either groups of proteins or single polypeptide chains[1, 2, 24, 49, 58, 61].

Antibodies directed against proteins of known function can be used to analyse polytene chromosomes more from a functional than a structural viewpoint. This had led to a re-evaluation of the original estimates of the number of total versus active genes in *Drosophila*, the interpretation of bands and interbands and, although still controversial, has led to the view that polytene chromosomes are more a structural consequence of the functional state of the genome rather than the result of a ubiquitous and highly ordered array of neatly packaged genes[27]. It should be remembered at this point that polytene nuclei represent interphase nuclei, which means that the experimenter views the chromosomes at a state in the cell cycle where diploid chromosomes are fully decondensed and no euchromatin can be detected as distinct structures.

If, as we like to believe, a single DNA molecule is organized in a chromosomal arm just as part of a bundle of 2^{10} DNA molecules is in a polytene chromosome, then such giant chromosomes offer indeed an unique tool to study chromatin organization.

In this chapter we shall first try to assess the structure and organization of polytene chromosomes as seen by phase contrast microscopy, then summarize the data on distribution of antigens of known and unknown functions on the polytene chromosomes and, where appropriate, introduce structural information on genes and the organization of genes as they are now available from gene cloning and DNA sequence analyses.

The constancy of banding pattern and the one band–one gene correlation

Beermann, in his work on banding patterns of *Chironomus tentans* polytene chromosomes, compared the banding patterns of the third chromosome in polytene

101

nuclei of four different organs and has come to the conclusion that, although there were differences as to the number of bands that could be counted in the different tissues, the overall patterns appeared to be preserved[8,9]. The described similarity has often been interpreted by many authors as meaning absolute identity, an interpretation Beermann has never subscribed to. The observed regularities naturally apply only to the patterns of alternating bands and interbands and not to patterns of puffs as the number, localization, and size of puffs vary drastically within one cell type with stage and development and are also different between two cell types of one developmental stage. As puffs have been found to show the highest rates of uridine incorporation they were rightly claimed to be the predominant sites of RNA synthesis, i.e. gene activity[47]. Puffing in many cases is hormonally induced, e.g. some puffs appear shortly after accumulation of the moulting hormone ecdysone[4,7].

The constancy of the banding pattern which was also described for different tissues of *D. melanogaster*[51,52] and the disappearance of individual bands during puffing has led to the early view of bands representing condensed inactive genes which, when activated, become decondensed assuming a large volume occupied by chromatin, matrix (non-histone) proteins and RNA product (for review see Ashburner and Berendes[5, 10, 22, 23]). This view implied that one band corresponds to one gene and that the number of bands one can observe should equal the number of genes of *Drosophila*. *Drosophila* geneticists generally concurred with this view and some set out to prove the one band–one gene hypothesis by 'saturation bombing' (saturation mutagenesis) of a limited chromosomal region and grouping the large number of mutations within this region into complementation groups. If nearly every complementation group is represented by more than one mutant isolated, the map should be pretty well saturated. The first results of such tedious but important studies indicated that the one band–one gene rule is correct[29]. The more information becomes available, however, both by strictly genetic analysis and by analysis of cloned gene clusters the more exceptions are found to this rule.

Citing examples on (a) genetic, (b) cytological and (c) molecular grounds, there are arguments which show that at least in some cases the number of genes might exceed the number of bands:

(a) Judd, in a continuing study on the region, has concluded that there could be more genes than bands, especially if one considers that there should be a proportion of genes that are neither homozygous lethal nor contributing to phenotype (Judd, personal communication).
(b) Sorsa et al.[64] in a detailed study on break points of inversions involving the white locus came to the conclusion that the white gene itself could only occupy a part of the chromosomal band the gene was designated to.
(c) In the haploid genome of *D. melanogaster* there are about 100 genes coding for each of the five histones[21]. The genes are clustered and found to be located exclusively in region 39 DE on chromosome 2L[45]. All 100×5 genes occupy a region comprising approximately five visible bands. There are now several more examples known where more than one cloned gene hybridizes to the same band[43].

There are other more simple, but nevertheless rather compelling reasons not to believe in a strict correlation of genes and bands. We find it hard to accept that *Drosophila melanogaster* should possess not more than 5000 genes (the maximal number of bands observable even in the electron microscope)[42], and, even if we are willing to accept that proposition, it becomes really hard to believe that other *Drosophila* species like *D. virilis* as well as the Chironomids should possess only 2000 genes[42] as their salivary gland chromosomes feature only this number of bands. While 5000 genes might be acceptable to some of us, 2000 surely is not enough for an organism of this complexity.

What, if not a single gene, is the chromomere? The fact that some bands at some stage of larval development dissipate into puffs indicates that these chromosomes represent units of transcription. As large puffs can cause the disappearance of not only one but several bands it could be argued that a transcriptional unit might be larger than one chromomere. This observation does not seriously threaten the chromomere concept, since in optimal preparations such additional bands can still be identified and two neighbouring transcriptional units may be activated independently but simultaneously[10, 12, 56]. What about the chromomeres whose band never puffs? While as early as 1962 Rudkin had noticed that some, albeit weak, incorporation of ³H-uridine occurs in regions other than puffs[53], only the tedious and painstaking counting of grains after uridine incorporation into unpuffed regions of salivary gland chromosomes of *Drosophila* by Zhimulev and his coworkers[60, 68], has led to the rather convincing result that: (a) there are many more sites showing low rates of uridine incorporation, and (b) these sites are interbands rather than puffs. Similar studies by Ananiev and Barsky[3] using highly stretched chromosomes to separate the bands more widely led these authors to the opposite conclusion, namely that there are many sites of RNA synthesis but that these are rather in bands than in interbands. Scrutinizing their data, however, one finds that some rather thick bands have no grains but that these are mostly at the interface between bands and interbands. The overstretching has probably led to separation of the interband region into two parts, the larger one with no grains, the band proximal one with grains.

The second line of evidence comes from measuring the distribution of the enzyme RNA polymerase B along polytene chromosomes (*Figure 7.1*). This enzyme is known to be responsible for the synthesis of all pre-mRNA. Having purified the enzyme to homogeneity[17] we were able in this laboratory to raise antibodies against RNA polymerase B and to decorate, by indirect immuno-fluorescence, the enzyme directly on fixed and squashed preparations of salivary glands[18, 27, 49]. As expected RNA polymerase was observed in puffs, but substantial amounts of enzyme were also seen at many sites along the chromosomes, Like the banding pattern seen on fixed and stained chromosomes a pattern of fluorescent bands was also observed which, on high enough magnification, revealed the fluorescence to be localizable in all interbands. Screening many chromosomal regions we never found one interband which was not decorated by the antibody; likewise we never observed a dark band to fluoresce. In contrast, staining the chromosomes for the presence of histone H1 and the core histones[14, 27, 35, 49] the fluorescence strictly followed the banding pattern, i.e. the amount of fluorescence was proportional to the amount of chromatin present. As the enzyme localization studies do not differentiate between active and inactive, i.e. transcribing and non-transcribing RNA polymerase molecules, such studies have to be complemented by ³H-uridine incorporation, preferably doing both immunofluorescence and incorporation studies on the same specimen. This has been done on isolated chromosomes of *Chironomus tentans* and the results indicated that qualitatively there is a good correlation between RNA synthesis and presence of enzyme molecules[57]. A third line of evidence for interband

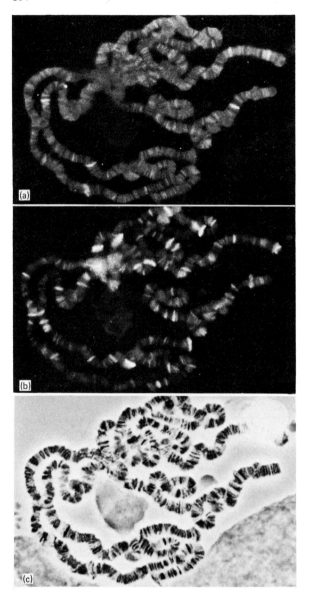

Figure 7.1. Double straining of polytene salivary gland
chromosomes of *Drosophila melanogaster* with monoclonal
antibodies against RNA polymerase B and a RNP-associated
antigen. (a) Chromosomes were incubated with monoclonal
anti-RNA polymerase B antibody and subsequently with
fluorescein-labelled rabbit-anti-mouse IgG. All interbands and
puffs show fluorescence. The chromocentre, which is mostly
transcriptional inactive shows only very weak fluorescence. (b)
The same chromosome spread was then incubated with
rhodamine-labelled monoclonal anti P11-antigen-antibody. The
fluorescence is distributed mainly in puffs but only in a subset of
interbands. (c) The chromosomes were stained after
fluorescence microscopy with orcein and photographed under
phase contrast.

transcription is provided by the work of Skaer[63] who described the presence of ribonucleoprotein (RNP) granules called perichromatin granules not only in puffs but also in interbands. He concluded that 33% of all interbands contained detectable amounts of granules and, unless it is assumed that the interband location of these granules is the result of trapping of nucleoplasmic RNPs during preparation, it is most likely that these interband granules represent *in situ* transcripts. Finally 5mC has been located by indirect immunofluorescence primarily in condensed bands as well as regressed puffs but not in interbands[13]. As the level of methylated DNA appears to be highest in inactive chromatin the data agree with the results already cited. It is of interest to note that 5mC has not been found in chromosomes of diploid cells of *Drosophila*.

The number of interbands in *Drosophila* polytene chromosomes can now be interpreted to reflect the number of distinctive transcriptional (or potentially transcriptional) units which are separated from each other by regions of condensed chromatin, the bands. As some of the basic cytogenetic observations are apparently contradictory to such a viewpoint the issue is still far from settled. For one, there are the two findings by Beermann, that the banding pattern of polytene chromosomes are similar in different tissues and that the interband DNA comprises only 5% of the total which could mean that only 5% of the genomic DNA might ever be transcriptionally accessible[9]. However, the latter measurement has been challenged recently by Laird[37, 38] who put the amount of interband DNA as high as 26%. More serious to the argument is the similarity of the banding pattern which forces one to the conclusion that in different cell types the same DNA sequences are transcribed at similar relative rates. If interband transcription is from genes coding for housekeeping functions (with puffed genes producing cell type and stage specific abundant pre-mRNA), a similarity of banding patterns in polytene chromosomes of different cell types is not impossible and might even be expected. Moreover, Beermann's finding of similarity might hold only for the observed cases and not for widely different stages of development as Ribbert[50] has reported rather dissimilar banding patterns in polytene nuclei of embryonic and larval tissue.

In addition, studies on pre-mRNA and mRNA sequence complexities in different species and developmental stages suggest that a low level transcription of several thousand genes at almost any time is not an unreasonable proposition[20, 26]. Therefore, the structural features of polytene chromosomes should be viewed as chromosomes in action, bands and interbands representing condensed, inactive and decondensed, active chromatin respectively. Does this mean that 95% of the *Drosophila* genome is not coding for any protein products (the proportion of DNA in bands according to Beermann)? Does it also mean that on puffing the bands which disappear on puff formation do not provide transcribable DNA? Or, alternatively, if band DNA on puffing becomes activated, do bands exclusively contain regulated genes, either totally inactive or totally active? In all these speculations we have to take into account the special feature of polyteny: every gene is represented by about 1000 copies. Does this mean that, as is usually the case for a haploid and, already with exceptions for a diploid nucleus, that all gene copies are either silent or active? Or could it rather be that depending on the level of activity, gene expression is regulated not only by the number of polymerase molecules started at every promoter signal per unit time, but also by the number of gene copies available for transcription? We have, some time ago, proposed this possibility as a model which implies that of the 1000 or so gene copies of housekeeping genes only 50 or 100 are available for transcription at the periphery of the band with the rest of the gene copies being tucked away in a condensed form still making up the band[27].

Final proof for the contention that the chromomere represents a functional rather than a structural unit with the consequence that silent genes are buried within condensed bands is expected to come from molecular cloning and transcription mapping of large contiguous chromosomal regions. One such transcription map of an approximately 200 kb region (encompassing segments 87D6 to 87E5) has already been completed and the results are rather clear that transcribed regions are clustered in chromosomal areas having thin bands alternating with interbands and a long stretch (150 kb) of DNA hybridizing to a thick band has largely failed to hybridize to mRNA (P. Spierer, personal communication). Thus, this type of evidence is in complete agreement with the above proposition[27].

Polytene chromosomes: The link between genetics and molecular biology

Just as the polytene chromosomes have helped *Drosophila* geneticists along for decades in allowing visual identification of deletions, inversions and other small scale aberrations which would otherwise only be accessible by tedious fine structure mapping, molecular biologists who barely (or not at all) understand what geneticists are doing but who are nevertheless intrigued by the little fruit fly are helped by the analytical potential of these interphase chromosomes. Firstly, regions of active and inactive chromatin can be distinguished by optical inspections; secondly, cloned pieces of DNA can, by *in situ* hybridization[15], be assigned to their original chromosomal location; thirdly with the use of antibodies the positioning of specific proteins and other antigens along the chromosomes and in reference to some specific landmark sites can be analysed[1, 49, 61]. If anybody were able to induce polytenization in HeLa cells this would deplete the ranks of molecular drosophilists in no time.

Thus, the uniqueness of Diptera providing easily visible and manageable polytene chromosomes has led to a concentration of chromatin studies on these species, especially on *Drosophila melanogaster*. We shall first list the type of uses polytene chromosomes offer in studies on the structural organization of chromatin, such as the chromosomal location of unique and repetitive sequences, proteins of known function, etc. Then we shall list a selected number of attempts to use the structural information polytene chromosomes can provide to assess possible functions of proteins of which we so far know nothing except that they can be localized in specific chromosomal areas.

Localization of DNA sequences

The technique of *in situ* hybridization, having been in use for almost a decade before gene cloning came into fashion, is an essential method for mapping cloned genes. Cloning any gene which can be genetically identified and mapped is now possible, if not via expression, then genetically via neighbouring transposons and/or by 'walking' i.e. by starting from an already cloned gene mapping nearby picking clones overlapping DNA sequences verifying continuously, by *in situ* hybridization of the cloned fragments, that the 'walk' is in the right direction[43].

There are, of course, complications primarily due to interspersed multigene families which, if hybridized to chromosomal DNA, will yield a number of chromosomal locations so that the investigator does not know whether the clone in question has sequences still from the right portion of the genome or from somewhere else. As hybridization of ^{3}H-nick-translated DNA requires at least a week or two for enough grains to develop in the radioautogram, walking as well as any screening of cloned DNA fragment is a tedious process. The use of biotinylated UTP in the nick translation reaction yields biotinylated DNA which hybridizes nearly as well as unsubstituted DNA[39]. This DNA can be made visible through antibiotin antibodies in combination with biotinylated anti-antibody plus avidine which is then stained by biotin-coupled peroxidase (double-sandwich technique – for some reason the single sandwich does not seem to do it). This method allows hybridization and detection of a DNA fragment within one day[40].

Should neither transposon nor already cloned closely linked gene be available there exists another possibility to come to the desired result, namely by micro-dissection of the band and cloning the isolated DNA[59]. This technique requires the ligation with vector DNA to be performed in nanolitre volumes as the concentration of DNA has to be high enough to obtain end to end ligation of two different molecules.

Localization of antigens of known and unknown functions

The use of polytene chromosomes for the localization of chromosome-associated nuclear proteins has, with availability of many antisera and monoclonal antibodies, become a powerful tool both for screening antisera and monoclonal antibodies for chromosomal localization of the antigens, as well as for obtaining some first hints as to the possible function of the unknown antigen. The assay of indirect immunofluorescence allows us, in a matter of hours, to see whether an antibody is directed against a protein associated with interphase chromatin and, if so, whether this protein is associated with active genes, i.e. puffs, and whether it is represented in all puffs or only a unique set of them. Alternatively, proteins can be assigned to inactive, i.e. condensed chromatin. This possibility of screening, if necessary, rather large numbers of antisera or, in the case of production of monoclonal antibodies, spent culture supernatants has prompted several laboratories to adopt the following experimental strategy to raise and to select antibodies directed against non-histone chromosomal (NHC) proteins of potentially interesting functions. The first strategy adopted was to extract and fractionate NHC proteins, inject the fractions into rabbits and test the resulting antisera on polytene chromosomes. Although a rather crude procedure (no fraction can be expected to consist of a homogeneously pure protein), it is surprising that in some cases antisera were obtained which appeared to be largely directed against a single antigen. Such antigens reported are the D1 protein of Alfageme et al.[1,2], the rho protein of Silver and Elgin[62], and the chromocentre associated protein of Will and Bautz[67]. Obviously we do not read as much about those antisera which give fluorescence patterns indicating a multitude of antigens reacting. For this reason the approach has its limits, especially since the antibody source (a single animal) is final, as one cannot expect to obtain the same monospecific antiserum twice on injection of the same fraction. With the availability of the monoclonal antibody technique[32] it became clear that this technique should be a method of choice as here rather crude fractions can be used as antigens and each antibody-producing clone recognizes only one antigenic determinant. The first large scale attempt to produce a series of monoclonal antibodies directed against NHC proteins of *Drosophila melanogaster* was made by Saumweber *et al.*[58] who came up with some 500 clones positive in a radioimmunoassay, of which we screened some 300 clones for immunofluorescence on polytene chromosomes. Of these some 60 showed a positive reaction with polytene nuclei, some reacting with the nuclear membrane, most antibodies, however, staining the chromosomes. The latter antibodies could be grouped into different classes, some reacting with bands, many reacting with interbands and puffs, some of which reacted primarily with puffs, etc.

Very few of the monoclonal antibodies (in fact only one so far) was found to give the same staining pattern as a monoclonal antibody directed against the largest subunit of RNA polymerase B. The majority of euchromatin-specific antibodies give a fluorescence pattern that favours large puffs. Here the fluorescence intensity is much greater than that observed with anti-RNA polymerase antibodies indicating that these antigens are either more accessible than RNA polymerase or that they are present in larger amounts[30]. Some of these puff specific antigens were meanwhile identified as being associated with RNP particles (Risau, personal communication)[25]. Comparing the distributions of different RNP-associated antigens with each other there are again families which show partial overlaps in fluorescence patterns with their respective antibodies. An example is T7 and P11 (the letters and numbers assigned to the antigens refer to the wells of the microtitre plates in which the original clones of antibody producing hybridoma cells were obtained). These two antigens are present in most puffs but in different amounts. Some puffs (e.g. 64A and 76A) are decorated preferentially by the T7 antibody whereas 63E and 76D are stained more strongly by the P11 antibody. In such studies the two antibodies to be compared were applied to the same squash preparation, one antibody labelled with one fluorescent dye (FITC) the other antibody with another dye (TRITC). The two dyes can be viewed separately in the fluorescence microscope using different filter combinations. In the case where the two antibodies are of the same immunoglobin subclass, they have to be coupled directly to the fluorescent dye; if, as in the case of P11 and T7, they represent different subclasses, indirect fluorescence can be applied using conventional subclass specific FITC and TRITC labelled antibodies.

Most antibodies crossreact with the corresponding antigens of other Diptera and even with nuclear antigens of mammals[25]. Some, however, do not crossreact with *D. hydei* or *D. virilis*[31]. Of four different hybridoma cell lines producing antibodies against apparently the same antigen, three (P11, Q14, Q16) fail to crossreact whereas the fourth (Q18) does[31]. *Drosophila* species of the melanogaster subgroup, like *D. yakuba* and *D. erecta* show crossreaction but chromosomes of the more distantly related species *D. simulans* which, however, still is classified as a member of the melanogaster family, give only weak signals with P11 antibodies. These data confirm rather nicely the family tree of *D. melanogaster* based on cytogenetic and morphological criteria.

Changes in the distribution of NHC proteins after heat shock

Raising the temperature of *Drosophila* larvae from 25°C to 37°C results in an immediate induction of nine puffs, some of which code for heat shock proteins which are synthesized almost exclusively at the elevated temperature[6]. While nothing is yet known about the function of these heat shock proteins (HSPs) except that they somehow contribute to the survival of the organism at the high temperatures, the induction of the heat shock puffs themselves represents the easiest and most reproducible system of gene activation. It is therefore interesting to follow the appearance of antigens associated with transcription in the developing heat shock puffs. RNA polymerase B was found to start accumulating at these sites almost immediately after raising the temperature[18]. Some RNP-associated antigens also accumulate within heat shock puffs with one noticeable exception: the P11 antigen is not found in all heat shock puffs except 93D. This latter puff apparently acquires huge amounts of P11 antigens after about 20 minutes of heat treatment, with non-heat shock puffs slowly becoming depleted of the antigen (Dangli, unpublished). This is rather surprising as it sets apart heat shock puff 93D from all the others. There were, however, other pieces of evidence that suggested 93D to be rather unique: the equivalent puff of *D. hydei* 48BC had been observed to be separately inducible by agents other than heat, such as vitamin B6 and some other chemical compounds[6]. cDNA of 48BD heat shock RNA was cloned by Peters *et al.*[48] and its sequence was found to feature several approximately 110 bp repeats having no open reading frames at all (Walldorf and Bautz, unpublished). In 48BC very large granules are seen by electronmicroscopy which are not observed in any other heat shock puff[11]. By immunoelectronmicroscopy these large granules were found to react with Q18 antibodies (which is the crossreacting homologue to P11). A search for the type of granules found in 48BC of *D. hydei* revealed the same in 93D, which were also found to bind antibodies Q18 and P11. Finally the 93 heat shock RNA (at least one of the two possibly occurring species) was found to remain in the nucleus and not to be transported into the cytoplasm as is the case for HSP70 mRNA, coded for by heat shock puffs 87A and 87C[41]. It appears then that the P11 antigen belongs to the special class RNPs whose function might be to keep the transcripts from being processed and/or transported into the cytoplasm. Indeed, there is now a reasonably good correlation between accumulation of e.g. ecdysone-induced RNA and presence of the P11 antigen, whereas those puffs whose RNA product is known to be exported rapidly out of the nucleus such as 3C, contain very little P11 antigen.

Another class of proteins known to be intimately associated with chromatin and associated with nucleosomes are the high mobility group (HMG) proteins[16], specifically HMG1 and 2 and HMG14 and 17, the latter have been implicated in gene activation[66]. Antibodies directed against calf thymus HMG1 have also been observed to decorate polytene chromosomes of *Chironomus tentans* by Kurth and Bustin[33]. These authors concluded from their studies that HMG1 is involved in the recondensation of puffs to form inactive bands again. In the same laboratory a rather unusual observation was made using antibodies directed against concanavalin A, a plant lectin specifically binding sugar moieties of glycoproteins and therefore usually used in the analysis of cell surfaces[34]. Concanavalin A was found to bind to Balbiani rings of *Chironomus tentans* chromosome IV, the amount of lectin and therefore glucose or mannose appeared to increase with the size of the puffs. It will be of interest to see whether this finding can be repeated with isolated and washed chromosomes.

Non-protein antigens

Polytene chromosomes have become a favoured assay system to see whether certain small molecules or whether specific types of DNA/RNA structures naturally occur in chromatin. Using antibodies against cyclic GMP, this small molecule was also found in puffs of *D. melanogaster* and thus it was implicated in gene expression[65]. Antibodies directed against DNA/RNA hybrids were also found to decorate polytene chromsomes[54,55]. The authors suggested that possibly hybrids of both regulatory RNA sequences and of nascent RNA chains might have been observed. Such speculation of course does not exclude or prove the existence of one or the other form. The most thrilling observation recently reported is the staining of interbands by antibodies directed against Z-DNA[44]. If correct, this result would mean that Z-DNA is a naturally occurring form of DNA. The recent finding of known autoimmunosera specifically reacting with Z-DNA supports this claim[36]. On the other hand attempts to confirm the interband localization of Z-DNA in different laboratories with different antibodies have yielded conflicting results: acid treatment of chromosomes prior to formaldehyde fixation leads to bright fluorescence in bands as well as in interbands (Kabisch, unpublished). One possible explanation to be considered is the popping of core histones out of nucleosomes on salt or acid treatment leading to supercoiled DNA, and it is known now that supercoiled DNA can easily flip into the Z conformation[46]. If this interpretation is correct the question whether free supercoiled DNA exists in nature becomes a very important issue, meaning that if free supercoiled DNA exists, Z-DNA must exist too.

A typical example of the wide applicability of the technique of immunological localization is the work of Gronemeyer and Pongs[19] who were able to localize the hormone ecdysone on polytene chromosomes. Small molecules such as steroid hormones are difficult to keep at their *in situ* localization as they are easily washed away under normal fixation conditions. The hormone was therefore crosslinked *in vivo* and although such treatment can lead to increased background levels, the results appeared rather convincing, namely that a number of potential puff sites had increased levels of fluorescence.

The examples given should suffice to document the usefulness of polytene chromosomes in studies of chromatin structure and function as well as chromosome organization. It is safe to predict that the number and types of antigens that can be assigned to active chromatin will increase considerably in the near future.

References

1. Alfageme, C. R., Rudkin, G. T. and Cohen, L. H. (1976). Locations of chromosomal proteins in polytene chromosomes. *Proceedings of the National Academy of Sciences, USA*, **73**, 2038–2042
2. Alfageme, C. R., Rudkin, G. T. and Cohen, L. H. (1980). Isolation, properties and cellular distribution of D1, a chromosomal protein of *Drosophila*. *Chromosoma*, **78**, 1–31
3. Ananiev, E. V. and Barsky, V. E. (1978). Localization of RNA synthesis sites in the 1B–3C region of the *Drosophila melanogaster* X chromosome.*Chromosoma*, **65**, 359–371
4. Ashburner, M. (1972). Puffing patterns in *Drosophila melanogaster* and related species. In *Results and Problems in Cell Differentiation Volume 4, Developmental Studies on Giant Chromosomes*. (Beermann, W., Ed.), pp. 101–151. Heidelberg, Springer
5. Ashburner, M. and Berendes, D. (1978). Puffing of polytene chromosomes. In *Genetics and Biology of* Drosophila, *Volume 26* (Ashburner, M. and Wright, T. E. D., Eds.), pp. 316–395. London and New York, Academic Press
6. Ashburner, M. and Bonner, J. J. (1979). The induction of gene activity in *Drosophila* by heat shock. *Cell*, **17**, 241–254
7. Becker, H. J. (1959). Die Puffs der Speicheldrüsenchromosomen von *Drosophila melanogaster*. I. Beobachtungen zum Verhalten des Puffmusters im Normalstamm und bei zwei Mutanten, giant and lethal-giant-larvae. *Chromosoma*, **10**, 654–678

8. Beermann, W. (1952). Chromomerenkonstanz und spezifische Modifikation der Chromosomen-struktur in der Entwicklung und Organdifferenzierung von *Chironomus tentans*. *Chromosoma*, **5**, 139–198

9. Beermann, W. (1962). *Riesenchromosomen: Protoplasmatologia*, Volume 4. Vienna, Springer

10. Beermann, W. (1973). Directed changes in the pattern of balbiani ring puffing in *Chironomus*: Effects of a sugar treatment. *Chromosoma*, **41**, 297–326

11. Derksen, J. and Willart, E. (1976). Cytochemical studies on RNP complexes produced by puff 2-48BC in *Drosophila hydei*. *Chromosoma*, **55**, 57–68

12. Derksen, J., Wieslander, L., van der Ploeg. M. and Daneholt, B. (1980). Identification of the balbiani ring 2 chromomere and determination of the content and compaction of its DNA *Chromosoma*, **81**, 65–84

13. Eastman, E. M., Goodman, R. M., Erlanger, B. F. and Miller, O. J. (1980). 5-Methylcytosine in the DNA of the polytene chromosomes of the Diptera *Sciara coprophila, Drosophila melanogaster* and *D. persimilis*. *Chromosoma*, **79**, 225–239

14. Elgin, S. C. R., Serunian, L. A. and Silver, L. M. (1978). Distribution patterns of *Drosophila* nonhistone chromosomal proteins. *Cold Spring Harbor Symposium on Quantitative Biology*, **42**, 839–850

15. Gall, J. G. and Pardue, M. L. (1969). Formation and detection of RNA–DNA hybrid molecules in cytological preparations. *Proceedings of the National Academy of Sciences, USA*, **63**, 378–383

16. Goodwin, G. H. and Johns, E. W. (1973). Isolation and characterisation of two calf thymus chromatin non-histone proteins with high contents of acidic and basic amino acids. *European Journal of Biochemistry*, **40**, 215–219

17. Greenleaf, A. L. and Bautz, E. K. F. (1975). RNA polymerase B from *Drosophila melanogaster* larvae, purification and partial characterization. *European Journal of Biochemistry*, **60**, 169–179

18. Greenleaf, A. L., Plagens, U., Jamrich, M. and Bautz, E. K. F. (1978). RNA polymerase B (or II) in heat induced puffs of *Drosophila* polytene chromosomes.*Chromosoma*, **65**, 127–136

19. Gronemeyer, H. and Pongs, O. (1980). Localization of ecdysterone on polytene chromosomes of *Drosophila melanogaster*. *Proceedings of the National Academy of Sciences, USA*, **77**, 2108–2112

20. Hastie, N. D. and Bishop, J. O. (1976). The expression of three abundance classes of messenger RNA in mouse tissues. *Cell*, **9**, 761–771

21. Hentschel, C. C. and Birnstiel, M. L. (1981). The organization and expression of histone gene families. *Cell*, **25**, 301–313

22. Holt, T. K. H. (1970). Local protein accumulation during gene activation. I. Quantitative measurements on dye binding capacity at subsequent stages of puff formation in *Drosophila hydei*. *Chromosoma*, **32**, 64–78

23. Holt, T. K. H. (1971). Local protein accumulation during gene activation. II. Interferometric measurements of the amount of solid material in temperature induced puffs of *Drosophila hydei*. *Chromosoma*, **32**, 428–435

24. Howard, G. C., Abmayr, S. M., Serunian, L. A., Sato, V. L. and Elgin, S. C. R. (1981). Monoclonal antibodies against a specific nonhistone chromosomal protein of *Drosophila* associated with active genes. *Journal of Cell Biology*, **88**, 219–225

25. Hügle, B., Guldner, H., Bautz, F. A. and Alonso, A. (1982). Cross-reaction of hnRNP-proteins of HeLa cells with nuclear proteins of *Drosophila melanogaster* demonstrated by a monoclonal antibody. *Experimental Cell Research*, **142**, 119–126

26. Izquierdo, M. and Bishop, J. O. (1979). An analysis of cytoplasmic RNA populations in *Drosophila melanogaster*, Oregon R. *Biochemistry Genetics*, **17**, 473–497

27. Jamrich, M., Greenleaf, A. L., Bautz, F. A. and Bautz, E. K. F. (1978). Functional organization of polytene chromosomes. *Cold Spring Harbor Symposium on Quantitative Biology*, **42**, 389–396

28. Jovin, T. M., van de Sande, J. H., Zarling, D. A., Arndt-Jovin, D. J., Eckstein, F., Füldner, H. H., Gréider, C., Grieger, I., Hamori, E., Kalisch, B., McIntosh, L. P. and Robert-Nicoud, M. (1982). Generation of left-handed Z-DNA in solution and visualization in polytene chromosomes by immunofluorescence. *Cold Spring Harbor Symposium on Quantitative Biology*, **47**, 143

29. Judd, B. H., Shen, M. W. and Kaufmann, T. C. (1972). The anatomy and function of a segment of a X chromosome of *D. melanogaster*. *Genetics*, **71**, 139–156

30. Kabisch, R. (1982). Differentielle Verteilung von Nichthistonprotein auf polytänen Chromosomen von *Drosophila melanogaster*. PhD Thesis, Heidelberg

31. Kabisch, R., Krause, J. and Bautz, E. K. F. (1982). Evolutionary changes in non-histone chromosomal proteins within the *Drosophila melanogaster* group revealed by monoclonal antibodies. *Chromosoma*, **85**, 531–538

32. Köhler, G. and Milstein, C. (1975). Continuous cultures of fused cells secreting antibody of predefined specificity. *Nature, London*, **256**, 495–497

33. Kurth, P. D. and Bustin, M. (1981). Localization of chromosomal protein HMG-1 in polytene chromosomes of *Chironomus thummi*. *Journal of Cell Biology*, **89**, 70–77
34. Kurth, P. D., Bustin, M. and E. N. Moudrianakis (1979). Concanavalin A binds to puffs in polytene chromosomes. *Nature, London*, **279**, 448–450
35. Kurth, P. D., Moudrianakis, E. N. and Bustin, M. (1978). Histone localization in polytene chromosomes by immunofluorescence. *Journal of Cell Biology*, **78**, 910–918
36. Lafer, E. M., Moller, A., Nordheim, A., Stollar, B. D. and Rich, A. (1981). Antibodies specific for left-handed Z-DNA. *Proceedings of the National Academy of Sciences, USA*, **78**, 3546–3550
37. Laird, C. D. (1980). Structural paradox of polytene chromosomes. *Cell*, **22**, 869–874
38. Laird, C. D., Ashburner, M. and Wilkinson, L. (1980). Relationship between relative dry mass and average band width in regions of polytene chromosomes of *Drosophila*. *Chromosoma*, **76**, 175–189
39. Langer, P. A., Waldrop, A. A. and Ward, D. C. (1981). Enzymatic synthesis of biotin-labeled polynucleotides: Novel nucleic acid affinity probes. *Proceedings of the National Academy of Sciences, USA*, **78**, 6633–6637
40. Langer-Safer, P. R., Levine, M. and Ward, D. C. (1982). Immunological method for mapping genes on *Drosophila* polytene chromosomes. *Proceedings of the National Academy of Sciences, USA*, **79**, 4381–4385
41. Lengyel, J. A., Ransom, L. J., Graham, M. L. and Pardue, M. L. (1980). Transcription and metabolism of RNA from the *Drosophila melanogaster* heat shock puff site 93D. *Chromosoma*, **80**, 237–252
42. Levefre, G. Jr. (1976). A photographic representation and interpretation of the polytene chromosomes of *Drosophila melanogaster* salivary glands. In *Genetics and Biology of Drosophila Volume 1a* (Ashburner, M. and Novitski, E., Eds.), pp. 32–60. New York, Academic Press
43. Meyerowitz, E. M. and Hogness, D. S. (1982). Molecular organization of a *Drosophila* puff site that responds to ecdysone. *Cell*, **28**, 165–176
44. Nordheim, A., Pardue, M. L., Lager, E. M., Moller, A., Stollar, B. D. and Rich, A. (1981). Antibodies to left-handed Z-DNA bind to interband regions of *Drosophila* polytene chromosomes. *Nature, London*, **294**, 417–422
45. Pardue, M. L., Kedes, L. H., Weinberg, E. S. and Birnstiel, M. L. (1977). Localization of sequences coding for histone messenger RNA on the chromosomes of *Drosophila melanogaster*. *Chromosoma*, **63**, 135–151
46. Peck, L., Nordheim, A., Rich, A. and Wang, J. C. (1982). Flipping of cloned d(pCpG)·d(pCpG)n DNA sequences from right to lefthanded helical structure by salt, Co(III), or negative supercoiling. *Proceedings of the National Academy of Sciences, USA*, **79**, 4560–4564
47. Pelling, C. (1964). Ribonucleinsäuresynthese der Riesenchromosomen. Autoradiographische Untersuchungen an *Chironomus tentans*. *Chromosoma*, **15**, 71–122
48. Peters, F. P. A. M. N., Grond, C. J., Sondermeijer, P. J. A. and Lubsen, N. H. (1982). Chromosomal arrangement of heat shock locus 2-48B in *Drosophila hydei*. *Chromosoma*, **85**, 237–249
49. Plagens, U., Greenleaf, A. L. and Bautz, E. K. F. (1976). Distribution of RNA polymerase on *Drosophila* polytene chromosomes as studied by indirect immunofluorescence. *Chromosoma*, **59**, 157–165
50. Ribbert, D. (1979). Chromomeres and puffing in experimentally induced polytene chromosomes of *Calliphora erythrocephala*. *Chromosoma*, **74**, 269–298
51. Richards, G. (1980). The polytene chromosomes in the fat body nuclei of *Drosophila melanogaster*. *Chromosoma*, **79**, 241–250
52. Richards, G. (1982). Sequential gene activation by ecdysteroids in polytene chromosomes of *Drosophila melanogaster*. VII. Tissue specific puffing. *Wilhelm Roux Archiv für Entwicklungsmechanik der Organismen*, **191**, 103–111
53. Rudkin, G. T. (1962). *Acta Histochimica, Supplement*, **2**, 77–84
54. Rudkin, G. T. and Stollar, B. D. (1977). Naturally occurring DNA/RNA hybrids. I. Normal patterns in polytene chromosomes. *ICN-UCLA Symposia on Molecular and Cellular Biology Volume 7*, pp. 257–269
55. Rudkin, G. T. and Stollar, B. D. (1977). High resolution detection of DNA–RNA hybrids *in situ* by indirect immunofluorescence. *Nature, London*, **265**, 472–473
56. Sass, H. (1981). Effects of DMSO on the structure and function of polytene chromosomes of *Chironomus*. *Chromosoma*, **83**, 619–643
57. Sass, H. (1982). RNA polymerase B in polytene chromosomes: Immunofluorescent and autoradiographic analysis during stimulated and repressed RNA synthesis. *Cell*, **28**, 269–278

58. Saumweber, H., Symmons, P., Kabisch, R., Will, H. and Bonhoeffer, F. (1980). Monoclonal antibodies against chromosomal proteins of *Drosophila melanogaster*. *Chromosoma*, **80**, 253–275

59. Scalenghe, F., Turco, E., Edström, J. E., Pirotta, V. and Melli, M. (1981). Microdissection and cloning of DNA from a specific region of *Drosophila melanogaster* polytene chromosomes. *Chromosoma*, **82**, 205–216

60. Semeshin, V. F., Zhimulev, I. F. and Belyaeva, E. S. (1979). Electron microscope autoradiographic study on transcriptional activity of *Drosophila melanogaster* polytene chromosomes. *Chromosoma*, **73**, 163–177

61. Silver, L. M. and Elgin, S. C. R. (1976). A method for determination of the *in situ* distribution of chromosomal proteins. *Proceedings of the National Academy of Sciences, USA*, **73**, 423–427

62. Silver, L. M. and Elgin, S. C. R. (1977). Distribution patterns of three subfractions of *Drosophila* nonhistone chromosomal proteins. Possible correlations with gene activity. *Cell*, **11**, 971–983

63. Skaer, R. J. (1977). Interband transcription in *Drosophila*. *Journal of Cell Sciences*, **26**, 251–266

64. Sorsa, V., Green, M. M. and Beermann, W. (1973). Cytogenetic fine structure and chromosomal localization of the white gene in *Drosophila melanogaster*. *Nature, London*, **245**, 34–37

65. Spruill, W. A., Hurwitz, D. R., Luccesi, J. C. and Steiner, A. L. (1978). Association of cyclic GMP with gene expression of polytene chromosomes of *Drosophila melanogaster*. *Proceedings of the National Academy of Sciences, USA*, **75**, 1480–1484

66. Weisbrod, S. and Weintraub, H. (1979). Isolation of a subclass of nuclear proteins responsible for conferring a DNase I-sensitive structure on globin chromatin. *Proceedings of the National Academy of Sciences, USA*, **76**, 630–634

67. Will, H. and Bautz, E. K. F. (1980). Immunological identification of a chromocenter-associated protein in polytene chromosomes of *Drosophila*. *Experimental Cell Research*, **125**, 401–410

68. Zhimulev, I. F. and Belyaeva, E. S. (1975). ^{3}H-Uridine labelling patterns in the *Drosophila melanogaster* salivary gland chromosomes X 2R and 3L. *Chromosoma*, **49**, 219–231

Experimental approaches

Technical advances at the molecular level have undoubtedly provided the impetus for the recent surge of progress in the field of molecular genetics. Yet it is when they are coupled with methodologies utilizing whole cells that these new technologies have proved most effective. The chapters in this section serve to outline some of the experimental approaches that can profitably be utilized at the cellular level. Their application to specific gene systems will be discussed in greater depth in the next section. Here, selected examples are used to illustrate the potential of these approaches in the study of gene function.

The technique of somatic cell fusion, covered in Chapter 8, has been in use for a number of years and has yielded valuable information concerning the regulation of gene activity at the cellular level. It has also proved to be a major tool in the assignment of genes to particular chromosomes. More recently, this approach has assumed even greater importance with the demonstration that fusions between myeloma cells and primary spleen cells can generate immortalized cell lines (hybridomas) which synthesize and secrete only a single (monoclonal) antibody species. Interpretation of data concerning gene expression in these fused cell systems is often complicated, however, by the presence of a hybrid cytoplasm and two nuclei. Cybrid constructions (enucleated cell/whole cell fusions) and reconstructed cells (cytoplast/karyoplast fusions) have confirmed that cytoplasmic factors are involved in the activation and/or repression of specific gene expression though the nature of these factors has still to be determined. In this respect, chromosome- and DNA-mediated gene transfers into tissue culture cells have largely superseded cell hybridization.

The injection of nuclei or cloned genes into cells is the subject matter discussed in Chapter 9. The nuclear transplantation experiments form a logical extension of the cell fusion studies and have principally utilized the large amphibian oocytes and eggs as recipient cells. The ability of early embryonic nuclei to replace the zygotic nucleus in the fertilized egg, and hence give rise to a normal adult, has provided strong evidence for the totipotent nature of these donor nuclei. In contrast, most adult nuclei injected into eggs are incapable of giving rise to normal offspring because they encounter chromosomal replication barriers. In oocytes, reprogramming of the nucleus often appears to take place, with genes being activated for those products that are normally found in oocytes and repressed for others which are not.

Cloned genes injected into oocyte nuclei are frequently accurately transcribed and processed, whether or not their products are normally found in oocytes. This occurs at high efficiency for those genes transcribed by RNA polymerase III and at lower efficiencies for protein-coding genes. There are some templates, such as chick ovalbumin and rabbit β-globin, which are transcribed poorly and/or inaccurately in amphibian oocytes: the reasons for this are not known. Injection of cloned sequences into eggs can lead to integration into the host nuclear DNA and the propagation of these sequences throughout the adult tissues. In some cases the injected genes can even be transmitted to the next generation via the germ line. The consequences of this are considered more fully in Chapter 11.

First, in Chapter 10, we provide a discussion of nuclear reactivation *in vitro*. The process of gene activation that occurs on transplantation of a transcriptionally inert nucleus into a foreign cytoplasm, be it through cell fusion or microinjection, can apparently be reproduced *in vitro* by incubating these nuclei with cytoplasmic extracts derived from transcriptionally active cells. Long-term effects cannot be monitored, due to the instability of nuclei *in vitro*, but the initial events behind the reactivation process can be studied. Moreover, with an *in vitro* assay system it should be possible to fractionate the cytoplasm and look for specific activating principles. A number of laboratories have adopted this approach in their search for gene-specific regulatory molecules, with varying degrees of success. The complexity of the system is high and it might seem appropriate to draw an analogy with attempts to fractionate *in vitro* transcription systems, where specificity of transcription is lost as soon as the initial fractionations are made. For at least some systems, however, this effect has not been observed and, in particular, this approach has proved eminently suited to the identification of specific second messengers of hormone action.

The final chapter in this section, Chapter 11, is concerned with the introduction of cloned genes, either as normal or mutated forms, or entire chromosomes into the nuclei of embryonic stem cells so that their expression can be studied during subsequent development and the process of cell differentiation. This has been achieved in mice through two different routes: microinjection into the pronuclei of fertilized eggs or introduction of teratocarcinoma stem cells (in a transformed or hybrid state) into the blastocyst. In the latter case one obtains chimaeric offspring whereas by microinjection all cells of the offspring should contain the foreign DNA, though not necessarily integrated at an expressing site. These techniques are still relatively new, yet they are already some way to fulfilling their potential. With the advent of site-directed integration and excision systems, they will no doubt provide the ultimate assay for gene function and bring a new meaning to the expression 'whole animal experiments'.

Chapter 8

Somatic cell hybridization

J. J. Lucas

Application of the techniques of cell fusion has resulted in dramatic advances in several areas of experimental biology, including animal virology, immunology, developmental biology and somatic cell genetics. Excellent reviews describing the historical development of these methods and the results of their use have been published[11, 18, 29, 58]. Perhaps the most recent important application of the powerful tool of cell hybridization is in the construction of hybridomas – antibody-producing cell lines made by fusing continuously growing myeloma cells to primary cells prepared from the spleens of animals immunized with particular antigens[39]. Hybridomas provide a continuous, large supply of monoclonal antibodies for use in a wide variety of studies[68].

Of special interest to readers of this volume may be fusion studies designed to elucidate the structural organization of eukaryotic genomes. Until very recently, somatic cell hybridization, along with more classical genetic analyses or family studies, was the primary method for the assignment of genes to particular chromosomes and to regions of chromosomes. The strategies used and results obtained from such analyses were described in detail by McKusick and Ruddle[49]. Newer methods for the transfer of genetic information from one cell to another, microcell-mediated gene transfer[15,21], chromosome-mediated transfer[47] and DNA-mediated transfer[4, 50, 55, 65], now complement the basic techniques of somatic cell fusion and promise to provide even greater resolution in the determination of gene maps[38]. Finally, as illustrated elsewhere in this book, the application of recombinant DNA technology[67] permits precise determination of many aspects of the organization of individual eukaryotic genes and gene families.

Somatic cell hybridization has also led to the identification of several regulatory mechanisms which apparently control expression of eukaryotic genomes. Some of the cell fusion studies in which these regulatory phenomena were first discovered and characterized are briefly described here. Then, some newly developed hybrid cell systems which promise to be particularly useful in determining the molecular mechanisms underlying these phenomena are discussed in detail.

Gene expression in hybrid cells

Development of cultured cell lines which retained traits characteristic of their tissues of origin permitted a new line of research: an investigation of the expression of such 'differentiation-specific' or facultative traits in cell hybrids. In these experiments, two cell types in different epigenetic states–each carrying a selectable genetic marker–were fused and cloned in a medium which allowed growth only of hybrid cells[45, 46]. Clonal lines–the progeny of the original fused cells, or heterokaryons–were analysed for their characteristic traits and, in many cases, were karyotyped to determine the contribution of each parental cell to the genetic constitution of the hybrids. As noted previously by others (see, for example, references 9 and 51), findings concerning the regulation of expression of facultative traits can be summarized by several general statements. As an excellent, detailed description of most of the experimental evidence leading to and supporting these statements has been published[58], the references cited here comprise a selective rather than an exhaustive list.

1. In hybrids formed by fusing somatic cells that express particular facultative traits with cells that do not, the specialized phenotype is, in most cases, extinguished[12, 13]. However, re-expression of the traits may occur upon loss of certain chromosomes contributed to the synkaryons by the parental cell not expressing the trait[37]. Also, the phenotype expressed by the hybrids may depend on the original gene dosage or balance. Thus, extinction was seen less frequently if the original 'differentiated' parental cell was tetraploid[10, 20]. That extinction resulted from decreased transcription of specific mRNAs was suggested by Deisseroth et al.[14] in their description of human fibroblast–mouse erythroleukemia (MEL) hybrids which had lost inducibility for globin synthesis, a characteristic of the MEL parent cell. Using a cDNA probe for globin sequences, they showed that mouse globin DNA sequences were present in the hybrid cell genome but were not transcribed in significant amounts.
2. In some hybrid cell types, however, a specialized phenotype may continue to be expressed[54]. It is yet unclear why certain facultative traits are not extinguished. Fougére and Weiss[19] have suggested the operation of an exclusivity principle: though a hybrid cell may have the potential to express two specialized phenotypes, only one will be expressed in a cell at any one time. Now in a hybridization experiment, the parental cell type not expressing the particular trait under study may yet possess other traits–not detected or assayed–which *are* extinguished. Alternatively, at least some exceptions to the more general rule of extinction may, in fact, reflect the gene dosage effects described above[58]. Many commonly used cell lines are, in fact, heteroploid and may possess aberrantly high numbers of the genes responsible for the traits under study. The work of Allan and Harrison[1], along with that of Axelrod et al.[3], also suggested that the manner in which hybrid cells are grown during selection might influence the observed expression of facultative traits: an adherent morphology was inconsistent with expression of the erythroid cell phenotype. Thus, when fibroblasts were fused to erythroid cells, globin inducibility was extinguished in cells which grew on a substrate but often expressed in cells which grew in suspension. That a cell's growth habit can dramatically affect its overall metabolic activity was also demonstrated by Ben-Ze'ev et al[5].
3. Cells that express a facultative trait may, upon fusion, induce or activate another cell to express a specialized function it normally does not. Excellent examples of this phenomenon are the synthesis of mouse or human albumin in some rat hepatoma–mouse fibroblast or mouse hepatoma–human leukocyte hybrids, respectively[9, 56]. As will be discussed in more detail below, the chromosomal

balance of the synkaryons may also play a role in the activation of facultative traits.

The findings summarized here are consistent with the notion that mammalian cells contain elements which can either extinguish or activate the expression of specific genes. Since the original discovery of these phenomena, however, little has been learned concerning the nature or mode of action of the putative regulatory elements. Most discussion implies that they act at the level of transcription of specific genes, although, for most cell systems studied, no evidence for such a mechanism has been provided. The major problem hindering progress in this area of research may be the complexity of the cell hybrids–bodies comprising mixed genomes, frequently undergoing chromosomal losses and other aberrations, within mixed cytoplasms. This problem has not, of course, prevented numerous useful applications of the basic technique of cell fusion followed by synkaryon selection. For example, Rankin and Darlington[57] have demonstrated that several liver-specific genes can be activated in human amniocytes by fusion to mouse hepatoma cell lines–a finding that could provide the basis for the development of methods for prenatal diagnosis of a variety of abnormalities. Also of considerable importance was the observation[8] that defects in cells from strains of mice carrying lethal albino deletions resulting in a variety of liver abnormalities[23] can be complemented by somatic cell fusion, thus strengthening the notion that the deletions are, in fact, in genes which regulate or control the expression of liver-specific traits and hepatic cell differentiation[24].

The problem noted above–that associated with chromosomal loss and rearrangement–can be alleviated, of course, by examining heterokaryons, the initial products of cell fusion. But such studies require that (a) the heterokaryons can be distinguished from homokaryons, and (b) the trait under investigation can be assayed at a single cell level. More recently, methods for the separation of heterokaryons from the mixed population of cells have been devised. Several of these techniques involve labeling of cells with non-toxic fluorescent dyes and separation by fluorescence-activated cell sorting[34,59].

Early analyses of heterokaryons indicated that extinction of facultative traits could also occur in these cells and was thus independent of chromosomal loss[27, 28, 62]. For example, within 24 hours after formation, heterokaryons formed by fusion of rat HTC cells (a line which is steroid-inducible for liver-specific tyrosine aminotransferase) and rat BRL-62 cells (a non-inducible line) had little or no enzyme and were unresponsive to steroid treatment.

More recently, Mével-Ninio and Weiss[52] presented a detailed analysis of the time course of phenotypic change in rat hepatoma–mouse fibroblast heterokaryons and cell hybrids. Heterokaryons were identified using Hoechst 33258, a fluorochrome which binds differentially to the chromatin of the two types of rodent nuclei; mouse and rat albumin were identified by immunofluorescence. Extinction of albumin synthesis occurred within 12 hours after fusion and was independent of gene dosage–that is, it also occurred when hypertetraploid hepatoma cells were used. By about 8–12 days after fusion, some hybrid colonies re-expressed rat albumin; several days later, some of these colonies also synthesized mouse albumin. The latter two events occurred more rapidly, however, if a hypertetraploid hepatoma cell was used and thus appeared to be dependent, to some degree, on gene dosage. The authors concluded that extinction was mediated by a diffusible factor from the fibroblast parent.

That cytoplasm might also contain factors which can activate gene expression in heterokaryons was suggested by two other recent studies. Wright[66] prepared heterokaryons from rat L6 myoblasts–a proliferating cell type which can be induced to differentiate and synthesize the embryonic form of skeletal myosin–and differentiated chick mononucleated muscle cells. Heterokaryons synthesized not only chick myosin but also the embryonic *and* adult light chains of rat skeletal myosin. As noted by the author, the results suggested the existence of a positive regulatory element in terminally differentiated chick cells which could apparently affect gene expression in a cell which was also determined for myogenesis. Linder *et al.*[42] described the synthesis of globin mRNA and the three normal forms of adult globin in cells formed from mature avian erythrocytes and several types of nonerythroid mammalian cells. No 'reprogramming' of the genome was detected, but the results showed that the cytoplasm of the mammalian cells could apparently induce reactivation of the dormant erythrocyte nuclei and re-expression of the genetic program expressed in erythroid cells before becoming dormant. These two studies are perhaps most noteworthy for the techniques employed. In the first, nearly pure populations of heterokaryons were prepared and the polypeptides synthesized by them were anlysed using two-dimensional gel electrophoresis. In the second, fusion was accomplished with a very high efficiency–about 50 to 100 erythrocyte nuclei were introduced per 100 mammalian cells–the life of the heterokaryons was prolonged by mitotic arrest using mitomycin C, and protein and RNA synthesis were monitored using gel electrophoresis techniques and molecular hybridization to globin probes, respectively. Application of powerful biochemical techniques to relatively pure population of hybrid cells promises to provide, in a relatively short time, information concerning the mechanisms underlying the various phenomena described above. In the next section, a set of new hybrid cell systems which may be particularly amenable to this approach are described.

Gene expression in cybrid and reconstructed cells

In order to investigate more directly the role of cytoplasmic–nuclear interactions in determining changes in nuclear gene expression, the techniques of 'cybridization'[7] and cell reconstruction or nuclear transplantation[15,63], have been employed. A cybrid is constructed by fusion of a cytoplast, or enucleated cell, to a whole cell; it initially contains a single nucleus within a mixed cytoplasm. Cybrids and their progeny can be selected from a mixed population of unfused parental bodies and homokaryons by combined use of a nuclear genetic marker, such as resistance to bromodeoxyuridine, and a mitochondrial marker, such as resistance to chloramphenicol[7]. A reconstructed cell is formed by fusion of a single cytoplast with a single karyoplast, a body containing a nucleus surrounded by a thin shell of cytoplasm and an outer cell membrane[17,61]. Reconstructed cells can likewise be isolated using genetic selection techniques. Furthermore, since methods have been devised for definitively identifying and isolating such cells, biochemical analysis of events occurring immediately after fusion is now feasible. (These techniques are reviewed in reference 59.) Several descriptions of what appear to be alterations in the phenotype of cells after exposure of a nucleus to a foreign cytoplasm, either by cybridization or nuclear transplantation, have been reported. Some noteworthy examples are:

1. Gopalakrishnan et al.[26] described the fusion of enucleated mouse neuroblastoma or fibroblast cells with mouse erythroleukemia cells, a cell line which could normally be stimulated to produce hemoglobin by treatment with dimethylsulfoxide. Many cybrid clones derived from these fusions could not be induced to synthesize hemoglobin for up to six months after their isolation. It thus appeared that a *transient* exposure to the cytoplasm of nonerythroid cells resulted in a long-term extinction of hemoglobin inducibility in the erythroleukemia cells. When such experiments were performed using two other independently isolated erythroleukemia cell lines, however, little or no extinction was observed. In a later report[53], the further characterization of one cybrid clone (E5-1B) which had been reported to be non-inducible was described. Now, when cultured under somewhat different conditions, the cells were found to be in fact heterogeneous. About 40% of a population of E5-1B cells could be induced to produce hemoglobin. Analysis of the globin gene sequences in the hybrid cells, using pancreatic DNase I as a probe for transcriptionally active genes, showed that about 40–45% of the globin gene sequences in a population of cells were preferentially resistant to nuclease digestion. Furthermore, subcloning the E5-1B cybrid line resulted in clones with varying degrees of inducibility. Thus, although these results demonstrated the ability of a foreign cytoplasm to affect inducibility in an erythroleukemia cell line, they did not, taken together, definitively demonstrate the stable long-term inheritance of the trait.

2. Albumin synthesis appeared to be extinguished in cybrids constructed by fusion of enucleated mouse fibroblasts with differentiated rat hepatoma cells[36]. Synthesis was suppressed in most cybrids within 10–20 hours after fusion. This phenomenon was clearly a transient one, in that albumin was again detected in large quantities by about 48 hours after fusion. It was concluded that extinction was mediated by a short-lived regulator which was not renewable in the absence of the fibroblast nucleus, a notion also consistent with the results of studies with heterokaryons described above.

3. A transient activation of another liver-specific function–steroid-inducible tyrosine aminotransferase activity–was seen in hybrid cells constructed by transplantation of mouse fibroblast nuclei into enucleated rat hepatoma cells[44]. Results of histochemical staining suggested that activation occurred in nearly all successfully reconstructed cells. Although the property was transmitted for many generations (as many as 100 in some experiments) it too was unstable and was eventually lost from all cells. An adequate explanation for such a long-term but ultimately unstable transmission has yet to be formulated. Transmission by cybridization but eventual loss of mitochondrial markers has also been described [35,64].

4. Synthesis of a second liver-specific enzyme–phenylalanine hydroxylase–was activated in cybrids constructed by fusion of enucleated rat hepatoma cells to mouse erythroleukemia cells[25]. In these experiments, cybrids were selected in medium lacking tyrosine; that is, only cells synthesizing the enzyme under investigation would grow. Cells containing the mouse genome, and synthesizing what was shown to be mouse phenylalanine hydroxylase, were obtained, albeit at a low frequency (10^{-5}–10^{-6}). In a later report[35] it appeared that the frequency of activation in subsequent experiments was even lower. Also described here was an analysis of cybrids for other traits characteristic of the cytoplast donor cell line. Of the assayed traits, only phenylalanine hydroxylase

was induced. The authors speculated that activation was dependent upon the selective pressure exerted, but the frequency of occurrence of enzyme-synthesizing cybrids in medium containing tyrosine was not in fact measured.

5. That the cytoplasm of actively growing cells contains the material necessary for reactivation of dormant nuclei was demonstrated by several groups. When transplanted into enucleated mammalian cells, avian erythrocyte nuclei increased in size, re-formed nucleoli and became transcriptionally active[16, 40, 43]. When placed into cytoplasts prepared from fibroblasts synchronized in the S phase of the cell cycle, DNA synthesis occurred in many reactivated nuclei[43]. Large amounts of globin were also synthesized in the hybrid cells[6]. As noted above, the three forms of normal adult globin (α^A, α^D and β-globin) were synthesized in heterokaryons formed from avian erythrocytes and mammalian cells[42]. In reconstructed cells, no adult β-globin was detected. Rather, a species having the characteristics of an early embryonic globin chain was synthesized[6]. Now it is unlikely that fibroblasts contain regulatory elements responsible for the specificity of globin gene expression in developing erythroid nuclei. However, the system may prove to be a uniquely useful one for probing the relationship of chromatin structure and globin gene expression. The techniques to be employed in such an analysis are described elsewhere[6].

6. The tumorigenicity of some transformed cells lines was partially suppressed by fusion to cytoplasts prepared from normal cell types[22, 33]. The phenomenon appeared not to be a general one. Thus, no suppression was seen when a particular SCV40-transformed cell line was fused to normal cell cytoplasts[33]. Similarly, the transformed phenotype–assessed by the ability to form vascularized tumors in the chorioallantoic membrane of fertilized chicken eggs or tumors in nude mice–was unchanged in hybrids prepared by fusion of karyoplasts from a transformed mouse line to cytoplasts from a normal human diploid fibroblast cell strain (Hightower, Klinger and Lucas, manuscript submitted for publication). Furthermore, since the molecular basis for expression of the transformed phenotype is largely unknown, it is yet unclear whether these instances of suppression represent examples of the phenomenon under consideration here.

   Another list of experiments could be appended to the above: those in which cytoplasts of one cell type were fused to whole cells or karyoplasts of another cell type and in which no evidence for a definitive change in nuclear gene expression was observed. For example, studies performed using embryonal carcinoma cells[41, 48], rat myoblasts[41], mouse adrenal tumor cells[60] (White and Lucas, unpublished results), or mouse neuroblastoma cells (White and Lucas, unpublished results) showed no indications of cytoplasmic regulation of nuclear encoded events. Several explanations for this lack of effect are feasible. For example, the function under study might be regulated, if at all, by mechanisms different from those controlling expression in the 'positive' studies listed above. Induction of some of the functions–for example, steroidogenesis in adrenal cell hybrids–would in fact require activation of a variety of enzymatic activities, some localized to the mitochrondria[2]. In other studies, transient effects may have been missed because cybrid or reconstructed cell clones were only assayed after many generations of growth in selective medium. Several of the effects discussed above–extinction of albumin synthesis or activation of tyrosine aminotransferase activity–lasted for only a limited period of time.

That numerous transient changes in gene expression occur after nuclear transplantation was suggested by a recent analysis of hybrid cells constructed by transplantation of nuclei from a transformed mouse cell line into cytoplasts from a normal human fibroblast strain (Hightower, Bruno and Lucas, manuscript in press). These cells can be produced in large quantities and definitely identified using a variety of morphological, genetic and immunological criteria[31,32]. An analysis of the proteins made in the hybrids by two-dimensional gel electrophoresis showed that nuclear-directed polypeptide synthesis was detectable as early as three hours after fusion. Several of the mouse polypeptides detected at such early times were rather minor species in the mouse parental cell type. The synthesis of some other polypeptides was not detectable until many hours after fusion. However, by about 48 hours after transplantation the pattern of proteins synthesized resembled almost exactly that of the nuclear donor parent cell. The overall morphology and detailed cytoskeletal structures of the cells likewise changed from those characteristic of the human cytoplast donor to those of the mouse nuclear donor.

Despite the rapidity of these changes, it is proposed that cytoplasmic–nuclear hybrid cells such as these provide an ideal opportunity to discover and investigate the mechanisms regulating gene expression in eukaryotic cells. As described in detail elsewhere[31] quantities of relatively pure hybrid cells suitable for many types of biochemical analysis can be readily prepared. Now under investigation in this laboratory are the changes in chromatin structure–methylation patterns, susceptibility to nuclease digestion, association with various chromosomal proteins–which accompany changes in expression of particular genes after nuclear transplantation.

Acknowledgements

Research performed in the author's laboratory was supported by PHS grant CA22302, awarded by the National Cancer Institute, and American Cancer Society grant MV-19. J. Lucas is an Established Investigator of the American Heart Association.

References

1. Allan, M. and Harrison, P. (1980). Co-expression of differentiation markers in hybrids between Friend cells and lymphoid cells and the influence of cell shape. *Cell*, **19**, 437–447
2. Asano, K. and Harding, B. W. (1976). Biosynthesis of adrenodoxin in mouse adrenal tumor cells. *Endocrinology*, **99**, 977–987
3. Axelrod, D. E., Gopalakrishnan, T. V., Willig, M. and Anderson, W. F. (1978). Maintenance of hemoglobin inducibility in somatic cell hybrids of tetraploid (2S) mouse erythroleukemia cells with mouse or human fibroblasts. *Somatic Cell Genetics*, **4**, 157–168
4. Bacchetti, S. and Graham, F. L. (1977). Transfer of the gene for thymidine kinase to thymidine kinase deficient human cells by purified herpes simplex viral DNA. *Proceedings of the National Academy of Sciences, USA*, **74**, 1590–1594
5. Ben-Ze'ev, A., Farmer, S. R. and Penman, S. (1980). Protein synthesis requires cell-surface contact while nuclear events respond to cell shape in anchorage-dependent fibroblasts. *Cell*, **21**, 365–372
6. Bruno, J., Reich, N. and Lucas, J. J. (1981). Globin synthesis in hybrid cells constructed by transplantation of dormant avian erythrocyte nuclei into enucleated fibroblasts. *Molecular Cell Biology*, **1**, 1163–1176
7. Bunn, C. L., Wallace, D. C. and Eisenstadt, J. M. (1974). Cytoplasmic inheritance of chloramphenicol resistance in mouse tissue culture cells. *Proceedings of the National Academy of Sciences, USA*, **71**, 1681–1685

8. Cori, C. F., Gluecksohn-Waelsch, S., Klinger, H. P., Pick, L., Schlagman, S. L., Teicher, L. S. and Wang-Chang, H.-F. (1981). Complementation of gene deletions by cell hybridization. *Proceedings of the National Academy of Sciences, USA*, **78**, 479–483

9. Darlington, G. J., Bernhard, H. P. and Ruddle, F. H. (1974). Human serum albumin phenotype activation in mouse hepatoma–human leukocyte cell hybrids. *Science*, **185**, 859–862

10. Davidson, R. L. (1972). Regulation of melanin synthesis in mammalian cells: Effect of gene dosage on the expression of differentiation. *Proceedings of the National Academy of Sciences, USA*, **69**, 951–955

11. Davidson, R. L. (1974). Gene expression in somatic cell hybrids. *Annual Review of Genetics*, **8**, 195–218

12. Davidson, R. L., Ephrussi, B. and Yamamoto, K. (1966). Regulation of pigment synthesis in mammalian cells as studied by somatic hybridization. *Proceedings of the National Academy of Sciences, USA*, **56**, 1437–1440

13. Davidson, R. L., Ephrussi, B. and Yamamoto, K. (1968). Regulation of melanin synthesis in mammalian cells, as studied by somatic hybridization I. Evidence for negative control. *Journal of Cell Physiology*, **72**, 115–127

14. Deisseroth, A., Burk, R., Pecciano, D., Mina, J., Anderson, W. F. and Nienhuis, A. (1975). Hemoglobin synthesis in somatic cell hybrids: globin gene expression in hybrids between mouse erthroleukemia and human marrow cells or fibroblasts. *Proceedings of the National Academy of Sciences, USA*, **72**, 1102–1106

15. Ege, T., Krondahl, U. and Ringertz, N. R. (1974a). Introduction of nuclei and micronuclei into cells and enucleated cytoplasms by Sendai virus induced fusion. *Experimental Cell Research*, **88**, 428–432

16. Ege, T., Zeuthen, J. and Ringertz, N. R. (1975). Reactivation of chick erythrocyte nuclei after fusion with enucleated cells. *Somatic Cell Genetics*, **1**, 65–80

17. Ege, T., Hamberg, H., Krondahl, U., Ericsson, J. and Ringertz, N. R. (1974b). Characterization of minicells (nuclei) obtained by cytochalasin enucleation. *Experimental Cell Research*, **87**, 365–377

18. Ephrussi, B. (1972). *Hybridization of Somatic Cells*. Princeton, Princeton University Press

19. Fougére, C. and Weiss, M. C. (1978). Phenotypic exclusion in mouse melanoma-rat hepatoma cells: pigment and albumin production are not reexpressed simultaneously. *Cell*, **15**, 843–854

20. Fougére, C., Ruiz, F. and Ephrussi, B. (1972). Gene dosage dependence of pigment synthesis in melanoma × fibroblast hybrids. *Proceedings of the National Academy of Sciences, USA*, **69**, 330–334

21. Fournier, R. E. K. and Ruddle, F. H. (1977). Microcell mediated transfer of murine chromosomes into mouse, Chinese hamster, and human somatic cells. *Proceedings of the National Academy of Sciences, USA*, **74**, 319–323

22. Giguére, L. and Morais, R. (1981). On suppression of tumorigenicity in hybrid and cybrid mouse cells. *Somatic Cell Genetics*, **7**, 457–471

23. Gluecksohn-Waelsch, S. (1979). Genetic control of morphogenetic and biochemical differentiation: lethal albino-deletions in the mouse. *Cell*, **16**, 225–237

24. Goldfield, A. E., Rubin, C. S., Siegel, T. W., Shaw, P. A., Schiffer, S. G. and Gluecksohn-Waelsch, S. (1981). Genetic control of insulin receptors. *Proceedings of the National Academy of Sciences, USA*, **78**, 6359–6361

25. Gopalakrishnan, T. V. and Anderson, W. F. (1979). Epigenetic activation of phenylalanine hydroxylase in mouse erythroleukemia cells by the cytoplast of rat hepatoma cells. *Proceedings of the National Academy of Sciences, USA*, **76**, 3932–3936

26. Gopalakrishnan, T. V., Thompson, E. B. and Anderson, W. F. (1977). Extinction of hemoglobin inducibility of Friend erythroleukemia cells by fusion with cytoplasm of enucleated mouse neuroblastoma or fibroblast cells. *Proceedings of the National Academy of Sciences, USA*, **74**, 1642–1646

27. Gordon, S. and Cohn, Z. (1970). Macrophage–melanocyte heterokaryons I. Preparation and properties. *Journal of Experimental Medicine*, **131**, 981–1003

28. Gordon, S. and Cohn, Z. (1971). Macrophage–melanoma cell heterokaryons IV. Unmasking the macrophage-specific membrane receptor. *Journal of Experimental Medicine*, **134**, 947–962

29. Harris, H. (1970). *Cell Fusion, The Dunham Lectures*. London, Oxford University Press

30. Hightower, M. J. and Lucas, J. J. (1980). Construction of viable mouse–human hybrid cells by nuclear transplantation. *Journal of Cell Physiology*, **105**, 93–103

31. Hightower, M. J. and Lucas, J. J. (1982). Nuclear transplantation with mammalian cells. In *Techniques in Somatic Cell Genetics* (Shay, J. W., Ed.), pp. 255–267. New York, Plenum Press

32. Hightower, M. J., Fairfield, E. F. and Lucas, J. J. (1981). A staining procedure for identifying viable cell hybrids constructed by somatic cell fusion, cybridization, or nuclear transplantation. *Somatic Cell Genetics*, **7**, 321–329

33. Howell, A. N. and Sager, R. (1978). Tumorigenicity and its suppression in cybrids of mouse and Chinese hamster cell lines. *Proceedings of the National Academy of Sciences, USA*, **75**, 2358–2362
34. Junker, S. (1981). A universally applicable method of isolating somatic cell hybrids by two-colour flow sorting. *Biochemical and Biophysical Research Communications*, **102**, 977–984
35. Kahn, C. R., Gopalakrishnan, T. V. and Weiss, M. C. (1981). Transfer of heritable properties by cell cybridization: specificity and the role of selective pressure. *Somatic Cell Genetics*, **7**, 547–565
36. Kahn, C. R., Bertolotti, R., Ninio, M. and Weiss, M. C. (1981). Short-lived cytoplasmic regulators of gene expression in cell hybrids. *Nature, London*, **290**, 717–720
37. Klebe, R. J., Chen, T. R. and Ruddle, F. H. (1970). Controlled production of proliferating somatic cell hybrids. *Journal Cell Biology*, **45**, 74–82
38. Klobutcher, L. A. and Ruddle, F. H. (1981). Chromosome mediated gene transfer. *Annual Review of Biochemistry*, **50**, 533–554
39. Köhler, G. and Milstein, C. (1975). Continuous cultures of fused cells secreting antibody of predefined specificity. *Nature, London*, **256**, 495–497
40. Ladda, R. L. and Estensen, R. D. (1970). Introduction of a heterologous nucleus into enucleated cytoplasms of cultured mouse L cells. *Proceedings of the National Academy of Sciences, USA*, **67**, 1528–1533
41. Linder, S., Brzeski, H. and Ringertz, N. R. (1979). Phenotypic expression in cybrids derived from teratocarcinoma cells fused with myoblast cytoplasms. *Experimental Cell Research*, **120**, 1–14
42. Linder, S., Zuckerman, S. H. and Ringertz, N. R. (1981). Reactivation of chicken erythrocyte nuclei in heterokaryons results in expression of adult globin genes. *Proceedings of the National Academy of Sciences, USA*, **78**, 6286–6289
43. Lipsich, L. A., Lucas, J. J. and Kates, J. R. (1978). Cell cycle dependence of the reactivation of chick erythrocyte nuclei after transplantation into mouse L929 cell cytoplasts. *Journal of Cell Physiology*, **97**, 199–208
44. Lipsich, L. A., Kates, J. R. and Lucas, J. J. (1979). Expression of a liver-specific function by mouse fibroblast nuclei transplanted into rat hepatoma cytoplasts. *Nature, London*, **281**, 74–76
45. Littlefield, J. W. (1964). Selection of hybrids from matings of fibroblasts *in vitro* and their presumed recombinants. *Science*, **145**, 709–710
46. Littlefield, J. W. (1966). The use of drug-resistant markers to study the hybridization of mouse fibroblasts. *Experimental Cell Research*, **41**, 190–196
47. McBride, O. W. and Ozer, H. L. (1973). Transfer of genetic information by purified metaphase chromosomes. *Proceedings of the National Academy of Sciences, USA*, **70**, 1258–1262
48. McBurney, M. W. and Strutt, B. (1979). Fusion of embryonal carcinoma cells to fibroblast cells, cytoplasts and karyoplasts. *Experimental Cell Research*, **124**, 171–180
49. McKusick, V. A. and Ruddle, F. H. (1977). The status of the gene map of the human chromosomes. *Science*, **196**, 390–405
50. Maitland, N. J. and McDougall, J. K. (1977). Biochemical transformation of mouse cells by fragments of herpes simplex virus DNA. *Cell*, **11**, 233–241
51. Malawista, S. E. and Weiss, M. C. (1974). Expression of differentiated functions in hepatoma cell hybrids. High frequency of induction of mouse albumin production in rat hepatoma × mouse lymphoblast hybrids. *Proceedings of the National Academy of Sciences, USA*, **34**, 333–361
52. Mével-Ninio, M. and Weiss, M. C. (1981). Immunofluorescence analysis of the time-course of extinction, reexpression, and activation of albumin production in rat hepatoma–mouse fibroblast heterokaryons and hybrids. *Journal of Cell Biology*, **90**, 339–350
53. Miller, D. M., Turner, P., Nienhius, A. W., Axelrod, D. E. and Gopalakrishnan, T. V. (1978). Active conformation of the globin genes in uninduced and induced mouse erythroleukemia cells. *Cell*, **14**, 511–521
54. Minna, J., Nelson, P., Peacock, J., Glazer, D. and Nirenberg, M. (1971). Genes for neuronal properties expressed in neuroblastoma × L cell hybrids. *Nature, New Biology*, **235**, 225–231
55. Minson, A. C., Wildy, P., Buchan, A. and Darby, G. (1978). Introduction of the Herpes simplex virus thymidine kinase gene into mouse cells using virus DNA or transformed cell DNA. *Cell*, **13**, 581–587
56. Peterson, J. A. and Weiss, M. C. (1972). Expression of differentiated functions in hepatoma cell hybrids: induction of mouse albumin production in rat hepatoma–mouse fibroblast hybrids. *Proceedings of the National Academy of Sciences, USA*, **69**, 571–575
57. Rankin, J. K. and Darlington, G. J. (1979). Expression of human hepatic genes in mouse hepatoma–human amniocyte hybrids. *Somatic Cell Genetics*, **5**, 1–10
58. Ringertz, N. R. and Savage, R. E. (1976). *Cell Hybrids*. New York, Academic Press
59. Shay, J. W. (Ed.) (1982). *Techniques in Somatic Cell Genetics*. New York, Plenum Press
60. Shay, J. W. and Clark, M. A. (1980). Alternative method for identifying reconstituted cells. *Proceedings of the National Academy of Sciences, USA*, **77**, 381–384

61. Shay, J. W., Porter, K. R. and Prescott, D. M. (1974). The surface morphology and fine structure of CHO (Chinese hamster ovary) cells following enucleation. *Proceedings of the National Academy of Sciences, USA*, **71**, 3059–3063
62. Thompson, E. B. and Gelehrter, T. D. (1971). Expression of tyrosine aminotransferase activity in somatic cell heterokaryons: evidence for negative control of enzyme expression. *Proceedings of the National Academy of Sciences, USA*, **68**, 2589–2593
63. Veomett, G., Prescott, D. M. Shay, J. and Porter, K. R. (1974). Reconstruction of mammalian cells from nuclear and cytoplasmic components separated by treatment with cytochalasen B. *Proceedings of the National Academy of Sciences, USA*, **71**, 1999–2002
64. Wallace, D. C., Pollack, Y., Bunn, C. L. and Eisenstadt (1976). Cytoplasmic inheritance in mammalian tissue culture cells. *In Vitro*, **12**, 758–776
65. Wigler, M., Silverstein, S., Lee, L.-S., Pellicer, A., Cheng, Y.-C. and Axel, R. (1977). Transfer of purified herpes virus thymidine kinase genes to cultured mouse cells. *Cell*, **11**, 223–232
66. Wright, W. E. (1981). Synthesis of rat myosin light chain in heterokaryons formed between undifferentiated rat myoblasts and chick skeletal monocytes. *Journal of Cell Biology*, **91**, 11–16
67. Wu, R. (Ed.) (1979). *Recombinant DNA. Vol. 68. Methods in Enzymology*. New York, Academic Press
68. Yelton, D. E. and Scharff, M. D. (1981). Monoclonal antibodies: A powerful new tool in biology and medicine. *Annual Review of Biochemistry*, **50**, 657–680

Chapter 9

Expression of nuclei and purified genes microinjected into oocytes and eggs

L. D. Etkin and M. A. DiBerardino

Introduction

The central problem in developmental biology concerns the process by which cell diversity is generated during embryogenesis. The problem is complex and may involve many facets such as reciprocal interactions between the nucleus and cytoplasm, alterations in the genome, sequential activation and repression of genes, and changes in the processing, transport and utilization of mRNAs. These processes have been examined by nuclear transplantation and gene injection into metazoan eggs and oocytes.

The concept of nuclear transplantation in metazoans was conceived by Spemann[118], who envisioned a unique experiment in which nuclei from somatic cells would be placed back into an egg. This would permit an analysis of the developmental potential of such nuclei which had already undergone many mitotic divisions and been exposed to different cellular environments during development and cell differentiation. This experiment was realized by Briggs and King[10] in their elegant nuclear transplantation studies in *Rana pipiens* and was extended mainly to other amphibians, and also to a limited degree to the invertebrate fruit fly, the protochordate, *Ciona*, and the vertebrate fish (reviewed in reference 33). More recently, nuclear transplantation tests of somatic cell nuclei have been initiated in the mammalian mouse by Illmensee and Hoppe[81] and McGrath and Solter[94a] who have perfected the technique with young embryonic cells and zygote pronuclei respectively. Results of these tests from the above species have shown that young embryonic nuclei are developmentally totipotent. However, nuclei from specialized cells tested so far in amphibians are developmentally restricted. The pertinent questions of whether irreversible changes occur in somatic cell nuclei during development and the nature of nuclear reprogramming, however, still remain unanswered.

One of the most promising techniques for studying the mechanisms of gene regulation at the transcriptional level as well as RNA processing and transport is the microinjection of purified DNA sequences into the nuclei of amphibian oocytes. Experiments of this nature were first performed by Mertz and Gurdon[105] who injected SV40 DNA into the nuclei (germinal vesicles) of *Xenopus* oocytes and detected transcripts which were produced under the direction of this DNA. To date a large number of genes have been injected into *Xenopus* oocytes and found to be transcribed, their RNAs processed, and mRNAs translated into protein (reviewed in references 47 and 68). While there is some discrepancy as to the accuracy and efficiency of transcription of some of these genes, the microinjection of purified genes into oocyte nuclei has served to answer several fundamental questions regarding the role of specific DNA sequences in transcriptional regulation, as well as the processing and transport of primary transcripts.

Recently, purified genes have been microinjected into fertilized eggs of the mouse, frog and fruit fly. These studies have shown that many eukaryotic genes are replicated and transcribed during development in these organisms. Evidence has also been presented demonstrating the integration of heterologous genes into the genome and passage to the next generation through the germ line, and by nuclear transplantation.

In this chapter we shall review the results of studies utilizing nuclear transplantation and injection of purified genes into oocytes and eggs in order to examine the questions of nuclear equivalence, nuclear reprogramming, mechanisms of nucleo–cytoplasmic interactions, processing and transport of RNA, function of putative regulatory sequences and the tissue specific and developmental regulation of gene expression.

Developmental potential of nuclei transplanted into eggs

The motivation for developing nuclear transplantation in amphibian eggs stemmed primarily from a fundamental problem posed by scientists around the turn of the century. They sought to answer whether or not cell specialization in multicellular organisms involves irreversible genetic changes. In the intervening years, various approaches (cytological, cell culture, genetic, molecular) have been used to focus on this problem. Although there are a few cases reported that involve loss and increase in DNA content and rearrangements of the sequence of nucleotides, the preponderance of evidence supports the concept that the gene content does not change. However, there are limitations to these studies; for example, (a) the techniques are not capable of testing the entire genome; (b) some procedures involve analysis of fixed and/or isolated cell components and we have to infer how living and organized cells function; and (c) the absolute level of resolution of these techniques, for example, molecular hybridization, is unknown. Thus far, the only technique available that theoretically tests the entire genome is nuclear transplantation. This procedure is performed within the confines of a living system. The ability to transplant a living nucleus permits the study of nucleo–cytoplasmic interactions and the analysis of a number of biological and biomedical problems, including normal cell differentiation, cancer, interspecific hybridization, immunobiology and aging[9, 12, 27, 33, 37, 65, 72, 83, 96]. Here we will be primarily concerned with the gene content of nuclei during cell differentiation and the cytoplasmic control of gene function.

Normal amphibian cell nuclei

The original nuclear transplantation procedure for metazoans was developed first in *Rana pipiens*, the leopard frog. It consisted of activation of the egg by pricking with a glass microneedle, manual removal of the egg nucleus with a glass microneedle, and finally transplantation of an embryonic somatic nucleus with its surrounding cytoplasm by means of a glass micropipette into the animal hemisphere of the enucleated egg[10]. The type of development that follows is then a

TABLE 9.1. Advanced development of nuclear transplants from larval cells

Source of cells	Total no. nuclei tested[a]	Post-neurula embryos (%)	Larvae		Adults		Reference
			Abnormal (%)	Normal (%)	Sterile (%)	Fertile (%)	
Intestine (X)	726	36 (5.0)	17 (2.3)	14 (1.9)	2 (0.3)	2 (0.3)	69
Ciliated epidermis (X)	174						85
Non-ciliated epidermis (X)	440	2 (0.4)	1 (0.2)	1 (0.2)		1 (0.2)	85
Melanophores (X)	200						85
Cultured melanophores (X)	257	2 (0.8)	2 (0.8)				85
Cell cultures (X)	3686	23 (0.6)[b]	6 (0.2)[b]	3[M] (0.1)			67
Intestine (X)							
stages 46–48	210	4 (1.9)	3 (1.4)	1 (0.5)[c]			100
stage 54	110	3 (2.7)	1 (0.9)	2 (1.8)[d]			100
stage 57	202	1 (0.5)		1 (0.5)[c]			100
Primordial germ (R)	410	35 (8.5)		31 (7.6)[e]			117

(a) Total number of nuclei tested includes results from serial transplantations
(b) Data have been estimated from graphs
(c) Died during early larval stage
(d) One died during early larval stage; the other metamorphosed
(e) Later development not reported
X – Xenopus laevis; R – Rana pipiens; M–Metamorphosed

reflection of the functional capacity of the introduced nucleus. Such tests revealed that many young embryonic nuclei promoted the eggs to develop into normal larvae which metamorphosed into juvenile frogs[10, 13] and if raised further, into sexually mature adults[95]. Thus, young embryonic nuclei can replace the zygote nucleus and are therefore genetically totipotent. These studies were confirmed in other amphibian species; for example, *Xenopus laevis*[51, 61], *Ambystoma mexicanum*[16], *Pleurodeles waltlii*[52] and others[96]. In the anuran *Xenopus* and the urodelens *Ambystoma* and *Pleurodeles* technical modifications were required. Activation of the egg is induced by either UV irradiation (*Xenopus*), or electrical stimulation (*Ambystoma, Pleurodeles*), and in the three species the egg nucleus is inactivated by UV irradiation and left in the egg. Extension of these tests to nuclei from determined regions of advanced stage embryos revealed that there was a progressive decrease in the number of eggs which developed normally[12, 33, 37, 52, 65, 83, 96]. The relative contributions made by the transplanted nucleus and the maternal templates of the egg to the formation of these nuclear transplants is unknown, but it is certain that a significant contribution is made by the transplanted nucleus because amphibian eggs lacking a functional nucleus form only partial blastulae[14].

The main interest today concerns the developmental potential of nuclei from specialized cell types, for these tests address most critically the genetic basis of cell specialization. Here we will summarize the results of nuclear transfers from larval and adult cells of amphibian species. In examining the genetic potential of these cells, we will consider only those studies in which post-neurula embryos, larvae, and adults developed (*Tables 9.1* and *9.2*). In the case of post-neurula embryos, the

TABLE 9.2. Advanced development of nuclear transplants from adult cells

Source of cells	Total no. nuclei tested[a]	Post-neurula embryos (%)	Larvae Abnormal (%)	Normal (%)	Reference
A-8 cell line (*X*)	365	2 (0.6)	2 (0.6)		85
Male germ (*R*)	116	4 (3.5)	1 (0.9)		36
Cell cultures (*X*)					
skin, lung, kidney	2322	26 (1.1)[b]	7 (0.3)[b]		91
Intestine (*X*)					
trough	564	5 (0.9)	3 (0.5)		94
crest	548	5 (0.9)	2 (0.4)	1 (0.2)[c]	94
Cell culture (*X*)					
skin	129	6 (4.6)	4 (3.1)		71
Spleen lymphocytes (*X*)	100	6 (6.0)	6 (6.0)		131
Erythrocytes (*X*)	440				22
Erythroblasts (*X*)	442		8 (2.0)		22

[a] Total number of nuclei tested includes results from serial transplantations
[b] Data have been estimated from graphs
[c] Died during an early larval stage
  X – *Xenopus laevis*; *R* – *Rana pipiens*

main organ systems are established in a rudimentary form, and in their latter stages they display muscle and nerve function. Larvae are free swimming tadpoles with differentiated cell types in various organ systems. Nuclei capable of providing genetic direction for this extent of development are considered pluripotent. In those cases when fertile adults develop, the genetic apparatus is totipotential.

Tests of somatic nuclei from larval and adult cells transplanted into *Xenopus* eggs have demonstrated that a small percentage of these nuclei can direct the differentiation of post-neurula and larval development, and in three cases fertile frogs developed from larval cell nuclei. These tests have been carried out on cell nuclei from a variety of tissues, namely, intestine[69, 100], cell cultures from minced tadpoles[67], an established cell line derived from the liver of an adult male[85] and cell cultures of skin, lung and kidney[91]. Whether these nuclear transplants developed from nuclei of stem cells or specialized cells is unknown. However, the studies do establish that differentiated organisms still contain a minor population of nuclei with the gene content capable of directing the differentiation of a wide variety of specialized cells.

The most critical attempts to test nuclei from defined cell types include those carried out on larval ciliated epidermis and melanophores, as well as adult cultured skin, lymphocytes, and erythroblasts. Ciliated epidermal cells and melanophores from young larvae exhibited discernible morphological characteristics, but their nuclei did not program for advanced development. However, nuclei of non-ciliated epidermal cells and cultured melanophores directed post-neurula and larval development and a single nucleus from a non-ciliated epidermal cells led to the formation of a fertile female frog[85].

Donor cells from cultured adult skin were judged to be specialized because 99.9% of the cells used for nuclear transplantation displayed immunoreactive keratin. Nuclear transfers from these skin cells led to four heart beat larvae (3.1%)[71]. In the case of lymphocytes derived from adult spleens, 96.1–98.7% of the donor cells contained immunoglobulins. Six per cent (6%) of their nuclei directed the formation of young larvae[131]. Finally, Brun[22] studied the developmental capacity of erythroblast nuclei. Adult frogs were first induced to a state of anemia in order to obtain erythroblasts. These cells are producing hemoglobin and therefore meet the criteria of a specialized cell type. Their nuclei directed recipient eggs to develop into 8 (2.0%) hatching larvae. In these three studies the use of the nucleolar marker (1NU) and determination of chromosome number established that development was due to the transplanted nucleus and not the egg nucleus left behind due to faulty enucleation.

Nuclear transplantation tests of germ cell nuclei are of particular importance, since their gene content remains totipotent except for occasional mutations. Such tests have been made in *Rana pipiens* with primordial germ cells of larvae and spermatogonial cells of adult males. Nuclei from primordial germ cells obtained from larvae at the initial stage of feeding promoted a much higher percentage of normal tadpoles[117] than have been obtained from somatic nuclei of larval donors. However, when spermatogonial nuclei from adult males were tested, a few post-neurulae developed and one abnormal larva resulted that fed for approximately a week and then died[36]. Although this larva is the most advanced nuclear transplant from adult cells, the inability to achieve complete and normal development from genetically totipotent nuclei indicated that the genetic restrictions in somatic nuclei may be due to incomplete expression in the cytoplasm of enucleated eggs (see sections on p. 134 and p. 135).

Amphibian cancer cell nuclei

In the preceding section we considered the genetic potential of nuclei from normal cells. So far, there has been only one malignant neoplasm, the renal adenocarcinoma of *R. pipiens*, that has been tested by nuclear transplantation into eggs. It has proved to be amenable to nuclear transplantation studies because primary and transplanted tumors as well as short term cultures of tumors are mainly diploid with very few deviations from the normal karyotype[34, 39, 40]. Nuclei of spontaneous renal carcinomas from adult frogs after injection singly into enucleated eggs promoted the formation of 3 (1.1%) abnormal larvae[39, 84]. Also, multiple nuclei from induced mesonephric carcinoma from metamorphosing tadpoles promoted 21 (6.5%) of the enucleated eggs to develop into abnormal larvae[97]. The best differentiated larvae from the two studies attained the feeding larval stage free of tumors, but failed to feed. Their inability to proceed further was due to chromosomal aberrations arising after transplantation into enucleated eggs[39]; (see also section on p. 134). Recovery of the egg nucleus in the exovate formed at the time of enucleation[84] and the use of a triploid nuclear marker in the donor cells[97] established that development was due to the transplanted nuclei. These studies demonstrated for the first time that *single* nuclei of *adult* frogs were capable of directing the differentiation of larvae, and suggested that carcinoma nuclei reversed their gene expression when transplanted into an embryonic environment. Although indirect fluorescent microscopic studies showed that 98.5% of the donor cells dissociated from comparable tumors are epithelial cells and not connective tissue cells[98], conclusive proof was still lacking. Since the carcinoma is a malignant neoplasm of the epithelial cells of the kidney tubules, the most direct approach to this problem would be to test nuclei from identifiable epithelial cells. Accordingly, a pronephric carcinoma was induced in a metamorphosing tadpole by injection of a herpes virus fraction into a triploid hatching larva. The parenchyma of the tumor was next grown in monolayer cultures which grew as epithelial cells. Triploid nuclei from these epithelial cells after transplantation into enucleated eggs promoted the development of triploid larvae free of tumors[41]. Since normal pronephric cells are preprogrammed to die, it is most likely that the surviving epithelial cells represented transformed carcinoma cells, whose genome has reversed its gene expression in the embryonic environment and specified cell types of larvae rather than the cancer phenotype.

Normal cell nuclei from other metazoan animals

Successful nuclear transplantation in amphibians suggested that similar experiments should be extended to other species of metazoan animals. So far, successful nuclear transfers of young embryonic nuclei have been accomplished in the eggs of the invertebrate *Drosophila* (reviewed in reference 80), the protochordate *Ciona intestinalis*[129], three species of the vertebrate teleost fish[54, 128] and the mammal mouse[81]. Substantial modifications were required in the procedures to meet the technical needs unique to each species (reviewed in reference 33). Nevertheless, these studies have demonstrated that the concept of genetic totipotency of young embryonic nuclei spans the Animal Kingdom from the invertebrate *Drosophila*, up to and including the mammalian mouse, *Mus musculus*.

Despite attempts to perform nuclear transplantations in mammals for over a decade, only recently did Illmensee and Hoppe[81] report success with nuclei from the inner cell mass of mouse blastocysts. Instead of enucleating the egg first, nuclei are injected singly into fertilized eggs and then the female and male pronuclei are removed with the micropipette. This operation involves only one penetration of the micropipette into the egg. Also, cytochalasin B is included in the operation medium to relax the cytoskeletal architecture of the egg. The fertilized egg and donor nuclei each bore different genetic markers and therefore development directed by the donor nuclei could be verified. When the developmental capacity of nuclei from inner cell mass and trophectoderm cells (predominantly mural) of the blastocyst were compared, only nuclei of the inner cell mass proved to be totipotent. Although 19% of the eggs injected with trophectoderm nuclei cleaved, all arrested during early preimplantation stages. However, 13% of the eggs injected with nuclei of the inner cell mass developed to late preimplantation stages. Of 16 blastocysts transferred into the uteri of pseudopregnant females, three (19%) live mice (two females and one male) were born. These mice bore the coat color, T6 translocation chromosomes and glucose phosphate isomerase (GPI) variant of the donor nuclei. One female and one male, mated with each other and also to non-nuclear transplant mice, have produced normal progeny bearing the coat color phenotype of the donor nuclei. Recently, McGrath and Solter[94a] microsurgically removed the zygote mouse pronuclei and introduced via cell fusion different genetically marked zygote pronuclei. Of 64 embryos transferred to the uteri 7 (10.9%) developed into adults. Thus, by blastocyst stage in the mouse, trophectoderm nuclei exhibit developmental restrictions, while innner cell mass nuclei, destined to form the cells of the embryo and pronuclei, remain totipotent.

Although amphibian species have proved technically amenable to nuclear transplantation studies of larval and adult cells, it is obviously important to extend these studies to nuclei of advanced cell types in the mammal and other non-amphibian forms in order to understand the genetic control of cell specialization among metazoan animals.

Cytoplasmic control of chromosome behavior

Sixty years, ago, Brachet[8] reported that, when sperm prematurely entered maturing oocytes of the sea urchin, the sperm chromatin condensed into chromosomes similar to those of the oocyte. Later, the same event was recorded in precociously inseminated oocytes of amphibians[4]. These early classical observations suggested that the state of the cytoplasm controlled nuclear behavior. Later, amphibian nuclear transplantations from advanced embryonic cell types, adult normal kidney and renal adenocarcinoma into *Rana* eggs indicated that the egg cytoplasm induced DNA synthesis in transplanted nuclei and changes in their chromosome morphology: (a) the number of cleaving eggs far exceeded the percentage of donor cells in mitosis; (b) metaphase chromosomes of nuclear transplant blastulae resembled blastula chromosomes and not the highly condensed metaphase chromosomes of the donor cells; and (c) nucleoli present in donor nuclei were absent in nuclear transplant blastulae but reappeared during their gastrulation[11, 15, 32, 39, 84]. Studies directed at understanding the initial behavior of transplanted nuclei were made[62, 122]. Soon after injection into eggs, transplanted embryonic nuclei behaved like pronuclei: they enlarged considerably, underwent chromatin decondensation, then entered the first mitosis in close synchrony with the events occurring in pronuclei of fertilized eggs. Thus, successful nuclear transplantations must depend on molecular substances present in the egg cytoplasm that control gene replication and ultimately gene transcription of transplanted nuclei.

The cytoplasmic control of nuclear and chromosomal behavior was further studied in *Bufo*, *Rana* and *Xenopus* eggs and oocytes. Mixed populations of cells were obtained from adult *Xenopus* liver, brain and blood. Multiple nuclei from each tissue type were injected into eggs and a large proportion of these nuclei synthesized DNA[57]. The percentage of transplanted nuclei synthesizing DNA in eggs, detected in autoradiograms, exceeded the percentage doing so in their own environment. Although these injected nuclei did not direct development, these studies established that some degree of reversibility is possible. Extension of these studies to diplotene oocytes and to those undergoing maturation demonstrated that transplanted nuclei conform to the nuclear characteristics of the resident nucleus. During the diplotene stage when the resident nucleus (germinal vesicle) is greatly enlarged, transplanted embryonic[121] and brain[64] nuclei enlarge, and the brain nuclei when presented with ^{3}H-uridine were shown to transcribe RNA. After germinal vesicle breakdown when the oocyte's own chromosomes condense and become aligned on the spindle of the first and second meiotic metaphases, transplanted embryonic[121], brain[21,138] and sperm nuclei[106,116] transformed into metaphase chromosomes on spindles. Recent studies have shown that erythroblast and even non-cycling and terminally differentiated erythrocyte nuclei transform into metaphase chromosomes aligned on spindles[1,93]. Thus, nuclear transplantation studies from different cell types in *Bufo*, *Rana* and *Xenopus* have firmly established the cytoplasmic control of nuclear and gene function. Some, aspects of these studies have now been confirmed in the mammalian mouse[3].

Cell fusion studies have also demonstrated the cytoplasmic control of nuclear behavior. When chick erythrocytes were fused with HeLa cells, the dormant erythrocyte nuclei initiated DNA synthesis and then RNA synthesis under the control of the metabolically active HeLa cells[74]. The same phenonemon has been observed in cell cultures when mammalian cells from different phases of the cell cycle are fused. Depending on whether the cells are in G1, S, G2 or M, three events are separately inducible, namely, induction of DNA synthesis, mitosis and chromosome condensation[102,110].

The main conclusion from these studies is that a new program of chromosome behavior and molecular synthesis is induced in foreign nuclei by cytoplasmic factors residing in the oocyte, egg or somatic cells. Among these factors three have now been identified and studied in the amphibian oocyte: maturation promoting factor (MPF) leading to dissolution of the nuclear membrane[101], chromosome condensing activity (CCA) responsible for condensing the chromosomes at metaphases I and II[101] and cytostatic factor (CSF) causing meiotic arrest at metaphase II[138,139]. MPF and CCA are operative in both meiosis and mitosis, and therefore may be universal regulatory substances controlling the cell cycle.

Cause of developmental restrictions

In the previous section it was seen that advanced embryonic and adult nuclei respond to the cytoplasmic signals in both oocytes and eggs, yet normal development has not occurred under the direction of adult nuclei. In order to determine the nature of the developmental restrictions displayed by nuclei from advanced cell types, an extensive series of studies was performed on abnormal embryos and larvae of *Rana* nuclear transplants (reviewed in reference 32). Most abnormal nuclear transplants examined exhibited chromosome abnormalities in number and/or structure, which in most cases arise during the first cell cycle of the egg[35]. These abnormalities are now known to be the cause of developmental arrest. The most severe chromosomal alterations cause developmental arrest of blastula embryos, whereas relatively minor karyotypic alterations permit development to early larval stages. Similar chromosome abnormalities have been observed in nuclear transplants of *Ambystoma mexicanum*[16], *Pleurodeles waltlii*[53] and *Xenopus laevis*[67]. Evidence has been presented previously that most of these numerical and structural changes in the chromosomes (a) involve genetic loss, (b) are a reflection of chromosomal differentiation acquired progressively through embryogenesis, and (c) are not a result mainly of technical damage[32, 33, 37]. It appears that the cytoplasmic cell cycle of the egg, which is much faster than the nuclear cycle of advanced cell types, induces premature changes in transplanted nuclei, resulting in incomplete DNA replication, chromosome breaks and rearrangements, and even hypoaneuploidy in the most severely affected embryos.

These results still leave unresolved whether or not the genome undergoes irreversible changes during cell differentiation. At least three hypotheses can be proposed to account for the developmental restrictions displayed by nuclei from advanced cell types: (a) irreversible genetic changes occur in the donor cells during cell differentiation; (b) the chromosome abnormalities are merely a result of insufficient time to reverse in the egg; and (c) the chromatin requires remodeling of its proteins that is not possible in the egg cytoplasm. At present, it is impossible to prove or disprove that the first hypothesis is a general event. With respect to the second proposition, one attempt has been made to lengthen the cell cycle in *Rana* eggs and this did improve development, but so far this success is limited to embryonic nuclei of tailbud endodermal cells[75]. The third hypothesis is tenable, especially when one considers the replacement of chromatin proteins by more basic proteins that occurs during spermatogenesis[42]. Also, after fertilization in the sea urchin, histone, specific to sperm chromatin, is replaced by embryonic histone, and additional protein replacements and modifications occur in the chromatin of male pronuclei[108]. Thus, it was hypothesized that conditioning of test nuclei in oocyte cytoplasm might lead to enhanced genetic potential[33].

Enhanced genetic potential of nuclei conditioned in oocytes

It was necessary first to determine whether somatic nuclei from embryonic cells could in fact direct embryogenesis after they resided in oocyte cytoplasm. For this test, blastula nuclei from undetermined cells and tailbud nuclei from determined endodermal cells were injected singly into maturing oocytes at first meiotic metaphase [78]. At this stage the oocytes are not activatable or fertilizable, but about 24 hours later (18°C), the oocytes can be activated by pricking with a glass microneedle. Approximately, ten minutes after activation the black dot indicative of the second meiotic metaphase of the egg nucleus of the oocyte is visible in the animal pole. This black dot was removed microsurgically with a glass microneedle. Both blastula and tailbud endodermal nuclei transformed into pronuclei and promoted development through embryogenesis (*Figure 9.1c*). In another series in which the nuclear transplants were fixed and studied cytologically, it was found that transplanted somatic nuclei conformed to the events displayed by the hosts' nuclei during maturation, i.e. they evolved metaphase chromosomes aligned on their own spindles (*Figure 9.1a* and *b*). Thus, somatic nuclei can reversibly respond to

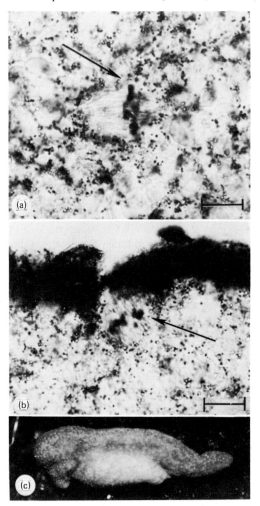

Figure 9.1. Oocyte at first meiotic metaphase, injected with a single somatic nucleus. (a and b) Oocyte fixed eight hours after nuclear transplantation. Scale bar 10 μm. (a) Injected nucleus has transformed into metaphase chromosomes aligned on a spindle located approximately one-third down from the animal pole. (b) Oocyte metaphase chromosomes are aligned on meiotic spindle located just beneath surface coat of animal hemisphere. Note difference in size of spindles in (a) and (b). The spindles of injected nuclei are 1.3 to 1.6 times longer (pole to pole) than the hosts' spindles. (c) Nuclear transplant derived from oocyte host (first meiotic metaphase) injected with a tailbud endodermal nucleus (stage 19). The nuclear transplant developed normally through neurulation and then exhibited slight abnormalities during the postneurula and larval stages. Photographed at stage 19, the animal later developed into a swimming larva (stage 22+). Length approximately 6 mm. From Hoffner and DiBerardino[78].

cytoplasms directing either meiotic or mitotic events and still retain the genetic potential to direct embryogenesis. Therefore, nuclei from differentiated cells could now be tested.

Recently, the effect of oocyte cytoplasm on the developmental potential of adult erythrocyte nuclei has been studied in *Rana*. This cell type is a non-cycling and terminally differentiated specialized cell type and provides a critical test for reversibility. When adult erythrocyte nuclei are injected into first meiotic metaphase oocytes and handled as described above for blastula and tailbud endodermal nuclei, swimming larvae result. But, if erythrocyte nuclei are injected into eggs, none of the nuclear transplants developed beyond the earliest stages of gastrulation[38]. In another study the ability of erythrocyte nuclei to synthesize DNA was studied in autoradiograms. Only 0.1% of erythrocyte nuclei taken directly from adult frogs synthesize DNA *in vitro* and 24% do so in a limited way after injection into eggs. However, over 75% synthesize DNA in activated eggs, if the nuclei have been previously incubated in diplotene and maturing oocytes[93]. These results with non-cycling and terminally differentiated adult erythrocytes demonstrate: (a) erythrocyte nuclei contain the genes to specify larval development; (b) dormant genes can be reactivated by conditioning in the cytoplasm of oocytes; (c) the means whereby genes are reactivated involve in some way the preparation of the chromosomes by oocyte cytoplasm to synthesize DNA in the cytoplasm of activated eggs; and (d) the formation of larvae attests to the fact that the DNA is replicated many times with a high degree of fidelity, and that the DNA is capable of widespread RNA synthesis required for protein synthesis in order to specify the cell, tissue and organ types present in larvae.

Control of the expression of specific genes from differentiated cell nuclei by oocyte components

The cytoplasmic influence on the molecular expression of nuclei has been examined by microinjection of somatic cell nuclei into amphibian oocytes. The results demonstrated that the general type of nucleic acid synthesized by the injected nuclei conforms to that being synthesized by the host cell[63, 64] (see p. 133). Whether the oocyte cytoplasm affects the expression of *specific* genes was investigated by injecting somatic cell nuclei into oocytes and analyzing the recipient oocytes for the presence of translational products of the injected nuclei. One technique which optimally resolves proteins is a two-dimensional polyacrylamide gel system involving isoelectric focusing of proteins in one dimension followed by SDS gel electrophoresis in the other dimension[107]. This, in conjunction with fluorographic techniques, permits the detection of newly synthesized proteins. The use of heterologous combinations such as HeLa or cultured *Xenopus* kidney cell nuclei injected into *Xenopus laevis* or *Pleurodeles waltlii* oocytes allows one to distinguish proteins synthesized under the direction of the injected nuclei from endogenous oocyte proteins. When HeLa cell nuclei were injected into *Xenopus* oocytes only a few of the HeLa specifc proteins were detected[28, 70]. This suggested that many of the genes which were active in the HeLa cells were not expressed when these nuclei were injected into oocytes. DeRobertis and Gurdon[28] also showed that a translational product which was not detectable in cultured *Xenopus* kidney cells was produced in detectable amounts by kidney cell nuclei following injection into oocytes of the urodele *Pleurodeles waltlii*. This product normally is synthesized in both *Xenopus* and *Pleurodeles* oocytes, the result implying, therefore, that the oocyte regulatory factors activated the silent gene in the injected kidney cell nuclei. One of the difficulties in interpreting this result is that the donor nuclei originated from cell lines which may possess abnormal chromosomal complements or exhibit aberrant regulation of gene expression.

The regulation of the expression of specific gene products with known function was carried out by Etkin[44,45]. The experimental design involved the use of two enzymes as markers of gene activity. Lactate dehydrogenase (LDH) is representative of a class of enzymes found in all cell types (housekeeping or ubiquitous functions), and alcohol dehydrogenase (ADH) is representative of a class of tissue-specific enzymes. These enzymes were selected because starch gel electrophoresis revealed species-specific differences between enzymes from two closely related amphibians, *Ambystoma mexicanum* and *Ambystoma texanum*. This permitted the detection of enzymes synthesized under the direction of *A. texanum* nuclei which had been transferred into *A. mexicanum* oocytes.

Adult liver cell nuclei of *A. texanum* (from cells producing a differentiated product, ADH, and a ubiquitous product, LDH) were transferred into *A. mexicanum* oocytes which synthesize LDH but not ADH. Recipients were cultured for periods ranging from one to three weeks. Three days after injection into the oocyte cytoplasm the *texanum* liver cell nuclei had increased their volume several fold (from 10 to 50 times). Closely associated with swelling was the incorporation of ³H-uridine into RNA in the injected liver cell nuclei. From these observations it was concluded that the injected *texanum* liver cell nuclei were responding to the oocyte cytoplasm and were indeed synthesizing RNA. *Mexicanum* oocytes were assayed for the presence of *texanum* LDH three weeks after receiving injections of *texanum* liver cell nuclei. *Texanum* LDH was detected in the recipient *mexicanum* oocytes. *Texanum* ADH, however, was never detected in any experiment. The appropriate controls, including the injection of *texanum* liver cell cytoplasm and cytoplasmic RNA, eliminated the possibility that the *texanum* LDH observed in the *mexicanum* oocytes was due to mRNA carryover with the injected nuclei. *Texanum* liver cell nuclei also were injected into enucleated *mexicanum* oocytes to determine if the maternal nucleus affects the expression of the injected nuclei. In all cases *texanum* LDH was detected suggesting that the presence or absence of the maternal nucleus did not affect the expression of LDH genes of the injected *texanum* liver cell nuclei.

These studies suggest that somatic cell nuclei injected into oocytes tend to produce products (i.e. ADH, LDH and those observed on two-dimensional gels) similar to those being produced by the oocyte. This implies that regulatory mechanisms exist which control the expression of the oocyte genome as well as any foreign genome introduced into the oocyte. In order to fully assess the question of the stability of the differentiated cellular phenotype, however, one must examine the regulation of many more differentiated products of injected somatic cell nuclei.

In the above experiments the expression of genes was analyzed at the translational level. While it is believed that most injected mRNAs are translated in *Xenopus* oocytes, it is possible that translational control mechanisms may affect the expression of some of the genes of the injected nuclei. Therefore, a more precise analysis would be to examine expression at the level of transcription. This was performed by analysing the regulation of the expression of genes coding for 5S ribosomal RNA, which of course do not produce any translational products.

Korn and Gurdon[86] microinjected nuclei from cultured *Xenopus* kidney cells and erythrocytes into *Xenopus* oocytes in order to assess the ability of the oocyte regulatory components to activate the oocyte specific 5S genes. As will be discussed in a later section (p. 143) *Xenopus* contains two sets of genes which code for 5S ribosomal RNA. These are somatic 5S genes and oocyte 5S genes. Both sets are active in oocytes, while only the somatic 5S set is predominantly active in somatic cells. The oocyte genes, however, are not completely inactive in somatic cells since they account for the production of up to 5% of the total 5S RNA. Korn and Gurdon[86] observed that when nuclei from both cultured kidney cells and erythrocytes were injected into oocytes the oocyte-specific gene products were detectable in half of the injected oocytes. On further examination they found that all oocytes from some females were able to activate the genes (or increase the rate of transcription of the 5S gene) while all oocytes from others were incapable. The reason for this difference is unclear. They also observed that treatment of nuclei with $0.35 \, \text{mol}/\ell$ NaCl prior to injection into oocytes from non-activating females resulted in an increase of the oocyte-specific 5S gene products produced by the injected nuclei. These results suggest that specific factors are involved in the regulation of the expression of these two sets of genes, and that it is possible (under certain conditions) to interfere with the action of these components. One could use this system to isolate molecular components involved in the regulation of the expression of specific genes. This could be accomplished by isolation and fractionation of components from cells in which specific genes are active and microinjecting these into recipient oocytes in an attempt to activate silent genes. This approach is currently underway in several laboratories.

Function of purified genes injected into nuclei of *Xenopus* oocytes

Transcription of injected DNA sequences

Microinjection of cloned sequences of DNA into the nuclei of amphibian oocytes has demonstrated that the genes are transcribed and the transcripts translated into protein. *Figure 9.2* illustrates a scheme used to test the expression of the sea urchin histone genes. One of the first successful attempts to examine the transcription of microinjected DNA in *Xenopus* oocytes was that by Colman[25]. He injected a synthetic polynucleotide poly(d(A–T)d(A–T)) into both eggs and oocytes of *Xenopus laevis* and demonstrated a stimulation of RNA synthesis, suggesting that the DNA was being transcribed. The use of the *Xenopus* oocyte as a transcriptional assay system was refined by Mertz and Gurdon[105]. They found that microinjected SV40 DNA was transcribed most efficiently when injected into the germinal vesicle (nucleus or GV) of *Xenopus* oocytes. Viral transcription continued for up to five days, and was proportional to the amount of DNA injected. They also observed transcription of adenovirus and several other DNAs. It has been observed that circular molecules are transcribed more efficiently than linear templates, when injected into the oocyte nucleus, and that both linear and circular molecules are degraded in the oocyte cytoplasm[135]. Circular DNA molecules are transcribed at a rate 500 times greater than linear DNA[73a]. This may be correlated with aberrant chromatin structure of injected linear DNA[104a].

Table 9.3 shows the diversity of genes which have been injected into *Xenopus* oocytes. Almost any DNA is transcribed when inserted into the nucleus, except in the case of several genes transcribed under the direction of polymerase II. Generally, 2–10 ng of purified DNA is injected into the germinal vesicle of each oocyte. This represents from 10^6–10^9 copies of a particular gene. Genes transcribed by RNA polymerase III (i.e. 5S and 4S RNA) are transcribed more efficiently (i.e. 10–20 transcripts/gene/hour) than either polymerase I or II type genes (0.1–1 transcripts/gene/hour). Several polymerase II type genes such as ovalbumin[126, 134], or rabbit globin[113] appear to be transcribed with very little efficiency in the oocyte,

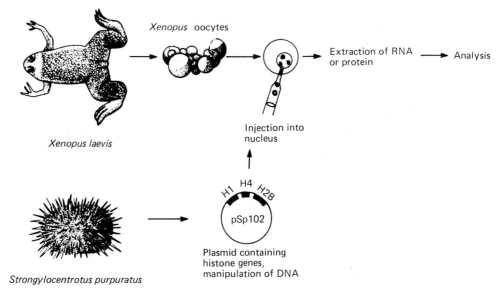

Figure 9.2. Strategy utilized in the microinjection of cloned genes into the nuclei of *Xenopus* oocytes. This is the scheme followed for genes which have been cloned from the sea urchin (*Strongylocentrotus purpuratus*). The eukaryotic DNA can be manipulated using restriction enzymes to delete specific sequences prior to recircularization and microinjection. Recipient oocytes are cultured for several hours or days. Newly synthesized proteins may be radiolabeled by incubating oocytes in radioactive amino acids, followed by extraction and analysis by gel electrophoresis and fluorography. Transcripts may be analyzed by either radiolabeling with uridine and fluorography, or by a variety of techniques including SI nuclease mapping of 5′ and 3′ ends or Northern blot analysis.

TABLE 9.3. Genes microinjected into *Xenopus* oocytes

Gene	Reference
SV40	29, 105, 112
4S tRNA	26a, 30, 79, 87
5S rRNA	19, 66
Adenovirus	105
Ovalbumin	90, 126, 134
Sea urchin histone	45, 48, 88. 109, Etkin and Childs
Early genes	(unpublished observations
Late genes	
18s and 28s rRNA	2, 125, 127
Mitochondrial DNA	137
Drosophila ADH	50
Polyoma *a* virus	112
Rabbit β-globin	113
Drosophila heat shock genes	130
Thymidine kinase	99
Chloramphenicol acetyl transferase	81a

while sea urchin histone[48, 109], viral thymidine kinase[99], *Drosophila* ADH[50] and viral genes[105, 112], are transcribed at significantly better rates. The precise reason for this discrepency is unclear. The expression of chicken ovalbumin gene is intriguing since its expression *in vivo* normally requires the presence of estrogen. The fidelity of expression in oocytes, however, is questionable in that the 5' and 3' termini of the transcripts have not been mapped and initiation of transcription may occur in the prokaryotic vector sequences.

Transcription of genes microinjected into *Xenopus* oocytes have been visualized under the electron microscope (*Figure 9.3*). Using this technique the RNA

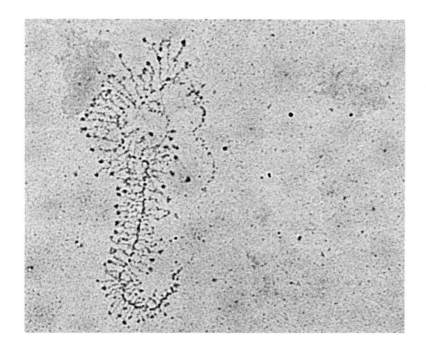

Figure 9.3. Visualization of transcription of microinjected *Xenopus* ribosomal RNA genes in *Xenopus* oocyte nuclei. This spread shows the apparently normal transcription occurring on microinjected molecule. Courtesy of Dr Aimee Bakken.

polymerase packing density and the mapping of transcription initiation and termination sites were determined[2, 125, 127]. Examination of DNA in the electron microscope has also revealed that microinjected DNA interacts with *Xenopus* chromosomal proteins resulting in the formation of minichromosome-like structures possessing nucleosomes[92, 136].

Processing of newly synthesized heterologous transcripts

One of the amazing features of the *Xenopus* oocyte is its ability to process correctly or modify the newly synthesized transcripts for many diverse genes. This is true for genes such as 4S tRNA genes from yeast which are transcribed under the direction of RNA polymerase III as well as for *Drosophila* ADH, globin, SV40, and adenovirus genes which are transcribed under the direction of RNA polymerase II. There are varying amounts of aberrant transcripts produced by many of these genes, but the oocyte machinery is capable of splicing out intervening sequences and performing a variety of post-transcriptional modifications on many of the RNAs. This points to the existence of enzyme systems in the *Xenopus* oocyte which perform a wide variety of functions on heterologous RNA.

An elegant example of this phenomenon is found in the processing of newly synthesized transcripts of yeast tRNA in *Xenopus* oocytes. DeRobertis and Olson[30] observed the appearance of a 108 nucleotide long RNA species following the microinjection of cloned yeast tRNA genes into the nuclei of *Xenopus* oocytes. This sequence was identified as the precursor to the 92 nucleotide long intermediate and the mature 78 nucleotide yeast tyrosine tRNA by gel analysis and nucleotide fingerprinting techniques. They also demonstrated that a 19 base pair 5' leader extension and an intervening sequence was removed from this precursor in the *Xenopus* oocyte. The final mature tRNA also had CCA added post-transcriptionally at the 3' end, and several of the bases were modified as they were in yeast. These results suggested that many of the frog enzymes involved in tRNA production were not species-specific and could recognize accurately and process yeast tRNA which had been transcribed by heterologous genes injected into the oocyte. The order and cellular location of these events coincided with events in yeast[103, 104].

RNAs synthesized under the direction of injected SV40 DNA were processed and their products translated into authentic protein products[29, 112, 113]. The intervening sequences appeared to be removed from *Drosophila* ADH mRNA[50]. This implies the existence of enzymes required for removal of intervening sequences and splicing of RNAs into functional transcripts. The ability of the *Xenopus* oocyte to perform the proper processing steps on diverse heterologous RNAs suggests that the mechanisms involved in these processes (i.e. mRNA splicing and modification) are widespread and highly conserved.

Identification of regulatory regions controlling transcription

4S tRNA genes

Repetitive genes coding for 4S RNA from various organisms have been isolated and well characterized with regard to their nucleotide composition. The use of cloning techniques and procedures for site-specific cleavage and ligation permitted the production of DNA sequences containing either complete coding regions or deletions of specific portions of these genes. These *in vitro* mutations have been used to analyze the role of specific sequences in the regulation of 4S genes by microinjection of manipulated DNA fragments into the oocyte and subsequent analysis of their transcriptional products.

Kressmann et al.[87, 88] microinjected genes coding for *Xenopus laevis* methionine tRNA into the nuclei of *Xenopus* oocytes. They demonstrated that 70% of the newly synthesized RNA in the injected oocytes was specific for this gene. Although these repetitive genes were organized as clusters *in vivo*, each individual gene was transcribed as a separate transcriptional unit. By coinjecting α-amanitin they demonstrated that the DNA was being transcribed under the direction of the proper RNA polymerase (polymerase III).

Telford et al.[124] and Kressmann et al.[89] used *in vitro* mutagenesis to analyze the regulatory regions of the *Xenopus* methionine tRNA. They cleaved a DNA fragment shown to be transcriptionally active into a 5' anterior and 3' posterior portion with restriction endonuclease treatment. The resultant DNA fragments were cloned and several million copies injected into *Xenopus* oocytes. Neither the 5' nor 3' portion alone was able to stimulate RNA transcription. They suggested that the promotion of transcription of *Xenopus* methionine tRNA gene required the presence of both the anterior and posterior portions of the gene. The anterior portion was the site of initiation of transcription while the posterior portion served a role such as a binding site for a specific regulatory protein. Recently, it was found that the control sequence which influences the rate of methionine tRNA synthesis is located at position 8–13 at the 5' end within the structural gene. A second regulatory sequence is located within a region between nucleotides 51–72[79]. These two regions can be moved apart by inserting pieces of DNA in between without affecting the rate or points of initiation and termination of transcription. Deletions within the region of these two sequences lead to inactivity of the gene.

5S ribosomal RNA genes

The genes coding for the 5S RNA component of ribosomes are repeated several thousand fold in *Xenopus laevis*. In this species there exists a dual 5S gene system consisting of two major sets of genes coding for 5S RNA. The more abundant set (oocyte 5S genes) codes for the oocyte type 5S RNA and is expressed only in the oocyte. The other set (the somatic 5S genes) is expressed in both oocytes and somatic cells. Besides the major oocyte component there are several minor oocyte type 5S RNA genes. The RNA products of each of the major somatic and oocyte gene sets differ by only a few nucleotides[18].

The transcription of isolated 5S genes from *Xenopus laevis* was examined by microinjecting cloned DNA fragments into the nuclei of *Xenopus* oocytes[19, 66]. DNA fragments containing either a single coding region or multiple coding regions produced accurate and faithful transcripts of the proper size and from the correct DNA strand. Co-injection of α-amanitin, an inhibitor of RNA synthesis, demonstrated that the genes were being transcribed by RNA polymerase III which normally transcribes these genes *in vivo*.

Recently, using an *in vitro* transcription system made from a *Xenopus* germinal vesicle extract, Bogenhagen et al.[7] and Sakonju et al.[115] were able to localize the control region of the 5S genes. By producing various deletion mutations in which they removed larger and larger portions of the 5' (upstream) prelude regions they found that transcription was still initiated even after removal of as many as 50 nucleotides into the coding region. Mutants deleted 55 or more nucleotides into the gene synthesized little or no 5S RNA. Further mapping localized the control region between nucleotides 55–83, in the middle of the coding sequence. This region was found to be essential for transcription of the genes. Engelke et al.[43] have isolated a protein from oocytes which is required for transcription of the 5S genes and binds to this internal control region. The belief that regulatory regions are always located at the 5' (upstream) end of the genes is therefore not correct, at least in the case of the 5S genes. In all probability this factor is a general transcription factor since its presence is required for transcription of both oocyte and somatic 5S genes. Therefore it is likely that there are other sequences, as well as other factors, necessary for the developmental regulation of the expression of oocyte and somatic 5S genes.

Genes coding for proteins

Histone genes of several species of sea urchin have essentially the same organization. The five histone genes which are expressed early in development are organized as a unit approximately 6.5 kb in length, which is repeated several hundredfold, while genes expressed later during development (late genes) are not tandemly clustered[24] (see Chapter 16 of this volume). The early genes are located on the same DNA strand and have the order H1, H4, H2B, H3 and H2A. The genes have been almost completely sequenced in two species *Psammechinus miliaris*[77] and *Strongylocentrotus purpuratus*[82, 123]. The availability of cloned DNA fragments containing portions of the complete repeat unit as well as complete nucleotide sequence information has made these genes useful in studies of transcriptional control.

The sea urchin histone genes are different from the 5S and 4S genes previously discussed, since they are transcribed under the direction of RNA polymerase II and their RNA products are translated into proteins. In this section we will consider how sea urchin histone genes and other genes transcribed by RNA polymerase II have been utilized to examine the role of specific nucleotide sequences in the transcriptional regulation of eukaryotic gene expression.

Probst *et al.*[109] and Hentschel *et al.*[76] using *Psammechinus miliaris*, and Etkin and Maxson[48] using *Strongylocentrotus purpuratus* demonstrated that accurate and functional transcripts were synthesized in recipient *Xenopus* oocytes following the injection of cloned fragments of the early sea urchin histone genes. Along with these apparently normal transcripts there were also a significant number of inaccurate transcripts. These included transcripts produced by inaccurate initiation and termination, improper strand selection[109] and 'read through' products of the eukaryotic portion of the injected plasmid[48]. Probst *et al.*[109] demonstrated that the inaccurate transcripts were limited to a nuclear location and were not detectable in the cytoplasm. Hentschel *et al.*[76] measured the levels of synthesis of the individual histone mRNAs produced from microinjected *P. miliaris* histone genes and found large quantitative variations in the levels of synthesis (or possible stability) of individual mRNAs. They observed that H2A and H2B mRNAs were very abundant, H1 showed intermediate concentration, and the H3 and H4 transcripts were very rare. This difference was due to differential rates of promotion in the case of H1 and H4, or defective termination in the case of the H3 gene. Recently, however, Stunnenberg and Birnstiel[120] were able to correct the defective termination of the sea urchin H3 gene transcription in *Xenopus* oocytes by coinjection of a chromosomal salt wash fraction from sea urchin embryos. The H3 late genes from *L. pictus* are transcribed *in vivo* utilizing a series of heterogeneous initiation sites[24]. A recircularized DNA sequence containing the *L. pictus* late H3 gene produces the same heterogeneous transcripts following injection into *Xenopus* oocytes (Childs and Etkin, unpublished observations).

Histone genes of both *S. purpuratus* and *P. miliaris* were sequenced and several highly conserved DNA sequences were found in the 5' upstream prelude or spacer sequences of the five genes[23, 58, 59, 77, 82, 123] (see *Figure 9.4*). These include the

Gene	CCAAT	GATCC	TAT	5'PyCATTCPu3'	spacing	coding
H1	—	28 CGGACGACCCGGG GCCTGCTGGGGCCC	CCTTATATTGAGCGTTGCC GGAATATAACTCGCAACGG	10 TTATTCGTTTTGTTAACC AATAAGCAAACAATTGG GpppAUUCG	23	ATG GCT TAC CGA AUG GCU ala
H2A	39 GACCAATGA CTGGTTACT	5 ATACGGATCCGGC TATGCCTAGGCCG	TGTATAAAAGGAAAGGTT ACATATTTTTCCTTTCCAA	8 CCATTCACAGTATCCAAA GGTAAGTGTCATAGGTTT GpppAUUCACA	62	ATC GCT TAG CGA AUG GCU ala
H2B	19 GACCAATCA CTGGTTAGT	3 ACCAGGATCCCGC TGGTCCTAGGGCG	CATATAAATAGCTGAAAAT GTATATTTATCGACTTTTA	12 TCATTCATCCCGTCACTC AGTAAGTAGGGCAGTGAG GpppAUUCA	39	ATG GCA TAC CGT AUG GCA ala
H3	24 GGACAATAG CCTGTTATC	6–8 CCGATCCCC GGCTAGGGG	GGTATAAAATAGCCACCAAA CCATATTTATCGGTGGTTT	14 CCATTCAAGTCATCGAAC GGTAAGTTCAGTAGCTTG GpppAUUCAAG	54	ATG TCT TAC AGA AUG UCU ser
H4	—	8–10 GATCC CTAGG	TGTAACAATACTCGGTGCA ACATTGTTATGAGCCACGT	14 TCATTCGCTTAGCGTAAT AGTAAGCGAATCGCATTA GpppAUUCGCUU	51	ATG TCA TAC AGT AUG UCA ser

Figure 9.4. Putative regulatory sequences located in the 5' prelude sequence of the sea urchin histone genes of *Strongylocentrotus purpuratus* and *Psammechinus miliaris*. The CCAAT, GATCC and TATA appear to be involved in the regulation of transcription of these genes (see text for discussion).

sequence 5'PyCATTCPu3' which has been identified as the putative cap signal marking the site of addition of the 7-methyl guanosine 5' cap sequence on most eukaryotic messenger RNAs. A second region located further upstream is an AT-rich sequence containing the consensus sequence TATA which has been identified as a putative promotor known as the Hogness-Goldberg box. Between 10–20 nucleotide pairs upstream there is a GC-rich region containing the sequence 5'GATCC3'. The GATCC consensus sequence located at this position may be unique to the histone genes. Further upstream is the consensus sequence CCAAT which has been identified in the 5' spacer sequences of many eukaryotic genes. These sequences or very close variations were identified in the 5' prelude regions of the five histone genes in both species of sea urchin. The H2A gene in *P. miliaris*, in addition to the conserved sequences mentioned above, also contains a 30 nucleotide long DNA block upstream from the GATCC and CCAAT motif. This is referred to as the H2A gene specific conserved sequence[58,59].

Utilizing site specific restriction enzymes and molecular cloning procedures Grosschedl and Birnstiel[58,59] constructed several deletion mutants of the H2A gene. They microinjected the modified DNA sequences into the nuclei of *Xenopus* oocytes. As an internal control the injected fragment of DNA contained an unmanipulated copy of the H2B gene. The deletion of the TATA sequence and the GATCC sequence resulted in the synthesis of a series of mRNAs containing several different initiation sites. Deletion of the cap sequence also resulted in the production of a transcript with a novel initiation site, but the site was the same distance downstream from the TATA as was the wild type mRNA. This suggested that the TATA motif may specify the site of initiation of transcription at a fixed position downstream. Therefore it appears that the presence of several of the conserved consensus sequences are required for specificity and accuracy of the initiation event.

In a further attempt to elucidate the function of the TATA sequence Grosschedl *et al.*[60] constructed a unique gene unit consisting of the sea urchin H2A and H2B protein coding regions and a mutated TATA (TAGA) sequence from the conalbumin gene. This unit was microinjected into the nuclei of *Xenopus* oocytes to determine the effect on the specificity and rate of initiation. Initiation of transcription occurred at a new site but always approximately 26 nucleotides downstream from the TATA or TAGA sequences. The amount of specific transcription was decreased five-fold with the mutant (TAGA) sequence.

Grosschedl and Birnstiel[59] have examined the role of specific upstream spacer sequences in the modulation of expression of the H2A histone gene. Deletion of a large segment of the 5' upstream spacer sequence (from positions −185 to −524 leaving the GATCC and TATA motif intact) reduced initiation of H2A transcription to about 1/15–1/20 of the wild type transcription. When they inverted this sequence they found a four- to five-fold increase in H2A gene expression. It was unexpected that a modulating segment would be located so far upstream from the site of transcriptional initiation. This segment of DNA contains several stretches of A and T as well as several inverted repeats in the region −345 to −184.

Similarly the 3' spacer sequence has been shown to play a role in the accurate termination of H2A gene transcription[6]. Preliminary oocyte injection experiments, using 3' deletion mutants, have highlighted the importance of a 23 bp conserved sequence found at the 3' end of all histone genes so far sequenced (with the exception of the yeast genes). This sequence, 5'AACGGC^C_TCTTTTCAG^G_AGCCACCA3', contains a 16 bp inverted repeat, and the ACCA motif; 3' termini of histone mRNAs localized to date fall within this ACCA region. A 12 bp deletion which removes the repeat elicits read-through of the polymerase into the downstream spacer DNA sequences. Removal of the coding sequence has no effect on termination, but elimination of the downstream spacer does.

Grosschedl and Birnstiel[58, 59] have classified the putative regulatory segments of the H2A histone gene of *P. miliaris* into three groups based on their possible function. Modulator segments are those sequences which control the rate of synthesis of H2A mRNAs and include the segment further upstream (-185 to -524) discussed above. This region is located far upstream from the initiation site and is distal to the four conserved sequences observed on the five sea urchin histone genes (the cap site, TATA, GATCC, and CCAAT). Selector sequences are specificity elements that select a unique or predominant 5' mRNA terminus, and include the TATA box. Finally, the intitiator elements are those which include the cap sequence that is found in the five sea urchin histone genes.

The specific function of the GATCC sequence has not been determined since the deletion of the TATA box region in Grosschedl and Birnstiel's experiments also involved the deletion of the GATCC motif. The only evidence relating to this point are the results of Etkin and Maxson[48] using cloned DNA segments of *S. purpuratus* histone genes. A DNA fragment containing the H1, H4 and H2B coding regions and all adjacent spacer sequences produced authentic transcriptional and translational products following microinjection into the nuclei of *Xenopus* oocytes. This DNA segment contained the three conserved sequences mentioned above. A second DNA fragment, however, which contained the H2B protein coding region and 115 base pairs of the upstream spacer did not produce transcriptional products in *Xenopus* oocytes. This DNA sequence was lacking a portion of the GC-rich region containing the GATCC and CCAAT consensus sequences. This DNA fragment contained a portion of the prokaryotic vector which may have influenced the expression of the H2B histone gene. This result, however, suggests that both the complete GATCC and TATA motif may be required for proper expression of the H2B gene.

In another series of experiments utilizing *in vitro* deletion mutations, McKnight *et al.*[99] analyzed the regulatory signals of the herpes simplex virus thymidine kinase gene. A cloned portion of DNA containing the thymidine kinase gene and 680 nucleotides of the 5' flanking region and 294 nucleotides of the 3' flanking region produced transcripts with accurate 5' and 3' termini following injection into nuclei of *Xenopus* oocytes. The mRNA was also translated into protein. When 10% of the nucleotides were deleted at the 5' end of the gene the microinjected DNA still produced accurate thymidine kinase (*tk*) mRNAs. If deleted to nucleotide 95 upstream from the coding region, the DNA directed synthesis of RNA at only 10% of the amount of *tk* mRNA synthesized by the complete DNA fragment even though the CCAAT and TATA sequences were intact. Deletion of nucleotides 85–32 resulted in a decrease of mRNA synthesis to 2% of the original level, while deletions retaining 18 or fewer nucleotides upstream from the coding region produced RNA with incorrect 5' termini. They found that the site of transcriptional initiation of the herpes simplex virus *tk* gene was specified by sequences residing 60–100 bp upstream from the gene. DNA sequencing data showed the presence of a variation of the consensus sequence TATAA (TATTAA) located 24 nucleotides upstream from the coding region and a variation of the consensus sequence CCAAT (CGAAT) located 80 nucleotides from the coding region. The region containing the TATTAA sequence was not essential for accurate initiation, but did enhance the *amount* of accurate initiation of transcription of the *tk* gene. Deletions lacking the CGAAT sequence showed reduced amounts of transcription.

The use of *in vitro* deletion mutants and the *Xenopus* oocyte as a functional assay system have demonstrated that the regulation of transcription of several eukaryotic genes is complex and probably involves the coordinated interaction of a number of specific DNA sequences. The evidence suggests that the TATA or Hogness-Goldberg box is probably an element determining specificity of initiation. The function of the CCAAT and the GATCC (which may be unique to histone genes) still remains unclear, but may involve quantitative aspects of the initiation process. Another consideration is the fact that there are many TATA or CCAAT sequences scattered throughout any given DNA sequence. It is perhaps important, therefore, to consider the spatial association between these sequences relative to one another as a means of regulating gene expression. This is made clear by Grosschedl and Birnstiel's analysis[59] of the H2A specific control region located many hundreds of nucleotides upstream from the sea urchin histone H2A gene in *P. miliaris*. As was pointed out previously, removal of this region decreased the rate of transcription of the H2A gene, but inversion of this sequence produced an increase in the rate of transcription. It is quite possible that this region may affect the higher order structure of the DNA (chromatin), thus influencing the relationship of the sequences downstream to one another.

Recently, Jones *et al.*[81a] showed that co-injection of a plasmid containing the gene coding for chloramphenicol acetyl transferase (CAT) linked to the adenovirus early promotor with a plasmid containing the gene coding for adenovirus E1A protein (a positive regulatory protein) produced a 5–10 fold stimulation in the expression of the CAT gene. This demonstrates that an exogenous factor can influence the rate of expression of a specific gene in the *Xenopus* oocyte system.

Fate and expression of cloned genes injected into eggs

The studies discussed so far involved the microinjection of purified segments of DNA into amphibian oocytes in order to assess the function of specific nucleotide sequences in transcription as well as to examine the transport and processing of RNA. An intriguing question concerns the nature of the mechanisms involved in the regulation of gene expression during development and in the differentiation of specific cell types. One approach to this question is the microinjection of purified genes into the developing embryo where stage- and cell-specific expression of these genes can be examined. This should include the injection of genes coding for ubiquitous products found in all cell types as well as genes coding for cell-specific products. It may then be possible to examine the fate and expression of these genes during development and in specific organs and tissues.

Experiments of this nature have been carried out in amphibians, mammals and insect species. In the amphibia cloned sequences of DNA were replicated and expressed during early development following microinjection into the fertilized egg of *Xenopus laevis*. These experiments were performed by microinjecting cloned genes for sea urchin histones[49, 50], rabbit β-globin[114], or *Drosophila* alcohol dehydrogenease[50] into fertilized eggs of *Xenopus laevis* within one hour after artificial insemination. The DNA was extracted at specific stages during development (i.e. immediately following injection, early cleavage, blastula, gastrula, etc.). Hybridization with the appropriate ^{32}P nick translated probe to Southern blots of the DNA showed an increase in the number of gene copies during these early stages. The number of DNA sequences declined after the gastrula stage of development; however, sequences were detected even in adult frogs[114]. A fraction of the DNA which hybridized with the probe was found migrating with the high molecular weight cellular DNA. The majority, however, migrated as the original plasmid. Treatment of isolated DNA with various restriction enzymes revealed changes in the pattern of restriction sites in the microinjected DNA[49, 50]. The presence of injected DNA sequences in the high molecular weight fraction and the change in restriction pattern for some of the DNA suggested that some of the heterologous sequences may be integrated into the frog genome. These exogenous DNA sequences can also be transmitted to clones of embryos by nuclear transplantation[49].

During the early cleavage stages transcripts from injected sea urchin histone genes and *Drosophila* alcohol dehydrogenase (ADH) genes were not detectable[50]. At the late blastula stage, however, discrete transcripts which co-migrated with authentic sea urchin histone and *Drosophila* ADH were detected. Sea urchin histone proteins were detected at the late blastula–early gastrula stage. This corresponded to the time of appearance of accurate transcriptional products from the injected genes. Bendig[5] observed sea urchin histone gene transcripts at the blastula stage which contained proper 5' and 3' termini, while Rusconi and Schaffner[114] observed the appearance of rabbit β-globin gene transcripts with proper 5' and 3' termini at this stage.

Newport and Kirschner[106a] have observed that genes for yeast tRNA were expressed at the mid-blastula stage in injected *Xenopus* embryos. The time of expression appears to be dependent upon the ratio of DNA to cytoplasm. They postulate the existence of a factor which inhibits transcription until the factor is titrated out by increasing amounts of DNA. Etkin (unpublished observations) found that the expression of the prokaryotic gene coding for chloramphenicol acetyl transferase (CAT) is expressed at the mid-blastula stage regardless of the type of promotor linked to the gene or if the gene was injected as a linear or circular molecule.

Experiments performed in the mouse demonstrated that genes for human β-globin, viral thymidine kinase[17, 107a, 132, 133], rabbit β-globin[26, 133], simian virus 40, herpes simplex virus[56] and Moloney leukemia virus[73] were replicated, incorporated into the mouse genome, and transcribed following microinjection into the pronuclei or cytoplasm of fertilized mouse eggs. There is also evidence that the DNA is transmitted to the next generation through the germ line[26, 55, 133]. In several cases translational products of the injected genes were detected[132, 133]. These results show that microinjected genes survive, replicate, are transcribed, and in some cases the products translated during development of both *Xenopus laevis* and mouse embryos. Recently, transformation was successfully performed in the fruit fly, *Drosophila melanogaster*[111, 119]. These investigators microinjected exogenous sequences of DNA inserted into the P transposable element. They observed that transformation occurred in 20–50% of the germ line cells. When the chromosomal DNA fragment containing the wild type rosy gene was inserted into the P transposable element and injected into mutant hosts, the recipient flies exhibited the wild type phenotype demonstrating that the injected gene was functional and its product corrected the mutant phenotype.

One of the primary goals of gene injection experiments centers around the analysis of cell- and tissue-specific expression of individual genes. To date studies have revealed that microinjected genes do not appear to be expressed in accordance with the developmental or tissue-specific pattern of expression of their endogenous counterparts. This is not surprising considering the fact that a relatively small piece of DNA is being introduced into an alien environment, far removed from the influence of its own neighboring genetic components. One of the possible reasons for this lack of tissue specificity in the expression of these genes, therefore, may be due to their abnormal location (site of integration) in the host genome. Evidence for this comes from experiments in which cloned retroviral genomes were microinjected into mouse zygotes[73]. These investigators observed that microinjected Moloney leukemia virus which was integrated at a chromosomal site different from the site of integration of normally infected mice was expressed in a different tissue-specific manner. This suggests that the expression of a developmentally regulated gene may be dependent upon its location in the chromosome. In most of the microinjection experiments in both amphibians and mice the injected genes are integrated at different sites in individual embryos and therefore do not appear to be expressed in a normal developmental or tissue-specific manner.

Acknowledgements

We thank N. J. Hoffner and Dr M. A. Handel for constructive comments on the manuscript. The experimental work from the author's laboratory reviewed in this article has been supported by research grants from the National Science Foundation (GB-41838 and previous ones), the National Institutes of Health (GM 23635) to M.D.B. and the National Science Foundation (PCM 80 23077), American Cancer Society and National Institutes of Health (GM 31479-01) to L.D. E.

References

1. Aimar, C. and Delarue, M. (1980). Changes in somatic nuclei exposed to meiotic stimulation in amphibian oocytes. *Biologie Cellulaire*, **38**, 37–42
2. Bakken, A., Morgan, G., Sollner-Webb, B., Roon, J., Busby, S. and Reeder, R. (1982). Mapping of transcription initiation and termination signals in *X. laevis* ribosomal DNA. *Proceedings of the National Academy of Sciences, USA*, **79**, 56–60
3. Balakier, H. and Czolowska, R. (1977). Cytoplasmic control of nuclear maturation in mouse oocytes. *Experimental Cell Research*, **110**, 466–469
4. Bataillon, E. and Tchou-Su (1934). L'analyse experimentale de la fecondation et sa definition par les processus cinetiques. *Annales des Sciences Naturelles, Zoologie*, **17**, 9–36
5. Bendig, M. (1981). Histone genes injected into *Xenopus* eggs persist and are expressed during early development. *Nature, London*, **292**, 65–67
6. Birchmeier, C., Grosschedl, R. and Birnstiel, M. L. (1982). Generation of authentic 3′ termini of an H2A mRNA *in vivo* is dependent on a short inverted DNA repeat and on spacer sequences. *Cell*, **28**, 739–745
7. Bogenhagen, D. F., Sakonju, S. and Brown, D. D. (1980). A control region in the center of the 5s RNA gene directs specific initiation of transcription II-3′ border of the region. *Cell*, **19**, 27–35
8. Brachet, A. (1922). Recherches sur la fecondation prematuree de l'oeuf d'oursin (*Paracentrotus lividus*). *Archives de Biologie, Paris*, **32**, 205–244
9. Briggs, R. (1979). Genetics of cell type determination. *International Review of Cytology, Supplement*, **9**, 107–127

10. Briggs, R. and King, T. J. (1952). Transplantation of living nuclei from blastula cells into enucleated frog's eggs. *Proceedings of the National Academy of Sciences, USA*, **38**, 455–457
11. Briggs, R. and King, T. J. (1953). Factors affecting the transplantability of nuclei of frog embryonic cells. *Journal of Experimental Zoology*, **122**, 485–506
12. Briggs, R. and King, T. J. (1959). Nucleocytoplasmic interactions in eggs and embryos. In *The Cell* (Brachet, J. and Mirsky, A. E., Eds.), Volume 1, pp. 527–617. New York, Academic Press
13. Briggs, R. and King, T. J. (1960). Nuclear transplantation studies on the early gastrula (*Rana pipiens*). *Developmental Biology*, **2**, 252–270
14. Briggs, R., Green E. U. and King, T. J. (1951). An investigation of the capacity for cleavage and differentiation in *Rana pipiens* eggs lacking 'functional' chromosomes. *Journal of Experimental Zoology*, **116**, 455–500
15. Briggs, R., King, T. J. and DiBerardino, M. A. (1961). Development of nuclear-transplant embryos. In *Symposium on the Germ Cells and Earliest Stages of Development* (Ranzi, S., Ed.), pp. 441–477. Instituto Lombardo Fond. A. Baselli, Milan
16. Briggs, R., Signoret, J. and Humphrey, R. R. (1964). Transplantation of nuclei of various cell types from neurulae of the Mexican axolotl (*Ambystoma mexicanum*). *Developmental Biology*, **10**, 233–246
17. Brinster, R. L., Chen, H. Y., Trumbauer, M., Senear, A. W., Warren, R. and Palmiter, R. D. (1981). Somatic expression of herpes thymidine kinase in mice following injection of a fusion gene into eggs. *Cell*, **27**, 223–231
18. Brown, D. D. (1979). Developmental genetics by gene isolation: the dual 5S RNA gene system in *Xenopus*. In *Mechanisms of Cell Change* (Ebert, J. D. and Okada, T. S., Eds.), pp. 65–70. New York, Wiley
19. Brown, D. and Gurdon, J. B. (1977). High fidelity transcription of 5S DNA injected into *Xenopus* oocytes. *Proceedings of the National Academy of Sciences, USA*, **74**, 2064–2068
20. Brown, D. and Gurdon J. B. (1978). Cloned single repeating units of 5S DNA direct accurate transcription of 5S RNA when injected into *Xenopus* oocytes. *Proceedings of the National Academy of Sciences, USA*, **75**, 2849–2853
21. Brun, R. (1974). The response of *Xenopus laevis* oocytes to injected somatic nuclei soon after germinal vesicle breakdown induced *in vitro*. *Experimental Cell Research*, **88**, 444–447
22. Brun, R. B. (1978). Developmental capacities of *Xenopus eggs*, provided with erythrocyte or erythroblast nuclei from adults. *Developmental Biology*, **65**, 271–284
23. Busslinger, M., Portmann, R., Irminger, J. C. and Birnstiel, M. L. (1979). Ubiquitous and gene specific regulatory 5′ sequences in a sea urchin DNA clone coding for histone protein variants. *Nucleic Acids Research*, **8**, 957–977
24. Childs, G., Nocente-McCrath, C., Lieber, T., Holt, C. and Knowles, J. (1982). Sea urchin late stage histone H3 and H4 genes: characterization and mapping of a clustered but nontandemly linked multigene family. *Cell*, **31**, 383–395
25. Colman, A. (1975). Transcription of DNAs of known sequence after injection into the eggs and oocytes of *Xenopus laevis*. *European Journal of Biochemistry*, **57**, 85–96
26. Constantini, F. and Lacy, E. (1981). Introduction of a rabbit β-globin gene into the mouse germ line. *Nature, London*, **294**, 92–94
26a. Cortese, A., Harland, R. and Melton, D. (1980). Transcription of tRNA gene *in vivo*: Single stranded compared to double stranded templates. *Proceedings of the National Academy of Sciences, USA*, **77**, 4147–4151
27. Danielli, J. F. and DiBerardino, M. A. (Eds.) (1979). Nuclear transplantation. *International Review of Cytology, Supplement*, **9**
28. DeRobertis, E. and J. B. Gurdon (1977). Gene activation in somatic nuclei after injection into amphibian oocytes. *Proceedings of the National Academy of Sciences, USA*, **74**, 2470–2474
29. DeRobertis, E. and Mertz, J. (1977). Coupled transcription–translation of DNA injected into *Xenopus* oocytes. *Cell*, **12**, 175–182
30. DeRobertis, E. and Olson, M. V. (1979). Transcription and processing of cloned yeast tyrosine tRNA genes microinjected into frog oocytes. *Nature, London*, **278**, 137–143
31. DeRobertis, E. M., Partington, G. A., Longthorne, R. and Gurdon, J. B. (1977). Somatic nuclei in amphibian oocytes: Evidence for selective gene expression. *Journal of Embryology and Experimental Morphology*, **40**, 199–214
32. DiBerardino, M. A. (1979). Nuclear and chromosomal behavior in amphibian nuclear transplants. *International Review of Cytology, Supplement*, **9**, 129–160
33. DiBerardino, M. A. (1980). Genetic stability and modulation of metazoan nuclei transplanted into eggs and oocytes. *Differentiation*, **17**, 13–30

34. DiBerardino, M. A. and Hoffner, N. (1969). Chromosome studies of primary renal carcinoma from Vermont *Rana pipiens*. In *Biology of Amphibian Tumors* (Mizell, M., Ed.), pp. 261–278. New York, Springer-Verlag
35. DiBerardino, M. A. and Hoffner, N. (1970). Origin of chromosomal abnormalities in nuclear transplants – A reevaluation of nuclear differentiation and nuclear equivalence in amphibians. *Developmental Biology*, **23**, 185–209
36. DiBerardino, M. A. and Hoffner, N. J. (1971). Development and chromosomal constitution of nuclear-transplants derived from male germ cells. *Journal of Experimental Zoology*, **176**, 61–72
37. DiBerardino, M. A. and Hoffner, N. J. (1980). The current status of cloning and nuclear reprogramming in amphibian eggs. In *Differentiation and Neoplasia*, Series: Results and Problems in Cell Differentiation Vol. II. (McKinnell, R. G. *et al.*, Eds.,) pp. 53–64. Berlin, Springer-Verlag
38. DiBerardino, M. A. and Hoffner, N. J. (1983). Gene reactivation in erythrocytes: Nuclear transplantation in oocytes and eggs of *Rana pipiens*. *Science,* **219**, 862–864
39. DiBerardino, M. A. and King, T. J. (1965). Transplantation of nuclei from the frog renal adenocarcinoma II. Chromosomal and histologic analysis of tumor nuclear-transplant embryos. *Developmental Biology*, **11**, 217–242
40. DiBerardino, M. A., King, T. J. and McKinnell, R. G. (1963). Chromosome studies of a frog adenocarcinoma line carried by serial intraocular transplantation. *Journal of the National Cancer Institute*, **31**, 769–789
41. DiBerardino, M. A., Mizell, M., Hoffner, N. J. and Friesendorf, D. G. (1982). Frog larvae cloned from nuclei of pronephric adenocarcinoma. *Differentiation*, **23**, 213–217
42. Dixon, G. H. (1972). The basic proteins of trout testis chromatin: aspects of their synthesis, post-synthetic modification and binding to DNA. *Acta Endocrinologica*, **168**, 130–154
43. Engelke, D. R., Sun-Yu, Ng., Shastery, B. S. and Roeder, R. G. (1980). Specific interaction of a purified transcription factor with an internal control region of 5S genes. *Cell*, **19**, 717–728
44. Etkin, L. D. (1976). Regulation of lactate dehydrogenase (LDH) and alcohol dehydrogenase (ADH) synthesis in liver nuclei, following their transfer into oocytes. *Developmental Biology*, **52**, 201–209
45. Etkin, L. D. (1978). Approaches to problems of gene regulation and cell type determination in amphibians. *American Zoologist*, **18**, 215–225
46. Etkin, L. D. (1981). Developmental regulation of sea urchin histone genes in *Xenopus* eggs and oocytes. *Journal of Supramolecular Structure and Cell Biochemistry Supplement*, **5**, 428
47. Etkin, L. D. (1982). Analysis of the mechanisms involved in gene regulation and cell differentiation by microinjection of purified genes and somatic cell nuclei into amphibian oocytes and eggs. *Differentiation*, **21**, 149–159
48. Etkin, L. D. and Maxson, R. E. Jr. (1980). The synthesis of authentic sea urchin transcriptional and translational products by sea urchin histone genes injected into *Xenopus laevis* oocytes. *Developmental Biology*, **75**, 13–25
49. Etkin, L. D. and Roberts, M. (1983). Transmission of integrated sea urchin histone genes by nuclear transplantation in *Xenopus laevis*. *Science*, **221**, 67–69
50. Etkin, L. D., Pearman, B. Roberts, M. and Bektesh, S. L. (1983). Replication, integration and expression of exogenous DNA injected into fertilized eggs of *Xenopus laevis*. Submitted for publication.
51. Fischberg, M., Gurdon, J. B. and Elsdale, T. R. (1958). Nuclear transplantation in *Xenopus laevis*. *Nature, London*, **181**, 424
52. Gallien, L. (1966). La greffe nucleaire chez les amphibians. *Annales de Biologie*, **5–6**, 241–269
53. Gallien, L. (1974). Signification des caryopathies induites dans la transplantation nucleaire chez les amphibians. *Année Biologie*, **13**, 285–292
54. Gasaryan, K. G., Hung, N. M., Neyfakh, A. A. and Ivanenkov, V. V. (1979). Nuclear transplantation in teleost *Misgurnus fossilis*. L. *Nature, London*, **280**, 585–587
55. Gordon, J. W. and Ruddle, F. H. (1981). Integraton and stable germ line transmission of genes injected into mouse pronuclei. *Science*, **214**, 1244–1246
56. Gordon, J. W., Scangus, G. A, Plotkin, D. J., Barbosa, J. A. and Ruddle, F. H. (1980). Genetic transformation of mouse embryos by microinjection of purified DNA. *Proceedings of the National Academy of Sciences, USA*, **77**, 7380–7384
57. Graham, C. F., Arms, K. and Gurdon, J. B. (1966). The induction of DNA synthesis by frog egg cytoplasm. *Developmental Biology*, **14**, 349–381
58. Grosschedl, R. and Birnstiel, M. L. (1980). Identification of regulatory sequences in the prelude sequences of an H2A histone gene by the study of specific deletion mutants *in vivo*. *Proceedings of the National Academy of Sciences, USA*, **77**, 1432–1436

59. Grosschedl, R. and Birnstiel, M. L. (1980). Spacer DNA sequences upstream of the TATAAATA sequence are essential for promotion of H2A histone gene transcription *in vivo*. *Proceedings of the National Academy of Sciences, USA*, **77**, 7102–7106

60. Grosschedl, R., Wasylyk, B., Chambon, P. and Birnstiel, M. L. (1981). Point mutation in the TATA box curtails expression of sea urchin H2A histone gene *in vivo*. *Nature, London*, **294**, 178–180

61. Gurdon, J. B. (1962). Adult frogs derived from the nuclei of single somatic cells. *Developmental Biology*, **4**, 256–273

62. Gurdon, J. B. (1964). The transplantation of living cell nuclei. *Advances in Morphogenesis*, **4**, 1–43

63. Gurdon, J. B. (1967). On the origin and persistence of a cytoplasmic state inducing nuclear DNA synthesis in frogs eggs. *Proceedings of the National Academy of Sciences, USA*, **58**, 545–552

64. Gurdon, J. B. (1968). Changes in somatic cell nuclei inserted into growing and maturing amphibian oocytes. *Journal of Embryology and Experimental Morphology*, **20**, 401–414

65. Gurdon, J. B. (1974). *The Control of Gene Expression in Animal Development*. Cambridge, Mass., Harvard University Press

66. Gurdon, J. B. and D. D. Brown, (1978). The transcription of 5S DNA injected into *Xenopus* oocytes. *Developmental Biology*, **67**, 346–356

67. Gurdon, J. B. and R. A. Laskey (1970). The transplantation of nuclei from single cultured cells into enucleate frogs' eggs. *Journal of Embryology and Experimental Morphology*, **24**, 227–248

68. Gurdon, J. B. and Melton, D. A. (1981). Gene transfer in amphibian eggs and oocytes. *Annual Review of Genetics*, **15**, 189–218

69. Gurdon, J. B. and Uehlinger, V. (1966). 'Fertile' intestine nuclei. *Nature, London*, **210**, 1240–1241

70. Gurdon, J. B., DeRobertis, E. M. and Partington, G. (1976). Injected nuclei in frog oocytes provide a living cell system for the study of transcriptional control. *Nature, London*, **260**, 116–120

71. Gurdon, J. B., Laskey, R. A., and Reeves, O. R. (1975). The developmental capacity of nuclei transplanted from keratinized cells of adult frogs. *Journal of Embryology and Experimental Morphology*, **34**, 93–112

72. Gurdon, J. B., Laskey, R. A., DeRobertis E. M. and Partington G. A. (1979). Reprogramming of transplanted nuclei in amphibia. *International Review of Cytology, Supplement*, **9**, 161–178

73. Harbers, K., Jahner, D. and Jaenisch, R. (1981). Microinjection of cloned retroviral genomes into mouse zygotes: integration and expression in the animal. *Nature, London*, **293**, 540–542

73a. Harland, R. M., Weintraub, H. and McKnight, S. L. (1983). Transcription of DNA injected into *Xenopus* oocytes is influenced by template topology. *Nature, London*, **302**, 38–43

74. Havris, H. (1967). The reactivation of the red cell nucleus. *Journal of Cell Science*, **2**, 23–32

75. Hennen, S. (1970). Influence of spermine and reduced temperature on the ability of transplanted nuclei to promote normal development in eggs of *Rana pipiens*. *Proceedings of the National Academy of Sciences, USA*, **66**, 630–637

76. Hentschel, C., Probst, E., Birnstiel, M. L. (1980). Transcriptional fidelity of histone genes injected into *Xenopus* oocyte nuclei. *Nature, London*, **288**, 100–102

77. Hentschel, C. C. and Birnstiel, M. L. (1981). The organization and expression of histone gene families. *Cell*, **25**, 301–313

78. Hoffner, N. J. and DiBerardino, M. A. (1980). Developmental potential of somatic nuclei transplanted into meiotic oocytes of *Rana pipiens*. *Science*, **209**, 517–519

79. Hofstetter, H., Kressmann, A. and Birnstiel, M. L. (1981). A split promoter for a eukaryotic tRNA gene. *Cell*, **24**, 573–585

80. Illmensee, K. (1976). Nuclear and cytoplasmic transplantation in *Drosophila*. In *The 8th Symposium of the Royal Entomological Society of London* (Lawrence, P. H., Ed.), pp. 76–96. Oxford, Blackwell Scientific Publications

81. Illmensee, K. and Hoppe, P. C. (1981). Nuclear transplantation in *Mus musculus*: Developmental potential of nuclei from preimplantation embryos. *Cell*, **23**, 9–18

81a. Jones, N. C., Richter, J. D., Weeks, D. L. and Smith, L. D. (1983). Regulation of adenovirus transcription by E1A in microinjected *Xenopus* oocytes. Submitted for publication

82. Kedes, L. H. (1979). Histones and histone messengers. *Annual Review of Biochemistry*, **48**, 837–870

83. King, T. J. (1966). Nuclear transplantation in amphibia. *Methods in Cell Physiology*, **2**, 1–36

84. King, T. J. and M. A. DiBerardino (1965). Transplantation of nuclei from the frog renal adenocarcinoma. I. Development of tumor nuclear-transplant embryos. *Annals of the New York Academy of Sciences*, **126**, 115–126

85. Kobel, H. R., Brun, R. B. and Fischberg, M. (1973). Nuclear transplantation with melanophores, ciliated epidermal cells, and the established cell line A-8 in *Xenopus laevis*. *Journal of Embryology and Experimental Morphology*, **29**, 539–547

86. Korn, L. J. and Gurdon, J.B. (1981). The reactivation of developmentally inert 5S genes in somatic nuclei injected into *Xenopus* oocytes. *Nature, London*, **289**, 461–465

87. Kressmann, A., Clarkson, S. G., Pirrotta, V. and Birnstiel, M. L. (1978). Transcription of cloned tRNA gene fragments and subfragments injected into the oocyte nucleus of *Xenopus laevis*. *Proceedings of the National Academy of Sciences, USA*, **75**, 1176–1180

88. Kressmann, A., Clarkson, S. G., Telford, J. L. and Birnstiel, M. L. (1978). Transcription of *Xenopus* tDNA met and sea urchin histone DNA injected into *Xenopus* oocyte nucleus. *Cold Spring Harbor Symposium on Quantitative Biology*, **42**, 1077–1082

89. Kressmann, A., Hofstetter, H., DiCapua, E., Grosschedl, R. and Birnstiel, M. L. (1979). A tRNA gene of *Xenopus laevis* contains at least two sites promoting transcription. *Nucleic Acids Research*, **7**, 1749–1763

90. Ladner, M., Chan, L. and O'Malley, B. W. (1979). Transcription of a cloned ovalbumin ds-cDNA in *Xenopus laevis* oocytes. *Biochemical and Biophysical Research Communications*, **86**, 1227–1229

91. Laskey, R. A. and Gurdon, J. B. (1970). Genetic content of adult somatic cells tested by nuclear transplantation from cultured cells. *Nature, London*, **228**, 1332–1334

92. Laskey, R. A., Honda, B., Mills, A. D., Morris, N. R., Wyllie, A., Mertz, J. E., DeRobertis, R. M. and Gurdon, J. B. (1977). Chromatin reconstitution and transcription of DNA in *Xenopus* eggs and oocytes. *Cold Spring Harbor Symposium on Quantitative Biology*, **42**, 171–178

93. Leonard, R. A., Hoffner, N. J. and DiBerardino, M. A. (1982). Induction of DNA synthesis in amphibian erythroid nuclei in *Rana* eggs following conditioning in meiotic oocytes. *Developmental Biology*, **92**, 343–355

94. McAvoy, J. W., Dixon, K. E. and Marshall, J. A. (1975). Effects of differences in mitotic activity, stage of cell cycle, and degree of specialization of donor cells on nuclear transplanation in *Xenopus laevis*. *Developmental Biology*, **45**, 330–339

94a. McGrath, J. and Solter, D. (1983). Nuclear transplantation in the mouse embryo by microsurgery and cell fusion. *Science*, **220**, 1300–1302

95. McKinnell, R. G. (1962). Intraspecific nuclear transplantation in frogs. *Journal of Heredity*, **53**, 199–207

96. McKinnell, R. G. (1978). *Cloning, Nuclear transplantation in Amphibia*. Minneapolis, University of Minnesota Press

97. McKinnell, R. G., Deggins, B. A. and Labat, D. D. (1969). Transplantation of pluripotential nuclei from triploid frog tumors. *Science*, **165**, 394–396

98. McKinnell, R. G., Steven, L. M. Jr. and Labat, D. D. (1976). Frog renal tumors are composed of stroma vascular elements and epithelial cells: What type nucleus programs for tadpoles with the cloning procedure? In *Progress in Differentiation Research* (Muller-Berat, N., Ed.), pp. 319–330. Amsterdam, North-Holland

99. McKnight, S. L., Gavis, E. R., Kingsbury, R. and Axel, R. (1981). Analysis of transcriptional regulatory signals of the HSV thymidine kinase gene: Identification of an upstream control region. *Cell*, **25**, 385–398

100. Marshall, J. A. and Dixon, K. E. (1977). Nuclear transplantation from intestinal epithelial cells of early and late *Xenopus laevis* tadpoles. *Journal of Embryology and Experimental Morphology*, **40**, 167–174

101. Masui, Y. and Markert, C. L. (1971). Cytoplasmic control of nuclear behavior during meiotic maturation of frog oocytes. *Journal of Experimental Zoology*, **177**, 129–146

102. Matsui, S.-I., Hiroshi, Y., Weinfeld, H. and Sandberg, A. A. (1972). Induction of prophase in interphase nuclei by fusion with metaphase cells. *Journal of Cell Biology*, **54**, 120–132

103. Melton, D. A. and Cortese, R. (1979). Transcription of cloned tRNA genes and the nuclear partitioning of a tRNA precursor. *Cell*, **18**, 1165–1172

104. Melton, D. A., DeRobertis, E. M. and Cortese, R. (1980). Order and intracellular location of the events involved in the maturation of a spliced tRNA. *Nature, London*, **284**, 143–148

104a. Mertz, J. (1982) Linear DNA does not form chromatin containing regularly spaced nucleosomes. *Molecular and Cellular Biology*, **2**, 1608–1618

105. Mertz, J. and Gurdon, J. B. (1977). Purified DNAs are transcribed after microinjection into *Xenopus* oocytes. *Proceedings of the National Academy of Sciences, USA*, **74**, 1502–1506

106. Moriya, M. and Katagiri, Ch. (1976). Microinjection of toad sperm into oocytes undergoing maturation division. *Development Growth and Differentiation*, **18**, 349–356

106a. Newport, J. and Kirschner, M. (1982) A major developmental transition in early *Xenopus* embryos. II. Control of the onset of transcription. *Cell*, **30**, 687–696

107. O'Farrell, P. (1975). High resolution two-dimensional electrophoresis of proteins. *Journal of Biological Chemistry*, **250**, 4007

107a. Palmiter, R. D., Chen, H. Y. and Brinster, R. L. (1982) Differential regulation of metallothionein–thymidine kinase fusion genes in transgenic mice and their offspring. *Cell*, **29**, 701–710

108. Poccia, D., Salik, J. and Krystal, G. (1981). Transitions in histone variants of the male pronucleus following fertilization and evidence for a maternal store of cleavage-stage histones in the sea urchin egg. *Developmental Biology*, **82**, 287–296

109. Probst, E., Kressmann, A. and Birnstiel, M. L. (1979). Expression of sea urchin histone genes in the oocyte of *Xenopus laevis*. *Journal of Molecular Biology*, **135**, 709–732

110. Rao, P. N. and Johnson, R. T. (1970). Mammalian cell fusion: Studies on the regulation of DNA synthesis and mitosis. *Nature, London*, **225**, 159–164

111. Rubin, G. and Spradling, A. (1982). Genetic transformation of *Drosophila* with transposable element vectors. *Science*, **218**, 348–353

112. Runnger, D. and Turler, H. (1978). RNAs of simian virus 40 and polyoma direct the synthesis of viral tumor antigens and capsid proteins in *Xenopus* oocytes. *Proceedings of the National Academy of Sciences, USA*, **75**, 6073–6077

113. Runnger, D., Matthias, P. D. and Huber, J. P. (1981). Transcription of complex structural genes in the *Xenopus* oocyte system. In *International Cell Biology 1980–1981* (Schweiger, H. G., Ed.), pp. 28–32

114. Rusconi, S. and Schaffner, W. (1981). Transformation of frog embryos with rabbit β-globin gene. *Proceedings of the National Academy of Sciences, USA*, **78**, 5051–5055

115. Sakonju, S., Bogenhagen, D. F. and Brown, D. D. (1980). A control region in the center of the 5S RNA gene directs specific initiation of transcription: 1. The 5′ border of the region. *Cell*, **19**, 13–25

116. Skoblina, M. N. (1976). Role of karyoplasm in the emergence of capacity of egg cytoplasm to induce DNA synthesis in transplanted sperm nuclei. *Journal of Embryology and Experimental Morphology*, **36**, 67–72

117. Smith, L. D. (1965). Transplantation of nuclei of primordial germ cells into enucleated eggs of *Rana pipiens*. *Proceedings of the National Academy of Sciences, USA*, **54**, 101–107

118. Spemann, H. (1938). *Embryonic Development and Induction*, p. 211. New Haven, Yale University Press

119. Spradling, A. and Rubin, G. (1982). Transposition of cloned P elements into *Drosophila* germ line chromosomes. *Science*, **218**, 341–347

120. Stunnenberg, H. G. and Birnstiel, M. L. (1982). Bioassay for components regulating eukaryotic gene expression: A chromosomal factor involved in the generation of histone mRNA 3′ termini. *Proceedings of the National Academy of Sciences, USA*, **79**, 6201–6204

121. Subtelny, S. (1968). Cytoplasmic influence of immature frog oocytes on nuclear behavior. *American Society of Cell Biology*, **39**, 131a

122. Subtelny, S. and Bradt, C. (1963). Cytological observations on the early developmental stages of activated *Rana pipiens* eggs receiving a transplanted blastula nucleus. *Journal of Morphology*, **112**, 45–59

123. Sures, I., Levy, S. and Kedes, L. (1980). Leader sequences of *Strongylocentrotus purpuratus* histone mRNAs start at a unique heptanucleotide common to all five histone genes. *Proceedings of the National Academy of Sciences, USA*, **77**, 1265–1269

124. Telford, J. L., Kressmann, A., Koski, R. A., Grosschedl, R., Muller, F., Clarkson, S. G. and Birnstiel, M. L. (1979). Delimitation of a promoter for RNA polymerase III by means of a functional test. *Proceedings of the National Academy of Sciences, USA*, **76**, 2590–2594

125. Trendelenburg, M. and Gurdon, J. B. (1978). Transcription of cloned *Xenopus* ribosomal genes visualized after injection into oocyte nuclei. *Nature, London*, **276**, 292–294

126. Trendelenburg, M. F., Mathis, D. and Oudet, P. (1980). Transcription units of chicken ovalbumin genes observed after injection of cloned complete genes into *Xenopus* oocyte nuclei. *Proceedings of the National Academy of Sciences, USA*, **77**, 5984–5988

127. Trendelenburg, M. H., Zentgraf, H., Franke, W. W. and Gurdon, J. B. (1978). Transcription patterns of amplified *Dytiscus* genes coding for ribosomal RNA after injection into *Xenopus* oocyte nuclei. *Proceedings of the National Academy of Sciences, USA*, **75**, 3791–3795

128. Tung, T. C., Wu, S. C., Tung, Y. Y. F., Yen, S. S. and Lu, T. Y. (1965). Nuclear transplantation in fishes. *Scientia Sinica*, **14**, 1244–1245

129. Tung, T. C., Wu, S. C., Yeh, Y. F., Li, K. S. and Hsu, M. C. (1977). Cell differentiation in ascidian studied by nuclear transplantation. *Scientia Sinica*, **20**, 222–233

130. Voellmy, R. and Runnger, D. (1982). Transcription of a *Drosophila* heat shock gene is heat induced in *Xenopus* oocytes. *Proceedings of the National Academy of Sciences, USA*, **79**, 1776–1780

131. Wabl, M. R., Brun, R. B. and Du Pasquier, L. (1975). Lymphocytes of the toad *Xenopus laevis* have the gene set for promoting tadpole development. *Science*, **190**, 1310–1312
132. Wagner, E. F., Stewart, T. A. and Mintz, B. (1981). The human β-globin gene and a functional viral thymidine kinase gene in developing mice. *Proceedings of the National Academy of Sciences, USA*, **78**, 5016–5020
133. Wagner, T. E., Hoppe, P. C., Jollick, J. D., Schull, D. R., Hudinka, R. L. and Gault, J. B. (1981). Microinjection of rabbit β-globin gene into zygotes and its subsequent expression in adult mice and their offspring.*Proceedings of the National Academy of Sciences, USA*, **78**, 6376
134. Wickens, M. P., Woo, S., O'Malley, B. W. and Gurdon, J. B. (1980). Expression of a chicken chromosomal ovalbumin gene injected into frog oocyte nuclei. *Nature, London*, **225**, 628–634
135. Wyllie, A. H., Gurdon, J. B. and Price, J. (1977). Nuclear localization of an oocyte component required for the stability of injected DNA. *Nature, London*, **268**, 150–152
136. Wyllie, A. H., Laskey, R. A., Finch, J. and Gurdon, J. B. (1978). Selective DNA conservation and chromatin assembly after injection of SV-40 DNA into *Xenopus* oocytes. *Developmental Biology*, **64**, 178–188
137. Zentgraf, H., Trendelenburg, M. F., Spring, H., Scheer, U. Franke, W. W., Muller, U., Drury, K. and Runnger, D. (1979). Mitochondrial DNA arranged into chromatin-like structures after injection into amphibian oocyte nuclei. *Experimental Cell Research*, **122**, 363–375
138. Ziegler, D. and Masui, Y. (1973). Control of chromosome behavior in amphibian oocytes. I. The activity of maturing oocytes inducing chromosome condensation in transplanted brain nuclei. *Developmental Biology*, **35**, 283–292
139. Ziegler, D. and Masui, Y. (1976). Control of chromosome behavior in amphibian oocytes. II. The effects of inhibitors of RNA and protein synthesis on the induction of chromosome condensation in transplanted brain nuclei by oocyte cytoplasm. *Journal of Cell Biology*, **68**, 620–628

Chapter 10

Gene activity in isolated nuclei

N. Maclean, S. P. Gregory and M. J. Pocklington

Introduction

The cloning of eukaryotic genes and the development of soluble cell-free systems
for their transcription has enabled DNA sequences, and in one case a specific
protein, to be identified as necessary for the correct transcription of certain
genes[70, 71]. However, these systems give very limited information about precise
sequence-specific regulation, since gene-specific transcription is not readily
reproduced in these situations. The difference in the transcription rates of different
genes, and of the same gene in different cell types, appears to be at least partly
determined by differences in the chromatin structure of the genes (reviewed in
reference 72).

It may eventually become possible, using chromatin proteins, to reconstruct
chromatin assembly events which lead to authentic differences in the transcription
of purified genes. Such systems would probably require defined histone variants
and modifications; nucleosome assembly factors[41]; modifying enzymes[5]; DNA of a
known methylation state; and DNA of a defined superhelical density, and probably
of a length comprising a whole chromatin 'domain'. Until the assembly of defined
chromatin structures becomes possible, we must learn about how chromatin
proteins affect transcription by investigating the structure of isolated chromatin and
nuclei from living cells with known transcriptional characteristics, and by observing
how experimental modifications of chromatin structure affect transcription.

In this chapter the transcriptional activity of isolated nuclei which appear to
maintain cell-specific patterns of transcription in cell-free systems, will be
discussed. The related topics of gene activity in nuclei after nuclear transplantation
and the activity of novel genes injected into nuclei have been covered in Chapters 8
and 9. Transcription from specific gene sequences in soluble systems using cloned
DNA sequences is mentioned in a number of other chapters.

Not only does the nucleosome organization impose on DNA some restriction to
its free transcription by RNA polymerases, but it is likely that the higher order
structure of chromatin also profoundly affects the transcriptional activity of the
DNA involved. Since some crucial aspects of higher order structure may be lost by
the isolation of chromatin from nuclei, it is likely that the transcriptional activity of
intact isolated nuclei more nearly reflects the *in vivo* state than does the activity of
isolated chromatin: studies on such activity may therefore be used to throw light on
mechanisms of gene regulation operative within living cells *in vivo*.

In this chapter it is proposed to discuss first some of the constraints and difficulties inherent in studying transcription in nuclear isolates, then to review a number of different experimental systems in which the transcriptional activity of isolated nuclei has been studied.

Studies on transcription with isolated nuclei are basically complex experimental systems. The nuclei must be held in an ionic environment which retains the nuclear envelope and internal chromatin in a relatively authentic condition and which also permits activity of the RNA polymerase enzymes. Magnesium and manganese ion concentrations are particularly critical[71]. It is also important to establish that the nuclear membrane is permeable to the additives such as polymerase and labelled nucleotides which may be needed, although in most nuclear isolates loss of nuclear constituents following exposure of nuclei to detergents during isolation is even more likely to pose problems.

Any of the eukaryotic nuclear RNA polymerase classes, or the *E. coli* RNA polymerase, can be used to transcribe isolated nuclei. Unfortunately, the eukaryotic class II RNA polymerase which transcribes pre-messenger RNA sequences and is therefore the most interesting as far as the control of transcription is concerned, will not initiate correctly unless an extract containing soluble factors from active cells is provided. It appears to be difficult to obtain authentic transcription by adding such an extract to isolated nuclei, because of its uncharacterized nature and the likelihood of non-specific effects. The authentic transcription of isolated nuclei by polymerase III, however, seems to be a possibility.

Although *E. coli* RNA polymerase does not give authentic transcription on eukaryotic chromatin, its very tendency to transcribe from exposed DNA, even if normal promoters are not present on that DNA, implies that this enzyme can be used as an indicator of 'DNA exposure'. A number of the experimental systems in this chapter have already exploited the *E. coli* enzyme to reveal specific changes in chromatin resulting from interactions with reactivating factors.

Perhaps the most serious problems are presented by the activity of nuclease enzymes whether within the nuclei and perhaps calcium activated, or introduced in cytoplasmic extracts used in the experimental protocol. Nucleases are likely to be responsible for rapid reduction in molecular weight of new RNA transcripts and for the introduction of nicks in the DNA template which may subsequently serve as pseudo-promotors for RNA polymerases.

Assuming that transcription is obtained and that adequate precautions have been taken to eliminate the possibility of spurious RNA synthesis, there yet remains the difficulty of identifying the transcripts, both as products of particular DNA sequences, and also as being of a size which would satisfy the demand for something more than short initiation transcripts or the completion of transcripts already in train before the experiment began.

As will become clear in the remaining sections of this chapter, it is now possible to come close to satisfying the demands of a really rigorous protocol so excluding the difficulties outlined above and also by the use of probes constructed from cloned genes, to identify transcripts in an unambiguous manner. It can be fairly claimed that the study of gene regulation by the use of isolated nuclei is now practicable, if not actually easy.

What are the special attractions of experimental systems involving isolated nuclei? If the procedures are complex and the results often ambiguous, why use such systems? The answer is surely that these experiments do offer the unique opportunity to expose genes in nuclei to purified regulatory molecules, RNA, protein, or whatever, thus providing an assay for factors suspected of acting as eukaryotic gene regulators.

Experimental systems

Rat liver nuclei exposed to rat liver cytoplasm

There are numerous reports in the literature of the last decade in which general RNA synthesis of isolated nuclei is recorded without particular regard to the sequence specificity of the transcripts. For example, Dvorkin et al.[22] have concluded that cytoplasm from rat hepatoma cells induces a reduction in the rate of RNA synthesis in nuclei isolated from normal rat liver. Bastian[2] has sought to compare the rate of RNA synthesis in isolated rat liver nuclei following exposure to cytoplasm from either regenerating or normal liver cells. Regenerating rat liver cytosol proved effective in enhancing the rate of RNA synthesis and a thermolabile protein was tentatively identified as the reactivating factor. Further work[3,4] has revealed changes in the relative transcription from repetitive and unique sequences in the rat liver nuclei, but the activity of specific unique sequences was not measured.

Responses of nuclei to low molecular weight RNA

RNA and especially low molecular weight RNA is a probable candidate for a role in gene regulation[20]. In the light of information now available about the many RNA sequences which are purely nuclear (see discussion in Chapter 25 of this volume) it is obviously sensible to consider some RNA as having a role in genetic control mechanisms. Prestayko and Busch[55] suggested that low molecular weight RNA might well have such a regulatory role and Lebedyeva and Platinov[43] have recorded some stimulation of transcription in rat liver nuclei following their exposure to fractions of nuclear RNA. But Kozlov et al.[38,39] have failed to find any stimulation of transcription of either RNA polymerase I or II, following exposure of rat liver nuclei to low molecular weight nuclear RNA.

Heat shock genes in *Drosophila*

It has become apparent over the last decade that most organisms are endowed with gene sequences which become specifically activated by heat shock. Presumably such genes code for proteins which help the organisms cope with the altered biochemical demands of elevated temperature.

It was established in 1962 by Ritossa[5] (see Ashburner and Bonner[1] for review) that cells from *Drosophila* larvae responded to a temperature elevation of 12 °C above the normal optimum of 25 °C by dramatic and fairly sudden changes in both protein and RNA synthesis. Much of the novel protein produced is the product of only a few gene sequences and these have come to be known as 'heat shock' genes. Heat shocked larvae also display a unique pattern of puffing in the polytene chromosomes of their salivary glands and at least some of these puffs are now known to represent activation of chromatin in the region of the heat shock genes.

Investigation and manipulation of the *Drosophila* heat shock situation has resulted in what is perhaps the most precise and convincing example of regulatory molecules being used to evoke a specific genetic response in isolated nuclei. Compton and Bonner[15] developed an *in vitro* assay by isolating nuclei from the salivary glands of *Drosophila* larvae and incubating their nuclei *in vitro*. It is known that if such nuclei are themselves exposed to heat shock, no activation of heat shock genes is detectable[6] but Compton and McCarthy[16] demonstrated that extracts of cytoplasm from heat shocked *Drosophila* tissue culture cells would evoke puffing of heat shock loci in isolated polytene nuclei exposed to such extracts. It therefore seemed that molecules were present in the cytoplasm of heat shocked cells which would enter nuclei and elicit specific genetic regulation within the chromatin. The system has been further exploited by Craine and Kornberg[19], and these authors have partially purified and characterized the cytoplasmic factor, proving it to be both heat labile and protease sensitive. They have further shown that while less than 0.01% of RNA synthesized by nuclei exposed to cytoplasm from non-heat shock cells hybridized to a heat shock gene probe (plasmid 123 E3 containing the gene for the 70 kD heat shock protein), between 1 and 2% of the RNA synthesized by nuclei exposed to heat shock cytoplasm will hybridize with the probe. Two further observations of Craine and Kornberg[19] deserve mention. The first is that, when the isolated nuclei were treated with heat shock factor but permitted only to transcribe with endogenous RNA polymerase, the RNA produced was predominantly homologous to the structural gene sequence. On the other hand, when exogenous *E. coli* derived RNA polymerase was provided, the RNA product was not homologous to the heat shock gene loci but hybridized primarily to sequences a short distance upstream from the 5' end of the heat shock loci. These loci only became accessible following activation of nuclei with heat shock factor, and did not depend on either presence or activity of histone H1 or endogenous RNA polymerase II. It therefore seems that the bacterial polymerase is revealing a specific chromatin decondensation event which extends upstream of the heat shock genes when these genes are activated by heat shock factor. This area of decondensation may be equivalent to DNase I-hypersensitive sites revealed in other systems. Two observations on this interesting series of experiments are appropriate. One is that the use of isolated nuclei *in vitro* has in this instance permitted partial purification of a protein factor which acts like a gene regulatory molecule. The second is that although bacterial RNA polymerase may not yield authentic transcription products when added to eukaryotic cell nuclei, the very tendency of the enzyme to initiate from non-coding sequences which become exposed (and the enzyme presumbly does not transcribe from non-exposed sequences in chromatin) permits its use as a tool to reveal specific changes in chromatin conformation and DNA accessibility which, though outside the structural gene seqence, may be significant in the process of gene activation in eukaryotic cells.

The reactivation of transcription in *Xenopus* erythrocyte nuclei

Isolated *Xenopus* erythrocyte nuclei have a very low transcriptional activity with *E. coli* RNA polymerase or homologous polymerases, but this can be increased by preincubating the nuclei in soluble extracts (S100s) from active cells (see *Figure 10.1*). This observation has formed the basis for studies on the control of transcription in these nuclei[29, 46].

    Two lines of investigation have been followed. Firstly, it was thought that both general and specific changes in chromatin structure resulting from incubation of the

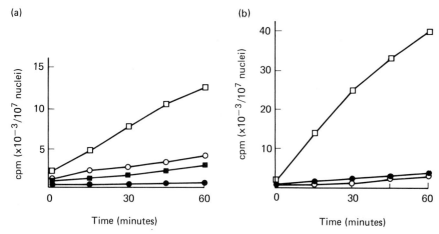

Figure 10.1. (a) Transcription of erythrocyte and erythroblast isolated nuclei with and without added *E. coli* RNA polymerase in NSM at 25°C. 2 U/ml *E. coli* RNA polymerase; 2×10^8 nuclei/ml; 600 μM ATP, CTP and GTP; 25 μM (5–^{3}H) UTP with a specific activity of 2 Ci/mmol; NSM was 70 mmol/ℓ KCl; 7 mmol/ℓ MgCl$_2$; 25 mmol/ℓ Tris-HCl pH 7.9; 10% v/v glycerol; 0.1 mmol/ℓ DTT. Erythrocytes were obtained by ventricular puncture to obtain the blood, followed by purification of the erythrocytes in a Ficoll gradient. The nuclei were prepared by lysing the red cells using 0.03% v/v Triton X100 in NSM, followed by extensive washing (5×1 ml per 10^8 nuclei) in NSM. Erythrocyte nuclei, ●; erythrocyte nuclei + *E. coli* RNA polymerase, ○; erythroblast nuclei, ■; erythroblast nuclei + *E. coli* RNA polymerase, □. (b) Transcription of isolated erythrocyte nuclei in the presence of *E. coli* RNA polymerase, following preincubation of the nuclei for 15 minutes in erythroblast S100 (□); erythrocyte S100 (●) or control (preincubated in NSM: ○). The nuclei were extensively washed prior to, and following preincubation, with 5×1 ml NSM per 10^8 nuclei.

nuclei in active cell S100s would be revealed by characterizing the changes in the transcription of sequences by *E. coli* RNA polymerase. Since it is known that *E. coli* RNA polymerase transcribes, in isolated chromatin and nuclei, DNA sequences which are transcribed *in vivo* at a greater frequency than those which are not[21], *E. coli* RNA polymerase can be used as a probe of chromatin structure. These studies were complemented by the use of DNase I and other DNases to probe for changes in chromatin structure resulting from the S100 treatment.

The second line of investigation which stemmed from the original reactivation observations was the attempted development of a fully homologous transcription system, involving isolated nuclei from *Xenopus* erythroid cells and purified RNA polymerase. Such a system would not have been feasible for the authentic transcription of genes by polymerase II, for the reasons already stated.

Changes in the specific sequence transcription by E. coli *RNA polymerase of* Xenopus *erythrocyte nuclei following 'reactivation' of the nuclei*

Given that the rate of total transcription in isolated erythrocyte nuclei by *E. coli* RNA polymerase changes after a brief treatment with heterologous S100 extracts, it is of great interest to determine whether particular gene sequences are activated. If this were so, this system could form the basis for the detection and purification of factors which cause gene-specific transcription by modifying the chromatin structure.

It is known that under certain conditions the *E. coli* polymerase will synthesize complementary RNA using an RNA template in media containing manganese[8, 53] and probably also involving RNA as a primer[60]. This problem can be overcome by omitting manganese from the transcription medium. Also, DNA dependent transcription is 80% inhibited by 10 µg/ml actinomycin D, whereas RNA dependent transcription is not, so the inclusion of 10 µg/ml actinomycin D in a reaction provides a suitable control[36, 56, 66].

The transcription of erythrocyte nuclei with *E. coli* RNA polymerase reveals that, following incubation with erythroblast S100, RNA hybridizing to a globin cDNA plasmid is produced in greater quantities than in nuclei which are not reactivated, while there is no detectable change in the amount of RNA transcribed from a tRNA gene cluster. This result suggests that the tRNA gene cluster of *Xenopus* may have a different chromatin structure from that of polymerase II transcribed genes, thus preventing its detectable transcription by *E. coli* RNA polymerase.

RNA sequences hybridizing to a vitellogenin cDNA clone (pXlvc 10) were also produced following reactivation and transcription of the erythrocyte nuclei with *E. coli* RNA polymerase. However, RNA hybridizing to a different vitellogenin cDNA clone (pXlvc 23) was not detected. This means that either the two vitellogenin genes have a different chromatin structure in erythrocyte nuclei, or pXlvc 10 contained a sequence which was present elsewhere in the *Xenopus* genome, and which was transcriptionally reactivated by the erythroblast S100. In short, evidence is found for the presence of transcripts in reactivated nuclei which will hybridize to a globin gene probe, one of two vitellogenin gene probes, but none which will hybidize to a tRNA gene probe.

For the analyses of changes in chromatin structure using the *E. coli* RNA polymerase, more information is required about what determines whether transcription by the polymerase takes place. It is certain that the *E. coli* RNA polymerase preferentially transcribes nucleosome-free regions in chromatin and this may explain the results of Craine and Kornberg[19]. *E. coli* RNA polymerase probably also transcribes the regions which are moderately DNase I sensitive, and which are probably associated with the non-histone proteins HMG14 and 17[31,36,56].

Other investigators have shown that *E. coli* RNA polymerase does not initiate at single sites around eukaryotic genes but indeed initiates at many sites on both strands of the DNA[42]. Therefore it is not possible to perform S1 mapping of the transcripts[7] which would localize initiation to a unique site.

Specific sequence digestion by DNase I after preincubation of erythrocyte nuclei with erythroblast S100

Since *E. coli* RNA polymerase cannot be used to give accurate information about chromatin structure, we have probed the structure of transcriptionally reactivated erythrocyte nuclei by digesting the nuclei to different extents with DNase I. This was followed by purification of the DNA, electrophoresis, blotting onto nitrocellulose, hybridization with nick-translated plasmid DNA and autoradiography. There was no visible difference in the overall digestion rates of control and preincubated nuclei by DNase I, as revealed by the appearance of ethidium bromide stained DNA smears. However, there were differences in the digestion of specific genes by the DNases, in the differently treated nuclei.

No difference in the digestion rate, or pattern of hypersensitive sites, was detected in the β-globin gene region. However, a slight difference in the digestion rate of the vitellogenin B gene was apparent following preincubation. In nuclei which had been treated with erythroblast S100, the hybridizing bands were almost completely removed by the action of 2 and 5 μg/ml DNase I on the preincubated nuclei, but not in the control nuclei. No hypersensitive sites were detected by the hybridization probe (pXlvc 10)[78].

Hybridization of a nick-translated plasmid which contained a *Xenopus* genomic tRNA gene cluster (pXt210)[14] to restriction enzyme-cleaved DNA prepared from DNase I-digested, S100-preincubated nuclei was also carried out to determine whether any changes in chromatin structure had been produced by the S100 treatment in the region of these tRNA genes.

1 2   3 4   5 6   7   8   9 10 11 12 13 14 15 16

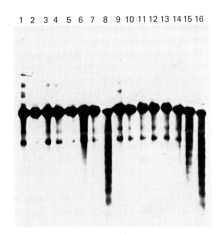

Figure 10.2. Hybridization of nick-translated (^{32}P-) pXt210 (tRNA gene cluster) to Hind III-cleaved, electrophoresed, filter-transferred DNA from treated erythrocyte nuclei. Lanes 1–8 nuclei preincubated in NSM; lanes 9–16 nuclei preincubated in whole ovary S100; DNase I concentrations: Lanes 1 and 9, 0 μg/ml; 2 and 10, 0.05 μg/ml; 3 and 11, 0.1 μg/ml; 4 and 12, 0.25 μg/ml; 5 and 13, 0.5 μg/ml; 6 and 14, 1 μg/ml; 7 and 15, 2.5 μg/ml; 8 and 16, 5 μg/ml.

The autoradiograph in *Figure 10.2* shows that there was a slight difference between the control and ovary S100-treated nuclei in terms of their digestibility by DNase I. Since lane 15 clearly shows pXt210-hybridizing sequences in a nucleosomal ladder while lane 7 does not, it appears that the tRNA sequences were digested more readily in nuclei which had been preincubated with ovary S100. No single DNase I-sensitive site within the repeated tRNA cluster was recognized since no sub-bands were generated.

Figure 10.3 shows an autoradiograph of Eco Rl digested RNA fron nuclei treated with NSM or erythroblast S100, followed by DNase I digestion and probed with nick-translated pXt210 as above. The Eco Rl digestion was partial which explains the appearance of the 'ladder'. The ladder corresponds to the monomer to 5 (or more) -mer units of the 3.18 kb genomic tRNA gene repeat sequence. It can clearly be seen that although the digestion of NSM-preincubated nuclei is what would be expected from the results above in *Figure 10.2*, the equivalently digested erythroblast S100-treated nuclei did not give the same pattern of digestion products. Lane 6 shows a smear of short hybridizing fragments rather than a nucleosomal ladder, and also a specific band at 4.5 kb, halfway between the monomer and dimer bands.

Clearly, a change in chromatin structure has been produced by the erythroblast S100 treatment on the erythrocyte nuclei. A DNase sensitive region appears to have been produced almost exactly half way between the Eco R1 sites which define

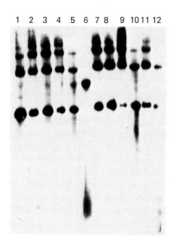

Figure 10.3. Hybridization of nick-translated pXt210 to partially Eco R1 cleaved, electrophoresed, filter-immobilized DNA from treated erythrocyte nuclei. Lanes 1–6 nuclei preincubated in erythroblast S100; lanes 7–12 nuclei preincubated in NSM; DNase I concentrations: Lanes 1 and 7, 0 µg/ml; 2 and 8, 0.25 µg/ml; 3 and 9, 0.5 µg/ml; 4 and 10, 1 µg/ml; 5 and 11, 2.5 µg/ml; 6 and 12, 5 µg/ml.

the repeating unit. Also, the nucleosomal structure of the region was altered by the erythroblast S100 treatment, since the nucleosomal ladder disappeared.

The arrangement of nucleosomes on tRNA genes has previously been shown to be random in active cells, but phased in erythrocytes[11]. A possible nucleosome-free region was detected in erythrocytes, in the region of the tRNA gene; however, this is not in a position which would account for the specific band detected in *Figure 10.3* above. Also, the possible nucleosome-free region was not detected by DNase I digestion of erythrocyte nuclei in the previous studies. Therefore, the putative nucleosome-free region detected in the previous work does not appear to correspond with the site of DNase I sensitivity produced by incubation of the nuclei in erythroblast S100. A DNase sensitive site produced by preincubation of the nuclei in erythroblast S100 was also detected in this work by a Ca^{2+} activated 'linker specific' nuclease (not shown). This was probably the same site as the one revealed by DNase I in the results above (Pocklington *et al.*, in preparation).

It is clear from the results presented above that alterations in the transcription of isolated nuclei can be produced by incubating them with S100 fractions from active cells. The sequence specificity of such alterations in transcriptional capacity were identified by transcription with *E. coli* RNA polymerase and digestion by DNase I. However, these results have not revealed whether the changes resemble authentic changes in chromatin structure. The reactivation of the globin and vitellogenin genes for transcription by *E. coli* RNA polymerase appear to be sequence-specific since the tRNA genes and another vitellogenin gene were not detectably activated, but it is difficult to reconcile the sequences which are reactivated with the known pattern of transcription in erythroblasts. The lack of detectable transcription of the multiple copy tRNA genes by the *E. coli* RNA polymerase, even after the S100 treatment which altered the structure of the tRNA gene chromatin, suggests that the *E. coli* RNA polymerase recognizes different aspects of chromatin structure than those revealed by DNase I digestion. One possible reason for this is that certain sequences are required for initiation by *E. coli* RNA polymerase which are not present in the DNase sensitive regions. Alternatively, the tRNA genes may assume a particular chromatin configuration which is refractory to transcription by *E. coli* RNA polymerase, whatever the DNase digestibility.

By extending the methods of analysis presented in this section, it may be possible to identify chromatin modifying activities from active cells which lead to the sequence-specific restructuring of chromatin. If changes in the transcriptional specificity of homologous RNA polymerases are also to be correlated with changes in chromatin structure, cell-free systems for the transcription of chromatin will need to be developed, and these are discussed briefly in the following section.

Intact chromatin systems for the cell-free transcription of specific genes

Originally it had been hoped that authentic transcription using homologous RNA polymerases could be established in suitably preincubated *Xenopus* erythrocyte nuclei, and this would have been the first step in the detection of factors which control transcription. Since specific soluble factors are required for the initiation of authentic transcription by RNA polymerase II, however, the transcription of isolated nuclei by this enzyme is not possible due to the lack of RNA polymerase II preparations which initiate correctly but which do not otherwise affect the nuclei. On the other hand we considered that authentic transcriptional reactivation of RNA polymerase III transcribed genes might be detectable in this system. Polymerase III transcribed genes are present in multiple copies in the genome and would be easy to detect because of their small size and well characterized nature. Efficient reinitiation of polymerase III transcribed genes has been detected in cell-free systems using isolated chromatin and RNA polymerase III. If this is a general property of polymerase III transcribed genes, the preincubation of isolated nuclei in appropriate S100 may lead to the authentic transcription of genes by polymerase III, and such an approach could result in the purification of the factors responsible for this specificity. At present, however, results in our laboratory have shown that isolated erythrocyte nuclei cannot be 'reactivated' for polymerase III gene transcription (5S RNA or tRNA) in a cell-free system using oocyte S100.

The work of other investigators has recently shown that 5S RNA gene chromatin assembled in the presence of 5S RNA transcription factor (called TF IIIA by R. G. Roeder and his colleagues[23]) is properly transcribed, while 5S RNA gene chromatin assembled in the absence of the factor is not transcribed, even when the transcription factor is added later [9,28]. This result is best understood in terms of proteins binding stably to the DNA. The transcription factor cannot replace nucleosomes after they have already formed on the 5S RNA genes. On the other hand, the assembly of nucleosomes on 5S genes will not dislodge the transcription factor if it is already bound to the controlling region. Therefore the activity of these genes appears to be determined by events which take place during chromatin construction[73].

If this is the case, it is unlikely that changes in gene expression can be brought about when transcription 'factors' are simply added to isolated nuclei. However, chromatin structures may be less stable in intact nuclei because the concentration of nucleoplasmin and other nucleoplasmic components might be expected to allow a more dynamic chromatin structure than is produced in the cell-free extracts used in the assembly of 5S RNA gene chromatin described above. Differences in the concentration of these components in the oocyte germinal vesicle of different female *Xenopus* individuals may explain why only some individuals yield oocytes which can reactivate the oocyte-type 5S RNA gene transcription in somatic cell nuclei when injected into oocyte germinal vesicles[37].

In the same study, the rearrangement of chromatin structure brought about by high salt (350 mmol/ℓ NaCl) treatment was sufficient to cause the reactivation of the oocyte-type 5S RNA gene in most isolated somatic cell nuclei when they were injected into oocyte germinal vesicles. This supports the idea that a moderate level of chromatin disruption which does not result in the loss of histones is required for the subsequent assembly, under the appropriate conditions, of transcriptionally active chromatin. A similar disruption of the chromatin mediated by nucleoplasmin or other nucleosome assembly/disassembly factors, may occur in intact cells.

Conclusions

Certain conclusions of general import and application can be drawn from these experiments with isolated *Xenopus* erythrocyte nuclei. They are:

1. Preparations of isolated nuclei from these and other cell types may make a unique contribution to our understanding of the regulation of transcription in eukaryotes. Gene regulatory proteins may be detected and identified by their effects on specific genes.
2. Deoxyribonucleases can be used as probes of chromatin structure following the treatment of isolated nuclei with extracts from other cells. In this way, treatments on factors which cause chromatin structural changes can be identified.
3. Although the *E. coli* RNA polymerase does not transcribe eukaryotic chromatin in a faithful manner, it does have a use in revealing sequences which are available for transcription. However, this information may simply represent the cumulative DNase sensitivity of particular sequences within the nucleus, so it is a matter of conjecture whether *E. coli* RNA polymerase can recognize chromatin structures which DNases, under appropriate conditions, cannot.
4. Terms such as 'open' and 'closed' chromatin conformation should not be used except in a physical context (i.e. when considering light and electron microscopical evidence). The transcription of isolated nuclei and their digestion with DNases, reveals strictly only whether particular sequences are transcribed or digested, not necessarily whether they are in an 'open' conformation or not.

Hormone-mediated transcription

In higher organisms there are a group of genes, known as the hormone-responsive genes, whose expression is under the direct control of the endocrine system. The products from such genes only accumulate in appreciable amounts in a limited number of cell types (target cells) and then only in response to specific hormone stimulation. In several cases this control has been demonstrated to act at the level of gene transcription[26, 40, 58, 63]; exposure of the target tissue to its hormone inducer causes the activation of a novel pattern of transcription specific to that cell type. Recent experiments have largely focused on two separate aspects of this process – the localization of those DNA sequences responsible for the hormonal regulation of specific gene transcription and the identification of other factors that may be required to elicit this response.

The transcription of hormone-responsive genes in isolated nuclei

Studies utilizing isolated nuclei have provided some of the strongest data in support of a direct hormonal effect on the transcription of a number of hormone-responsive genes. Comparisons have been made *in vitro* between the transcriptional activity of nuclei that have been isolated from target cells either before or after treatment with the relevant activating hormone(s). Transcription in such nuclear preparations, at least initially, reflects their activity *in vivo*.

Thus, of the ^{3}H-RNA synthesized by oviduct nuclei isolated from chickens that had been chronically stimulated with the oestrogen diethylstilboestrol, 0.23% hybridized to filters containing immobilized cloned ovalbumin cDNA[63]. An additional 0.17% of the total RNA hybridized to filters containing DNA derived from intervening sequences present within the ovalbumin gene. No ovalbumin gene transcripts were detectable in RNA synthesized by spleen or liver nuclei, or in RNA synthesized by oviduct nuclei isolated from chickens that had been withdrawn from hormone treatment. These results complement those changes already observed in intact oviduct cells following the administration of oestrogen[51] and are consistent with regulation of ovalbumin mRNA production at the transcriptional level. Stabilization of the resultant mRNA is also likely to occur during the period of hormone stimulation[18], but the primary factor in accounting for this accumulation of ovalbumin mRNA sequences is the dramatic increase in transcription of the ovalbumin gene[63].

Hormone-responsive genes like ovalbumin are generally protein coding and thus transcribed by RNA polymerase II. Increases in the activity of this enzyme have been reported for nuclei isolated from various tissues stimulated by steroid and/or peptide hormones. Prostate nuclei from rats show greater polymerase II activity following testosterone treatment[69], as do isolated dog thyroid nuclei after previous exposure of the cells to thyrotropin[35]. However, the definitive evidence in support of a direct hormonal effect on transcription has to come from studies that monitor the expression of a specific hormone-responsive gene. A number of genes in addition to ovalbumin have shown greater activities in nuclei isolated from target cells after hormone treatment than they do in nuclei isolated from unstimulated tissue. These include the casein[26] and mouse mammary tumour virus (MMTV) genes[61], both of which are activated by glucocorticoids, and the vitellogenin genes of chicken liver nuclei which respond to oestradiol treatment[52]. The α_{2u}-globulin genes are also expressed in liver nuclei isolated from male rats, but not in those isolated from female rats[13].

In all these cases it has been necessary to detect a low level of specific transcription amidst a high background of endogenous RNA. The levels of free RNA polymerases in isolated nuclei are generally low[45], so that most of the RNA synthesis one observes is due to the elongation of *in vivo* initiated chains. Prolonged rates of transcription, approaching those encountered *in vivo*, are thus rather infrequent in nuclear systems. Isolated nuclei are more generally used as transient expression systems, to define the pattern of nuclear activity at a particular stage in a cell's life history. In order to determine whether a specific gene sequence is still active in a preparation of isolated nuclei it is necessary to circumvent the problems posed by the bulk of this pre-formed RNA.

Hybridization of RNA radioactively labelled *in vitro* to immobilized cDNA sequences has been used successfully in monitoring transcription of the ovalbumin gene. However, in other systems where the rate of specific transcription is much lower, there may be excessive 'noise' from the rest of the labelled transcripts or a high background of unlabelled transcripts that will distort the hybridization data. To overcome this, it is possible to incorporate a mercurated precursor into the newly synthesized RNA such that this material can be purified away from the endogenous RNA using a thiol-based affinity column. This RNA can then be probed with an isotopically labelled cDNA and the identity of its sequences confirmed by retention of the hybrids on their passage through a second thiol-containing column. The precautions that have to be exercised whilst using mercurated nucleotides have been well documented (see discussions in reference 13 and references quoted in reference 45). Using rigorous experimental protocols it has been possible to demonstrate that these precursors do not affect the preferential transcription of hormone-responsive genes in nuclei which have been isolated from stimulated target cells[25, 48, 50].

Using isolated nuclei to identify hormone receptors/second messengers

Having established that the transcription of a particular gene is under hormonal control, the next step is to identify the causative agents of the gene activation.

For steroid hormones there is a large body of evidence to suggest that their effects are mediated by specific, high-affinity receptor proteins (reviewed in reference 30). These receptors are present in significant amounts only in target cells where they occur in the cytosol and, at least in some cases, on the cell surface. Binding of steroid by its receptor causes a change in protein conformation or enzymic activity, which eventually leads to its translocation into the cell nucleus. Whereas the native receptor is unable to bind to nuclei or DNA, the hormone–receptor complex has a high affinity for selected 'acceptor' sites in the target cell chromatin. Some of these are presumably located at or near those DNA sequences activated by the hormones and it is conceivable that binding at these sites facilitates polymerase entry to the relevant hormone-responsive genes. Cascade effects cannot be ruled out, however, and the mechanisms by which the receptor–hormone complex facilitates transcription remain unknown. The availability of cloned DNA fragments is at present allowing an assessment of the sequence specificity of receptor–complex binding to DNA[17, 49].

In steroid receptor studies, isolated nuclei have been used primarily as assay systems for receptor–complex binding[27]. Identification of the receptor protein is not difficult since it will bind labelled steroid with high affinity. Movement of the receptor–complex can also be followed by simply monitoring the movement of this specific tag. The situation as regards peptide hormones is somewhat more complicated. These hormones exert their effects without the apparent need to enter the cell. Thus, labelled peptide hormones remain extracellular whilst still inducing the activation of specific hormone-responsive genes. In these cases it has been postulated that a second messenger is responsible for conveying the signal for gene activation from the membrane to the nucleus. This messenger must be released into the cytosol as a direct consequence of the interaction between the peptide hormone and the cell membrane. It is thus this factor that is initially responsible for the alterations in cellular activity.

The role of cyclic AMP as a hormone secondary messenger has long been established[62]. However, it is clear that a number of peptide hormones do not function via this route. Isolated nuclei are now providing a means for identifying the nature of these other second messengers. For instance, a low molecular weight factor induced in liver cells following insulin treatment is capable of stimulating transcription in nuclei isolated from untreated cells[32].

More recently, the prolactin-induced activation of β-casein gene transcription has been subjected to this type of examination[65]. When microsomes from lactating rabbit mammary gland were treated with prolactin and subsequently removed by centrifugation, there remained in the supernatant a factor which was capable of stimulating in isolated mammary cell nuclei a five- to ten-fold increase in the rate of β-casein gene transcription. 28S rRNA transcription was not affected. Moreover, other lactogenic hormones could induce the same effect, although not when added directly to the nuclei. The factor stimulating β-casein gene transcription was also released from membranes derived from other tissues containing prolactin receptors, such as those from liver, ovary and brain, but not by those membranes from heart, lung and muscle which are known not to bind prolactin. Treatment of isolated rabbit liver or reticulocyte nuclei caused no stimulation of either β-casein gene transcription or globin gene transcription. These data are suggestive, then, of a specific role for this factor as a mediator of prolactin activity[64]. Preliminary characterization of the nature of this factor suggests that it is thermostable (10 minutes, 100 °C), trypsin sensitive and has a molecular weight of around 1000 daltons[65]. The striking similarity with the putative insulin second messenger[34, 59] may be suggestive of a common mechanism of action for second messengers. An identical route is not employed here, though, since insulin is incapable of releasing the β-casein gene activating factor even though mammary gland membranes do contain insulin receptors.

Localization of those DNA sequences required for hormonal regulation of gene expression

The ability to manipulate DNA sequences *in vitro*, to reintroduce these modified templates into living cells, and then to chart their effect on the cellular phenotype ('reverse genetics') has revolutionized molecular biology. It is now possible to identify those regions of the genome that are implicitly involved in the induction of hormone-responsive gene transcription.

Most extensively studied, to date, has been the mouse mammary tumour virus (MMTV), a single stranded RNA virus that replicates via a double stranded DNA intermediate (provirus). The provirus can integrate at random in the mouse genome and since viral transcription is inducible by glucocortricoid hormones, the signals for hormone regulation presumably reside within its length. This has been confirmed by analysing the rates of transcription of viral and flanking cellular DNA at proviral integration sites following hormone stimulation[67]. Co-transfection of mouse Ltk⁻ cells with a cloned MMTV provirus and the thymidine kinase gene of herpes simplex virus leads to integration of the viral DNA into the genome and the retention of its hormone responsiveness[12, 33]. Crucial to these experiments, of course, is the fact that Ltk⁻ cells contain receptors for glucocorticoid hormones.

Recent experiments using fragments of the MMTV genome suggest that the sequences necessary for hormone inducibility lie near the 5' end of the provirus close to the viral promoter[24]. Transference of this fragment to the 5' end of a dihydrofolate reductase cDNA brings this non-hormone-responsive gene under the regulation of glucocorticoids[44].

16 kb fragments of cloned rat genomic sequence DNA, each containing single copies of $\alpha_{2\mu}$-globulin genes, have similarly been transfected into mouse Ltk⁻ cells and they also retain enough information to respond to dexamethasone treatment[40]. In this case, there are approximately 10 kb of DNA flanking the 5' end of the gene. For human growth hormone (HGH) genes co-transformed into mouse fibroblasts there is again a three- to five-fold induction of HGH mRNA upon glucocorticoid stimulation. This time the sequences required for induction have been localized to

the region within 500 nucleotides 5' of the mRNA start site[58]. Fusion of this segment of flanking DNA to the hormone-insensitive structural gene sequence of thymidine kinase also renders the hybrid gene responsive to glucocorticoid induction[58].

Comparisons can be drawn here with the induction of α-interferon gene transcription on exposure to Newcastle disease virus[47] and the production of mouse metallothionein I in response to cadmium treatment[10]. Indeed, the chicken ovalbumin gene is regulated like a viral gene when placed under the influence of a herpes simplex virus promoter[54].

In conclusion, these experiments serve to emphasize the fact that the primary level of hormone action is at the transcriptional level. Moreover, it is those sequences immediately flanking the 5' end of the gene that are most important in deciding whether a gene is to be transcribed or not and, if so, at what levels.

References

1. Ashburner, M. and Bonner, J. J. (1979). The induction of gene activity in *Drosophila* by heat shock. *Cell*, **17**, 241–254
2. Bastian, C. (1977). Comparative studies of transcription in isolated nuclei. Effect of homologous and of heterologous cytosol. *Biochemical and Biophysical Research Communications*, **74**, 1109–1116
3. Bastian, C. (1978). Comparative studies on molecular hybridization of nuclear RNA synthesized by isolated rat liver nuclei: effect of homologous and heterologous cytosol. *Biochemical and Biophysical Research Communications*, **83**, 893–900
4. Bastian, C. (1980). The effect of cytosol from regenerating rat liver on the *in vitro* RNA synthesis of isolated cell nuclei from a Morris hepatoma: comparative studies on molecular hybridization of nuclear RNA. *Biochemical and Biophysical Research Communications*, **92**, 80–88
5. Belikoff, E, Wong, L.-J. and Alberts, B. M. (1980). Extensive purification of histone acetylase A, the major histone N-acetyl transferase activity detected in mammalian cell nuclei. *Journal of Biological Chemistry*, **255**, 11448–11453
6. Berendes, H. D. and Boyd, S. B. (1969). Structural and functional properties of polytene nuclei isolated from salivary glands of *Drosophila hydei*. *Journal of Cell Biology*, **41**, 591–599
7. Berk, A. J. and Sharp, P. A. (1977). Sizing and mapping of early adenovirus RNA by gel electrophoresis of Sl endonuclease-digested hybrids. *Cell*, **12**, 721–732
8. Bernard, O., Cory, S. and Adams, J. M. (1977). Synthesis of complementary RNA on RNA templates using the DNA-dependent RNA polymerase of *E. coli*. *Biochimica et Biophysica Acta*, **478**, 407–416
9. Bogenhagen, D. F., Wormington, W. M. and Brown, D. D. (1982). Stable transcription complexes of *Xenopus* 5S RNA genes: a means to maintain the differentiated state. *Cell*, **28**, 413–421
10. Brinster, R. L., Chen, H. Y., Warren, R., Sarthy, A. and Palmiter, R. D. (1982). Regulation of Metallothionein–Thymidine Kinase fusion plasmids injected into mouse eggs. *Nature, London*, **296**, 39–42
11. Bryan, P. N., Hofstetter, H. and Birnstiel, M. L. (1981). Nucleosome arrangement on tRNA genes of *Xenopus laevis*. *Cell*, **27**, 459–466
12. Buetti, E. and Diggelmann, H. (1981). Cloned Mouse Mammary Tumor Virus DNA is biologically active in transfected mouse cells and its expression is stimulated by glucocorticoid hormones. *Cell*, **23**, 335–345
13. Chan, K. M., Kurtz, D. T. and Feigelson, P. (1979). Transcription of the α_{2u}-globulin gene in male rat liver nuclei *in vitro*. *Biochemistry*, **17**, 3092–3096
14. Clarkson, S. G., Kurer, V. and Smith, H. O. (1978). Sequence organisation of a cloned tDNA^{met} fragment from *Xenopus laevis*. *Cell*, **14**, 713–724
15. Compton, J. L. and Bonner, J. J. (1978). An *in vitro* assay for the specific induction and regression of puffs in isolated polytene nuclei of *Drosophila*. *Cold Spring Harbor Symposium on Quantitative Biology*, **42**, 835–838
16. Compton, J. L. and McCarthy, B. J. (1978). Induction of the *Drosophila* heat shock response in isolated polytene nuclei. *Cell*, **14**, 191–201

17. Compton, J. G., Schrader, W. T. and O'Malley, B. W. (1982). Selective binding of chicken progesterone receptor A subunit to a DNA fragment containing ovalbumin gene sequences. *Biochemical and Biophysical Research Communications*, **105**, 96–104

18. Cox, R. F. (1977). Estrogen withdrawal in chick oviduct. Selective loss of high abundance classes of polyadenylated messenger RNA. *Biochemistry*, **16**, 3433–3443

19. Craine, B. L. and Kornberg, T. (1981). Activation of the major *Drosophila* heat-shock genes *in vitro*. *Cell*, **25**, 671–681

20. Davidson, E. H. and Britten, R. J. (1973). Organisation, transcription and regulation in the animal genome. *Quarterly Review of Biology*, **48**, 565–613

21. Dobson, M. E. and Ingram, V. M. (1980). *In vitro* transcription of chromatin containing histones hyperacetylated *in vitro*. *Nucleic Acids Research*, **8**, 4201–4219

22. Dvorkin, G. A., Efimova, V. F. and Koganitskaya, L. I. (1973). Effect of cytoplasm on RNA synthesis in isolated nuclei of rat liver cells. *Doklady Akademii nauk SSSR, Series Biology*, **210**, 461–463

23. Engelke, D. R., Ng, S.-Y., Shastry, B. S. and Roeder, R. G. (1980). Specific interaction of a purified transcription factor with an internal control region of 5S genes. *Cell*, **19**, 717–728

24. Fasel, N., Pearson, K., Buetti, E. and Diggelmann, H. (1982). The region of Mouse Mammary Tumour Virus DNA containing the long terminal repeat includes a long coding sequence and signals for hormonally regulated transcription. *EMBO*, **1**, 3–7

25. Ganguly, R. and Banerjee, M. R. (1978). RNA synthesis in isolated nuclei of lactating mammary cells in presence of unmodified and mercury-labelled CTP. *Nucleic Acids Research*, **5**, 4463–4477

26. Ganguly, R., Mehta, N. M., Ganguly, N. and Banerjee, M. R. (1979). Glucocorticoid modulation of casein gene transcription in mouse mammary gland. *Proceedings of the National Academy of Sciences, USA*, **76**, 6466–6470

27. Garroway, N. W., Orth, D. N. and Harrison, R. W. (1976). Binding of cytosol receptor–glucocorticoid complexes by isolated nuclei of glucocorticoid-responsive and non-responsive cultured cells. *Endocrinology*, **98**, 1092–1100

28. Gottesfeld, J. and Bloomer, L. J. (1982). Assembly of transcriptionally active 5S RNA genes chromatin *in vitro*. *Cell*, **28**, 781–791

29. Gregory, S. P., Hilder, V. A. and Maclean, N. (1977). Transcriptional reactivation of isolated *Xenopus* erythrocyte nuclei: patterns of RNA synthesis. *Journal of Cell Science*, **28**, 49–60

30. Grody, W. W., Schrader, W. T. and O'Malley, B. W. (1982). Activation, transformation, and sub-unit structure of steroid hormone receptors. *Endocrine Reviews*, **3**, 141–163

31. Hodd III, H. G., Sahasrabuddhe, C. G., Plishker, M. F. and Saunders, G. F. (1980). Use of RNA polymerase as an enzymatic probe of nucleosomal structure. *Nucleic Acids Research*, **8**, 3851–3864

32. Horvat, A. (1980). Stimulation of RNA synthesis in isolated nuclei by an insulin-induced factor in liver. *Nature, London*, **286**, 906–908

33. Hynes, N. E., Kennedy, N., Rahmsdorf, V. and Groner, B. (1981). Hormone responsive expression of an endogenous proviral gene of mouse mammary tumor virus after molecular cloning and gene transfer into cultured cells. *Proceedings of the National Academy of Sciences, USA*, **78**, 2038–2042

34. Kiechle, F. L., Jarett, L., Kotogal, N. and Popp, D. A. (1981). Partial purification from rat adipocyte plasma membranes of a chemical mediator which stimulates the action of insulin on pyruvate dehydrogenase. *Journal of Biological Chemistry*, **256**, 2945–2951

35. Kleiman De Pisarev, D., Pisarev, M. A. and Spaulding, S. W. (1979). Early effects of thyrotropin on ribonucleic acid transmission in the thyroid. *Endocrinology*, **104**, 693–696

36. Konkel, D. A. and Ingram, V. M. (1978). Is there specific transcription from isolated chromatin? *Nucleic Acids Research*, **5**, 1237–1252

37. Korn, L. J. and Gurdon, J. B. (1981). The reactivation of developmentally inert 5S genes in somatic nuclei injected into *Xenopus* oocytes. *Nature, London*, **289**, 461

38. Kozlov, A. P., Plyusnin, A. Z., Evtushenko, V. I. and Seits, I. F. (1979). Studies of low molecular weight nuclear RNA synthesis in isolated nuclei. *Molecular Biology, Moscow*, **12**, 91–99

39. Kozlov, A. P., Plyusnin, A. Z., Evtushenko, V. I. and Seits, I. F. (1979). Effect of low molecular weight and nuclear RNA on activity of RNA polymerase II in nuclei of rat liver. *Biokhimiya*, **44**, 27–32

40. Kurtz, D. T. (1981). Hormonal inducibility of rat α_{2u}-globulin genes in transfected mouse cells. *Nature, London*, **291**, 629–631

41. Laskey, R. A., Honda, B. M., Mills, A. D. and Finch, J. T. (1978). Nucleosomes are assembled by a specific protein which binds histones and transfers them to DNA. *Nature, London*, **275**, 416–420

42. Lavialle, L., Reuveni, Y., Thoren, M. and Salzman, N. P. (1982). Molecular interaction between simian virus 40 DNA and *Escherichia coli* RNA polymerase. *Journal of Biological Chemistry*, **257**, 1549–1557

43. Lebedyeva, L. M. and Platinov, O. M. (1977). Effect of nuclear RNA from normal and regenerating rat liver on RNA synthesising ability of isolated nuclei. *Dopovidi Akademiyi Nauk USSR*, **7**, 642–644
44. Lee, F., Mulligan, R., Berg, P. and Ringold, G. (1981). Glucocorticoids regulate expression of dihydrofolate reductase cDNA in mouse mammary tumour virus chimaeric plasmids. *Nature, London*, **294**, 228–232
45. Maclean, N. and Gregory, S. P. (1981). Transcription in isolated nuclei. In *The Cell Nucleus*, Volume VIII (Busch, H., Ed.), pp. 139–191. New York, Academic Press
46. Maclean, N. and Hilder, V. A. (1977). Studies on the template activity of 'isolated' *Xenopus* erythrocyte nuclei. II. The effect of cytoplasmic extracts. *Journal of Cell Science*, **24**, 119–129
47. Mantei, N. and Weissmann, C. (1982). Controlled transcription of a human interferon gene introduced into mouse L cells. *Nature, London*, **297**, 128–132
48. Mizuno, S., Tallmann, N. A. and Cox, R. F. (1978). Estrogen withdrawal in chick oviduct: Characterization of RNA synthesized in isolated nuclei using a mercurated precursor. *Biochimica et Biophysica Acta*, **520**, 184–202
49. Mulvihill, E. R., Le Pennec, J.-P. and Chambon, P. (1982). Chicken oviduct progesterone receptor: Location of specific regions of high-affinity binding in cloned DNA fragments of hormone-responsive genes. *Cell*, **28**, 621–632
50. Nguyen-Huu, M. C., Sippel, A. H., Hynes, N. E., Groner, B. and Schutz, G. (1978). Preferential transcription of the ovalbumin gene in isolated hen oviduct nuclei by RNA polymerase B. *Proceedings of the National Academy of Sciences, USA*, **75**, 686–690
51. O'Malley, B. W. and Means, A. R. (1974). Female steroid hormones and target cell nuclei. *Science*, **183**, 610–620
52. Panyim, S., Ohno, T. and Jost, J. P. (1978). *In vitro* RNA synthesis and expression of vitellogenin gene in isolated chicken liver nuclei. *Nucleic Acids Research*, **5**, 1353–1370
53. Pays, E., Donaldson, D. and Gilmour, R. S. (1979). Specificity of chromatin transcription *in vitro*: anomalies due to aberrant DNA synthesis. *Biochimica et Biophysica Acta*, **562**, 12–130
54. Post, L. E., Norrild, B., Simpson, T. and Roizman, B. (1982). Chicken ovalbumin gene fused to a HSV promotor and linked to a thymidine kinase gene is regulated like a viral gene. *Molecular Cell Biology*, **2**, 233–240
55. Prestayko, A. W. and Busch, H. (1968). Low molecular weight RNA of the chromatin fraction from Novikoff hepatoma and rat liver nuclei. *Biochimica et Biophysica Acta*, **169**, 327–337
56. Reff, M. E. and Davidson, R. L. (1979). *In vitro* DNA dependent synthesis of globin RNA sequences from erythroleukemic cell chromatin. *Nucleic Acids Research*, **6**, 275–287
57. Ritossa, F. (1962). A new puffing pattern induced by heat shock and DNA in *Drosophila*. *Experientia*, **18**, 571–573
58. Robins, D. M., Paek, I., Seeburg, P. M. and Axel, R. (1982). Regulated expression of human growth hormone genes in mouse cells. *Cell*, **29**, 623–631
59. Seals, J. R. and Czech, M. P. (1980). Evidence that insulin activates an intrinsic plasma membrane protease in generating a secondary chemical mediator. *Journal of Biological Chemistry*, **255**, 6529–6531
60. Shih, T. Y., Young, H. A., Parks, W. P. and Scolnick, E. M. (1977). *In vitro* transcription of Moloney leukemia virus genes in infected cell nuclei and chromatin: Elongation of chromatin associated ribonucleic acid by *Escherichia coli* ribonucleic acid polymerase. *Biochemistry*, **16**, 1795–1801
61. Stallcup, M. R., Ring, J. and Yamamoto, K. R. (1978). Synthesis of mouse mammary tumor virus ribonucleic acid in isolated nuclei from cultured mammary tumor cells. *Biochemistry*, **17**, 1515–1521
62. Sutherland, E. W. and Rall, T. W. (1960). The relation of adenosine 3′ 5′-phosphate and phosphorylase to the actions of catecholamines and other hormones. *Pharmacological Reviews*, **12**, 265–299
63. Swaneck, G. E., Nordstrom, J. L., Kreuzaler, F., Tsai, M.-J. and O'Malley, B. W. (1979). Effect of estrogen on gene expression in chicken oviduct: Evidence for transcriptional control of ovalbumin gene. *Proceedings of the National Academy of Sciences, USA*, **76**, 1049–1053
64. Teyssot, B., Houdebine, L.-M. and Djiane, J. (1981). Prolactin induces release of factor from membranes capable of stimulating β-casein gene transcription in isolated mammary cell nuclei. *Proceedings of the National Academy of Sciences, USA*, **78**, 6729–6733
65. Teyssot, B., Djiane, J., Kelly, P. A. and Houdebine, L.-M. (1982). Identification of the putative prolactin second messenger activating β-casein gene transcription. *Biology of the Cell*, **43**, 81–88
66. Tsai, M. J., Tsai, S. Y., Chang, C. W. and O'Malley, B. W. (1978). Effect of estrogen on gene expression in the chick oviduct. *In vitro* transcriptions of the ovalbumin gene. *Biochimica et Biophysica Acta*, **521**, 689–707

67. Ucker, D. S., Ross, S. R. and Yamamoto, K. R. (1981). Mammary tumor virus DNA contains sequences required for its hormone-regulated transcription. *Cell, 27,* 257–266

68. Wahli, W., Ryffel, G. U., Wyler, T., Jaggi, R. R., Weber, R. and David, I. B. (1978). Cloning and characterization of synthetic sequences from the *Xenopus laevis* vitellogenin structural gene. *Developmental Biology, 67,* 371–383

69. Wang, T. Y. and Loor, R. M. (1979). Testosterone-activated RNA synthesis in isolated prostate nuclei. *Journal of Steroid Biochemistry, 10,* 299–304

70. Weil, P. A., Segall, J., Harris, B., Ng, S.-Y. and Roeder, R. G. (1979). Faithful transcription of eukaryotic genes by RNA polymerase III in systems reconstituted with purified DNA templates. *Journal of Biological Chemistry, 254,* 6163–6173

71. Weil, P. A., Luse, D. S., Segall, J. and Roeder, R. G. (1979). Selective and accurate initiation of transcription at the Ad2 major late promoter in a soluble system dependent on purified RNA polymerase II and DNA. *Cell, 18,* 469–484

72. Weisbrod, S. (1982). Active chromatin. *Nature, London, 297,* 289–295

73. Woodland, H. R. (1982). Stable gene expression *in vitro. Nature, London, 297,* 457–458

Chapter 11

Whole animal systems

C. M. Croce

Introduction

In order to understand the molecular basis of regulation of gene expression during embryonal development and during the process of cell differentiation, we need to be able to introduce specific genes and their mutant derivatives in the developing animals and possibly, to develop *in vitro* systems in which stem cells, when exposed to specific inducers, can differentiate into different more differentiated or terminally differentiated cell types. It is necessary, therefore, to develop *in vivo* and *in vitro* systems that allow the introduction of either cloned DNA molecules or entire chromosomes into the nuclei of stem cells.

Murine teratocarcinoma-derived stem cells, also called *embryonal carcinoma cells*, have the ability to differentiate into a large variety of different tissues when injected into syngenic mice[40] and are able to participate in the normal development of chimeric mice following their injection into the mouse blastocyst[3, 18, 33] (5–6 days old embryo). Therefore they are equivalent to cells of the inner mass of the mouse blastocyst. Not only can the teratocarcinoma stem cells participate in the normal development and differentiate into germ cells, but they can also be cultured *in vitro*, maintaining the ability to differentiate spontaneously *in vitro* and to participate in the normal development of chimeric mice following their injection into the mouse blastocyst[11, 36]. Cell lines of teratocarcinoma stem cells that are incapable of differentiating spontaneously[1, 41] have also been isolated. These cell lines are formed by uniform populations of stem cells that can be used for the study of the expression of a large variety of gene products and for the transcription and modification of a large number of genes[8, 15, 24, 25]. In addition, such nullipotent stem cells can be induced to differentiate by the use of different inducers such as retinoic acid[41]. Therefore, it is possible to obtain practically pure cultures of stem and differentiated cells that can be investigated by biochemical and molecular means.

Teratocarcinoma

Teratocarcinomas are tumors that are formed by a mixture of a large variety of differentiated tissues and by a population of malignant stem cells that maintain the ability to differentiate into a large variety of different cell types. Teratocarcinomas can be experimentally obtained by placing early embryos into ectopic places such as under the kidney capsule[9]. An important characteristic of the malignant stem cells called embryonal carcinoma cells, is that they 'normalize' following differentiation[21]. Since the injection of single stem cells into syngenic mice results in the appearance of tumors containing a large array of differentiated tissues, the teratocarcinoma-derived stem cells are considered to be pluripotent[21]. More recently Brinster[3] at the University of Pennsylvania and Illmensee and Mintz[18,33] at the Institute for Cancer Research in Fox Chance have injected teratocarcinoma stem cells into mouse blastocysts and have obtained chimeric mice in which their tissues derived at least in part from the teratocarcinoma stem cells. In some of the chimeras even the germ line was derived from the teratocarcinoma stem cells[18,33]. These results indicate that teratocarcinoma stem cells can participate in normal development and that they are truly totipotent since they can differentiate into cells of many tissues of the body including the germ line.

Since it is possible to obtain pure cultures of teratocarcinoma stem cells that can be induced to differentiate, it becomes possible to ask questions regarding the molecular basis of differential gene expression in stem versus differentiated cells.

Teratocarcinoma stem cell hybrids

The availability of continuous cell lines of teratocarcinoma stem cells has made it possible to select for drug resistant mutants that can be used for hybridization experiments.

The phenotype of hybrids between teratocarcinoma stem cells and more differentiated cells differs considerably depending on the partner cells with which the stem cells have been hybridized. Quite often, interspecific and intraspecific hybrids resemble the teratocarcinoma stem cell parent in morphology, biochemical properties and their ability to differentiate[5, 29, 32]. Interestingly, somatic cell hybrids that behave like teratocarcinoma stem cells still fail to express the major histocompatibility antigens and β_2-microglobulin of the stem cells but express the histocompatibility antigen and β_2-microglobulin of the differentiated parental cells[5]. We have hybridized a teratocarcinoma-derived stem cell line deficient in thymidine kinase and capable of differentiating *in vitro* and *in vivo* and of participating in the normal development of chimeric mice when injected into the mouse blastocyst[19] with either mouse myeloma cells[5] or with human fibroblasts[19] or with rat hepatoma cells[19, 27].

The hybrids between mouse teratocarcinoma and myeloma cells were morphologically indistinguishable from the teratocarcinoma stem cell parent, were capable of differentiating into a large number of different tissues when injected into nude mice and did not produce and secrete immunoglobulin chains[5]. They expressed, however, the histocompatibility antigens of the mouse myeloma parent[5]. The hybrids between the mouse teratocarcinoma stem cell mutants and human fibroblasts have lost almost all of the human chromosomes and most of them have retained only human chromosome 17 that carries the gene for thymidine kinase in addition to the entire mouse chromosome complement[19]. Similar findings were obtained using hybrids between a different teratocarcinoma stem cell line and either Friend cells (erythroleukemia)[32] or mouse thymus derived cells[32]. Some of the hybrids between Friend's cells and teratocarcinoma stem cells resembled,

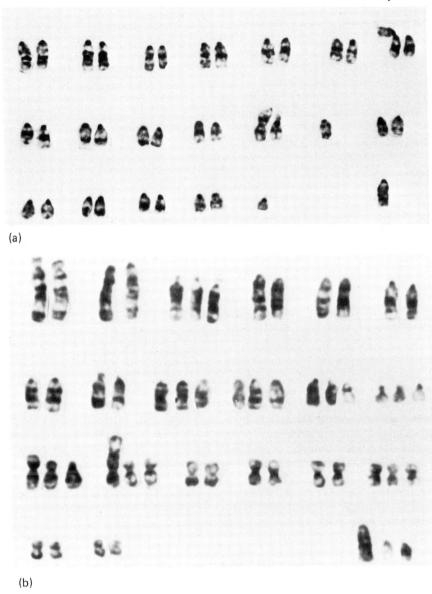

(a)

(b)

Figure 11.1. Karyotypes of (a) parental TK⁻ OTT6050 mouse teratocarcinoma and (b)
HPRT⁻ Fu5AH rat hepatoma cells after trypsin-Giemsa banding standing.

however, the erythroleukemic cell parent and not the teratocarcinoma stem cells.
A completely different result, however, was obtained in the case of the hybrids
between teratocarcinoma stem cells and rat hepatoma cells. The karyotypes of the
parental cells are shown in *Figure 11.1*. These hybrids were extremely heteroploid
(*Figure 11.2*) and were, in fact, morphologically different from both parental cells.
They formed tumors with no differentiation in phenotypically normal tissues and
did not express any of the differentiated functions of the hepatoma parent[27].

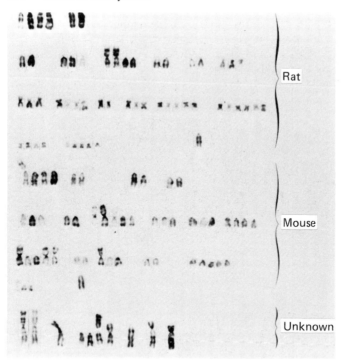

*Figure 11.2.*Karyotype of hybrid As3 between OTT6050 and Fu5AH cells.

Therefore it can be concluded that in most cases hybrids between teratocarcinoma stem cells and differentiated cells resemble the stem cell parent. Hybrids that resemble the differentiated parent or that do not resemble any of the parental cells, however, can also be obtained.

Injection of teratocarcinoma stem cell hybrids into the mouse blastocysts

To determine whether somatic cell hybrids between teratocarcinoma stem cells and differentiated cells are able to participate in the normal development of a chimeric mouse and whether it is possible to introduce foreign genes into chimeric mice we attempted to introduce foreign genetic information into chimeric mice via somatic cell hybridization and microinjection of well characterized hybrid cells into the mouse blastocyst. For this purpose we have hybridized pluripotent teratocarcinoma stem cells with human fibroblasts and we obtained somatic cell hybrids that have retained only human chromosome 17[19]. These hybrids expressed human thymidine kinase and galactokinase activity. In collaboration with Illmensee of the University of Geneva and Hoppe at the Jackson Laboratory, we have injected single hybrid cells into 103 experimental blastocysts of a different mouse strain (C57BL/6). The teratocarcinoma stem cells were derived from the 129 mouse strain. Since the C57BL/6 and the 129 strains express different electrophoretic forms of the enzyme glucose phosphate isomerase (GPI), it is possible to determine which mouse tissues are chimeric by looking at the expression of GPI gel electrophoresis[19]. Following the injection of the hybrids into the C56BL/6 blastocysts, the blastocysts were surgically transferred to pseudopregnant foster mothers. Forty-six mice were born and two of them showed coat mosaicism (*Figure 11.3a* and *b*). Therefore the results of this experiment show that the hybrids are capable of differentiating into

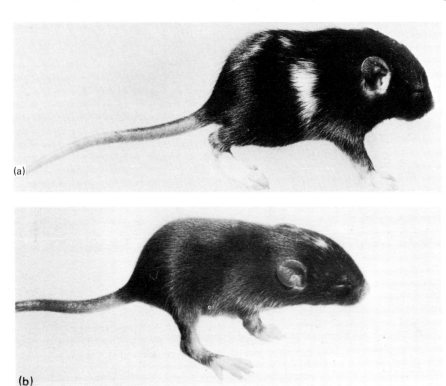

Figure 11.3. (a) Coat mosaicism in chimeric mouse A developed from a BL6 blastocyst after microinjection of a human–mouse hybrid cell. In addition to the expected black coat phenotype of the recipient, this mouse also exhibited several white patches on the back, tail and feet and a large white clone extending from the midlateral to the ventral side, all of which derived from the injected hybrid cell and comprised about 20% of the total coat. (b) Coat mosaicism of an additional chimeric mouse.

phenotypically normal skin. We have also sacrificed some of the mice and have studied them for 129 teratocarcinoma cell hybrid contribution to the formation of internal organs by GPI analysis. As shown in *Figure 11.4*, some of the mice were chimeric in several different tissues. Both chimeric mice shown in *Figure 11.4*, for example, were chimeric in the brain, heart, kidney and pancreas. The different tissues were morphologically and histologically normal. Interestingly, the muscles of the chimeric mice instead of showing a GPI two-band pattern like the other chimeric tissues, revealed a GPI three-band pattern due to the fact that muscles are syncitia that contain in the same cytoplasm nuclei of 129 and C57BL/6 origin (*Figure 11.4*). Since the GPI is a dimer of identical subunits, the muscles express a band formed by a dimer of the 129 and C57BL/6 enzyme subunits.

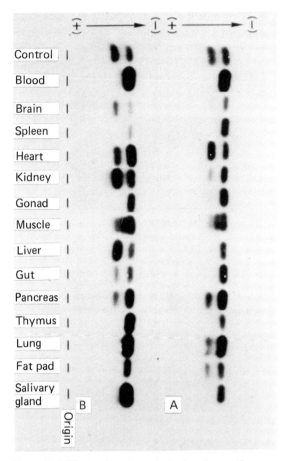

Figure 11.4. Electrophoretic analysis of strain-specific allevic variants of GPI in blood cell lysates and tissue homogenates from two chimeric mice (A and B). Control represents a 50:50 mixture of slow-migrating 129 and fast-migrating BL6 types of GPI in blood cell lysates. In each genetically mosaic mouse, clonal descendants of the injected human–mouse hybrid cell contributed substantially to several internal organs and also fused normally with myoblasts of the host, as judged from heterodimeric enzyme expression in skeletal muscle. In addition, mouse A showed conspicuous coat mosaicism (*Figure 11.3*) whereas chimeric mouse B did not.

Since the hybrid cells we have injected into the mouse blastocysts contained human chromosome 17 and expressed human galactokinase (GLK) activity, we have tested the chimeric tissues for the expression of the human enzyme. A weak human specific GLK activity was observed only in the heart of mouse A and in the kidneys of mouse B[19]. The factor likely to be responsible for the lack of human GLK expression in the majority of the chimeric tissues is that chromosome 17 might be lost from the hybrid cells during *in vivo* growth in the absence of HAT[26] selection. These results were quite exciting, however, because they showed the feasibility of introducing foreign genetic information into living experimental animals, but also indicated that we should look for more stable somatic cell hybrids to obtain consistent foreign gene expression in the chimeric mice.

Since rat × mouse hybrid cells tend to lose rat chromosomes very slowly[7], it was decided to attempt to inject the interspecies hybrids between pluripotent mouse teratocarcinoma stem cells and rat hepatoma cells into the mouse blastocyst[27]. Single rat–mouse hybrid cells from the As3 hybrid clone were injected into 179 blastocysts, of which 144 were surgically implanted into pseudopregnant foster mothers to allow development to term (*Figure 11.5*). A total of 62 experimental mice were delivered, 35 females and 27 males. None of these mice showed any detectable coat mosaics, nor did they reveal in the breeding that germ line derived from the injected cells. However, after biochemical analysis for specific allelic GPI variants, three of the 62 mice (about 5%) were found to have hybrid cell-derived contributions to a limited number of internal organs of endomesodermal origin (*Figure 11.6*)[17]. In chimeric mouse A, a male 10 months of age, the liver, lung, kidneys, intestinal tract and adipose tissue (fat pad) contained a substantial number of cell progeny from the injected rat–mouse hybrid. Only two organs, the liver and lung, showed the teratocarcinoma strain-specific GPI variant in mouse B (*Figure 11.6*). Chimeric mouse C, a male four months of age, revealed at autopsy an abnormally enlarged liver with globular irregularities of the first lobe. The particular area as well as the remaining liver expressed both the mouse- and

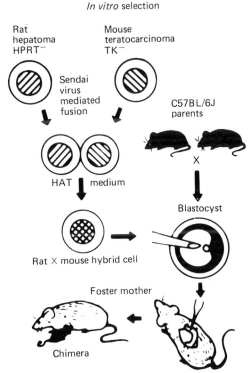

In vitro selection

Rat hepatoma HPRT⁻

Mouse teratocarcinoma TK⁻

Sendai virus mediated fusion

C57BL/6J parents

HAT medium

×

Blastocyst

Rat × mouse hybrid cell

Foster mother

Chimera

In vivo differentiation

Figure 11.5. Experimental scheme of cycling rat–mouse hybrid cells through mice via blastocyst injection. TK⁻ mouse teratocarcinoma cells and HPRT⁻ rat hepatoma cells were fused by use of inactivated Sendai virus and selected in HAT medium, in which only the hybrid cells could survive. These cells were then injected singly into BL6 blastocysts carrying several genetic markers, thereby enabling any *in situ* differentiation of cell hybrids. After microinjection, the blastocysts had to be surgically implanted in pseudopregnant foster mothers to allow development to term. The experimental offspring were analyzed for hybrid cell contributions to the coat, internal organs, and germ line.

182

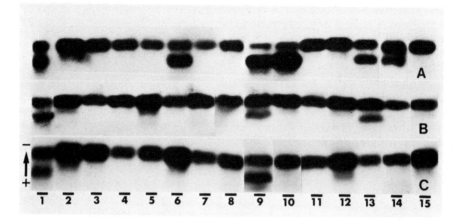

Figure 11.6. Cellulose acetate electrophoresis of mouse strain-specific enzyme variants of GPI in cell extracts from various organs of chimeric mice A, B and C. 1, 50:50 control mixture of slow-migrating strain 129 and fast-migrating strain BL6 types; 2, blood; 3, brain; 4, spleen; 5, heart; 6, kidneys; 7, gonads; 8, skeletal muscle; 9, liver; 10, intestinal tracts; 11, pancreas; 12, thymus; 13, lung; 14, adipose tissue; and 15, salivary glands. In chimeric mouse A, the kidneys, liver, intestinal tracts, lung and adipose tissue; in chimeric mouse B, the liver and lung; and in chimeric mouse C, only the liver showed teratocarcinoma strain-specific GPI activity derived from the hybrid cell progeny.

TABLE 11.1. Rat- and mouse-specific enzyme expression in several tissues of chimeric mice A, B, and C derived from injections of single rat–mouse hybrid cells into BL6 blastocysts

	Enzyme																		
	GPI		*ENO-1*		*TRI-1*		*LDH*		*GPD*		*GLK*		*NP*		*ALD*		*ADH*	
Tissue	*R*	*M*	*R*	*M*	*R*	*M*	*R*	*M*	*R*	*M*	*R*	*M*	*R*	*M*	*R*	*M*	*R*	*M*
FU5AH	+		+		+		+		+		+		+		+		+	
OTT6050		+		+		+		+		−		+		+		−		+
As3	+	+	+	+	+	+	+	+	−	+	+	+	+	+	−	−	−	+
Liver R	+		+		+		+		+		+		+		+		+	
Liver M		+		+		+		+		+		+		+		+		+
Liver C	(+)	+	(+)	+	−	+	+	+	−	+	−	+	−	+	(+)	+	+	+
Tumor C	+	+	−	+	(+)	+	(+)	+	−	+	−	+	−	+	+	+	+	+
Liver B	+	+	−	+	(+)	+	+	+	+	+	−	+	+	+	+	−	+	+
Liver A	+	+	−	+	+	+	+	+	+	+	+	+	+	+	−	+	−	+
Gut A	+	+	+	+	(+)	+	−	+	+	+	−	+	−	+	−	+	−	+
Kidney A	+	+	(+)	+	+	+	+	(+)	−	+	+	+	+	+	−	+	−	+

FU5AH, rat hepatoma; OTT6050, mouse teratocarcinoma; As3, rat–mouse hybrid cell line; R, rat; M, mouse; +, strong enzyme activity; (+), weak enzyme activity; −, no enzyme detectable. GPI, glucose phosphate isomerase; ENO-1, enolase-1; TRI-1, tripeptidase-1; LDH, lactate dehydrogenase; GPD, glycerol-3-phosphate dehydrogenase; GLK, galactokinase; NP, nucleoside phosphorylase; ALD, aldolase; ADH, alcohol dehydrogenase.
Note the difference in enzyme expression within a given mosaic organ and among the different mosaic organs. Note also in liver B the presence of rat ALD only, whereas liver A, intestinal tract, A, and kidney A express only the mouse ALD, and liver C and its tumor both show rat and mouse ALD. Although the mosaicism in the liver, intestinal tract, and kidneys of the chimeric mouse A is derived from the cell progeny of one injected hybrid cell, the clonal origin is only apparent from the GPI, ALD, and ADH data, whereas the variations concerning the other enzyme data likely result from chromosomal segregation or dosage compensation during development *in vivo*.

rat-specific GPI patterns. No other tissue was found to be mosaic, as judged from enzyme analysis (*Figure 11.6*). The abnormal liver lobe showed morphological features typical of a hemangio-endothelioma. All the other mosaic organs of the three chimeric mice did not reveal any kind of morphological abnormality when examined in serial sections[17].

After the GPI test had been completed, only the mosaic tissues were further analyzed for a number of different enzymes that allowed us to distinguish between rat and mouse gene products. The electrophoretic analysis of these enzymes showed that, in addition to GPI, eight rat-specific variants could be detected in the different mosaic tissues (*Table 11.1*). However, these rat enzymes were not always present in all of the mosaic organs; rather, there was considerable variation observed concerning the presence and activity of the rat-specific gene products (*Figured 11.7*). The absence of the enzymes GPD, ADH and ALD in the hybrid cell line and their appearance in some of the mosaic tissues demonstrated that these rat genes, inactive *in vitro*, became activated *in vivo* (*Figure 11.6*)[17].

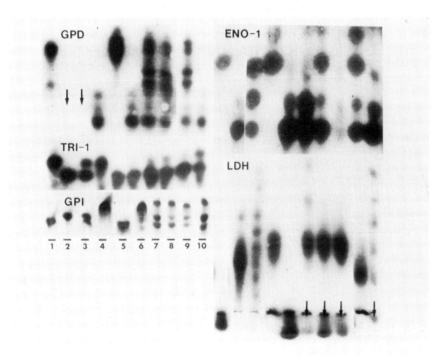

Figure 11.7. Starch gel electrophoresis of mouse- and rat-specific enzyme variants of GPD, TRI-1, GPI, ENO-1, and LDH in cell extracts from those organs of chimeric mice, A, B, and C that contained hybrid cell-derived tissue participation as judged from the GPI analysis (see *Figure 11.4*). Slots 1–5 are controls: 1, Fu5AH rat hepatoma; 2, OTT6050 mouse teratocarcinoma; 3, As3 rat–mouse hybrid; 4, mouse liver; 5, rat liver. For ENO-1 and TRI-1: 4, rat liver; 5, mouse liver. Slots 6–10 represent chimeric tissues: 6, liver C; 7, liver B; 8, liver A; 9, intestinal tract A; 10, kidneys A. Note the absence of GPD activity (arrow) in the hybrid cells and its appearance in livers B and A and in intestinal tract A. The formation of mouse–rat heteropolymers in these tissues documents the functional cooperation of the interspecific gene products. Note the appearance of adult rat LDH-5 (arrow) not detectable in the hybrid cells.

These results indicate that somatic cell hybrids between teratocarcinoma stem cells and differentiated cancer cells, that have retained the great majority of the chromosome of both parents, are still capable of differentiating into phenotypically normal tissues and that euploidy is not a *sine qua non* for normal differentiation and development in chimeric mice. Since all chimeric mice which were obtained were chimeric in the liver, it seems possible that the phenotype of the differentiated parent might play a role in determining the type of tissues that can derived from the teratocarcinoma stem cell hybrids. Finally, these results indicate that foreign genes can be expressed during the differentiation of teratocarcinoma stem cell hybrids *in vivo*. It is of considerable interest that enzymes such as GPD, ADH and ALD which are not expressed in the injected hybrid line A_3 were expressed in some of the mosaic tissues demonstrating that these rat genes, inactive *in vitro*, became activated during development *in vivo*.

Experiments similar to those carried out with hybrid cells can be carried out with teratocarcinoma stem cells transformed by recombinant DNA vectors carrying selectable markers[24, 25]. The resulting chimeric mice can then be studied for the regulation of the expression of the inserted genes in different tissues. Differentiation of the teratocarcinoma stem cell transformants into germ cells might also result in the development of mouse strains carrying specific foreign genes. In order to attempt to obtain transmission through the germ line it seems likely, however, that we need euploid teratocarcinoma-derived stem cell lines.

Microinjection of cloned DNA molecules into fertilized eggs

An alternative way to introduce specific genes into mice is to microinject the genes of interest into fertilized eggs which are then implanted into pseudopregnant females[4, 6, 13]. Nucleic acid hybridization procedures are used to detect the presence of the introduced molecules in the offspring[35]. Up to 40% of the mice developed from the injected eggs contain the injected sequences integrated in the chromosomal DNA. Tandem duplication of the inserted sequences often occurs. Integration of the duplicated sequences into a single mouse chromosome has been reported by Costantini and Lacy[6] who have also shown vertical transmission of the inserted sequences to the progeny. In some cases, expression of the introduced genes has been observed[4]. This method seems quite powerful in order to introduce specific genes into the mouse germ line and to investigate the regulation of the expression of the inserted sequences in the different mouse tissues[35]. Recently Palmiter *et al.*[35] have carried out some beautiful experiments by inserting a herpes simplex virus thymidine kinase gene under the control of the promoter/regulatory region of the mouse metallothionein-1 gene. Interestingly, the enzyme was induced by heavy metals. Exposure to glucocorticoid, however, did not result in the induction of the enzyme[35]. This approach should result in a better understanding of the regulatory sequences involved in the control of gene expression in the different mouse tissues. We have used the same approach to introduce the genomic sequences of the embryonal human μ-immunoglobulin heavy chain gene in order to determine whether it undergoes rearrangement during β-cell differentiation. The human sequences were detected in approximately 20% of the mice derived from the injected eggs and were found to be transmitted vertically to their progeny. Analysis of the β-cells derived from these mice carrying the human μ chain gene is in progress.

DNA-transformed murine teratocarcinoma-derived stem cells: regulation of expression of transfecting genomes

Developmentally pluripotent teratocarcinoma stem cells are able to restrict expression of oncogenic viral genes[14, 43, 44] which are fully expressed in teratocarcinoma-derived differentiated cells. Since molecular mechanisms operative in the regulation of expression of viral genes in this differentiating model system may be analogous to mechanisms involved in gene regulation during embryogenesis, it is important to define the molecular basis for suppression of expression of viral genes in teratocarcinoma stem cells. A series of studies has been conducted by various investigators in an attempt to pinpoint the block to expression of papovaviruses in infected stem cells[37, 38]. When F9[1] stem cells were infected with Simian virus 40 (SV40), low levels of unspliced early viral RNA were detected[38]. Wild type polyoma virus infection of PCC4 aza stem cells[34] resulted in production of small amounts of viral RNA[47] suggesting that polyoma virus early transcription is blocked in these cells. Polyoma virus host range mutants which produce infectious virus in PCC4 aza stem cells[20, 47] and F9 stem cells[12, 39] have been isolated. The nature of the nucleotide changes permitting polyoma virus infection of F9 cells differs from the rearrangements present in the genomes of polyoma virus mutants that infect PCC2 aza cells, but the changes in both types of mutants occur in the same region of the viral genome, just to the late side of the origin of DNA replication. It is possible that the polyoma host range mutations affect early transcriptional initiation in stem cells[20, 47]. We have isolated and characterized[24, 25] F9 derived stem and differentiated cell clones carrying single integrated copies of the SV40 genome and have compared deoxyribonuclease (DNase I) sensitivity[15], methylation[16], and transcription[25] of the SV40 genome and the linked pBR322 and HSV-1 thymidine kinase (HSV-1 *tk*) genomes in the stem and differentiated cells[16]. DNase I probing of chromatin structure has provided evidence that transcriptionally active chromatin has a more accessible configuration than transcriptionally inactive chromatin; that is, actively transcribed genes are preferentially digested by DNase I[15, 16]. Since nucleases have proved to be a useful tool in investigation of chromatin structure, we have initiated DNase I sensitivity studies of chromatin structure in teratocarcinoma-derived stem and differentiated cells[15, 16].

Recently, restriction enzyme analysis has been used to compare the state of methylation of specific genes (or sites near specific genes) in cells in which the genes are silent compared with cells in which they are active, and striking differences have been found[31, 48]. In particular, certain genes (or sequences near the genes) seem to be relatively undermethylated in cells in which that gene is expressed. This correlation has been observed in the globin gene[30, 46], the ovalbumin gene[23, 31] and several viral genes[10, 42]. These findings have suggested that methylation of certain residues may be one of the functions that is correlated with inactivity of a specific gene. Thus, we also investigated the state of methylation of specific cytosine residues in the plasmid genomes within the transfected F9 stem and differentiated cell clones.

Finally, we have compared the transcriptional state of the pBR322, HSV-1 *tk*, and SV40 genomes in stem and differentiated cells.

Characterization of the F9/SV and F9/PY transfectants

Transfections of F9 TK⁻ cells with the SV-containing plasmid pC6 (*Figure 11.8*) resulted in the isolation of two independent stem cell clones, 12-1 and 14-2, which

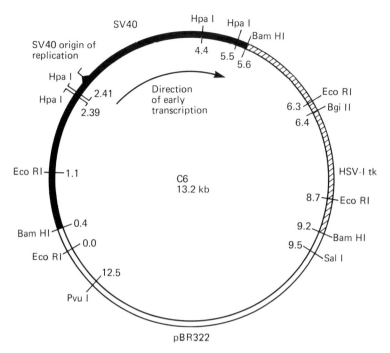

Figure 11.8. Restriction map of 13.2 kb plasmid C6. One copy of Bam HI-digested SV40 DNA is inserted into the pBR322/HSV-1 *tk* vector, pHSV-106.

carry one copy each of the pC6 plasmid genome integrated into the cellular genome through a site on the pBR322 genome. The organization of the pC6 plasmid genome within 12-1 stem cells has been described[25] and is shown in *Figures 11.9* and *11.10* and a similar strategy was used to determine the organization of the plasmid genome in 14-2 cells. The 15-1, another stem cell clone, has been isolated by transfection of F9 TK⁻ cells with the PyC6 plasmid which contains a polyoma virus genome which was inserted into the pBR322/HSV-1 *tk* vector using the same method described for construction of the pC6 plasmid[24]. Preliminary results indicate that 15-1 cells contain one copy of the PyC6 plasmid genome integrated into the cellular genome through a site on the polyoma virus genome.

Differentiated cell clones 12-1a and 14-2a have been isolated after retinoic acid induction of the homologous stem cell clones, 12-1 (*Figure 11.11*) and 14-2. Differentiated cell clones can be readily isolated from the F9/SV40 stem cell clones after retinoic acid treatment probably because, upon differentiation, these cells behave like SV40-transformed cells and have a high cloning efficiency. The 15-1 cells, on the other hand, behave like F9 cells after retinoic acid treatment; that is, they differentiate readily but the differentiated cells do not clone efficiently, and we have not been able to establish a continuously growing differentiated line from these cells. This is probably due to the fact that the PyC6 plasmid genome is integrated into the cellular DNA through a site in the polyoma sequences. If the integration event has interrupted the polyoma early region genes, then transforming gene products would not be produced and differentiated cells would thus not became transformed. It will be important in these cells to determine the precise integration site of the plasmid genome within the polyoma virus sequences.

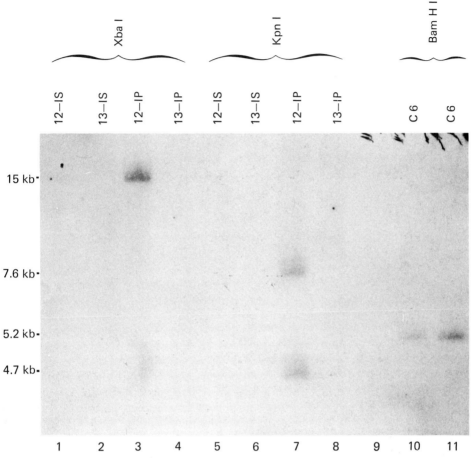

Figure 11.9. Hybridization of ^{32}P-labeled SV40 DNA to 13-1 and 12-1 cellular DNA after restriction of endonuclease digestion and transfer from agarose gel. Hirt supernatant (S) and pellet (P) DNA (10 μg/lane) from 12-1 and 13-1 stem cells were cleaved with Xba I (lanes 1–4) and Kpn I (lanes 5–8); electrophoresed in a 0.7% agarose slab gel; denatured; transferred to a nitrocellulose sheet; and hybridized to ^{32}P-labeled SV40 DNA. Bam HI-cleaved and similarly treated pC6 DNA (lane 10, 12.5 pg; lane 11, 6.25 pg) was included as a marker.

The 12-1 stem cells express the stage-specific embryonic antigen, SSEA-1[22], which is normally found on murine pre- and post-implantation embryos and on teratocarcinoma-derived stem cells but not on differentiated derivatives[22], nor are H2 antigen and β_2-microglobulin detected on 12-1 stem cells[8]. These stem cells are also negative for SV40 T antigen and SV40 specific tumor associated surface antigen (TASA)[22, 24, 25]. Conversely, 12-1a cells no longer express SSEA-1[22], but synthesize SV40 T antigen and TASA and the major histocompatibility antigens, H2[8] and β_2-microglobulin[8]. The 14-2 and 14-2a cells express the same surface and viral antigens as their respective counterpart cells, 12-1 and12-1a. Thus by morphological and immunological criteria, the 12-1 and 14-2 cells are identical to their parent stem cell, F9; the differentiated cells 12-1a and 14-2a are, on the other hand, not identical to the F9-derived differentiated parietal endoderm cell,

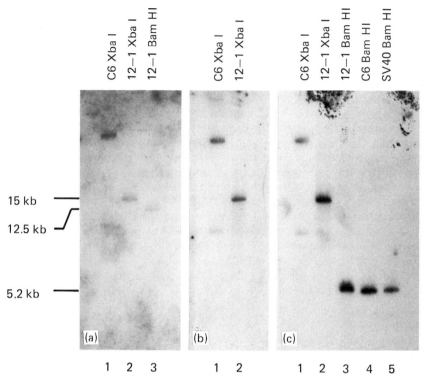

Figure 11.10. Hybridization of (a) ³²P-labeled pBR322 DNA, (b) HSV-1 *tk* DNA, and (c) SV40 to Xba I- and Bam HI-cleaved 12-1 cellular DNA. Hirt pellet DNA (10 µg/lane) from 12-1 cells was cleaved with Xba I and Bam HI restriction endonucleases and applied to an agarose slab gel as indicated above each lane in this figure. After electrophoresis, the gel was cut into three parts, and the DNA in each gel was denatured and transferred to nitrocellulose sheets. Each of the three nitrocellulose sheets was hybridized to a different ³²P-labeled probe: (a) Hybridized to ³²P-labeled pBR322 DNA; (b) hybridized to ³²P-labeled HSV-1 *tk* DNA; and (c) hybridized to ³²P-labeled SV40 DNA. pC6 DNA, Bam HI-cleaved pC6 DNA, and Bam HI-cleaved SV40 DNA are included as markers.

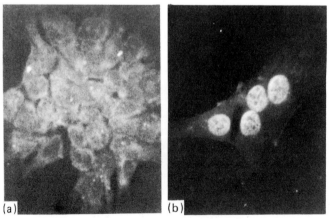

Figure 11.11. Indirect immunofluorescence of SV40 T antigen in transformant 12-1 cells (a) before and (b) after induction of differentiation with 0.1 µmol/ℓ retinoic acid.

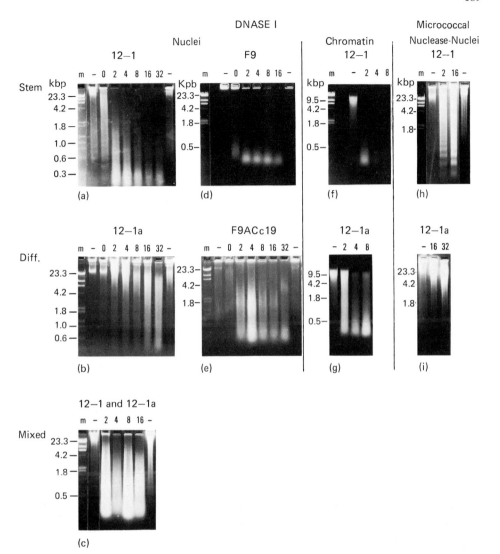

Figure 11.12. Nuclease digestion of stem and differentiated cell nuclei and chromatin. DNase I digestions were performed as described[16], using 10 µg of DNase I/mg of DNA. Micrococcal nuclease digestions were performed as described[16], using 10 units of micrococcal nuclease/mg of DNA. DNA was isolated from each sample and subjected to electrophoresis (10 µg/lane) on 1.0% agarose gels for various periods of time (7–15 hours). Numbers across the top of each figure indicate the time (minutes) of digestion with nuclease and those on the left indicate sizes of fragments (kb pairs); minus signs above the left lanes indicate nuclei or chromatin samples incubated at 37°C for two minutes without nuclease; minus signs above the rightmost lanes in some figures indicate nuclei samples incubated for 32 minutes at 37°C without nuclease. (a) and (d) 12-1 and F9 stem cell nuclei digested with DNase I; (b) and (e) 12-1a and F9ACc19 differentiated cell nuclei digested with DNase I; (c) equal numbers of 12-1 and 12-1a cells mixed before isolation of nuclei and then digested with DNase I; (f) and (g) 12-1 and 12-1a chromatin digested with DNase I; the 12-1 four minute and eight minute lanes (f) contained only 4 µg of DNA. (h) and (i) 12-1 and 12-1a nuclei digested with micrococcal nuclease.

F9ACc19[15]. They are morphologically dissimilar to F9ACc19 cells, behaving like SV40-transformed fibroblasts and they express the SV40 specific tumor antigens, T and t (detected by immunoprecipitation, data not shown) and 12-1a (presumably 14-2a also) expressed SV40 specific TASA[22].

The F9/Py cell, 15-1, like the F9/SV stem cells, expresses SSEA-1 and lacks H2 and β₂-microglobulin; preliminary results of indirect immunofluorescent staining for polyoma T antigen suggest that polyoma virus coded early antigens may not be detectable in 15-1 stem cells.

DNase I sensitivity of plasmid genomes in stem and differentiated cells

We have previously shown that chromatin of teratocarcinoma-derived stem cells is more sensitive to DNase I digestion than is chromatin of differentiated cells derived from the homologous stem cell[15]. Under the same conditions of digestion, less DNA remains undigested in stem cells than in differentiated cells (*Figure 11.12*) and the average molecular weight of resistant DNA fragments is much lower in stem than in differentiated cells (*Figure 11.12*), indicating that there are more DNase I accessible sites in stem cell chromatin. The DNase I hypersensitivity of stem cell chromosomal DNA is not due to endogenous nuclease activity, since purified chromatin of stem cells is hypersensitive to DNase I digestion and isolated

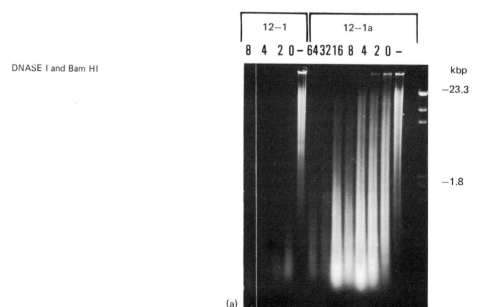

Figure 11.13. DNase I sensitivity of plasmid genomes in stem and differentiated cells. 12-1 and 12-1a nuclei were digested with DNase I (10 µg/mg of DNA) (numbers across the top of each figure indicate the time (minutes) of digestion and those on the right indicate size of fragments (kb pairs); minus signs indicate that nuclease was not added), and DNAs were isolated at each time point and digested with Bam HI. Agarose gels (a) were blotted, and nitrocellulose filters were hybridized with (*see facing page*) (b) nick-translated ^{32}P-labeled HSV-1 *tk*, (c) SV40, or (d) pBR322 DNA.

DNA from 12-1 nuclei incubated at 37°C for 32 minutes without DNase I is not significantly degraded (*Figure 11.12*).

To study DNase I sensitivity of plasmid genomes within 12-1 and 12-1a cells, DNase I digested DNAs of 12-1 and 12-1a cells were cleaved with Bam HI, electrophoresed, transferred to nitrocellulose paper, and the filters were hybridized to ^{32}P-labeled SV40, HSV-1 *tk*, or pBR322 DNA[15]. Results of such an experiment are shown in *Figure 11.13*. The ethidium bromide stained agarose gel (*Figure 11.13*) shows the distribution of DNase I/BAM HI-digested DNA fragments from the two

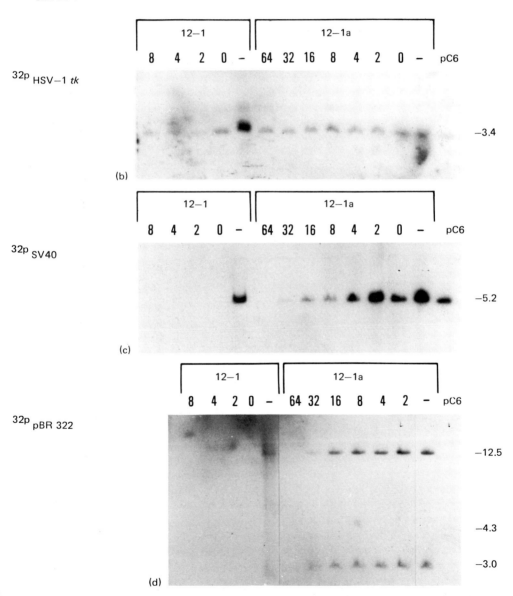

Figure 11.13 (cont.)

cell types. The 5.2 kb SV40 genome is still detectable in 12-1a cells after 16 minutes of DNase I digestion (*Figure 11.13*). When labeled pBR322 DNA is used as the probe, the results are similar to those obtained for the SV40 genome. The pBR322 genome is not detected as a single 4.3 kb band, since pC6 plasmid is integrated into cellular DNA through a site on the pBR322 genome; when the cellular DNA is cleaved with Bam HI and analyzed by electrophoresis, Southern transfer and hybridization with labeled pBR322 DNA, two bands are detected (12.5 kb and 3.0 kb) which contain pBR322 DNA and host flanking sequences. Thus, in *Figure 11.12* the disappearance of the 12.5 kb and 3.0 kb bands with increasing time of DNase I digestion could be due to DNase I sensitivity sites of host flanking sequences, pBR322 sequences, or both. *Figure 11.13* shows that the HSV-1 *tk* 3.4 kb fragment is still detectable in 12-1a nuclei after 64 minutes of DNase I digestion (10 μg DNase I/mg DNA) and is still detectable in 12-1 nuclei after 8 minutes of DNase I digestion. Thus purified cellular DNA from the two cells was digested with DNase I (0.1 μg/ml) and DNase I sensitivity of HSV-1 *tk* and SV40 genomes within naked cellular DNAs were compared. The HSV-1 *tk* 3.4 kb fragment was still intact at 32 minutes of DNase I digestion, whereas the SV40 genome was not detectable at the first time point (0 minutes) after addition of DNase I (data not shown). The HSV-1 *tk* gene, within naked cellular DNA, is insensitive to DNase I digestion. It will be important to determine the basis for this DNase I insensitivity.

Comparison of methylation of plasmid genomes within stem and differentiated cells

Since more than 90% of the methylated cytosine residues in eukaryotic DNA are present in the dinucleotide, CpG, restriction enzymes which contain this dinucleotide within their recognition sequence can be used to locate methyl groups in eukaryotic DNA. Hpa II, which cleaves the SV40 genome at only one site, recognizes the tetranucleotide sequence CCGG but will not cut this sequence if the internal C is methylated (CMGG). Msp I recognizes the same tetranucleotide sequence but cuts either CCGG or CMGG; Hha I, which cleaves the SV40 genome at two sites, also does not cut its recognition sequence (GCGC) if the internal C is methylated. When 12-1 stem and 12-1a differentiated cellular DNAs were cleaved with Hpa II and Msp I and hybridized to ^{32}P-labeled SV40 DNA after Southern transfer, the pattern of hybridization for all four DNA samples (12-1/Hpa II, 12-1/Msp I, 12-1a/Hpa II and 12-1a/Msp I) was the same (data not shown) indicating that the cytosine residue in the Hpa II site is not methylated in either 12-1 or 12-1a cells. Similarly, the cytosine residues within the Hha I sites on the SV40 genome are not methylated in either 12-1 or 12-1a cells.

The restriction endonuclease Ava I, which cleaves the pBR322 genome at only one site, does not cut its recognition sequence, CPyCGPuG, if the internal CG is methylated. If 12-1 and 12-1a cellular DNAs are cleaved with Ava I and hybridized to ^{32}P-labeled pBR322 DNA after Southern transfer, the hybridization patterns for the two cells are not the same (*Figure 11.14*). The high molecular weight bands which are seen in Ava I cut 12-1a DNA but not in Ava I cut 12-1 DNA indicate a methylated cytosine(s) either at the pBR322 Ava I site(s) or in host flanking sequences.

Transcription of plasmid genomes in stem and differentiated cells

Total RNA was extracted from 12-1 and 12-1a cells and assayed for presence of pBR322 and HSV-1 *tk* specific RNA by transfer to nitrocellulose filters[45] and

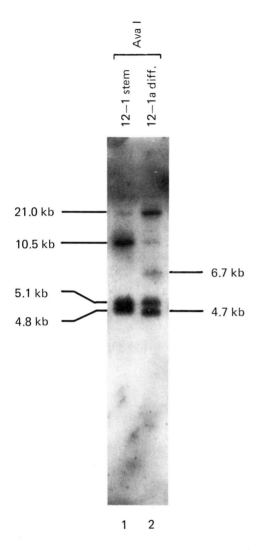

Figure 11.14. Differential methylation of cytosine residues in pBR322/host sequences. Total cellular DNA from 12-1 stem and 12-1a differentiated cells was digested with Ava I, a restriction enzyme which does not cut its recognition sequence (CPyCGPuG) if the internal cytosine residue is methylated. The cleaved DNA was electrophoresed on a 0.7% agarose gel, blotted to nitrocellulose paper, and hybridized to a nick-translated pBR322 probe. Since the pattern of hybridization is different from the two cell lines, we conclude that the distribution of 5-methylcytosine residues is different in the two cells with the differentiated cell DNA probably containing more methylated cytosine in the Ava I recognition sequence contained within pBR322 or host flanking sequences.

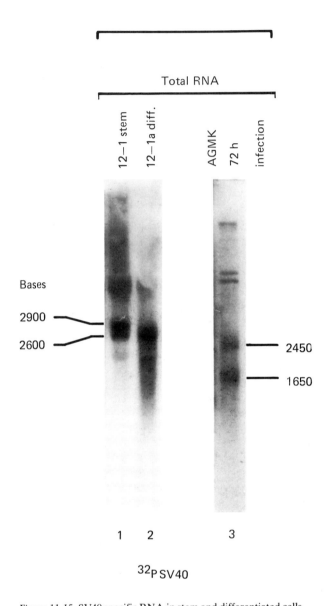

Total RNA

12–1 stem

12–1a diff.

AGMK 72 h infection

Bases

2900

2600

2450

1650

1   2   3

^{32}P SV40

Figure 11.15. SV40 specific RNA in stem and differentiated cells.
Total cellular RNA from 12-1 stem and 12-1a differentiated cells
was denatured, electrophoresed, transferred to a nitrocellulose
filter, and hybridized to ^{32}P-labeled SV40 DNA to detect SV40
specific RNA in the two cell types. Polyadenylated fractions of
total RNA and of cytoplasmic RNA from stem and differentiated
cells contained the same SV40 specific RNA species (2900 and
2600 bases) seen above in total cellular RNA from the two cell
lines.

hybridization to ^{32}P-labeled pBR322 or HSV-1 *tk* probes. The pBR322 specific RNA was not present in either cell, whereas HSV-1 *tk* specific RNA was present in both (data not shown).

Total RNA and oligo dT-cellulose selected polyadenylated (poly A$^+$) RNA from 12-1 and 12-1a cells were also hybridized to ^{32}P-labeled SV40 DNA. A total RNA fraction from both 12-1 and 12-1a cells (*Figure 11.15*), as well as the poly A$^+$ fractions from both cells (not shown) contained the 2900 base SV40 small t mRNA and the 2600 base large T mRNA. Results of S1 nuclease analysis[2] of SV40 transcripts from 12-1 stem cells indicate that these messages are spliced and transported to the cytoplasm of stem cells (not shown)[25].

Conclusion of the *in vitro* studies

To monitor the molecular events associated with the expression of the SV40 and polyoma genomes on differentiation, we have constructed teratocarcinoma stem cell lines, F9/SV and F9/Py, containing single integrated copies of the SV40 or polyoma genomes. This was accomplished by introducing a recombinant plasmid consisting of pBR322 linked to the HSV-1 *tk* gene and the SV40 polyoma genome into the F9 TK$^-$ cell. F9/Py stem cells are SSEA-1 positive, H2 and β_2-microglobulin negative. The F9/SV differentiated clones have been established.

In two F9/SV stem cell clones, 12-1 and 14-2, there is a single copy per cell of the entire plasmid genome integrated through a site of the pBR322 genome; the plasmid genome does not rearrange during differentiation and is the same in the homologous differentiated daughter clones, 12-1a and 14-1a. In the F9/Py stem cell clone, the plasmid genome is integrated through a site on the polyoma virus genome which may interrupt the polyoma virus early region, precluding expression of polyoma T antigens in differentiated cells.

One stem/differentiated F9/SV40 cell pair, 12-1 and 12-1a, was examined for DNase I sensitivity, methylation and transcription of plasmid genomes in the two cell types, with the intention of correlating DNase I insensitivity and hypomethylation of specific plasmid genomes with transcriptional activity of that plasmid genome in the two cell types.

Unexpectedly, we observed that at the same DNase I concentration, total stem cell DNA within isolated nuclei was more sensitive to DNase I digestion than total 12-1a DNA treated in the same way. The DNase I hypersensitivity of stem cell chromatin is not due to higher endogenous nuclease activity in stem cells nor does it seem to be due to better permeability of stem cell nuclei, since purified stem cell chromatin is more DNase I sensitive than purified 12-1a chromatin. The SV40 and pBR322 genomes are at least ten-fold more sensitive to DNase I digestion in 12-1 than in 12-1a cells; the H2 and β-microglobulin genes, which are not transcribed in 12-1 cells[8] are also more sensitive to DNase I digestion in 12-1 than in 12-1a cells[8], in which they are fully expreseed. Thus, we favor the conclusion that DNase I hypersensitivity of stem cell chromatin is not related to its transcriptional state; that is, DNase I hypersensitivity of stem cell chromatin has a different configurational basis than the DNase I hypersensitivity of active genes in somatic cells. Alternatively, it is possible that all genes in stem cells are in an 'active' or 'open' configuration regardless of the transcriptional state. It should be possible to determine if the latter hypothesis is correct by comparing DNase I sensitivity of two regions of the stem cell genome, such as the early (transcribed) and late (non-transcribed) regions of the SV40 genome, at low DNase I concentrations

$(0.1\,\mu g/ml\,DNA)^{15}$; if the entire stem cell genome is in the 'active' configuration, there should be no difference in DNase I sensitivity of differentially transcribed regions of the stem cell genome.

The SV40 genome is not differentially methylated in 12-1 and 12-1a cells, at least at the cytosine residues cleaved by Hpa II, and Hha II and Hha I restriction sites are within the late region of the SV40 genome, which is not trascribed in either 12-1 or 12-1a cells; in addition, the lack of methylation of these cytosine residues may simply reflect lack of methylation of the transfecting plasmid genome. Since differential methylation of Ava I recognition site(s) of the pBR322 genome and/or host flanking sequences was observed in these two cells, it is possible that *de novo* methylation of this cytosine residue(s) has occurred during the transition from stem to differentiated cell.

Since pBR322 specific RNA is not detected in stem or differentiated cells, the simplest interpretation of these results is that pBR322 genome is not transcribed in either cell type. The HSV-1 *tk* gene is transcribed in both cell types which is, of course, expected since both cells depend for their survival on expression of HSV-1 thymidine kinase[24, 25]. The SV40 early gene product, T antigen, is not detected in 12-1 (or 14-2) stem cells by immunofluorescence or immunoprecipitation using monoclonal anti-T antibody[25]. Since polyadenylated, spliced, cytoplasmic mRNAs for both large T and small t are present in stem cells in amounts comparable to the same mRNAs found in 12-1a cells, the block to expression of SV40 T antigens may be post-transcriptional; it is possible, however, that there is some defect in the stem cell SV40 mRNAs that is not detectable by S1 nuclease analysis. It will be important to determine if the SV40 mRNAs produced in stem cells are translatable *in vitro* in order to define further the molecular basis for lack of production of SV40 T antigen in teratocarcinoma-derived stem cells.

Teratocarcinoma stem cell hybrids to study X chromosome inactivation

Inactivation of one of the two X chromosomes occurs in females during early embryonal development[28]. As a result females are mosaics of cells that express only one of the two X chromosomes[28]. In order to understand the molecular basis of X chromosome inactivation it was felt necessary to develop an *in vitro* cell system in which X chromosome inactivation may occur following the differentiation of stem cells. For this purpose a teratocarcinoma stem cell line was constructed that carries an integrated plasmid vector containing an intact SV40 genome[24, 25] and is deficient in hypoxanthine phosphoribosyltransferase (HPRT)[6a]: it was also reasoned that in order to study the molecular basis of X chromosome inactivation it is necessary to obtain clones of differentiated cells expressing either one of the X chromosomes. Since the F9 teratocarcinoma stem cells terminally differentiate following retinoic acid induction[24, 41] it is extremely hard to obtain continuous cell lines. On the contrary, if the stem cells contain an integrated SV40 genome, these differentiated derivatives are able to grow indefinitely[16, 24, 25]. Therefore, we have hybridized the 12-1-6TG stem cells (SV40$^+$ HPRT) with spleen cells derived from male *Mus caroli* which express a different electrophoretic form of glucose-6-phosphate dehydrogenase (G6PD). Both differentiated and stem cell hybrids were obtained. Interestingly, the stem cell hybrids express both parental G6PD and G6PD heteropolymer. Following induction with retinoic acid one of the two X chromosomes became inactivated in the differentiated cell clones (Croce, 1982). Therefore we have constructed a cell system that allows the study of X chromosome inactivation *in vitro*. Since cloned DNA probes of X-linked HPRT and G6PD genes that undergo to inactivation are available it becomes possible to investigate the molecular basis of X chromosome inactivation in mammalian cells.

Conclusions

Two different systems are presently available to insert genes of interest into living mice: the introduction of teratocarcinoma stem cell hybrids or DNA transformed teratocarcinoma stem cells into the mouse blastocyst or the introduction of cloned DNA molecules into the nuclei of fertilized eggs. The advantage of the teratocarcinoma stem cell system is that it is possible to inject hybrids or transformants that have been already well characterized. The major disadvantage of the system, however, is that it is very hard to obtain chimeras the germ line of which derive from the stem cells. The microinjection of DNA molecules has the advantage that a high fraction of the progeny has the introduced DNA sequences integrated in the chromosomal DNA of every cell of the body of the animal, including its germ cells. Therefore, it is easy to obtain vertical transmission of the foreign sequences. The disadvantage is that the sequences might integrate into sites that do not allow the normal expression of the gene of interest. This system, however, will become enormously powerful when it will be possible to direct the microinjected DNA sequences to specific sites within the genome.

The availability of teratocarcinoma-derived stem cell lines that can be induced to differentiate when exposed to the appropriate inducers make it possible to investigate the molecular basis of differential gene expression in stem versus differentiated cells. Since the teratocarcinoma stem cells correspond to cells of a five-day old embryo, the information we derive from these investigations might be relevant to the mechanisms of gene control during early embryogenesis.

The development and the improvements of these different approaches to the study of developmental biology should improve our understanding of the genetic mechanisms resulting in differential gene expression during differentiation and embryonal development.

References

1. Artz, K. and Jacobs, F. (1974). Detection of H-2 antigens on the cells of the early mouse embryo. *Transplantation*, **16**, 675–677
2. Berk, A. J. and Sharp, P. A. (1977). Sizing the mapping of early adenovirus mRNAs by gel electrophoresis of S1 endonuclease-digested hybrids. *Cell*, **12**, 721–732
3. Brinster, R. L. (1974). The effect of cells transferred into the mouse blastocyst on subsequent development. *Journal of Experimental Medicine*, **140**, 1049–1056
4. Brinster, R. L., Chen, H. Y., Trumbauer, M., Senear, A. W., Warren, R. and Palmiter, R. D. (1981). Somatic expression of herpes thymidine kinase in mice following injection of a fusion gene into eggs. *Cell*, **27**, 223–231
5. Correani, A. and Croce, C. M. (1980). Expression of the teratocarcinoma phenotype in hybrids between totipotent mouse teratocarcinoma and myeloma cells. *Journal of Cellular and Comparative Physiology*, **105**, 73–79
6. Costantini, F. and Lacy, E. (1981). Introduction of a rabbit β-globin gene into the mouse germ line. *Nature, London*, **294**, 92–94

6a. Croce, C. M., (1980). Cancer genes in cell hybrids. *Biochimica et Biophysica Acta,* **605,** 411–430

7. Croce, C. M., Koprowski, H. and Litwack, G. (1974). Regulation of the corticosteriod inducibility of tyrosinc aminotransferase in interspecific hybrid cells. *Nature, London,* **249,** 839–841

8. Croce, C. M., Linnenbach, A., Huebner, K., Parnes, J. R., Margulies, D., Appella, E. and Seidman, J. (1981). Control of expression of histocompatibility antigens (H2) and β_2-microglobulin in F9 teratocarcinoma stem cells. *Proceedings of the National Academy of Sciences, USA,* **78,** 5754–5758

9. Damjanov, I., Solter, D., Belicza, M. and Skreb, N. (1971). Teratomas obtained through extrauterine growth of seven-day mouse embryos. *Journal of the National Cancer Institute,* **46,** 471–475

10. Desrosiers, R. C., Mulder, C. and Fleckenstein, B. (1979). Methylation of herpes virus saimiri DNA in lymphoid tumor. *Proceedings of the National Academy of Sciences, USA,* **76,** 3839–3843

11. Dewey, M., Martin, D. W., Martin, G. R. and Mintz, B. (1977). Mosaic mice with teratocarcinoma-derived mutant cells deficient in hypoxanthine phosphoribosyltransferase. *Proceedings of the National Academy of Sciences, USA,* **74,** 5564–5568

12. Fujimura, F. K., Deininger, P. L., Friedmann, T. and Linney, E. (1981). Mutation near the polyoma DNA replication origin permits productive infection of F9 embryonal carcinoma cells. *Cell,* **23,** 809

13. Gordon, J. W., Scangos, G. A., Plotkin, D. J., Barbosa, J. A. and Ruddle, F. H. (1980). Genetic transformation of mouse embryos by microinjection of purified DNA. *Proceedings of the National Academy of Sciences, USA,* **77,** 7380–7384

14. Huebner, K., Tsuchida, N., Green, C. and Croce, C. M. (1979). A murine teratocarcinoma stem cell line carries suppressed oncogenic virus genomes. *Journal of Experimental Medicine,* **150,** 392–405

15. Huebner, K., Linnenbach, A., Weidner, S., Glenn, G. and Croce, C. M. (1981). Deoxyribonuclease I sensitivity of plasmid genomes in teratocarcinoma stem and differentiated cells. *Proceedings of the National Academy of Sciences, USA,* **78,** 5071–5075

16. Huebner, K., Linnenbach, A., Glenn, G., Weidner, S. and Croce, C. M. (1982). In *Mechanisms of Chemical Carcinogenesis,* pp. 503–515. Alan Liss, Inc.

17. Illmensee, K. and Croce, C. M. (1979). Xenogenic gene expression in chimeric mice derived from rat-mouse hybrid cells. *Proceedings of the National Academy of Sciences, USA,* **76,** 879–883

18. Illmensee, K. and Mintz, B. (1976). Totipotency and normal differentiation of single teratocarcinoma cells cloned by injection into blastocysts. *Proceedings of the National Academy of Sciences, USA,* **73,** 549–553

19. Illmensee, K., Hoppe, P. C. and Croce, C. M. (1978). Chimeric mice derived from human-mouse hybrid cells. *Proceedings of the National Academy of Sciences, USA,* **75,** 1914–1918

20. Katinka, M., Yaniv, M., Vasseur, M. and Blangy, D. (1980). Expression of polyoma early functions in mouse embryonal carcinoma cells depends on sequence rearrangements in the beginning of the late region. *Cell,* **20,** 393–399

21. Kleinsmith, L. J. and Pierce, G. B. (1964). Multipotentiality of single embryonal carcinoma cells. *Cancer Research,* **24,** 1544–1551

22. Knowles, B. B., Pan, S., Solter, D., Linnenbach, A., Croce, C. M. and Huebner, K. (1980). Expression of H-2, laminin and SV40 T and TASA on differentiation of transformed murine teratocarcinoma cells. *Nature, London,* **288,** 615

23. Kuo, M. T., Mandel, J. L. and Chambon, P. (1979). DNA methylation: correlation with DNase I sensitivity of chicken ovalbumin and conalbumin chromatin. *Nucleic Acids Research,* **7,** 2105

24. Linnenbach, A., Huebner, K. and Croce, C. M. (1980). DNA-transformed murine teratocarcinoma cells: regulation of expression of simian virus 40 tumor antigen in stem versus differentiated cells. *Proceedings of the National Academy of Sciences, USA,* **77,** 4875–4879

25. Linnenbach, A., Huebner, K. and Croce, C. M. (1981). Transcription of the SV40 genome in DNA-transformed murine teratocarcinoma stem cells. *Proceedings of the National Academy of Sciences, USA,* **78,** 6386–6390

26. Littlefield, J. W. (1964). Selection of hybrids from matings of fibroblasts *in vitro* and their presumed recombinants. *Science,* **145,** 709–710

27. Litwack, G. and Croce, C. M. (1979). Somatic cell hybrids between totipotent mouse teratocarcinoma and rat hepatoma cells. *Journal of Cellular and Comparative Physiology,* **101,** 1–8

28. Lyon, M. F. (1961). Gene action in the X-chromosome of the mouse (*Mus musculus* L.) *Nature, London,* **190,** 372

29. McBurney, M. W. (1976). Clonal lines of teratocarcinoma cells *in vitro*–differentiation and cytogenetic characteristics. *Journal of Cellular and Comparative Physiology,* **89,** 441–456

30. McGhee, J. G. and Ginder, G. D. (1979). Specific DNA methylation sites in the vicinity of the chicken β-globin genes. *Nature, London*, **280**, 419–420

31. Mandel, J. L. and Chambon, P. (1979). DNA methylation: organ specific variations in the methylation pattern within and around ovalbumin and other chicken genes. *Nucleic Acids Research*, **7**, 2081–2103

32. Miller, R. A. and Ruddle, F. H. (1976). Pluripotent teratocarcinoma–thymus somatic cell hybrids. *Cell*, **9**, 45–55

33. Mintz, B. and Illmensee, K. (1975). Normal genetically mosaic mice produced from malignant teratocarcinoma cells. *Proceedings of the National Academy of Sciences, USA*, **72**, 3585–3589

34. Nicholas, J. F., Avner, P., Gaillard, J., Guenet, J. L., Jakob, H. and Jacob, F. (1976). Cell lines derived from teratocarcinomas. *Cancer Research*, **36**, 4224–4231

35. Palmiter, R. D., Chen, H. Y. and Brinster, R. L. (1982). Differential regulation of metallothionein–thymidine kinase fusion genes in transgenic mice and their offspring. *Cell*, **29**, 701–710

36. Papioannou, V., McBurney, M., Gardner, R. and Evans, M. (1975). Fate of teratocarcinoma cells injected into early mouse embryos. *Nature, London*, **258**, 70–73

37. Segal, S. and Khoury, G. (1979). Differentiation as a requirement for simian virus 40 gene expression in F9 embryonal carcinoma cells. *Proceedings of the National Academy of Sciences, USA*, **76**, 5611–5615

38. Segal, S., Levine, A. J. and Khoury, G. (1979). Evidence for non-spliced SV40 RNA in undifferentiated murine teratocarcinoma stem cells. *Nature, London*, **280**, 335–337

39. Sekikawa, K. and Levine, A. J. (1981). Isolation and characterization of polyoma host range mutants that replicate in nullipotenial embryonal carcinoma cells. *Proceedings of the National Academy of Sciences, USA*, **78**, 1100–1104

40. Stevens, L. C. (1970). The development of transplantable teratocarcinomas from intratesticular grafts of pre- and postimplantation mouse embryos. *Developmental Biology*, **21**, 364–382

41. Strickland, S. and Mahdavi, V. (1978). The induction of differentiation in teratocarcinoma stem cells by retinoic acid. *Cell*, **15**, 393–403

42. Sutter, D. and Doerfler, W. (1980). Methylation of integrated adenovirus type 12 DNA sequences in transformed cells is inversely correlated with viral gene expression. *Proceedings of the National Academy of Sciences, USA*, **77**, 253–256

43. Swartzendruber, D. E. and Lehman, J. M. (1975). Neoplastic differentiation: Interaction of simian virus 40 and polyoma virus with murine teratocarcinoma cells *in vitro*. *Journal of Cellular and Comparative Physiology*, **85**, 179–188

44. Swartzendruber, D. E., Friedrich, T. D. and Lehman, J. M. (1977). Resistance of teratocarcinoma stem cells to infection with simian virus 40: early events. *Journal of Cellular and Comparative Physiology*, **93**, 25–30

45. Thomas, P. S. (1980). Hybridization of denatured RNA and small DNA fragments transferred to nitrocellulose. *Proceedings of the National Academy of Sciences, USA*, **77**, 5201

46. Van der Ploeg, L. H. T. and Flavell, R. A. (1980). DNA methylation in the human γδβ-globin locus in erythroid and nonerythroid tissues. *Cell*, **19**, 947–958

47. Vasseur, M., Kress, C., Montreau, N. and Blangy, ?. (1980). Isolation and characterization of polyoma virus mutants able to develop in embryonal carcinoma cells *Proceedings of the National Academy of Sciences, USA*, **77**, 1068–1072

48. Waalwijk, C. and Flavell, R. A. (1978). DNA methylation at a CCGG sequence in the large intron of the rabbit β-globin gene: tissue-specific variations. *Nucleic Acids Research*, **5**, 4631–4661

Expression of specific genes

Our understanding of eukaryotic genes has ripened in the last six years as a direct result of the development of new technologies. The first, and most important, is the recombinant DNA methodology which permits the molecular cloning in a bacterial host of essentially any segment of DNA. These methods were first exploited using bacterial plasmids, autonomously replicating circular DNAs which frequently carry genes for resistance to antibiotics. The engineering of these DNAs to cause them to have single cleavage sites for a given restriction enzyme made it possible to insert a foreign DNA segment into that site by cleavage of both DNAs and subsequent ligation. Needless to say, it was and is a prerequisite of this technology that the DNA enzymology is up to the same standard as the DNA technology itself. Such plasmid vectors have been elaborated to carry single cleavage sites for a large number of restriction enzymes to facilitate their general use. Thus, pBR322, or its derivatives has single sites for Eco RI, Hind III, Bam HI, Sal I, Sph I, Pvu II, Pst I, Pvu I and Eco RV which can be used for the insertion of foreign DNA segments. Plasmid cloning technology permits the easy introduction of DNA inserts of up to 10–15 kb in length. Unfortunately the size of the plasmid influences the efficiency of transformation of *E. coli*; large plasmids are difficult to introduce into *E. coli* so that for cloning of large DNA segments alternative phage or cosmid vectors must be used. Finally, plasmid cloning systems have been developed for a number of other bacterial species, e.g. gram positive bacteria such as *B. subtilis* which are of particular interest for the production and secretion of foreign proteins in this host; and broad host range plasmids such as pSF1010 allow cloning in a number of other gram negative bacteria such as *Pseudomonas*. Despite its limitations, plasmid cloning is the method of choice for the fine-scale dissection of a eukaryotic gene once it has been cloned from its host on a large DNA segment as will be discussed below. It is historically the first system, and was the first system used to construct a repository of random DNA segments in a bacterial host. In this approach, DNA from a given source is cleaved in one way or other to generate a random set of overlapping DNA fragments. In the first experiments shear was used to generate these random DNA fragments; more recently partial cleavage with one or more restriction enzymes is used. The random fragments are then cloned in the vector of choice to generate a collection of DNA segments which has been called either a library (for the intellectuals) or bank (for the businessmen).

If the number of transformed bacteria is large enough statistically the entire genome of the organism is represented; the number of clones required depends on the size of the inserts and the size of the genome studied. Screening of this library with a probe for a given DNA region allows the isolation of overlapping clones which, in turn, permit the reconstruction of the way the gene region is built up. In *Figure 1*, for example, a series of overlapping phage clones of the human β-globin locus is depicted.

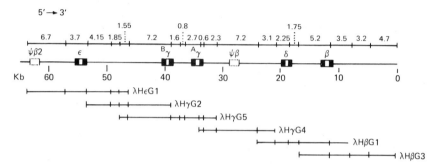

Figure 1. The linkage map of the human β-globin gene locus as shown by the structural analysis of overlapping phage λ clones. The genomic segments of the clones isolated are shown together with the cleavage sites for the enzyme Eco RI (redrawn from reference 6). The numbers indicate the size of fragments in kb.

As stated above, cloning of large DNA segments requires vectors adapted for this purpose. In practice, phage vectors are used for DNA inserts up to 25 kb in size. For larger fragments up to 40 kb cosmids based on the λ system are used.

Cosmids are cloning vehicles derived from plasmids which also contain the λ phage DNA cohesive ends (cos). Only a small region near the cos site is recognized by the λ phage packaging system, both *in vivo* and *in vitro*. A DNA containing two cos sites in the same orientation separated by 35–53 kb can therefore be packaged *in vitro* as described below and can be transduced into *E. coli* at high efficiency. Inside the bacteria the cosmid circularizes (in the same way as does λ DNA), but because the cosmid does not contain all the essential λ phage genes, it is not able to go through a lytic cycle. Thereafter, cosmids replicate as plasmids and are therefore constructed with suitable selective markers such as resistance to antibiotics; this is used for phenotypic selection. Since cosmids only need to contain a replication origin, selective marker(s) and cos site, they can be small, frequently less than 5 kb in size. The upper size limit of DNA fragments that can be inserted in cosmids and packaged into λ phage particles is therefore approximately 45 kb, much larger than it would be possible to clone in λ or plasmid vectors. The size selection imposed by the *in vitro* packaging system also leads to a minimum size for cloned DNA: if the cosmid to be packaged is 5 kb, the inserted DNA must be at least 30 kb long. The cosmid cloning system therefore offers an efficient system for cloning large DNA fragments, with a low background of non-recombinant colonies, since cosmid dimers are too small to be packaged.

A standard cloning expedition for a eukaryotic gene will usually start with a given mRNA. This will either be purified (if it is an abundant mRNA species and used directly as a probe) or will first be cloned as a double stranded DNA by first copying the mRNA with reverse transcriptase converting the cDNA copy to a double stranded DNA and cloning in a plasmid vector. The cDNA or mRNA is then used as a radioactively labelled probe to screen a phage or cosmid library. Clones are picked and characterized by restriction enzyme digestion. The DNA segments of interest are then subcloned onto a plasmid to facilitate detailed analysis.

The second important technical breakthrough in the last five years has been the development of routine methods for determining DNA sequences by either chemical or enzymatic approaches. Once a gene is cloned, the sequence of several thousand base pairs can be obtained within weeks or at most months which obviously permits detailed data to be obtained of the type described in the chapters in this section.

In addition, the development of DNA blotting technology has made possible the analysis of even single copy mammalian genes directly in chromosomal DNA without the necessity of cloning the DNA segment of interest. This allows a number of experiments that would be tedious by cloning; for example, the screening of large numbers of individuals of a given species for a given DNA segment. More importantly, it can be used for the diagnosis of disease at the DNA level and should become a powerful addition to the repertoire in the clinic.

A final but vital new approach which has made possible many of the interesting conclusions of the last few years has its firm base in the new DNA technology and is called Reversed Genetics. Our understanding of the prokaryote *E. coli* and its viruses has depended to a great extent on the application of classical genetics to this bacterial system. In this approach mutants with a given phenotype are obtained and the molecular basis of the phenotype is then determined. This approach could never be applied in such detail to a higher eukaryote such as a mammal. For a rare mutant to be obtained requires searching among millions of progeny – clearly not a feasible proposition for a mouse. Reversed Genetics is exactly as its name suggests – a mutation is created at a given known site in or near a gene and its phenotypic effects are subsequently studied. This approach was first applied to the RNA phage Qβ in the pre-cloning era[6a] and has since been applied to countless cloned genes. The mutations generated can range from the gross deletion of large amounts of DNA sequence to single point mutations. These will be referred to in many of the chapters of this section. The Reversed Genetics approach requires, of course, methods for the analysis of phenotypic expression of the mutated genes. Fortunately, eukaryotic genes can be transcribed *in vitro* in the nuclei of *Xenopus* oocytes after microinjection and in animal cells such as mouse L cells or HeLa cells after the coprecipitation of DNA with calcium phosphate and its (presumptive) uptake by phagocytosis. In addition, genes of lower eukaryotes such as yeast and *Dictyostelium* can be expressed in homologous systems after introducing the DNA into these cells by analogous methods.

Looking back over this period we should ask if what we have learned has justified all the effort. Fortunately, dramatic new insights into gene organization have been obtained. Of these the most startling must be that both viral[1, 3, 5, 9] and cellular genes[2, 4, 8, 12] are frequently interrupted by introns. The position of the introns correlates well with the separation of functional domains in the protein molecule which supports the idea that exons reflect the evolutionary history of functional units of proteins, and that these units must have been shuffled to generate novel combinations of these units which, in turn, generate diversity in protein function[7]. The way in which split genes are expressed is understood in outline. Split genes are transcribed from a promoter and RNA synthesis initiates at the cap site of the mRNA. The entire gene (including introns) is transcribed and in at least some cases transcription terminates downstream from the 3′ end (polyA site) of the mRNA. Cleavage occurs at the polyA site and polyA is added post-transcriptionally. Removal of the introns by RNA splicing occurs by a mechanism which is as yet only poorly understood. DNA sequence analysis of splice sites shows that 5′ splice sites have a ↓ GT (↓ indicates the cleavage point) and 3′ splice sites an AG ↓ cleavage (see Chapter 17). It is also known that pre-mRNA exists as pre-mRNP particles and there are some suggestions that small nuclear or cytoplasmic RNAs are involved[11]. The enzymology of this process, is, however, not clear although we have some understanding of how the process works for yeast tRNA.

Promotion of RNA synthesis is at least partially understood, at least at the level of the DNA sequences which constitute the promoter. There are at least three RNA polymerases in higher eukaryotes: RNA pol I (for the large rRNAs), pol II (for the mRNAs) and pol III (for tRNAs and 5SRNA). While the promoter sequences for RNA pol II are located in the 5′ flanking regions of the gene (see e.g. Chapter 17), those for RNA pol III are found within the gene itself (see Chapter 14). As in bacteria, specific DNA sequences are required for promotion and we presume that specific DNA–protein interactions are involved in an analogous way to that shown for *E. coli*.

The fact that transcription, initiation, termination and RNA processing must occur for eukaryotic messengers raises the possibility that each step can be used for the regulation of gene expression. Thus, it is likely that transcription is one of the major control sites for gene expression. It is not possible to detect transcripts of, say, a globin gene in normal non-erythroid cells and this is generally true for most genes. In addition, the chromatin structure of active genes differs from that of inactive genes (see Chapter 2). Is differential transcription used to effect quantitative changes in the level of mRNA? It is likely that this occurs for the mouse α-amylase gene: liver amylase mRNA is produced from a promoter 206

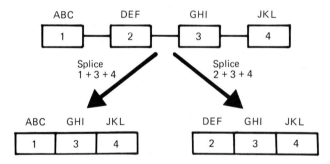

Figure 2. A schematic illustration of the differential splicing events in the mouse α-amylase genes. In the salivary gland the left-hand splicing scheme is followed; in the liver the right-hand scheme.

nucleotides upstream from the AUG, whereas in the salivary gland one hundred-fold more mRNA is produced from a promoter 95 nucleotides upstream from the AUG. The implication is, therefore, that the liver promoter is weak and the salivary gland promoter is strong and that this is at least one of the mechanisms whereby the dosage of α-amylase is kept different in the two tissues. The two mRNAs are also spliced differently as shown schematically in *Figure 2*. It is possible that these two mRNAs are translated with different efficiencies and that this too is used to regulate the level of α-amylase. Finally, there is micro-heterogeneity in the 3′ and 5′ ends of the RNA for α-amylase and a number of other genes. The latter presumably results from flexibility in the choice of site of initiation of RNA synthesis. The former is caused by transcription through more than one polyA addition site. The processing machinery then must choose to cleave at one or other of these sites and this will generate mRNAs with 3′ untranslated regions of different lengths (seen, for example, for dihydrofolate reductase mRNAs). It is possible that this length difference affects translational efficiency and also the efficiency of splicing.

Differential splicing and the use of different polyA addition sites is certainly used to generate different proteins from a single mosaic gene. First seen for the DNA viruses like SV40, it is a favourite method used in the immunoglobulins to generate protein forms which are secreted, or which remain hooked-up in the membrane by virtue of a hydrophobic transmembrane segment. The exon for the latter is eliminated by differential splicing in the mRNA for the secreting protein (see Chapter 20). Despite our understanding of these few titbits of mRNA regulation we are quite ignorant of the factors which regulate tissue-specific expression. It is, however, now possible to obtain regulated transcription in a number of cell systems; thus, globin genes respond to differentiation signals in erythroleukaemia cells after the introduction of these genes by DNA-mediated gene transfer (see Chapter 17); the transcription of interferon genes is stimulated in fibroblasts in response to virus infection (U. Weidle, H. Ragg and C. Weissmann, personal communication, 1982); and the metallothionin gene responds to Cd^{2+} in fibroblasts. In some cases the DNA sequences required for the specific response have been shown to lie in the 'promoter region' (using the term loosely here) as shown by deletion mapping. Thus, sequences from −115 (that is, 115 nucleotides to the 5′ side of the cap site) to the cap site are required for the response of the interferon gene, and linking the interferon promoter to the body of the globin gene allows induction of globin mRNA by virus infection! (U. Weidle, H. Ragg and C. Weissmann, personal communication). In these, and therefore presumably many other cases, it is likely that a specific DNA–protein interaction causes the response.

Transcriptional regulation is not likely to be this simple however since there are a number of experiments that suggest that DNA sequences far from genes influence expression (see, for examples, Chapters 17 and 18). It is possible that chromatin structure (see Chapter 2) and DNA methylation (Chapter 4) are involved with these as yet poorly understood effects.

The recombinant DNA revolution has given us a vast amount of data on the way gene families are organized. Related genes are clustered in one or a few short regions of chromosomes and gene–gene distances are of the order of a few to tens of kilobases. The intergenic DNA contains repeated sequences but these repeated sequences are sometimes also present in introns or the untranslated portion of mRNAs (Chapters 18 and 19). One of the most interesting discoveries to come out of the study of repeated sequences is the fact that these may have the property of mobile genetic elements (see Chapter 15). Originally restricted to retroviruses, this group of DNAs now includes the long repeated sequences of *Drosophila* such as copia and '412') and it is likely that many repeated sequences in mammals are mobile. Thus, Alu repeats are flanked by short direct repeats and an Alu repeat has been found inserted in a tubulin pseudogene[10]; in this case, since the sequence of the tubulin gene is known, it is possible to see that the short duplication is of the tubulin gene and therefore that it is of the same nature as seen for bacterial transposons and retroviruses.

In this section a series of experts have contributed chapters on specific gene systems. As has been traditional in biological sciences, basic principles have emerged from the study of suitable model systems. One has only to remember the impact of the *E. coli lac* operon on molecular biology. Likewise, a number of eukaryotic genes have served as models. The expression of genes during terminal differentiation has been studied in erythropoiesis (globin genes), myogenesis (actin, etc.), oogenesis, silk synthesis and so on. The role of hormones in gene expression has been studied for the avian egg white proteins and vitellogenin genes. Although most genes are present in a single copy, certain gene products are needed in large amounts at certain times in the life cycle of organisms. The way nature copes with this is to have multiple gene copies as is seen for the genes encoding the ribosomal RNAs (Chapter 12), tRNAs (Chapter 14) and histones (Chapter 16). This poses its own particular problems and is therefore of special interest. A number of gene systems, particularly the immunoglobulins, have their own intrinsic interest in addition to their role as model systems and are included in this section.

Finally, the size of any book is finite and we have had to leave out a number of interesting genes and restrict the length of the chapters.

References

1. Berget, S., Moore, C. and Sharp, P. A. (1977). Spliced segments at the 5′ terminus of adenovirus 2 in RNA. *Proceedings of the National Academy of Sciences, USA*, **74**, 3171–3175
2. Breathnach, R., Mandel, J. L. and Chambon, P. (1977). Ovalbumin gene is split in chicken DNA. *Nature, London*, **270**, 314–319
3. Chow, L. T., Gelinas, R. E., Broker, T. R. and Roberts, R. J. (1977). An amazing arrangement at the 5′ ends of adenovirus 2 messenger RNA. *Cell*, **12**, 1–8
4. Doel, M. T., Houghton, M., Cook, E. A. and Carey, N. H. (1977). The presence of ovalbumin mRNA coding sequences in multiple restriction fragments of chicken DNA. *Nucleic Acids Research*, **4**, 3701–3703
5. Dunn, A. R. and Hassell, J. A. (1977). A novel method to map transcripts: Evidence for homology between an adenovirus mRNA and discrete multiple regions of the viral genome. *Cell*, **12**, 23–36
6. Fritsch, E. F., Lawn, R. M. and Maniatis, T. (1980). Molecular cloning and characterization of the human α-like globin gene cluster. *Cell*, **19**, 959–972
6a. Flavell, R. A., Sabo, D., Bandle, E. and Weissmann, C. (1974). Site-directed mutagenesis: Generation of an extracistronic mutation in bacteriophage QB RNA. *Journal of Molecular Biology*, **89**, 255–272
7. Gilbert, W. (1978). Why Genes in Pieces? *Nature, London*, **271**, 501
8. Jeffreys, A. J. and Flavell, R. A. (1977). A physical map of the DNA regions flanking the rabbit β-globin gene. *Cell*, **12**, 429–439
9. Klessig, D. F. (1977). Two adenovirus m-RNAs have a common 5′ terminal leader sequence encoded at least 10 kb upstream from their main coding regions. *Cell*, **12**, 9–12
10. Lemischka, I. and Sharp, P. A. (1982). The sequences of an expressed rat tubulin gene and a pseudogene with an inserted repetitive element. *Nature, London*, **300**, 330–335
11. Lerner, M. R., Boyle, J. A., Mount, S. Wolin, S. L. and Steiz, J. (1980) Are snRNPs involved in splicing? *Nature, London*, **283**, 220–244
12. Tilghman, S. M., Tiemeire, D. C., Seidman, J. G., Peterlin, B. M., Sullivan, M., Maizel, J. V. and Leder, P. (1978). Intervening sequence of DNA identified in the structural portion of a mouse β-globin gene. *Proceedings of the National Academy of Sciences, USA*, **75**, 725–729

Chapter 12

Genes for ribosomal RNA

D. M. Glover

Introduction

Ribosomal RNA represents the major transcriptional product in cells. In the higher eukaryotes the demand for rRNA is met by having multiple copies of its genes, which are physically located in the nucleolus. In most higher eukaryotes, the genes are tandemly arranged, and can be visualized frozen in transcription by electron microscopy. One such example of rDNA transcription matrices is shown in *Figure 12.1*. Each unit of transcription is separated from the next by a non-transcribed spacer (NTS) region. The transcribed regions are densely packed with polymerase molecules and nascent chains of the primary transcripts. The nascent RNA chains associate with proteins and generate pre-ribosomal particles. In the human tumour cell-line, HeLa, for example, these particles have an 80S sedimentation coefficient, comparable with mature ribosomes[26, 48]. The pre-ribosomal particles contain some of the proteins found in mature ribosomal sub-units together with some additional non-ribosomal nucleolar proteins, which apparently have enzymic activities for the endonucleolytic processing of the primary transcript of the rDNA.

In contrast to the enzymology of rDNA processing, the pathway by which the primary transcript is cleaved has been well characterized for a number of organisms. The primary transcript is rapidly methylated, predominantly in those regions which correspond to the mature rRNA sequences. This has been shown by following the incorporation of labelled methionine into the primary transcript, processing intermediates and the mature rRNA[52]. If methylation is used to identify oligonucleotides in fingerprinting experiments, the same pattern is seen for the primary 45S HeLa cell transcript as the mature 18S and 28S rRNAs[31]. In human cells, about 48% of the primary transcript is discarded and these sequences are not methylated. The processing pathways have been elucidated by the electrophoretic analysis of pulse labelled RNA. It has also been possible to visualize the processing intermediates by electron microscopy, and determine their relationships by a comparison of their secondary structures[53]. Ribosomal RNA has a high degree of secondary structure which is apparent as a complex pattern of duplex hairpin regions when partially denatured. The denaturation patterns for 28S rRNA and 20S (the direct precursor to 18S) RNA are shown in *Figure 12.2,* together with the denaturation map for the 45S RNA from which they are derived. Such an examination of each of the intermediates allows them to be positioned in the pathway, also shown in *Figure 12.2*.

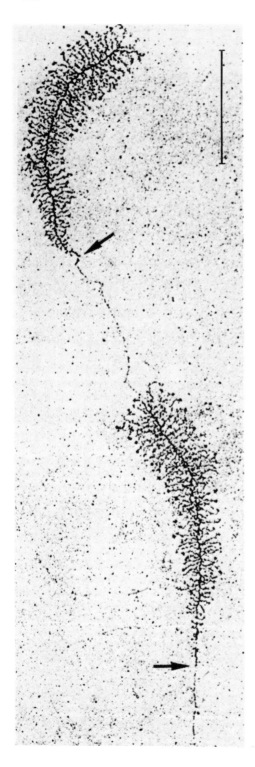

Figure 12.1. Transcription matrices of ribosomal genes from *Xenopus* oocyte nucleoli. The arrows point to the start points of transcription. The bar represents 1 μm. From Scheer *et al.*[47]

(a)

(b)

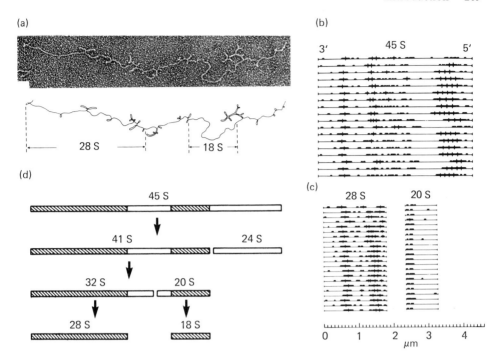

(d)

(c)

Figure 12.2. Secondary structure maps of rRNA from HeLa cells. The micrograph (a) shows the secondary structure of the 45S primary ribosomal RNA from HeLa cells. Structure maps can be drawn from contour length measurements of such molecules and are shown for (b) the 45S RNA and (c) the 28S and 20S RNA molecules. The structural relationships between these molecules can readily be seen, and a comparison of such data for each of the processing intermediates permits a scheme to be drawn for the processing pathway as shown in (d). (This figure is taken from the data of Wellauer and Dawid[53]. In the *original* version of this figure, incorrect assignments of 5' and 3' ends of the RNA have been given. This was because the ascites exonuclease used to digest the RNAs has the opposite polarity to that believed at the time. The figure has therefore been changed and the correct polarities are now shown.)

The basic features of this processing pathway are seen in all eukaryotes. The ribosomal genes have, in fact, been extremely highly conserved and show extensive cross species homology. During the evolution of eukaryotes the transcription unit has, however, become larger. In yeast, for example, the primary transcript is about 7.2 kb whereas in man it is bout 13 kb. This is largely due to an increase in the proportion of the unit that is discarded in processing, but in addition the RNA of the large ribosomal subunit has increased in length (3.8 kb to 4.9 kb for these two organisms) and there has been a slight shortening of the 18S RNA (2 kb in yeast; 1.75 kb in man). Molecular cloning techniques make it a relatively trivial exercise to clone the rDNA from any eukaryote, and consequently the preliminary characterization of rDNA genes has been reported for many organisms. This chapter will concentrate upon describing the rDNA from just three organisms which show different patterns of rDNA organization and expression. Two of these organisms have rDNA in which the gene for the large rRNA is interrupted by an 'insertion' or 'intervening sequence'. Such sequence interruptions are found in the rDNA of a diverse range of organisms including yeast mitochondrial rDNA,

Chlamydomonas chloroplast rDNA, *Physarum* rDNA, *Tetrahymena* rDNA and the rDNA of many insects. We will examine here the 28S rRNA genes of *Drosophila* and *Tetrahymena* which have sequence interruptions differing markedly in several respects. First of all, though, we will look at the rDNA of *Xenopus laevis*, which does not have such interruptions and is perhaps the best characterized.

The rDNA of *Xenopus laevis*

The rDNA of *Xenopus laevis* was the first from any eukaryotic organism to be studied. This was because it has a much greater G + C content than the rest of the genome, and could therefore be purified as a density satellite upon equilibrium centrifugation in cesium chloride[6]. The tandemly repeating pattern of organization of this DNA was obvious from an examination of partially denatured DNA in the electron microscope[57].

Heterogeneity of the non-transcribed spacer (NTS)

The NTS region has a higher G + C content than the transcribed region, and therefore remains double stranded under conditions in which the transcribed regions are partially melted. Measurements of the double stranded region containing the NTS suggested that the spacer regions in the tandem array were homogeneous in length, but subsequent molecular characterization of this region has shown this early conclusion to be erroneous. *Xenopus* rDNA was the first eukaryotic DNA to be propagated in *E. coli*, in experiments in which Eco RI fragments of gradient purified rDNA were cloned in the plasmid vector pSC101[33]. The Eco RI fragments fell into two size classes: a homogeneous set of fragments 4.5 kb in length and a heterogenous set of fragments ranging from 6.0 kb to 6.9 kb. The homogeneous set of fragments contain most of the 28S gene, and the heterogenous set the 18S gene and NTS.

The heterogeneity results from variation in the number of sub-repetitive elements within the NTS itself. This was first evident from a set of heteroduplex experiments carried out by Wellauer *et al.*[55]. In the first type of experiment, the DNA from a single cloned segment containing the NTS was denatured and allowed to reassociate. A fully reannealed structure was not always seen in the electron microscope, but sometimes two single stranded loops of equal length were seen. This is indicative of a tandemly repeating structure annealing out of register. When two clones with NTSs of different lengths were denatured and then mixed and reannealed, heteroduplex structures were formed in which in addition to two loops of the kind described above there was a further loop corresponding in length to the length difference between the two NTSs. The limits within which the loops occur define the extremes of the sub-repetitive sequences within the NTS. Two repetitive regions were mapped in these experiments, but subsequently detailed restriction mapping and sequencing experiments have shown that one of these regions defined by electron microscopy is itself comprised of two repetitive areas separated by a non-repetitive sequence. The restriction mapping experiments elegantly demonstrate three repetitive regions within the NTS. In these experiments restriction fragments from within the NTS were labelled at only one of their ends and then partially digested with enzymes which cleave within the repetitive elements.

211

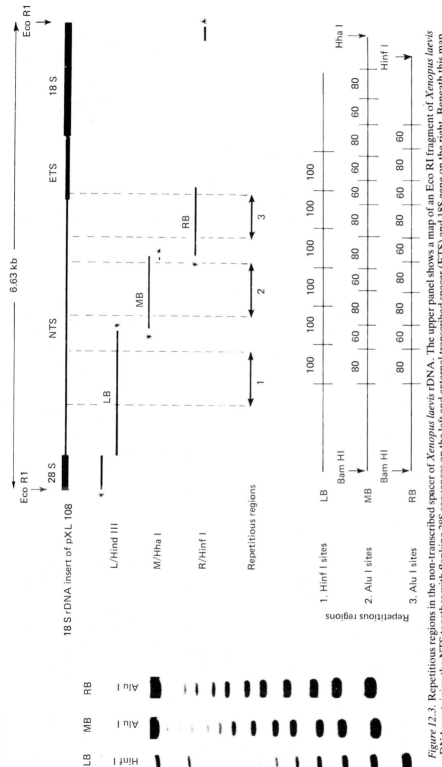

Figure 12.3. Repetitious regions in the non-transcribed spacer of *Xenopus laevis* rDNA. The upper panel shows a map of an Eco RI fragment of *Xenopus laevis* rDNA containing the NTS together with flanking 28S sequences on the left and external transcribed spacer (ETS) and 18S gene on the right. Beneath this map are shown a set of restriction fragments labelled at one end (*) and which cover three repetitious regions. The repetitive elements within three of these fragments, LB, MB and RB, are revealed by autoradiography (left) following partial digestion and electrophoresis as a regularly spaced series of bands. The cleavage maps for these three restriction fragments are shown in the lower part of the figure. DNA sequencing studies have subsequently shown that the repeating units in regions 2 and 3 are 60 and 81 nucleotide pairs. Sequencing has also detected another repetitious region, 0, between region 1 and the 3′ end of the 28S gene, and having weak homology with region 1.

Following gel electrophoresis and autoradiography, a 'ladder' of labelled fragments is seen (*Figure 12.3*). Each 'rung' on this 'ladder' is a fragment corresponding in length to the distance between the labelled terminus and successive cleavage sites[5]. The three repetitious regions are shown in *Figure 12.3*: region 1 has a repeating unit of 100 base pairs whereas regions 2 and 3 are almost identical and have alternating 81/60 base pair elements. DNA sequencing shows that these 60 and 81 base pair arrangements are identical excepting a run of 21 nucleotides. These repetitious sequences within the NTS allow recombination to take place between rDNA units that are not in perfect alignment. The length heterogeneity of the spacers indicates that such unequal exchanges do occur. On an evolutionary time scale, this presents one mechanism whereby rDNA variants could be moved about between different chromatids. This, together with selection forces, could then represent one way by which desirable changes could be spread through the tandemly arranged genes or, reciprocally, eliminated from a population.

This extent of NTS heterogeneity is not seen in all tissues. During oogenesis the number of rRNA genes is increased by up to 5000-fold in order to provide for the large numbers of ribosomes that are accumulated in the oocyte. The structure of the rDNA in somatic and oocyte tissue is the same, but usually only a subset of NTS lengths are seen in the oocyte rDNA indicating that a fraction of the chromosomal sequences are selectively amplified[56]. This is thought to occur by rDNA units becoming extrachromosomal and then replicating as rolling circles. rDNA isolated by density gradient centrifugation from oocytes has up to 9% circular molecules corresponding in length from 1–20 rDNA units[21]. Furthermore rolling circle structures can be seen by autoradiography following pulse labelling with ^{3}H-thymidine[44].

Studies on transcription

The three repetitious elements in the NTS are separated from each other by non-repetitious regions which contain cleavage sites for Bam HI and have acquired the name 'Bam islands' These two regions have considerable homology to the sequences immediately 5' to the transcription start point[34, 36]. This has led to a scheme to account for the evolution of the NTS by a series of reduplication events of an ancestral sequence postulated to be at the boundary of the external transcribed spacer and the NTS. This leads to the interesting idea that by duplicating the promoter region, a sequence has been created that is capable of binding RNA polymerase molecules at a high density. In support of this speculation is the observation that occasionally transcription does initiate in the spacer, possibly at the Bam islands[47].

The development of *in vitro* systems for studying transcription by RNA polymerase I has lagged behind those for RNA polymerases II and III. Such systems are now being developed[20, 25, 32] and these soon should lead to the analyses of the types described in Chapter 13. Microinjection of cloned rDNA into the nuclei of *Xenopus laevis* oocytes has served as an assay for rDNA expression. This can be followed by electron micrsocopy which reveals that about 10% of the injected circular DNA molecules are associated with nascent chains of ribonucleoprotein[51]. The high packing density of transcripts and smooth appearance of the NTS resemble the appearance of the endogenous transcription units and a higher density of nucleosome particles is seen on the plasmid vector sequences (*Figure 12.4*). Bakken et al.[1] have used this system to monitor the effect

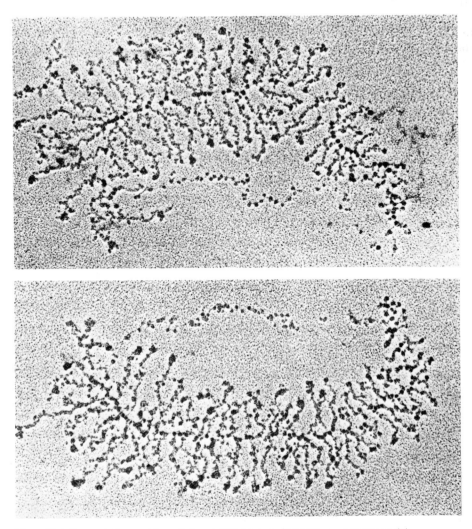

Figure 12.4. Transcription matrices of cloned rDNA injected into *Xenopus* oocyte nuclei.
The spacer sequences which directly precede the transcript have an 'unblocked'
appearance unlike the plasmid vector DNA. From Trendelenberg and Gurdon[51].

of deletions generated *in vitro* on cloned rDNA transcription units. They conclude
that the promoter lies between 320 nucleotides upstream and 113 nucleotides
downstream from the transcription initiation site, and that a cluster of four T
residues at the 3′ end of the transcription unit are required for the termination of
transcription. If only two Ts are present then termination does not occur correctly.
A similar approach has been taken by Moss and Birnstiel[35]. They have
microinjected clones of *X. laevis* rDNA into oocyte nuclei of *X. borealis*. There are
sufficient nucleotide differences between the 5′ sequences of the 40S rRNA of
these two species, that transcripts of the injected gene can be recognized in an
excess of *X. borealis* rRNA by the S1 protection mapping technique of Berk and
Sharp[3]. In these experiments they were able to confirm that transcription can

initiate at the Bam island sequences in the spacer. More recent experiments show that the major promoter element is in the region mapped by Bakken *et al.*[1]. Furthermore when two rDNA clones are injected one of which has no spacer, and the other having a spacer containing Bam islands, then the latter is preferentially transcribed (Moss, personal communication). Further studies on these interesting sequences should prove to be most illuminating.

The rDNA of *Drosophila melanogaster*

The two types of insertion sequence

The rRNA genes of *Drosophila melanogaster* were shown to be clustered at the nucleolar organizer (NO) regions of the X and Y chromosomes by Ritossa and Spiegelman[43], who were able to correlate rDNA copy number with the number of NOs in the genotype. The genes were first cloned by Glover *et al.*[18] and subsequently a high proportion shown to have a 5 kb 'insertion' in their 28S genes[17, 39, 54, 58]. This was the first example of a split gene to be found in eukaryotes, but subsequently such discontinuities were discovered in many eukaryotic genes and given the name 'intervening sequences' or 'introns'. The term 'insertion' has remained with the *Drosophila* rDNA discontinuity and is perhaps a fortunate choice since unlike other genes having intervening sequences, the *Drosophila* rDNA units that contain insertions are rarely transcribed, and it is units that do not contain insertions that serve as templates for rRNA synthesis[24, 29]. In this and also other respects the rDNA insertions could resemble the prokaryotic insertion sequences (IS elements) which cause strongly polar mutations in the operons into which they are inserted. There are two types of rDNA insertion: the type I insertions are the most common and only occur in the rDNA of the X chromosome where they are found in 60–70% of rDNA units. The type II insertions are found in about 14% of the rDNA units on both X and Y chromosomes. The two types of insertion are non-homologous; they occur at sites which are 60 nucleotide pairs apart in the uninterrupted 28S gene; they differ in their patterns of transcription.

   The most commonly found type I insertion is 5 kb in length, but there are in addition a set of shorter insertions of up to about 1 kb (see *Figure 12.5*). These represent a co-terminal sub-set of sequences from the right hand side of the 5 kb sequence. These shorter insertions are flanked on both sides by a sequence present only once in the uninterrupted rDNA sequence[11, 45]. This is a structural feature commonly found at the insertion sites of many transposable elements. Unlike other transposable elements, however, the type I insertions are always found at this site in rDNA. Furthermore, the length of the duplicated sequence flanking the insertions varies from 7 to 15 nucleotides (*Figure 12.6*), whereas with all other transposable elements that have been described a fixed length duplication is generated characteristic of the element. The sequences of all the shorter elements are contained within the 5 kb insertion suggesting that they were derived from this longer form. The 5 kb insertions, however, are not flanked by a duplication of an rDNA sequence, but to their left there is a deletion of nine nucleotides of rDNA[46]. This feature is also reminiscent of the deletions that can be generated alongside other transposable elements.

   In situ hybridization experiments show that the type I sequences are found not only in the nucleolus, but also in the chromocentral heterochromatin of salivary

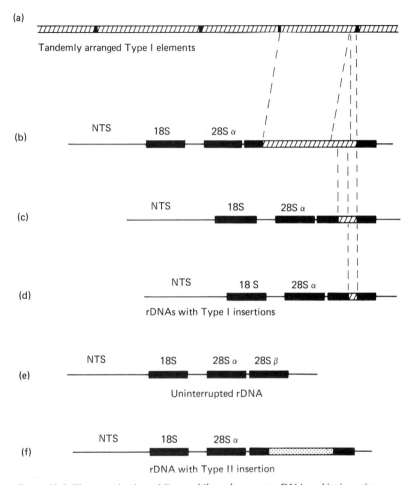

Figure 12.5. The organization of *Drosophila melanogaster* rDNA and its insertion elements. (a) A representation of tandemly arranged insertion elements as occur in the chromocentral heterochromatin[24]. These elements may have sequence deletions, as does the one shown here, or also may contain sequence insertions (not shown in this diagram). (b) A map of an rDNA unit containing a 5 kb type I insertion; (c) and (d) two units containing shorter insertions of approximately 1 kb and 0.5 kb. These shorter insertions are co-terminal subsets of the 5 kb insertion. (e) Uninterrupted rDNA unit. The 28S gene encodes four RNAs which are hydrogen bonded within the 28S rRNA: the 5.8S and 2S RNAs are not shown in these diagrams and map in the internal transcribed spacer; the other two polynucleotide components have been called the α and β chains; the insertions map within the gene for the β chain. (f) An rDNA unit with a type II insertion. This insertion is not homologous to the type I insertion and occurs at a position 60 nucleotides upsteam in the direction of rDNA transcription from the type I insertions[46].

gland chromosomes and also a band, 102C, on chromosome 4[38]. Type II insertions, on the other hand, are only found in the nucleolus. The majority of the 'non-nucleolar' type I insertion elements occur in tandem arrays[10, 23]. These tandemly arranged type I elements are more heterogeneous in length and sequence than the insertions in the rDNA, perhaps indicating that the mechanisms which operate to maintain homogeneity of rDNA also work to some extent upon the

216

Insertions flanked by duplications:

LEFT JUNCTION   RIGHT JUNCTION

RI10

                              insertion 0.525(0.530) kb

GAATGGATTAACGAGAGATTCCTACTGTCCCTATCTACTACCACAGTTCCGCTG----------ATCCGAAAAGCATACATTGTCCCTATCTACTATCTAGCAA
                              ******************   ***************

MB27

                              insertion 1.006(1.007) kb

GAATGGATTAACGAGAGATTCCTACTGTCCCTGTCTTAGCTGGGAGCAG----------ATCCGAAAAGCATACATTGTCCTATCTACTATCTAGCAA
                              ********   ********

RI9

                              insertion 0.047kb

GAATGGATTAACGAGAGATTCCTACTGTCCCTATCTACTTGGAACACGCCACGTAAAAT----------ATCCGAAAAGCATACATTGTCCTATCTACTATCTAGCAA
                              **************   **************

Uninterrupted rDNA:

GAATGGATTAACGAGAGATTCCTACTGTCCCTATCTACTATCTAGCGAA

Figure 12.6. Duplicated sequences flanking the short type I insertions in *D. melanogaster* rDNA. The short type I insertions in rDNA are flanked by sequences present only as a single copy within the uninterrupted units (nucleotides marked by asterisks). The 5 kb insertion is not flanked by a duplication but has a deletion of rDNA sequences at its left junction. RI9, RI10 and MB27 refer to three clones of rDNA. rDNA sequences are underlined and insertion sequences are overlined. The right hand junction is identical in all clones that have been examined. From Roiha and Glover[45].

rDNA insertion elements. In one cloned segment containing five tandemly arranged type I elements (see *Figure 12.5*), each unit was shown to be flanked on the right by 14 nucleotides of rDNA sequence and on the left by a 20 nucleotide rDNA sequence identical to the sequences contiguous to the 5 kb type I insertion in a full rDNA unit[46]. This tells us that these insertion elements were once present in the nucleolus. Their transposition to the chromocentral sites cannot have occurred recently since the sequences at the two sites show considerable sequence divergence. Furthermore, the event generating the deletion of rDNA by the 5 kb insertion must have occurred a considerable time ago, since this deletion is evident within the small rDNA segment on the left flank of each type I element in the tandem array.

Insertions have now been detected in the rDNA from many species of Dipteran flies. The site of the insertion in *D. virilis* rDNA is identical to the site in *D. melanogaster*[41], and the insertions, which are either 5 or 10 kb in length are flanked on either side by a duplication of 14 nucleotides of rDNA. The 10 kb insertion consists of two copies of the 5 kb unit separated by a third copy of the 14 nucleotide rDNA segment, suggesting that it has arisen by unequal crossing-over between units with 5 kb insertions[40]. There is complete sequence divergence in the terminal regions of the insertions from the two species, and yet there are internal regions that show some homology[2]. The existence of insertions at the same site in these Dipteran species might still indicate that insertions originally arose in an ancestral species and have been stably maintained at this location. Alternatively, site-specific recombination might be an ongoing process in the rDNA of *Drosophila* species. In this case the duplication of rDNA elements might be analogous to the duplication of the *att* site sequences that results from homologous recombination between the genomes of *E. coli* and its lambdoid phages.

Transcription of units containing insertions

The low levels of transcription of rDNA units containing insertions are intriguing. The deletion of rDNA sequences adjacent to the 5 kb type I insertions would preclude these rDNA units from acting as a template for functional rDNA. One might still have expected abortive transcription of these units, especially since they have identical sequences to uninterrupted units in the putative promoter region at the 5′ end of the transcription unit[28]. Restriction fragments of such regions from both interrupted and uninterrupted genes can also be transcribed equally well in an *in vitro* system for RNA polymerase I[25]. Long and Dawid[29], however, could only detect *in vivo* on the order of 6 to 10 molecules of high molecular weight transcripts containing type I sequences per nucleus and about fifty 1 kb RNA molecules in the cytoplasm of embryonic cells. This should be compared with about 1300 molecules per nucleus of the 8 kb primary transcript from uninterrupted units. These results are consistent with the observation by electron microscopy of long silent regions interspersed with active transcription units in nucleolar chromatin from embryonic cells[30]. Transcription of the insertions does not even occur at a significant level in fly strains that exhibit the bobbed phenotype[28]. These flies show shorter bristles, slow development and other pleiotropic effects as a consequence of deletions of more than 50% of their rDNA, and must be under strong pressure to use all their available rDNA. A similar situation probably pertains with *D. hydei*, in which the severity of the bobbed phenotype has been correlated with the number of uninterrupted rDNA units and not with the total number of rDNA units[16].

rDNA units containing type II insertions are transcribed at comparably low levels. In this case the transcripts are only found in the nucleus, the most abundant corresponding in length to the most common type II insertion $(3.4\,\mathrm{kb})$[24]. There is in addition a set of higher molecular weight transcripts linked mainly to rDNA upstream in the rDNA transcription unit from the insertion. This different pattern of transcription of the two types of insertion might reflect the different structures of their junctions with the rDNA sequence. There is no duplication of rDNA sequences on either side of the insertion, but a striking tract of about $20\,\mathrm{dA}$ residues occurs at the right-hand junction[46]. The most abundant transcripts containing type II sequences seem to have an endpoint at this dA tract, suggesting either that transcription terminates here or that newly synthesized RNA is rapidly cleaved at this site. The concentration of insertion transcripts varies by up to three orders of magnitude between strains, the highest levels having been observed in flies carrying the y^+Y chromosome[24]. This observation may go some way towards explaining the electron microscopic observations of Chooi[9] that nurse cells of flies carrying the y^+Y chromosome have a significant proportion of long transcription units.

Variations in rRNA gene number

The heterogeneity of the insertion sequences and also of the non-transcribed spacer region gives a complex, chromosome-specific restriction pattern, which has been used to follow the inheritance of rDNA. Many *Drosophila* tissues contain polytene chromosomes in which the rDNA undergoes fewer rounds of replication relative to sequences in euchromatic regions. Spear and Gall[49] showed that the rDNA content of diploid tissues is proportional to the number of nucleolar organizers in the genome. The rDNA content of the polytene salivary gland chromosomes is proportionally lower, by a factor of two to eight, than in diploid cells and reaches a roughly constant level irrespective of the number of nucleolar organizers. The restriction patterns of rDNA from these tissues show that ribosomal genes from only one nucleolar organizer are replicated during polytenization, and within this nucleolar organizer not all genes are replicated equally[14]. The replication of rDNA from only one of the X chromosomes in females explains the finding that rDNA levels are the same in X/X polytene cells as in X/O polytene cells, but the molecular mechanism of their differential replication is not understood[13].

In addition to this tissue-specific, developmental variation in the rDNA content of cells, the numbers of rDNA units can undergo inheritable variation in germ line cells. It is beyond the scope of this chapter to give a full review of the genetic investigation of this phenomenon of 'rDNA magnification'. This together with other aspects of rDNA amplification in *Drosophila* have been recently reviewed by Kunz *et al.*[27]. Magnification occurs when rDNA deficient chromosomes are maintained together in male flies. Chromosomes having large rDNA deletions can be satisfactorily maintained in fly stocks for many generations as long as they are placed against a chromosome having a full complement of rDNA. If, for example, an rDNA deficient chromosome *Xbb* is placed against the severely rDNA deficient *Ybb*[-] chromosome, then this male fly exhibits an extreme 'bobbed' phenotype. If these two chromosomes are maintained together for several generations, then one sees a stepwise increase in the rDNA content of the X chromosome and in parallel a reversion of the phenotype back to wild type. Two models for magnification have been put forward. Tartof[50] has proposed that the occurrence of unequal crossing over between sister chromatids generates cells with increased numbers of rDNA units, which have a selective advantage. In support of this hypothesis is the appearance during magnification not only of revertants of the bobbed mutation but also more severely bobbed mutants, presumably the reciprocal products of such exchange. Secondly, there is some correlation of the frequency of magnification with somatic mosaicism of the bobbed phenotype, which would result from the crossing over being a mitotic event. Finally magnification does not occur with a ring chromosome where unequal sister chromatid exchange would give rise to inviable zygotes. However, it is also possible that putative regulatory regions are modified in such a chromosome.

Ritossa[42], on the other hand, proposes synthesis of extrachromosomal copies of rDNA without phenotypic effect, in the first generation that rDNA deficient chromosomes are placed together. Furthermore he argues that the integration of such extrachromosomal DNA into both X and Y chromosomes could explain the increase in *meiotic* crossing over during certain magnification crosses.

As yet there is insufficient evidence at the molecular level to favour one of these models over the other. Whichever model is closer to reality, both require recombination events to take place in the rDNA. Tandemly arranged sequences within the NTS of *Drosophila* rDNA serve as an indication that unequal exchanges can occur in *Drosophila* just as in *Xenopus*. There is also considerable evidence that the type I insertion can be involved in site-specific recombination. The elements themselves or large blocks of rDNA flanked by insertion elements could therefore be involved in this process. We have recently looked for changes in the gross organizational patterns of these sequences during magnification process. There seems to be no selective magnification of individual rDNA units, as one might expect if small extrachromosomal elements were produced as in Ritossa's model. There is, however, in some crosses, an increase not only in rDNA content but also of tandemly arranged type I elements from the heterochromatic blocks which flank the nucleolus organizer[11a]. It seems, therefore, as if a large segment of the chromosome is amplified as a response to the severe bobbed phenotype.

The rDNA of *Tetrahymena*

Tetrahymena like other ciliated protozoans has two types of nucleus: the germinal micronucleus which is not expressed during the vegetative life of the cell, and the macronucleus which is formed during conjugation. The macronucleus is transcriptionally active and is highly polyploid. In the germ line micronucleus there is a single copy of a rRNA gene[60]. Upon conjugation this is excised from the chromosome to give a single 11 kb free rDNA copy[37]. Molecules of this kind persist only for a few generations during which time they undergo replication to form about 100 palindromic dimeric repeats[15,23]. A short repeating sequence (CCCCAA) is found occupying about 400 base pairs close to the end of the palindrome[4]. Within this repeating unit there are single stranded breaks in the DNA, and the end of the molecule either has a hairpin structure or is attached to a protein. This repeated sequence element is found at many different genomic locations in the micronucleus and forms the end points of many linear genomic segments in the macronucleus.

Some strains of *Tetrahymena* possess an intervening sequence (IVS) in all their copies of rDNA[7, 12, 59]. In this case the use of the term 'intervening sequence' seems appropriate, since in a manner analogous to that found for IVS elements in genes encoding polypeptides, the *Tetrahymena* rDNA IVS is spliced out of the primary transcript. Curiously other strains do not have an intervening sequence. The palindromic DNA is transcribed bidirectionally to give 35S primary transcripts. For those species containing rDNA IVSs, the first step in processing which can be seen *in vivo* by pulse labelling experiments, is the excision of the IVS. This is followed by other endonucleolytic processing steps. Zaug and Cech[61] have developed an *in vitro* system to study the splicing event. Isolated nuclei are incubated in the presence of α-amanitin to inhibit non-rRNA synthesis. A discrete 0.4 kb RNA molecule corresponding to the length of the intervening sequence is produced. Both linear and circular forms of this RNA are observed (*Figure 12.7*), the circular form apparently being derived from the linear form[19]. If the

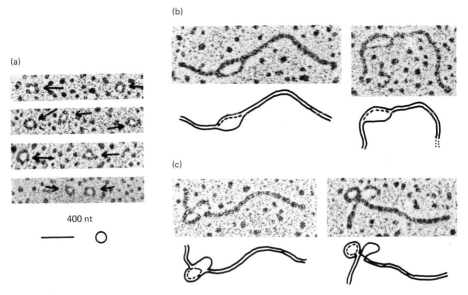

Figure 12.7. Circular RNAs homologous to the IVS of *Tetrahymena* rDNA. The intervening sequence excised from the primary transcript of *Tetrahymena* rDNA occurs as both linear and circular RNA molecules, 400 nucleotides long. (a) Electron micrographs of the circular form. On the right are the R loop structures formed when either the linear (b) or circular (c) forms of this excised RNA are annealed with rDNA. From Grabowski *et al.*[19]

transcription is allowed to take place in a low concentration of monovalent cation, then splicing is inhibited and an intermediate accumulates[8]. The reaction continues if monovalent and divalent cations are added together with a guanosine compound. This can either be GTP, GDP, GMP or guanosine. During the reaction the 3'-OH of the guanosine becomes linked to the 5' end of the IVS RNA via a phosphodiester bond. When guanosine molecules with 5'-phosphoryl groups are used, the phosphoryl groups are not removed. These observations have led to the splicing model shown in *Figure 12.8*.

    The splicing activity is associated with the primary transcript and is not destroyed by treatment with SDS and phenol, boiling in the presence of SDS and

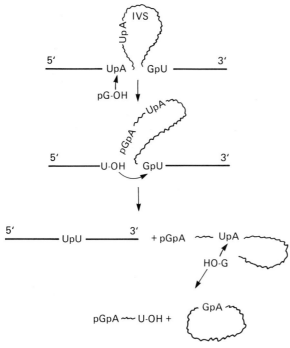

Figure 12.8. Model for splicing of the *Tetrahymena* primary rRNA transcript. In the presence of a guanosine nucleotide and monovalent cations, the IVS excises itself from the precursor. The guanosine cofactor is added to the 5′ end of the linear IVS RNA through a 3′–5′ phosphodiester bond. The linear IVS RNA is cleaved at a point 15 nucleotides from its 5′ end and the original 3′ end of the molecule is joined to the newly created 5′-phosphate. None of these reactions requires a protein and it would therefore be seen that the RNA itself has many characteristics expected for a protein enzyme[62].

mercaptoethanol or protease treatment. A protein cannot be the catalytic agent, unless it is protected from the above treatment by virtue of its association with the RNA. An alternative would be that the RNA itself is folded in such a way that it is endowed with the necessary catalytic abilities. Circular RNA molecules have been observed in yeast mitochondria where they could arise from the processing of the mitochrondrial rDNA transcripts. This contrasts with the processing of yeast tRNA in which the IVS is excised as a linear molecule. The generality of this splicing mechanism therefore remains to be determined.

References

1. Bakken, A., Morgan, G., Sollner-Webb, B., Roan, J., Busby, S. and Reeder, R. H. (1982). Mapping of transcription initiation and termination signals on *Xenopus laevis* ribosomal DNA. *Proceedings of the National Academy of Sciences, USA*, **79**, 56–60
2. Barnett, T. and Rae, P. M. M. (1979). A 9.6 kb intervening sequence in *D. virilis* rDNA, and sequence homology in rDNA interruptions of diverse species of *Drosophila* and other diptera. *Cell*, **16**, 763–775

3. Berk, A. J. and Sharp, P. A. (1977). Sizing and mapping of early adenovirus mRNAs by gel electrophoresis of S1 endonuclease resistant hybrids. *Cell*, **12**, 721–732
4. Blackburn, E. H. and Gall, J. G. (1978). A tandemly repeated sequence at the termini of the extrachromosomal ribosomal RNA genes in *Tetrahymena*. *Journal of Molecular Biology*, **120**, 33–53
5. Boseley, P., Moss, T., Machler, M., Portmann, R. and Birnsteil, M. (1979). Sequence organisation of the spacer DNA in a ribosomal gene unit of *X. laevis*. *Cell*, **17**, 19–32
6. Brown, D. O. and Weber, C. S. (1968). Gene linkage by RNA–DNA hybridisation. *Journal of Molecular Biology*, **34**, 661–698
7. Cech, T. R. and Rio, D. (1979). Localisation of transcribed regions on extrachromosmal ribosomal RNA genes of *Tetrahymena thermophila* by R-loop mapping. *Proceedings of the National Academy of Sciences, USA*, **76**, 5051–5055
8. Cech, T. R., Zaug, A. J. and Grabowski, P. J. (1981). *In vitro* splicing of the ribosomal RNA precursor of *Tetrahymena*: Involvement of a guanosine nucleotide in the excision of the intervening sequence. *Cell*, **27**, 487–496
9. Chooi, W. Y. (1979). The occurrence of long transcription units among the X and Y ribosomal genes of *Drosophila melanogaster*: Transcription of insertion sequences. *Chromosoma*, **74**, 57–81
10. Dawid, I. B., Long, E. O., Di Nocera, D. P. and Pardue, M. L. (1981). Ribosomal insertion-like elements in *Drosophila melanogaster* are interspersed with mobile sequences. *Cell*, **25**, 399–408
11. Dawid, I. B. and Rebbert, M. L. (1981). Nucleotide sequences at the boundaries between gene and insertion regions in the rDNA of *Drosophila melanogaster*. *Nucleic Acids Research*, **9**, 5011–5020
11a. de Cicco and Glover, D. M. (1983). Amplification of rDNA and type I sequences in *Drosophila* males deficient in rDNA. *Cell*, **32**, 1217–1225
12. Din, N., Engberg, J., Kaffenberger, W. and Eckert, W. A. (1979). The intervening sequence in the 26S rRNA coding region of *T. thermophila* is transcribed within the largest stable precursor for rRNA. *Cell*, **18**, 525–532
13. Endow, S. A. (1980). On ribosomal gene compensation in *Drosophila*. *Cell*, **22**, 149–156
14. Endow, S. A. and Glover, D. M. (1979). Differential replication of ribosomal gene repeats in polytene nuclei of *Drosophila*. *Cell*, **17**, 597–605
15. Engberg, J., Andersson, A., Leick, V. and Collins, J. (1976). Free ribosomal DNA molecules from *Tetrahymena pyriformis* GL are giant palindromes. *Journal of Molecular Biology*, **104**, 455–470
16. Franz, G. and Kunz, W. (1981). Intervening sequences in ribosomal RNA genes and bobbed phenotype in *Drosophila hydei*. *Nature, London*, **290**, 638–640
17. Glover, D. M. and Hogness, D. S. (1977). A novel arrangement of the 18S and 28S sequences in a repeating unit of *Drosophila melanogaster* rDNA. *Cell*, **10**, 167–176
18. Glover, D. M., White, R. L., Finnegan, D. J. and Hogness, D. S. (1975). Characterisation of six cloned DNAs from *Drosophila melanogaster*, including one that contains the genes of rRNA. *Cell*, **5**, 149–157
19. Grabowski, P. J., Zaug, A. J. and Cech, T. R. (1981). The intervening sequence of the ribosomal RNA precursor is converted to a circular RNA in isolated nuclei of *Tetrahymena*. *Cell*, **23**, 467–476
20. Grummt, I. (1981). Specific transcription of mouse ribosomal DNA in a cell-free system that mimics control *in vivo*. *Proceedings of the National Academy of Sciences, USA*, **78**, 727–731
21. Hourcade, D., Dressler, D. and Wolfson, J. (1973). The amplification of ribosomal RNA genes involves a rolling circle intermediate. *Proceedings of the National Academy of Sciences, USA*, **70**, 2926–2930
22. Karrer, K. M. and Gall, J. G. (1976). The macronuclear ribosomal DNA of *Tetrahymena puriformis* is a palindrome. *Journal of Molecular Biology*, **104**, 421–453
23. Kidd, S. J. and Glover, D. M. (1980). A DNA segment from the chromocentral heterochromatin of *D. melanogaster* which contains five tandemly repeating units homologous to the major rDNA insertion. *Cell*, **19**, 103–119
24. Kidd, S. J. and Glover, D. M. (1981). *Drosophila melanogaster* ribosomal DNA containing type II insertions is variably transcribed in different strains and tissues. *Journal of Molecular Biology*, **151**, 645–662
25. Kohorn, B. D. and Rae, P. M. M. (1982). Accurate transcription of truncated ribosomal templates in a *Drosophila* cell-free system. *Proceedings of the National Academy of Sciences, USA*, **79**, 1501–1505
26. Kumar, A. and Warner, J. R. (1972). Characterisation of ribosomal precursor particles from HeLa cell nucleoli. *Journal of Molecular Biology*, **63**, 233–246
27. Kunz, W., Grimm, C. and Franz, G. (1982). Amplification and synthesis of rDNA: *Drosophila*. In *The Cell Nucleus*, Volume 10, (Busch, H., Ed.), Academic Press, in press

28. Long, E. O., Collins, M., Kiefer, B. I. and Dawid, I. B. (1981). Expression of the ribosomal DNA insertions in bobbed mutants of *Drosophila melanogaster*. *Molecular Genetics*, **182**, 377–384
29. Long, E. O. and Dawid, I. B. (1979). Expression of ribosomal DNA insertions in *Drosophila melanogaster*. *Cell*, **18**, 1185–1196
30. McKnight, S. L. and Miller, O. L. (1976). Ultrastructural patterns of RNA synthesis during early embryogenesis of *Drosophila melanogaster*. *Cell*, **8**, 305–319
31. Maden, B. E. H., Salim, M. and Summers, D. F. (1972). Maturation pathway for rRNA in the HeLa cell nucleus. *Nature, London*, **237**, 5–9
32. Miller, K. G. and Sollner-Webb, B. (1981). Transcription of mouse rRNA genes by RNA polymerase I: *in vitro* and *in vivo* initiation and processing sites. *Cell*, **27**, 165–174
33. Morrow, J. F., Cohen, S. N., Chang, A. C. Y., Boyer, H. W., Goodman, H. M. and Helling, R. B. (1974). Replication and transcription of eukaryotic DNA in *E. coli*. *Proceedings of the National Academy of Sciences, USA*, **71**, 1743–1747
34. Moss, T. and Birnstiel, M. (1979). The putative promoter of a *Xenopus laevis* ribosomal gene is reduplicated. *Nucleic Acids Research*, **6**, 3733–3743
35. Moss, T. and Birnstiel, M. (1982). The structure and function of the ribosomal gene spacer. In *The Nucleolus*, (Jordan and Callis, Eds.), Cambridge University Press
36. Moss, T., Boseley, P. G. and Birnstiel, M. (1980). More ribosomal spacer sequences from *Xenopus laevis*. *Nucleic Acids Research*, **8**, 467–485
37. Pan, W. C. and Blackburn, E. M. (1981). Single extrachromosomal ribosomal RNA gene copies are synthesised during amplification of the rDNA of *Tetrahymena*. *Cell*, **23**, 459–466
38. Peacock, W. J., Appels, R., Endow, S. and Glover, D. M. (1981). Chromosomal distribution of the major insert in *Drosophila melanogaster* 28S rRNA genes. *Genetical Research*, **37**, 209–214
39. Pellegrini, M., Manning, J. and Davidson, N. (1977). Sequence arrangement of the rDNA of *Drosophila melanogaster*. *Cell*, **10**, 213–224
40. Rae, P. M. M. (1982). Unequal crossing-over accounts for the organisation of *Drosophila virilis* rDNA insertions and the integrity of flanking 28S genes. *Nature, London*, **296**, 579–581
41. Rae, P. M. M., Kohorn, B. D. and Wade, R. P. (1980). The 10 kb *Drosophila virilis* 28S rDNA intervening sequence is flanked by a direct repeat of 14 base pairs of coding sequence. *Nucleic Acids Research*, **8**, 3491–3504
42. Ritossa, F. (1972). Procedure for magnification of lethal deletions of genes for ribosomal RNA. *Nature, London*, **240**, 109–111
43. Ritossa, F. and Spiegelman, S. (1965). Localisation of DNA complementary to ribosomal RNA in the nucleolus organiser region of *Drosophila melanogaster*. *Proceedings of the National Academy of Sciences, USA*, **53**, 737–745
44. Rochaix, J. D., Bird, A. and Bakken, A. (1974). Ribosomal RNA gene amplification by rolling circles. *Journal of Molecular Biology*, **87**, 473–488
45. Roiha, H. and Glover, D. M. (1981). Duplicated rDNA sequences of variable lengths flanking the short type I insertions in the rDNA of *Drosophila melanogaster*. *Nucleic Acids Research*, **9**, 5521–5532
46. Roiha, H., Miller, J. R., Woods, L. C. and Glover, D. M. (1981). Arrangement and rearrangement of sequences flanking the two types of rDNA insertion in *D. melanogaster*. *Nature, London*, **290**, 749–753
47. Scheer, U., Trendelenberg, M., Krohne, G. and Franke, W. (1977). Lengths and patterns of transcriptional units in the amplified nucleoli of oocytes of *Xenopus laevis*. *Chromosoma*, **60**, 147–167
48. Shepherd, J. and Maden, B. E. H. (1972). Ribosomal assembly in HeLa cells. *Nature, London*, **236**, 211–214
49. Spear, B. B. and Gall, J. G. (1973). Independent control of ribosomal gene replication in polytene chromosomes of *Drosophila melanogaster*. *Proceedings of the National Academy of Sciences, USA*, **70**, 1359–1363
50. Tartof, K. D. (1973). Unequal mitotic sister chromatid exchange and disproportionate replication as mechanisms regulating ribosomal gene redundancy. *Cold Spring Harbor Symposium on Quantitative Biology*, **38**, 491–500
51. Trendelenberg, M. F. and Gurdon, J. B. (1978). Transcription of cloned *Xenopus* ribosomal genes visualised after injection into oocyte nuclei. *Nature, London*, **276**, 292–294
52. Weinberg, R. A. and Penman, S. (1970). Processing of 45S nucleolar RNA. *Journal of Molecular Biology*, **47**, 169–178
53. Wellauer, P. K. and Dawid, I. B. (1973). Secondary structure maps of ribosomal RNA: Processing of HeLa rRNA. *Proceedings of the National Academy of Sciences, USA*, **70**, 2827–2831

54. Wellauer, P. K. and Dawid, I. B. (1977). The structural organisation of ribosomal DNA in *Drosophila melanogaster*. *Cell*, **10**, 193–212
55. Wellauer, P. K., Dawid, I. B., Brown, D. D. and Reeder, R. H. (1976). The molecular basis for length heterogeneity in ribosomal DNA from *X. laevis. Journal of Molecular Biology*, **105**, 461–486
56. Wellauer, P. K., Reeder, R. H., Dawid, I. B. and Brown, D. D. (1976). The arrangements of length heterogeneity in repeating units from amplified and chromosomal rDNA from *X. laevis. Journal of Molecular Biology*, **105**, 487–506
57. Wensink, P. C. and Brown, D. D. (1971). Denaturation map of the rDNA of *X. laevis. Journal of Molecular Biology*, **60**, 235–248
58. White, R. L. and Hogness, D. S. (1977). R loop mapping of the 18S and 28S sequences in the long and short repeating units of *Drosophila melanogaster* rDNA. *Cell*, **10**, 177–192
59. Wild, M. A. and Gall, J. G. (1979). An intervening sequence in the gene coding for 25S ribosomal RNA of *Tetrahymena pigmentosa. Cell*, **16**, 565–573
60. Yao, M. C. and Gall, J. G. (1977). A single integrated gene for ribosomal RNA in a eukaryote, *Tetrahymena pyriformis. Cell*, **12**, 121–132
61. Zaug, A. J. and Cech, T. R. (1980). *In vitro* splicing of the ribosomal RNA precursor in nuclei of *Tetrahymena. Cell*, **19**, 331–338
62. Zaug, A. J., Grabowski, P. J. and Cech, T. R. (1983). Autocatalytic cyclization of an excised intervening sequence RNA is a cleavage–ligation reaction. *Nature, London*, **301**, 578–583

Chapter 13

5S Ribosomal RNA genes

J. R. Miller

Introduction

5S RNA is ubiquitous in evolution, being associated with the large subunit of both eukaryotic and prokaryotic ribosomes. All known species of 5S RNA are approximately 120 nucleotides long and sequence conservation is such that a general secondary structure model can be applied with minor variations to all sequenced 5S RNAs.

Both prokaryotes and eukaryotes possess multiple copies of the gene coding for 5S RNA. In prokaryotic genomes the genes for 5S RNA are often linked to those for the larger ribosomal RNA (16S + 23S) molecules. In eukaryotes 5S RNA genes are as a rule separate from the major ribosomal RNA genes (18S + 28S), the exceptions being some lower eukaryotes such as *Dictyostelium* and yeast, where 5S genes are interspersed amongst rRNA genes. In all cases so far studied transcription of eukaryotic 5S genes has been demonstrated by the use of the fungal toxin α-amanitin, to be mediated by RNA polymerase III.

This chapter will deal predominantly with the structure and expression of the 5S RNA genes of *Xenopus*. Justification for this choice lies in the detailed knowledge of the structure of these genes and more importantly in the increased understanding of their regulation[17].

Structure of 5S genes

The 5S RNA genes in *Neurospora* appear to be scattered singly throughout the genome[33]. However, in most eukaryotes the 5S genes occur as tandem repeats, with each 5S RNA coding sequence being contained within a larger region of DNA termed the 5S repeat unit. In this article such repeat units containing a 5S gene will be termed 5S DNA. In all cases the 5S DNA coding sequence has been demonstrated to be contiguous to the 5S RNA sequence; therefore these genes do not possess intervening sequences. There is thought to be no processing of the primary transcript at the 5′ end, since 5S RNA can be isolated with a 5′ terminal triphosphate group, but short extensions at the 3′ end have been observed in rat liver and HeLa nuclei and in *Drosophila* (see below). These are apparently removed by rapid processing events. Termination of transcription in *Xenopus* is mediated by a sequence of four or more T residues in the non-coding strand to produce, without processing, a mature RNA of length 120 nucleotides.

Yeast 5S DNA

In the yeast *Saccharomyces cerevisiae* there are 140–150 copies of the 5S RNA gene/haploid genome. The genes for 5S RNA and the precursor for the three ribosomal RNA molecules, 28S, 18S and 5.8S RNAs, have been demonstrated to be contained on the same repeat units. 5S RNA and the rRNA precursor molecule are made as separate transcripts from opposite strands of the repeat unit[21].

Drosophila 5S DNA

Drosophila melanogaster has been demonstrated to possess approximately 160 copies of the 5S gene/haploid genome. These are located at a single site 56F, on the right arm of chromosome 2. The repeat unit is approximately 375 nucleotides in length with a coding region of constant length, and a spacer region which occurs in size classes differing in length by 6–8 nucleotides[34]. The primary transcript has been shown to be a 135 nucleotide long molecule, which is processed back at its 3′ end to the mature 120 nucleotide form[15]. It is interesting to note that a cluster of 5 T residues occurs on the non-coding strand at a position 10 residues downstream from the corresponding 3′ end of the mature RNA. The structure of this gene and the various *Xenopus* 5S genes described is illustrated schematically in *Figure 13.1*.

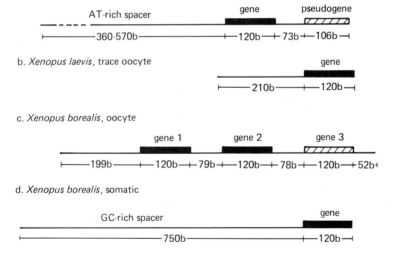

Figure 13.1. Schematic diagram illustrating the structure of various 5S RNA genes. Reproduced from Fedoroff and Brown[9] by kind permission of the author and the MIT Press.

Xenopus 5S genes

During oogenesis in *Xenopus* and other amphibia very large quantities of ribosomes are accumulated. Consequently large amounts of 28S, 18S and 5S RNA must be synthesized. In *Xenopus* the genes coding for the 40S precursor to 28S, 18S and 5.8S RNA are amplified as much as 5000-fold during oogenesis, to permit the high level of RNA synthesis required. An equally large amount of 5S RNA must be synthesized by the developing oocyte. The genes coding for 5S RNA are unlinked to the genes for 18S and 28S RNA and in contrast the 5S genes are not amplified during oogenesis. Instead in both *X. laevis* and *X. borealis* there are two different 5S gene families each encoding 5S RNA molecules of a distinct sequence class. Both frogs have many thousands of genes coding for oocyte-specific 5S RNA. Although these genes are a permanent component of the genome in all tissues, the 5S RNA they encode is synthesized only during oogenesis, and thus is found only in developing oocytes. In addition each species possesses several hundred genes which encode somatic 5S RNA; these genes are expressed in both oocytes and somatic tissues. However, in oocytes the somatic 5S RNA is present as a minor component, due to the very high level of oocyte 5S RNA synthesis.

The genes for 5S RNA have been shown by *in situ* hybridization to be located at or near the telomere region of the long arm of most if not all the chromosomes of *Xenopus laevis*. Presumably this DNA is oocyte-specific, therefore the location of the somatic type 5S DNA is still unclear.

Xenopus laevis *oocyte 5S RNA genes*

The oocyte-specific 5S RNA genes of *X. laevis* consist of two distinct gene families, the oocyte 5S DNA and the trace oocyte 5S DNA. There are approximately 24 000 tandemly arrayed repeat units of oocyte 5S DNA/haploid genome. The repeat units, which are heterogeneous in length, have been shown to contain two distinct features: an A + T-rich spacer region and a G + C-rich region which contains the 5S gene and also a 'pseudogene' sequence. The A + T-rich spacer contains a number of 15 bp long repeated sequences and it is the frequency of reiteration of these sequences which accounts for the observed length heterogeneity between 5S repeat units[9]. The G + C-rich region contains a duplication of 174 nucleotides of sequence, commencing 73 nucleotides before the 5' end of the 5S gene and terminating at position 101 of the gene. The duplicated portion of the 5S gene was termed the 'pseudogene'[16] since it shared 85% sequence homology with the intact 5S gene[24]. The functional significance of the pseudogene is unknown, although recent studies have shown that it is capable of being transcribed. A comparison of the sequences of two cloned repeat units demonstrated the presence of micro-heterogeneities in both spacer, gene and pseudogene; neither cloned gene sequence corresponded exactly to the bulk 5S RNA sequence. However, the observed level of mismatch was lower than the expected random value, suggesting a correction mechanism was in operation[22].

A second family of oocyte-specific 5S genes, the *X. laevis* trace oocyte 5S DNA, was isolated as a minor satellite component of the *X. laevis* genome by Brown *et al.*[7]. This gene family was shown to have a repeat unit length of 350 bp, with no detectable length heterogeneity between individual repeat units. The 5S RNA coded for by this gene family differs from both the oocyte and somatic sequences. Low levels have been detected in oocyte 5S RNA preparations, but not in somatic

5S RNA, and thus this RNA has been termed oocyte-specific. Sequencing of three repeat units revealed that the spacer DNA was A + T-rich and contained a partial duplication, and that micro-heterogeneities existed in both spacer and gene between repeat units. There were estimated to be approximately 1300 repeat units/haploid genome[29].

Xenopus laevis *somatic 5S RNA genes*

The DNA coding for *X. laevis* somatic 5S RNA was isolated by Peterson *et al.*[29] by a combination of ultracentrifugation and gel electrophoresis. Sequencing of a cloned repeat unit demonstrated that the 5S gene was surrounded by a G + C-rich spacer, with a repeat length of approximately 880 nucleotides. There were estimated to be 400 copies of this repeat unit/haploid genome, and the repeats were demonstrated to be homogeneous in length. Both gene and spacer of this repeat unit have been shown to be extensively homologous to a cloned *X. borealis* somatic 5S DNA repeat (see below).

Xenopus borealis *oocyte 5S DNA*

Korn and Brown [18] demonstrated that some regions of *X. borealis* oocyte 5S DNA consist of clusters of 5S RNA genes separated by A + T-rich spacers of variable length. One 755 nucleotide segment of DNA was sequenced and shown to contain three 5S RNA coding sequences arranged in tandem head to tail manner. Two of the 5S coding sequences were very similar to each other and to the bulk *X. borealis* oocyte 5S RNA sequence. The third gene differed in 15 positions from the dominant 5S RNA sequence, including a G to C transition at the initiation nucleotide. A genomic restriction fragment corresponding to the 5' half of this gene was isolated and found to give an identical sequence to the cloned DNA. Therefore a substantial proportion of the 5S genes differ from the norm, perhaps coming into the category of pseudogenes.

All three cloned genes were transcribed in *Xenopus* germinal vesicle extracts, although gene 3 was only transcribed at half the efficiency of the other two. The product from gene 3 was 121 nucleotides long and started with an adenine at position −1 relative to other 5S genes[20]. This differs from the other 5S gene transcripts which initiate at guanine residues and is presumably due to the sequence deviation around the start site. However, no transcripts corresponding to the diverged gene have been detected amongst the *in vivo* 5S RNA population. Length heterogeneity between gene clusters was demonstrated to be due to differing numbers of a tandemly repeated 21 nucleotide long sequence. There were estimated to be 9000 *X. borealis* oocyte 5S genes/haploid genome.

Xenopus borealis *somatic 5S DNA*

A cloned repeat unit of *X. borealis* somatic 5S DNA was sequenced by Peterson *et al.*[29]. The 5S gene was present in a tandem repeat, surrounded by G + C-rich spacer. Individual repeat units were found to be homogeneous in length and there were shown to be approximately 700 repeats/haploid genome. Sequencing revealed a close homology between *X. borealis* and *X. laevis* somatic 5S repeat units. The respective 5S coding sequences differed at only two positions, and the spacers showed homologous regions interspersed with non-homologous segments. One general feature to emerge from these studies is that all three oocyte-specific 5S DNAs, *X. laevis* oocyte, *X. laevis* trace oocyte and *X. borealis* oocyte, have A + T-rich spacer sequences, (76%, 60%, 61% respectively) with the major species having repeat units of heterogeneous length, whereas the two somatic 5S DNAs have G + C-rich spacer sequences (*X. laevis* 57%, *X. borealis* 61%) and are homogeneous in length.

Expression and regulation of 5S RNA genes

Accurate transcription of *Xenopus* 5S genes

Parker and Roeder[26] demonstrated that when *X. laevis* 5S DNA in the form of chromatin was transcribed by a purified RNA polymerase III preparation, as much as 50% of the total RNA synthesized was a transcript of the coding strand of the 5S gene region. In contrast when a native *X. laevis* 5S DNA template was used, the same enzyme was found to transcribe both strands of the 5S repeat unit with only 1.3% of the total transcripts being 5S RNA. Similar results with this native template were obtained using *X. laevis* RNA polymerase I and *E. coli* RNA polymerase, thus demonstrating a complete lack of specificity. Therefore accurate transcription of 5S DNA was shown to require at least one chromatin associated component, in addition to the template and a purified RNA polymerase III preparation. Subsequently Ng *et al.*[25] discovered that correct and specific 5S RNA synthesis was obtained with an *X. borealis* oocyte 5S DNA clone, exogenous purified RNA polymerase III and unknown components from an immature *X. laevis* oocyte chromatin supernatant fraction. If the chromatin supernatant fraction was omitted, completely non-specific RNA synthesis was observed. In parallel Brown and Gurdon[5] demonstrated that injection of genomic *X. borealis* oocyte and *X. laevis* oocyte 5S DNAs into *X. laevis* oocyte germinal vesicles would generate 5S RNA molecules of correct length and sequence. Cloned single repeating units of *X. laevis* and *X. borealis* 5S DNA were also later shown to support accurate synthesis of 5S RNA when injected into the oocyte germinal vesicle[6]. These experiments demonstrated that a single cloned 5S repeat unit contained all the information required for 5S RNA synthesis in the nucleus of an injected oocyte. This was also shown to apply to a linear 5S repeat unit excised from its plasmid vector. In addition the 5S transcription observed in oocyte injection experiments was shown to be mediated by RNA polymerase III[12]. Similar results were obtained by Birkenmeier *et al.*, using a lysate of manually isolated *X. laevis* oocyte nuclei[1].

Location of an intragenic control region

Brown and colleagues used 5' and 3' deletion mutants of the 5S gene to demonstrate that within the 5S gene, between positions 50 and 83, there exists a region of DNA which directs RNA polymerase III to initiate transcription approximately 50 nucleotides upstream from the 5' border of the region[3,31]. Two series of deletions extending into the *X. borealis* somatic 5S gene from either the 5' or 3' side were used in these experiments. They were constructed by cutting a cloned *X. borealis* somatic 5S gene within the vector sequence. In each case a restriction enzyme was used which would leave a large segment of vector sequence at one end of the 5S gene and a very small segment of vector DNA at the other end of the 5S gene. The insert fragment was then treated with exonuclease III over a time course, the reaction terminated and the products made flush ended using S1 nuclease. Subsequently the blunt ended fragments containing a range of deletions were recloned. The end points of the various deletions could be estimated by looking at the size of various diagnostic restriction fragments within the 5S gene. Precise determination of the end points was accomplished by DNA sequencing. The ability of the deleted genes to support transcription was then assayed using an oocyte nuclear extract. These experiments are summarized in *Figure 13.2*.

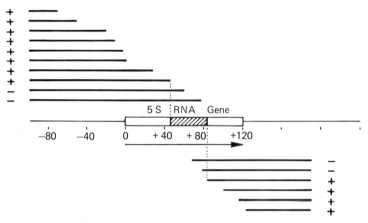

Figure 13.2. Location of an intragenic control region in the *X. borealis* somatic 5S gene. The extent of the 5' and 3' deletions constructed and their effects on transcription are summarized. The hatched segment is the intragenic control region. Reproduced from Bogenhagen *et al.*[3] by kind permission of the authors and the MIT Press.

Recombinant plasmids with deletions extending as far as the last nucleotide of the 5' flanking region were found to give specific initiation of transcription of the 5S gene although in some cases the exact position of initiation was altered by a few nucleotides. The study of cloned genes containing deletions extending into the 5' portion of the 5S gene, revealed that the control region responsible for initiation of transcription is located downstream from nucleotide +50 of the 5S gene. Deletion mutants which retain the sequence downstream from position 50 were found to synthesize a molecule of approximately 5S RNA size. RNA fingerprinting showed these molecules to be part 5S sequence and part vector sequences, the 5' proximal 5S sequence being replaced by vector sequence. Initiation was shown to have occurred at a variety of sites in the vector, always at a G nucleoside.

Two recombinants were constructed which contained an insertion of either 6 or 20 nucleotides of sequence between positions 40–41 of an otherwise intact 5S gene. Transcription of these 'maxigenes' demonstrated that each supported the synthesis of an RNA species corresponding in size to normal 5S RNA. Fingerprinting of the maxigene transcripts revealed that initiation of transcription had occurred downstream from the normal site. In the case of the 20 nucleotide insertion, transcription had initiated between 21 and 24 nucleotides downstream from the normal position. This experiment indicated that initiation of 5S gene transcription occurred a fixed distance upstream from the internal control region. By a procedure similar to the one described above, the 3' boundary of the internal control region was shown to lie between nucleotides 80–83. Deletion from the 3′ end past nucleotide 120 abolishes the termination site of the 5S gene, therefore a normal pattern of transcripts will not appear on a gel. This problem was solved by using cordycepin triphosphate, an ATP analogue which causes termination of transcription when incorporated in place of ATP. Thus partial incorporation of this nucleotide resulted in partial termination of 5S transcription, and a reproducible pattern of transcripts upon gel electrophoresis.

In this way the *X. borealis* somatic 5S gene was shown to contain an intragenic region responsible for the initiation of transcription located between nucelotides 50–83. That this region alone is responsible for initiation was demonstrated by the cloning of a fragment spanning nucleotides 41–87 of the 5S gene into the BstI site of pBR322. The recombinant clone produced was shown to initiate transcription of vector sequences at the correct distance upstream from the 5S insert.

Isolation of a transcription factor specific for 5S genes

A transcription factor which is required for 5S RNA synthesis *in vitro* was isolated from an *X. laevis* ovarian extract by Engelke *et al.*[8] This protein was detected and isolated by its ability to facilitate transcription of exogenous *X. borealis* oocyte 5S genes in an unfertilized egg extract, which although known to contain high levels of RNA polymerase III, would not transcribe added 5S genes. Fractionation of the ovarian extract by column chromatography showed that a 37 000 d protein consistently copurified with the transcription complementation activity. This protein was shown also to be required for the transcription of a cloned *X. borealis* somatic 5S gene in the egg extract. However a cloned *Xenopus* tRNA[Met] gene was correctly transcribed and processed by the egg extract in the absence of the factor.

Possible binding of this transcription factor to the 5S gene was studied by the 'footprinting' technique[8]. In this method a restriction fragment, labelled at one 5′ or 3′ end, is partially digested with DNase I. When the products are analysed on a denaturing polyacrylamide gel, a ladder of discrete fragment sizes differing by a single base is observed. However, if a protein is bound to a defined site on the DNA molecule, it will protect that segment of DNA from the DNase. Thus gel electrophoresis will reveal a blank section of the ladder, corresponding to the region of DNA protected by the protein. Roeder and colleagues discovered that the transcription factor they had isolated would specifically interact with both the *X. borealis* somatic and oocyte 5S genes and in each case an intragenic region was protected from DNase cleavage. The regions involved were positions 47–96 of the *X. borealis* somatic 5S gene, positions 45–92 of the *X. borealis* oocyte 5S gene 1 and positions 75–96 of the *X. borealis* oocyte gene 3. The first two regions are almost identical to the 5S internal control region described by Brown and colleagues, but the region protected in the third *X. borealis* oocyte gene represents only the 3′ half of this region. Why this should be so is not clear since this gene is apparently transcribed normally *in vitro*, although at half the efficiency of the other *X. borealis* oocyte genes[18, 20].

These interactions were observed in the presence of purified factor and DNA only. Neither RNA polymerase III nor egg extract mediate any specific DNase protection of the 5S gene, nor do they alter or abolish the protection conferred by this factor. Moreover, loss of specific binding to mutant genes deleted beyond +83 from the 3′ direction correlates exactly with the loss of their ability to initiate transcription[32]. Therefore it seems very likely that through its interaction with the internal control region, this factor is involved in the initiation of 5S RNA transcription.

The 5S transcription factor is identical to a 5S RNA storage protein

Immature *Xenopus* oocytes possess no detectable free 5S RNA. All the 5S RNA has been shown to be present in either 7S or 42S ribonucleoprotein particles. The 7S particle was recently shown by Picard and Wegnez[30] to consist of one molecule of 5S RNA complexed with one protein molecule of 45 000 d molecular weight. The 5S transcription factor previously isolated by Engelke *et al.*[8] had an estimated molecular weight of 37 000 d; hereafter this factor will be referred to as TF IIIA. Subsequently TF IIIA was found by Honda and Roeder[14] to comigrate on denaturing gel electrophoresis with both a protein from the 7S RNP particle and one from the 42S RNP particle. The proteins were transferred from the gel onto chemical affinity paper, and subsequently challenged with radioactive antibody raised against TF IIIA. Only the protein from the 7S RNP particle was found to cross-react with the TF IIIA antisera; neither the 42S RNP particle-derived protein, nor any *Xenopus* ribosomal proteins gave a reaction. Both TF IIIA and the 7S protein were cleaved with cyanogen bromide, and the products separated by SDS polyacrylamide gel electrophoresis. Both proteins gave an identical cleavage pattern of four daughter peptides. Thus TF IIIA and the 7S particle-derived protein seemed structurally similar if not identical. This observation therefore raised the question as to whether the two proteins were functionally identical.

Results indicate that both the 7S RNP particle and the purified 7S protein can substitute for TF IIIA in complementing a soluble *Xenopus* egg extract for 5S RNA transcription. However, neither the 42S RNP particle nor any of the proteins purified from it can complement the egg extract for transcription[14, 27]. The binding of both the 7S RNP particle and the 7S derived protein to 5S DNA was then studied by Pelham and Brown using a cloned *X. borealis* somatic 5S gene and the DNase protection technique[27]. Briefly the findings were that the 7S RNP particle does not give any protection of the 5S gene. However, if the 7S particle was first treated with pancreatic ribonuclease, the preparation was then observed to protect the 45–96 region of the 5S gene from DNase. Purified protein from the 7S particle gave protection against DNase. Preincubation of the protein with *Xenopus* 5S RNA abolished the DNase protection, whereas preincubation of the protein with *E. coli* tRNA had no effect on the protection mediated by the protein. These experiments demonstrated that the protein from the 7S RNP particle was functionally identical to TF IIIA. Furthermore this protein was capable of binding either to nucleotides 45–96 of the 5S gene, or to 5S RNA, but not simultaneously to both.

5S RNA can inhibit its own synthesis

Pelham and Brown demonstrated that transcription of a cloned *X. borealis* somatic 5S gene in an oocyte nuclear lysate was inhibited upon addition of exogenous *Xenopus* 5S RNA to the lysate[27]. When exogenous 5S RNA was added to a lysate containing a mixture of two templates, 5S DNA and the genes coding for adenovirus VA RNA, only 5S RNA synthesis was reduced; VA RNA synthesis was unaffected. Similarly, transcription of an *X. borealis* oocyte 5S clone in a mammalian whole cell extract was specifically inhibited by HeLa 5S RNA, whereas tRNA gene transcription was unaffected[11]. Thus 5S RNA was demonstrated to be capable of specifically inhibiting its own synthesis. Further experiments demonstrated that the inhibition was dependent on the 5S DNA concentration[27]. As the 5S DNA concentration was lowered, a constant level of exogenous 5S RNA produced increasing inhibition of 5S RNA synthesis. When 5S RNA was added to a lysate which had been preincubated with 5S DNA, there was a lag time of 40 minutes before any reduction in the rate of 5S RNA synthesis was observed. Thus once an active transcription complex had formed, it was not immediately susceptible to inhibition by exogenous 5S RNA. The production of this stable transcription complex is a rapid event, occurring within five minutes[4]. Once formed the complex is resistant to dilution and high concentrations of competitor DNA. Moreover, it is stable for at least 40 rounds of transcription suggesting that associated factors must either remain bound or, if released, selectively reassociate with this structure. TF IIIA is required for the assembly of the active complex and must be highly bound judging by its resistance to adsorption by competing 5S RNA. However, alone it is not sufficient to generate an active complex. The identity of the other factors required is not known but there is some suggestion that they may be necessary for the formation of active complexes on other RNA polymerase III transcription units. Preincubation with histones in TF IIIA-depleted extracts can give rise to a stable transcriptionally inactive complex on *X. borealis* somatic 5S DNA which is subsequently refractory to TF IIIA stimulation. Under these conditions tRNA gene transcription is still active. Such experiments are thus providing us with an insight into how the stability of the differentiated state is maintained.

Does 5S RNA autoregulate its own synthesis in oocytes?

The experiments described above indicate that TF IIIA, a 5S transcription factor present in oocytes which binds to the internal control region of the 5S gene, is structurally and functionally identical to a 5S RNA storage protein found in 7S RNP particles. The observation that exogenous 5S RNA will inhibit the transcription of the 5S gene can therefore be explained by binding of 5S RNA to TF IIIA and the subsequent production of 7S RNP particles, thus making TF IIIA unavailable for binding to the control region of the 5S gene.

If this scheme of events operates *in vivo*, regulation of 5S RNA synthesis in oocytes would be accomplished by regulation of the level of transcription factor. A given quantity of transcription factor would result in a certain level of 5S RNA synthesis. The newly synthesized RNA would then sequester the transcription factor, and lead to a decrease in the rate of 5S RNA synthesis in the absence of fresh synthesis of transcription factor.

A transcription factor related to TF IIIA exists in *Xenopus* somatic cells

Pelham *et al.* examined somatic cells of *X. laevis* for the presence or absence of a molecule similar to the oocyte transcription factor TF IIIA[28]. Initially they discovered that in an *X. laevis* kidney cell extract, transcription of a cloned *X. borealis* somatic 5S gene required an intact intragenic region extending from nucleotides 50–83. Moreover, synthesis from this template was inhibited by addition of exogenous 5S RNA from either oocytes or somatic cells. These experiments demonstrated the presence in somatic cells of a 5S transcription factor with similar properties to the oocyte factor, namely the capability of binding either to an intragenic control region or to 5S RNA. Antibodies were raised against purified oocyte transcription factor and were shown to cause approximately 95% inhibition of transcription in an oocyte nuclear extract and approximately 80% inhibition in a somatic cell extract using the same cloned *X. borealis* somatic 5S DNA substrate. Thus 5S RNA synthesis in *Xenopus* somatic cells required a protein which was antigenically similar to the oocyte transcription factor. To detect this protein, nuclear extracts were subjected to denaturing gel electrophoresis and transferred to nitrocellulose filters which were then probed with radioactively labelled antisera against oocyte factor and autoradiographed. The somatic cell extract was shown to contain a protein which reacted specifically to antisera against oocyte transcription factor, but which migrated more slowly on the gel than the oocyte factor. When a more sensitive experiment was performed, extracts of *Xenopus* tissue culture cells and of adult *Xenopus* liver were both found to contain an immunoprecipitated protein which comigrated on a gel with the oocyte factor. In each extract there was 10 times more somatic than oocyte factor present. A quantitative experiment demonstrated that the small amount of oocyte factor persisting in somatic cells was insufficient to support the observed levels of somatic 5S RNA transcription. Therefore, by inference, the somatic protein must be active as a transcription factor in somatic cells. Levels of this protein are too low to permit, thus far, the direct experiments which revealed the properties of TF IIIA.

The somatic cell extract was also shown to be capable of efficient transcription of oocyte-type 5S genes. Therefore the absence of transcription of oocyte 5S genes in somatic cells is not simply due to the presence of different transcription factors in oocytes and somatic cells.

Developmental regulation of 5S RNA synthesis

The genes coding for somatic 5S RNA are known to be expressed in both oocytes and somatic cells, whereas the oocyte-type 5S genes are expressed only in developing oocytes[10]. Accurate transcription of cloned 5S genes has been obtained in injected oocytes and cell-free systems derived from both oocytes and somatic cells. None of these systems however demonstrated any developmental regulation, in that the oocyte-type genes are always transcribed. Korn and Gurdon attempted to obtain developmentally specific expression of 5S genes by a series of experiments involving the injection of *Xenopus* somatic cell nuclei into *Xenopus* oocyte germinal vesicles[19]. This procedure allowed them to ask whether the oocyte could reprogram the somatic cell nucleus to express oocyte-type 5S genes.

Initially *X. laevis* kidney tissue culture cell nuclei were injected into individual oocytes from two different *X. laevis* females. The oocytes of one female were found to activate the oocyte-type 5S genes, whereas in the oocytes of the other female there was no increase in synthesis of oocyte 5S RNA over the background value. A total of 50 *X. laevis* females were then tested for their ability to activate oocyte 5S RNA synthesis in somatic cell nuclei. Twenty-seven were found to be activating, 16 were non-activating and seven gave intermediate values. In several cases oocytes were taken from a given female at intervals over a few months. Each female always demonstrated the same activating or non-activating pattern. The possibility existed that the oocyte 5S RNA synthesis observed could have been due to stimulation of the endogenous oocyte 5S genes in the host. This was eliminated by injecting *X.*

laevis kidney culture cell nuclei into *X. borealis* oocytes, and observing that the oocyte 5S RNA synthesized was of the *X. laevis* type and not specific to *X. borealis*. Erythrocyte nuclei were also shown to be capable of activation to synthesize oocyte 5S RNA, demonstrating that this effect was not peculiar to nuclei from a culture cell line. Korn and Gurdon also discovered that a short preincubation of the nuclei in 0.35 mol/ℓ NaCl, which will remove non-histone proteins but not histones, followed by injection into the oocytes of a non-activating female resulted in activation of the oocyte 5S RNA genes in 75% of the assays performed[19]. A further experiment demonstrated that neither the 5S transcription factor TF IIIA nor the 7S RNP particle would stimulate transcription of oocyte 5S genes, when mixed with somatic cell nuclei prior to injection of the mixture into the germinal vesicles of non-activating oocytes.

Similar results have been reported from *in vitro* studies[4]. Chromatin isolated from *X. laevis* erythrocytes or tissue culture cells synthesized somatic-type 5S RNA and germinal vesicle chromatin synthesized oocyte-type 5S RNA regardless of whether the incubation was performed in extracts derived from oocytes (presumably non-activating) or from somatic cells. Developmental control would thus appear to be a function of the template rather than the extract. Again, pre-washing somatic cell chromatin with 0.6 mol/ℓ NaCl can activate oocyte 5S RNA expression.

The nature of the transition of the oocyte genes from the inactive to the active state remains unclear. One may speculate that it has to do with the presence or absence of one or more non-histone chromosomal proteins. As nuclei injected into oocytes do not divide, these experiments demonstrated that a gene which was once inert could itself become active.

Termination of *Xenopus* 5S gene transcription

A feature common to *Xenopus* 5S genes is the existence of two or three T clusters on the non-coding strand, adjacent and downstream to the site of transcription termination. The mature 5S transcript normally has its 3' end in the middle of the first T cluster.

Bogenhagen and Brown constructed an *X. borealis* somatic 5S deletion mutant with the entire 3' flanking region after the first T cluster replaced by vector DNA[2]. They observed that this mutant gave correct termination of transcription *in vitro* and concluded that termination was mediated by the first TTTT sequence, and that no 3' flanking sequence was required. Deletion of an additional five nucleotides of 5S DNA, leaving only one T at position 119, followed by vector sequence, resulted in the production of high molecular weight RNA with no detectable trace of 5S RNA. Mutant genes were constructed by either inserting a piece of foreign DNA or deleting a segment of 5S DNA from the sequence between the initiation and termination regions. Transcription of these mutants produced a range of different sized '5S' molecules. This demonstrated that the site of termination was independent of the distance from either the 5' end of the RNA or the internal control region. Further studies demonstrated that a T4 sequence was an efficient terminator if surrounded by a G + C-rich region. However, if the T4 sequence was surrounded by A-rich sequences, then termination efficiency was reduced. Substitution of a C for a T to generate TTTC in place of the normal first T cluster, caused a large decrease in efficiency of termination.

The *X. laevis* 5S pseudogene can be transcribed

Recent experiments have demonstrated that the 5S pseudogene can be transcribed, both by an oocyte nuclear extract[2] and by microinjected oocytes[23]. The results indicated that pseudogene transcription was accurate, and initiated at the first nucleotide of the pseudogene. Transcription was mediated by RNA polymerase III. When injected into oocytes in the presence of the 5S gene, the pseudogene was found to be transcribed 27% as efficiently as the 5S gene. However, when injected in the absence of the 5S gene, the level of pseudogene transcription could be as high as 85% of that of the 5S gene. This suggested a competition effect between the 5S gene and pseudogene promoters. Both sets of experiments demonstrated that the TTTT sequence at position 103–106 of the pseudogene is inefficient as a terminator. Studies with a cloned molecule containing the pseudogene plus the adjacent A + T-rich spacer downstream, showed that the bulk of the pseudogene transcripts terminated at a T6 sequence in the spacer. The predominant transcript produced comigrated with 5S RNA upon gel electrophoresis. Thus the apparent transcriptional inactivity of the pseudogene *in vivo* could be accounted for either by competition from the 5S gene or comigration of a pseudogene transcript with 5S RNA on gel electrophoresis.

Acknowledgements

I am grateful to Don Brown, Nina Fedoroff and the MIT Press for allowing reproduction of their figures and to Glyn Millhouse for photography.

References

1. Birkenmeier, E. H., Brown, D. D. and Jordan, E. (1978). A nuclear extract of *Xenopus laevis* oocytes that accurately transcribes 5S RNA genes. *Cell*, **15,** 1077–1086
2. Bogenhagen, D. F. and Brown, D. D. (1981). Nucleotide sequences in *Xenopus* 5S DNA required for transcription termination. *Cell*, **24,** 261–270
3. Bogenhagen, D. F., Sakonju, S. and Brown, D. D. (1980). A control region in the center of the 5S RNA gene directs specific initiation of transcription: II. The 3′ border of the region. *Cell*, **19,** 27–35
4. Bogenhagen, D. F., Wormington, W. M. and Brown, D. D. (1982). Stable transcription complexes of *Xenopus* 5S RNA genes: a means to maintain the differentiated state. *Cell*, **28,** 413–421
5. Brown, D. D. and Gurdon, J. B. (1977). High-fidelity transcription of 5S DNA injected into *Xenopus* oocytes. *Proceedings of the National Academy of Sciences, USA*, **74,** 2064–2068
6. Brown, D. D. and Gurdon, J. B. (1978). Cloned single repeating units of 5S DNA direct accurate transcription of 5S RNA when injected into *Xenopus* oocytes. *Proceedings of the National Academy of Sciences, USA*, **75,** 2849–2853
7. Brown, D. D., Carroll, D. and Brown, R. D. (1977). The isolation and characterization of a second oocyte 5S DNA from *Xenopus laevis*. *Cell*, **12,** 1045–1056
8. Engelke, D. R., Ng, S. Y., Shastry, B. S. and Roeder, R. G. (1980). Specific interaction of a purified transcription factor with an internal control region of 5S RNA genes. *Cell*, **19,** 717–728
9. Fedoroff, N. V. and Brown, D. D. (1978). The nucleotide sequence of oocyte 5S DNA in *Xenopus laevis*. 1. The AT-rich spacer. *Cell*, **13,** 701–716
10. Ford, P. J. and Southern, E. M. (1973). Different sequences for 5S RNA in kidney cells and ovaries of *Xenopus laevis*. *Nature, New Biology*, **241,** 7–12
11. Gruisseum, W. and Seifart, K. H. (1982). Transcription of 5S RNA genes *in vitro* is feedback-inhibited by HeLa 5S RNA. *Journal of Biological Chemistry*, **257,** 1468–1472
12. Gurdon, J. B. and Brown, D. D. (1978). The transcription of 5S DNA injected into *Xenopus* oocytes. *Developmental Biology*, **67,** 346–356

13. Hamada, H., Muramatsu, M., Urano, Y., Onishi, T. and Kominami, R. (1979). *In vitro* synthesis of a 5S RNA precursor by isolated nuclei of rat liver and HeLa cells. *Cell*, **17**, 163–173

14. Honda, B. M. and Roeder, R. G. (1980). Association of a 5S gene transcription factor with 5S RNA and altered levels of the factor during cell differentiation. *Cell*, **22**, 119–126

15. Jacq, B., Jourdan, R. and Jordan, B. R. (1977). Structure and processing of precursor 5S RNA in *Drosophila melanogaster*. *Journal of Molecular Biology*, **117**, 785–795

16. Jacq, C., Miller, J. R. and Brownlee, G. G. (1977). A pseudogene structure in 5S DNA of *Xenopus laevis*. *Cell*, **12**, 109–120

17. Korn, L. J. (1982). Transcription of *Xenopus* 5S ribosomal RNA genes. *Nature, London*, **295**, 101–105

18. Korn, L. J. and Brown, D. D. (1978). Nucleotide sequence of *Xenopus borealis* oocyte 5S DNA: Comparison of sequences that flank several related eukaryotic genes. *Cell*, **15**, 1145–1156

19. Korn, L. J. and Gurdon, J. B. (1981). The reactivation of developmentally inert 5S genes in somatic nuclei injected into *Xenopus* oocytes. *Nature, London*, **289**, 461–465

20. Korn, L. J., Birkenmeier, E. H. and Brown, D. D. (1979). Transcription initiation of *Xenopus* 5S ribosomal RNA genes *in vitro*. *Nucleic Acids Research*, **7**, 947–958

21. Kramer, R. A., Philippsen, P. and Davis, R. W. (1978). Divergent transcription in the yeast ribosomal RNA coding region as shown by hybridisation to separated strands and sequence analysis of cloned DNA. *Journal of Molecular Biology*, **123**, 405–416

22. Miller, J. R. and Brownlee, G. G. (1978). Is there a correction mechanism in the 5S multigene system? *Nature, London*, **257**, 556–558

23. Miller, J. R. and Melton, D. A. (1981). A transcriptionally active pseudogene in *Xenopus laevis* oocyte 5S DNA. *Cell*, **24**, 829–835

24. Miller, J. R., Cartwright, E. M., Brownlee, G. G., Fedoroff, N. V. and Brown, D. D. (1978). The nucleotide sequence of oocyte 5S DNA in *Xenopus laevis*. II. The GC-rich region. *Cell*, **13**, 717–725

25. Ng, S. Y., Parker, C. S. and Roeder, R. G. (1979). Transcription of cloned *Xenopus* 5S RNA genes by *X. laevis* RNA polymerase III in reconstituted systems. *Proceedings of the National Academy of Sciences, USA*, **76**, 136–140

26. Parker, C. S. and Roeder, R. G. (1977). Selective and accurate transcription of the *Xenopus laevis* 5S RNA genes in isolated chromatin by purified RNA polymerase III. *Proceedings of the National Academy of Sciences, USA*, **74**, 44–48

27. Pelham, H. R. B. and Brown, D. D. (1980). A specific transcription factor that can bind either the 5S RNA gene or 5S RNA. *Proceedings of the National Academy of Sciences, USA*, **77**, 4170–4174

28. Pelham, H. R. B., Wormington, W. M. and Brown, D. D. (1981). Related 5S RNA transcription factors in *Xenopus* oocytes and somatic cells. *Proceedings of the National Academy of Sciences, USA*, **78**, 1760–1764

29. Peterson, R. C., Doering, J. L. and Brown, D. D. (1980). Characterization of two *Xenopus* somatic 5S DNAs and one minor oocyte-specific 5S DNA. *Cell*, **20**, 131–141

30. Picard, B. and Wegnez, M. (1979). Isolation of a 7S particle from *Xenopus laevis* oocytes: a 5S RNA–protein complex. *Proceedings of the National Academy of Sciences, USA*, **76**, 241–245

31. Sakonju, S., Bogenhagen, D. F. and Brown, D. D. (1980). A control region in the center of the 5S RNA gene directs specific initiation of transcription: 1 The 5' border of the region. *Cell*, **19**, 13–25

32. Sakonju, S., Brown, D. D., Engelke, D., Ng, S.-Y., Shastry, B. S. and Roeder, R. G. (1981). The binding of a transcription factor to deletion mutants of a 5S ribosomal RNA gene. *Cell*, **23**, 665–669

33. Selker, E. U., Yanofsky, C., Driftmier, K., Metzenberg, R. L., Alzner-DeWeerd, B. and Rajbhandary, U. L. (1981). Dispersed 5S RNA genes in *N. crassa*: structure, expression and evolution. *Cell*, **24**, 819–828

34. Tschudi, C. and Pirrotta, V. (1980). Sequence and heterogeneity in the 5S RNA gene cluster of *Drosophila melanogaster*. *Nucleic Acids Research*, **8**, 441–451

Chapter 14

Transfer RNA genes

S. G. Clarkson

Transfer RNAs are small but major constituents of every cell. As a group, they are universally important as the donors of amino acids to polypeptide chains. Individual species may also have crucial roles in other processes[99]. In *E. coli* attenuation of the tryptophan biosynthetic operon depends on the level of charged $tRNA^{Trp}$, while cell wall synthesis in staphylococci involves a novel $tRNA^{Gly}$ species. The presence of tRNA-like structures at the 3' end of viral RNA genomes and the use of certain specific tRNAs as primers for reverse transcriptase indicate a similar diversity of tRNA function in higher organisms. Presumably in response to these varied needs, the amounts of different kinds of tRNAs change appreciably during normal development and neoplasia. For these reasons, the genes that encode these small and relatively easily isolated molecules have been extensively studied.

Some hints of the complexity of tRNA gene arrangement in eukaryotes were evident from hybridization studies almost a decade ago. More recent DNA sequence analyses have revealed that tRNA coding sequences may be highly conserved or diverged, interrupted, set far apart or close together, on the same or opposite DNA strands. Despite this structural diversity, an unexpected common functional pattern has emerged – namely, that much of the information needed for tRNA gene transcription by one kind of polymerase, RNA polymerase III, is contained within the genes themselves.

Before discussing these topics in more detail, it is important to note that these are features of eukaryotic nuclear tRNA genes and that few of them are shared by the tRNA genes of chloroplasts and mitochondria. Spectacular advances have been made in the structural analyses of these cell organelles, the most notable being the sequence elucidation of the entire bovine and murine mitochondrial genomes[2, 6]. These mammalian organelles use a different genetic code with a simpler set of tRNAs, most of which lack some of the characteristic features of tRNAs from all other sources. In addition, the tRNAs are derived from transcripts of both strands of the mitochondrial genome. Post-transcriptional processing is also important in the biosynthesis of nuclear encoded tRNAs, but the primary transcripts are very much shorter and they usually contain only a single tRNA species. Because of this fundamental difference in the mode of transcription, the remainder of this review will be concerned with eukaryotic nuclear tRNA genes.

tRNA gene numbers

Estimates for the total number of tRNA genes in eukaryotes at present range from a few hundred in the nematode *Caenorhabditis elegans* to several thousand in the frog *Xenopus laevis* (*Table 14.1*). In contrast, *E. coli* contains about 60 tRNA genes and thus only one or a few copies of a gene for a particular tRNA species[7, 10]. Hybridization kinetics and sensitive chromatographic techniques suggest, however, that the number of individual tRNA species is very similar in prokaryotes and eukaryotes. *Drosophila melanogaster*, for example, contains about 60 kinetically distinct[110] and about 100 chromatographically distinct[42] tRNA species. Some of the latter are identical in primary sequence and differ only in the extent of their post-transcriptional modification. It is likely, then, that eukaryotes contain multiple copies of relatively few kinds of tRNA genes.

TABLE 14.1. Repetition of eukaryotic tRNA genes

| Species | Number of genes/haploid genome | | | |
	Total tRNA[a]	Single tRNA species		Reference
Dictyostelium discoideum		$tRNA^{Trp}$	~6	84
Physarum polycephalum	1050			
Neurospora crassa	2640			
Saccharomyces cerevisiae	360	$tRNA^{Tyr}$	8	78
		$tRNA^{Ser}_{UCA}$	3	11, 31, 80
		$tRNA^{Ser}_{UCG}$	1	31, 32, 80
		$tRNA^{Ser}_2$	~11	81
Euglena gracilis	740			
Tetrahymena pyriformis	800–1450			
Caenorhabditis elegans	300			
Bombyx mori		$tRNA^{Ala}$	~20	44
Drosophila melanogaster	590–900	$tRNA^{Lys}_2$	~12	39
		$tRNA^{Tyr}$	~23	63
Xenopus laevis	6500–7800	$tRNA^{Met}_1$	~310	18
		$tRNA^{Met}_2$	~170	18
		$tRNA^{Val}$	~240	18
Rattus norvegicus	6500			
Homo sapiens	1310	$tRNA^{Met}_1$	~12	90

[a] Total number of genes for all tRNA species (from reference 66).

It should be borne in mind that the gene number estimates are derived from DNA–RNA hybridization experiments with crude 4S RNA and are subject to several possible errors. Contamination by fragments of longer RNAs encoded by more highly repeated genes can lead to overestimates. Conversely, some tRNA genes will not be detected if their products are at low concentration in the pool of cellular RNA. In addition, we now appreciate that several features of tRNA gene organization – the occasional presence of intervening sequences, inverted repeats, and incomplete genes – can influence the outcome of these experiments. Some, but not all, of these limitations can be overcome by the use of more specific hybridization probes, such as aminoacyl-labelled tRNAs[18], purified tRNAs, particularly if iodinated at a specific modified nucleotide[78], and labelled restriction fragments of cloned tRNA genes. These more specific hybridization assays have confirmed that tRNA genes are reiterated, at least in higher eukaryotes, and have also shown that the number of genes coding for an individual tRNA species can vary by more than ten-fold (*Table 14.1*).

The hybridization conditions that have been used up to now are too insensitive to detect minor sequence variants and they may not always discriminate between genes coding for different tRNAs that share significant sequence homology. However, when combined with independent genetic evidence, they can give reliable estimates of the number of functional tRNA genes. In yeast, for example, three restriction fragments that hybridize with $tRNA_{UCA}^{Ser}$ have been correlated with the three known loci that can mutate to serine-inserting *ochre* suppressors[11, 31, 80]. A similar combination of molecular and genetic evidence has shown that yeast contains a single $tRNA_{UCG}^{Ser}$ gene which can mutate to a recessive lethal *amber* suppressor[11, 31, 32, 80]. This represents the first example of a eukaryotic single copy tRNA gene.

The simplest interpretation of the widespread repetition of tRNA genes is that it reflects the need for a large quantity of tRNA, amounting to ~20% of the mass of cellular RNA. Specialized cells can produce comparable amounts of proteins from single genes by the repeated translation of stable mRNAs. This option is not open to tRNA genes because the needed products are the mature transcripts themselves. Another option, gene amplification, has not been described for this class of gene. We might therefore expect to find a positive correlation between the abundance of a particular tRNA species and the number of genes that encode it. Such a correlation exists for the 'UCN' family of serine tRNAs and their genes in yeast. Thus, $tRNA_{UCG}^{Ser}$ and $tRNA_{UCA}^{Ser}$, encoded by 1 and 3 genes, respectively, are minor tRNA species whereas $tRNA_2^{Ser}$, which recognizes the codons UCC and UCU, is by far the most abundant serine tRNA isoacceptor and is encoded by at least 11 genes[80, 81].

An extension of this teleological argument may partly account for the rather poor correlation between the number of tRNA genes in various eukaryotes and their genetic complexity (*Table 14.1*). The very high number of *X. laevis* tRNA genes, for example, may be dictated by the unusual synthetic demands of oogenesis during which single oocytes accumulate as much as 40 ng tRNA. Calculations suggest that these tetraploid cells would indeed need at least 30 000 genes to synthesize this amount of tRNA in the 9–12 months of oogenesis[18]. Hence the high abundance of some tRNA genes may imply that their expression is restricted to certain developmental periods.

tRNA gene arrangements

The almost universal repetition of tRNA genes in eukaryotes raises several interesting questions concerning their organization. Are identical tRNA genes widely scattered throughout the genome? If not, are they clustered solely with themselves or with other tRNA genes? Is there a functional logic to their arrangement? Answers to the first two questions are accumulating rapidly but the third is proving more intractable. Indeed, as the following examples demonstrate, variability is the most striking feature of tRNA gene organization in eukaryotes.

Yeast

Most yeast tRNA genes are dispersed throughout the genome. This was first recognized for the eight tRNA^{Tyr} genes which are unlinked and map onto six different chromosomes[79]. Each of these genes is contained within a distinct restriction fragment, seven of which have been cloned[78]. Only one of these fragments contains an additional tRNA gene, which is close to the expectation for a random distribution[78]. That this is the rule rather than the exception is suggested by statistical analysis of a random collection of yeast tRNA gene clones[4], and by structural analyses of several tRNA gene families. The tRNA^{Phe} genes, for example, are found on at least 10 different restriction fragments ranging in size from 3–20 kb[107]. Similarly, the tRNA^{Leu} genes are located on at least 14 other fragments of comparable size range[108]. Of the 11 restriction fragments which hybridize with tRNA$^{Ser}_2$, only one contains an additional tRNA gene[81]. The individual members of other tRNA gene families that have been cloned also exhibit the same scattered arrangement. Thus, single genes for tRNA$^{Met}_3$[77], tRNA$^{Glu}_3$[29,33], tRNA$^{Arg}_2$ and tRNA$^{Val}_1$[3] have been found on separate DNA fragments of 3–7 kb which contain no other tRNA genes. Information is so far limited on the nature and exact location of other kinds of nearby genes, although the cytochrome *c* gene has been mapped to within 2.2 kb of one of the tRNA$^{Ser}_2$ genes[81], and an origin of replication is thought to reside on a 1.9 kb fragment containing a tRNA$^{Glu}_3$ gene[33]. In neither case, however, is there any compelling reason to believe that the association is functionally significant.

Yeast does contain, however, a few tightly clustered tRNA genes. In *S. cerevisiae*, single copies of tRNA$^{Arg}_3$ and tRNA^{Asp} genes are separated by an identical 10 kb of DNA at two or more sites in the genome. The genes share the same polarity, with the tRNA$^{Arg}_3$ gene upstream, and they are co-transcribed *in vitro* as a dimeric tRNA precursor[94]. A similar arrangement of two other genes has been found in *S. pombe*. In this case, a tRNA$^{Ser}_{UCG}$ gene containing a 16 bp intervening sequence is followed by 7 bp of DNA and then an uninterrupted initiator tRNA^{Met} gene. *In vitro*, these two genes are also transcribed together as a single precursor[67].

Drosophila melanogaster

More than 15% of the tRNA genes of *D. melanogaster* have been localized by *in situ* hybridization of labelled tRNAs to squashes of salivary gland polytene chromosomes[63]. No common pattern to their arrangement is yet evident but several interesting features have emerged: the tRNA genes are widely distributed over most chromosomes except the fourth; the 60 or so loci that have been detected contain quite different numbers of tRNA genes; any one locus may contain genes coding for tRNAs which accept different amino acids; and genes for a single tRNA species may be found at more than one locus. By now, DNA fragments from several of these loci have been cloned. Their structural analyses have confirmed the varied complexity of tRNA gene arrangements in *Drosophila*. In one well studied case, overlapping clones have been obtained that span 94 kb of DNA from region 42A of chromosome arm 2R[39,55,115,116]. Within this 94 kb, a central 46 kb region contains eight tRNA^{Asn} genes, four tRNA$^{Arg}_2$ genes, five tRNA$^{Lys}_2$ genes and a single tRNA^{Ile} gene. The genes are irregularly spaced, although to some extent sub-clustered, and they are transcribed from both DNA strands (*Figure 14.1*). Some of the genes with opposite polarity were orginally detected by electron

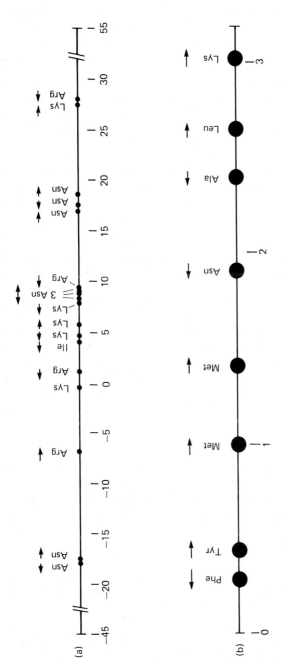

Figure 14.1. Two complex tRNA gene clusters. (a) Arrangement of the 18 tRNA genes located at region 42A of *Drosophila melanogaster*. No additional tRNA genes have been found from 30 to 55 or from −20 to −45. (b) Arrangement of the 8 tRNA genes of a 3.18 kb fragment of *Xenopus laevis* DNA. The fragment is tandemly repeated ∼150 times at a single chromosomal locus. Distances are given in kilobases. Arrows indicate the polarity of the genes. Compiled from references 21, 39, 55, 70, 71, 115, 116.

microscopy as inverted repeat structures[116] and only later were recognized as tRNA genes by DNA sequencing and by more sensitive hybridization assays[55, 115].

DNA sequencing has also directly confirmed the presence of identical tRNA genes at different chromosomal loci. Region 50AB on chromosome arm 2R, for example, contains a cluster of five tRNA[Ile] genes, all having the same sequence as the one found at 42A. These genes are irregularly spaced, distributed over both DNA strands, and are intermingled with two tRNA[Leu] genes containing related 38 bp and 45 bp intervening sequences[85].

The initiator tRNA genes are also located at two regions, in this case 61D and 70DE of chromosome arm 3L. Two tRNA[Met] genes from 61D are not tightly clustered, however, nor are they mixed with other tRNA genes. Instead, they occur as single copies within repeated DNA segments of 415 bp that are separated from each other by many kb of DNA[97]. To add to the diversity, *Drosophila* also contains some tRNA genes within closely-linked repeated structures. For example, single copies of a tRNA[Gly] gene are found within two direct repeats of at least 1.1 kb at chromosomal region 56F. The repeats are separated by no more than 0.9 kb of DNA and may even be contiguous[49].

Xenopus laevis

Most *X. laevis* tRNA genes are well separated from the rest of the genomic DNA by buoyant density centrifugation[17, 20]. The DNA segments containing genes for tRNA$_1^{Met}$, tRNA$_2^{Met}$ and tRNA^{Val} can also be distinguished from each other by their different buoyant densities[17, 20]. These genes therefore lie within separate gene clusters. One of these clusters has been studied in some detail. Its novel feature is that it comprises about 150 copies of a tandemly repeated 3.18 kb fragment[21]. From DNA sequencing, the major repeat unit contains two tRNA$_1^{Met}$ genes and single genes for six other tRNA species[70, 71]. The genes are irregularly spaced, they are found on both DNA strands, and one of them, a tRNA[Tyr] gene, contains a 13 bp intervening sequence (*Figure 14.1*). This cluster is located at or near a telomere on one short chromosome[34]. *X. laevis* contains other 3.18 kb fragments which are partially homologous to the major repeat unit but they lack the tRNA$_1^{Met}$ genes[88]. These fragments are tandemly repeated about 20-fold but their chromosomal location is not yet known.

Mammals

The mammalian tRNA genes that have been examined to date exhibit three different patterns of organization. The ~12 human initiator tRNA genes are found on separate restriction fragments of 1.2–11 kb[90]. They thus resemble the dispersed tRNA gene families of yeast. In contrast, at least three tRNA[His] genes are contained within a 15 kb fragment of mouse DNA. Their 3' flanking regions are identical for 62 bp, suggesting that they may once have been part of a tandemly repeated structure[46]. A tRNA gene cluster has also been found within a 2.1 kb fragment of rat DNA. In this case, however, single genes for tRNA[Asp], tRNA[Gly] and tRNA[Glu] are separated by DNA segments of about 450 bp which show no extensive sequence homologies[92].

This brief survey illustrates the various ways that tRNA genes can be organized within even a single species. No eukaryotic counterparts have yet been found, however, for two arrangements characteristic of *E. coli* – the location of certain tRNA genes in the spacers between ribosomal RNA genes, and the tight clustering of several different kinds of tRNA genes that are co-transcribed as large multimeric precursors[99]. The patterns of gene arrangement that have emerged, together with the results of direct functional tests, suggest instead that most eukaryotic tRNA genes are transcribed individually. At present the only known exceptions are the pairs of yeast genes noted earlier.

tRNA gene stability

RNA sequencing has shown that tRNAs are very highly conserved between distantly related eukaryotes. The $tRNA^{Lys}_2$ sequences, for example, are identical in *Drosophila* and mammals while their $tRNA^{Gly}_1$ sequences are 95% homologous[49, 55]. Despite this phylogenetic stability, DNA sequencing has made it clear that the members of tRNA multigene families are not always identical. Such variant tRNA genes, which differ from the expected DNA sequence by one or more nucleotides, have been found amongst the $tRNA^{Glu}$ and $tRNA^{Val}_4$ genes of *D. melanogaster*[1, 54] and the initiator tRNA genes of *D. melanogaster*, *X. laevis* and humans[71, 90, 97]. The tRNA sequences encoded by these variants have not been detected *in vivo*, suggesting that the genes are transcribed infrequently or not at all, or that their transcripts are unstable. One of these variants, a human $tRNA^{Met}_1$ gene containing a G → T substitution within the conserved TΨCGA sequence, has been assayed in a homologous *in vitro* transcription system. It has also been introduced into intact cells via a viral vector. In both situations the variant gene is efficiently transcribed but the precursor RNA is not processed to mature tRNA[117, 118]. It is possible, however, that functional tRNAs may be produced from some variant genes, particularly from those that replace a non-conserved nucleotide in a loop region or from those that substitute a complementary base pair in a stem region.

In addition to minor sequence variation, some tRNA genes have been extensively disrupted so that only short regions of the coding sequence remain intact. Examples of such incomplete genes, or tRNA pseudogenes, have been found recently in a *D. melanogaster* fragment which contains four regions of homology with the initiator tRNA, the longest of which is only 34 bp. The hybridization of this fragment to about 30 dispersed chromosomal sites suggests that these pseudogenes may have arisen by the repeated insertion and excision of a transposable element into an intact tRNA gene[97].

At present there are very few examples of variant and incomplete tRNA genes. More impressive is the evolutionary stability of tRNA coding sequences and their maintenance at widely scattered genomic sites. The mechanism by which this is achieved is unknown but some plausible schemes have been proposed for the evolutionary history of certain tRNA gene clusters. The two *Drosophila* $tRNA^{Gly}$ genes at region 56F, for example, are contained within sequences that are identical for at least 280 bp; they thus appear to be the result of a simple duplication[49]. Some long range sequence homologies can also be detected around the two $tRNA^{Met}$ genes on the *X. laevis* 3.18 kb repeat unit, although their immediate 5' flanking sequences are not conserved and they also differ by a single nucleotide within their coding sequences. Hence these two genes may also have arisen through an early duplication event which was then followed by sequence divergence; the tRNA gene cluster may then have been generated by duplication of the entire 3.18 kb

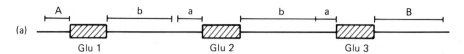

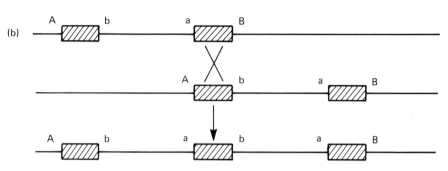

Figure 14.2. Model for the evolution of a tRNA gene cluster. (a) Sequence homologies around the three tRNA^{Glu} genes located at region 62A of *Drosophila melanogaster*. (b) Generation of a gene triplet from a hypothetical ancestral gene pair by unequal crossing over. The C → T transition in gene 2 happened after the crossover event because it is found only in this gene. Adapted from reference 54.

fragment[71]. A third factor, unequal crossing-over, has been invoked to explain the structural features of a *D. melanogaster* cluster of three tRNA^{Glu} genes[54]. In this scheme, an ancestral gene first gave rise to two genes, genes 1 and 3 (*Figure 14.2*), by a duplication event which was then followed by divergence of the flanking regions. An unequal crossover between these two genes then led to a central third gene, gene 2, which subsequently mutated by a single base transition.

Intervening sequences

Eukaryotic tRNA genes are sometimes interrupted by DNA sequences that do not code for the mature tRNA. These intervening sequences first detected within the tRNA^{Tyr} and tRNA^{Phe} genes of yeast[41, 107], one of which, the *SUP4* tRNA^{Tyr} gene, encodes a functional *ochre* suppressor and thus remains active despite its discontinuity. The way this is done is now clear. The intervening sequence is transcribed as part of the tRNA precursor and then, in two well coordinated reactions collectively known as splicing, it is excised from the precursor by an endonuclease while the two halves of the tRNA are joined together by an RNA ligase[57, 58, 76, 83]. A conditional lethal mutant, *rna1*, has been found in yeast in which these splicing reactions are blocked. At the non-permissive temperature this mutant accumulates ~12 unspliced precursor tRNAs, thereby suggesting that only a few kinds of yeast tRNA genes are split by intervening sequences[53, 57]. They are similarly restricted to a subset of tRNA genes in other eukaryotes. The tRNA^{Leu} genes, for example, are the only known split tRNA genes of *Drosophila*[85]. Of the eight tRNA genes located on the *X. laevis* 3.18 kb fragment, only the tRNA^{Tyr} gene contains an intervening sequence[70, 71]. The five kinds of tRNA genes that have been sequenced from mammals all lack intervening sequences yet mammalian cells contain the requisite splicing enzymes[64, 103].

The intervening sequences within tRNA genes vary significantly both in length, from 13 to ~60 bp, and in sequence. Although some sequence homologies around their 3′ boundaries have been noted[85], they are not particularly strong nor are they universal[84]. No extensive homologies exist either in the middle of intervening sequences or at their 5′ splice junctions, suggesting that the fidelity of splicing is not sequence-dependent. Nucleotides at the boundaries can, however, influence splicing efficiency. Thus an A → G transition at the 5′ splice junction of a mutant yeast SUP4 tRNA^{Tyr} gene results in a slow but nevertheless accurate excision of the intervening sequence from the precursor[22].

A striking feature of these intervening sequences is their common location: in all cases they are found one nucleotide 3′ to the anticodon. In apparent contradiction, it was earlier suggested that the intervening sequence immediately abuts the anticodon of a X. laevis tRNA^{Tyr} gene[70]. This proposal was based on the nature of the modified base next to the anticodon of this tRNA. Recent evidence[64] suggests, however, that this is a modified G rather than t^{6}A and this, in turn, would place the intervening sequence one nucleotide further in the 3′ direction (Figure 14.3) in keeping with all other split tRNA genes.

An interesting consequence of this location is that unspliced precursor tRNAs retain many of the characteristic features of the tRNA cloverleaf. In particular, the aminoacyl, D and TΨ stems are conserved while the intervening sequence alters only the secondary structure of the anticodon stem and loop region. In most unspliced yeast tRNA precursors the anticodon triplet can base pair with complementary nucleotides of the intervening sequence[57, 107] (Figure 14.3), thereby protecting the anticodon from cleavage by S1 nuclease[76]. A similar secondary structure can also be drawn for an unspliced transcript of a Neurospora crassa tRNA^{Phe} gene[93] but this is not a feature of every split tRNA gene. For example, the intervening sequence is capable of base pairing with only one nucleotide of the anticodon of a tRNA^{Tyr} from X. laevis[70], and not at all with the anticodon of a tRNA^{Trp} from Dictyostelium discoideum[84]. The SUP4 mutation in yeast, which converts a wild type tRNA^{Tyr} to a functional ochre suppressor, reduces the complementarity between the anticodon and the intervening sequence (Figure 14.3). Moreover, in X. laevis oocytes the SUP4 precursor is spliced more efficiently than its wild type counterpart[75], presumably because its enlarged anticodon loop makes it resemble more the frog precursor (Figure 14.3). This result further demonstrates that accurate splicing does not require the 3′ splice junction to be in a loop because another consequence of the SUP4 mutation is that it moves the 3′ splice junction to a base-paired region (Figure 14.3). An A → G transition at position 37 of the SUP4 gene restores the 3′ splice junction to a loop but places the 5′ splice junction within a region that can exist in a fully base-paired form (Figure 14.3). The precursor of this mutant is also accurately processed, although very slowly, both in yeast and in X. laevis oocytes[22, 75]. Hence the 5′ splice junction may need to be in a loop for efficient splicing but the accuracy of this process does not depend on a particular secondary structure of the intervening sequence and anticodon stem and loop region. Instead, the splicing enzymes appear to recognize certain secondary and tertiary structural features of the mature tRNA within the unspliced precursor.

Elaborate mechanisms thus exist to remove intervening sequences from the transcripts of certain eukaryotic tRNA genes. But why they are included in these transcriptional units in the first place is far from clear. One possibility is that they contribute sequences that are essential for transcription. This now seems unlikely,

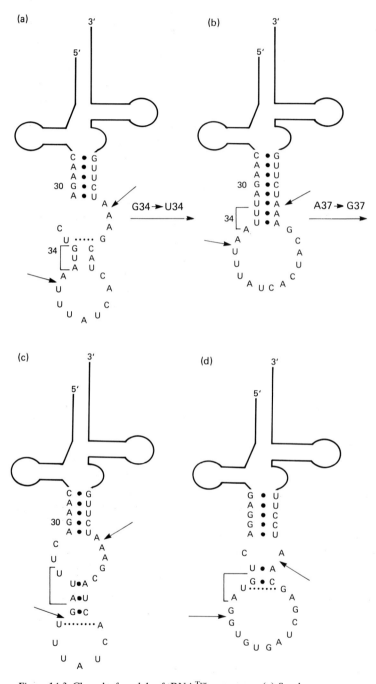

Figure 14.3. Cloverleaf models of tRNA^Tyr precursors. (a) *Saccharomyces cerevisiae sup⁺*. (b) *SUP4* single mutant. (c) *SUP4-G37* double mutant. (d) *Xenopus laevis sup⁺*. Only nucleotides in the anticodon arm and intervening sequence are shown, in each case in unmodified form. The anticodon is indicated by brackets. Sites for removal of the intervening sequence are indicated by arrows. Compiled from references 22, 64, 70.

however, for the intervening sequences within some yeast tRNA genes have been both enlarged with heterologous DNA[14,56] and, more important, entirely deleted[109] without the loss of gene expression. The strong conservation of eukaryotic tRNA coding sequences makes it similarly unlikely that they accelerate the evolution of new gene products, a role proposed for the intervening sequences within structural genes[40]. Indeed, a case may be made for the exact opposite – namely, that intervening sequences have been introduced into certain tRNA genes to reduce the chance of their illegitimate recombination with other related tRNA genes. The widely scattered tRNA^{Tyr} and tRNA^{Phe} genes of yeast, for example, possess very different amino acid acceptor stems but 69% homology between the rest of their coding sequences[41,107]. Recombination between these genes presumably is hindered by their very dissimilar intervening sequences. The *X. laevis* tRNA^{Tyr} and tRNA^{Phe} coding sequences are also closely related (70% overall homology) but only the tRNA^{Tyr} gene contains an intervening sequence[70]. These genes are set close together but on opposite DNA strands, an arrangement that may sufficiently reduce the chance of their recombination that there is no need for an intervening sequence within the tRNA^{Phe} gene. The yeast tRNA$^{Ser}_{UCG}$ and tRNA$^{Ser}_{UCA}$ coding sequences provide examples of even more closely related but functionally distinct genes: their 96% homology reflects only a 3 bp difference[11,80]. The probability of their gene conversion would seem to be particularly high, except that an intervening sequence is present in the single tRNA$^{Ser}_{UCG}$ gene but absent from the three tRNA$^{Ser}_{UCA}$ genes.

Transcription initiation and termination sites

The nuclear tRNA genes of eukaryotes are transcribed by RNA polymerase III, a complex enzyme of ~700 000 molecular weight which is composed of at least 10 different subunits[86]. The enzyme by itself does not transcribe purified tRNA genes with fidelity but requires additional components whose number, nature and modes of action have not yet been characterized. Specific transcription of these genes results in the synthesis of tRNA precursors containing extra 5′ and 3′ nucleotides and, in a few cases, intervening sequences. Although these tRNA precursors were first detected in mammalian cells some 15 years ago[13] the steady state levels of the primary transcripts are so low in normal cells that it has not yet been possible to precisely identify the *in vivo* transcription initiation and termination points of a single eukaryotic tRNA gene. This problem has been circumvented in recent years by the microinjection of cloned tRNA genes into the nucleus of frog oocytes[23,61] or by their incubation in a variety of cell-free extracts[28, 68, 73, 95, 101, 111]. Both approaches yield large amounts of tRNA precursors and, as a result, the 5′ termini of the primary transcripts of several eukaryotic tRNA genes have been characterized by the combined use of DNA sequencing and either S1 nuclease mapping or direct RNA sequencing (*Table 14.2*).

Most tRNA genes possess a single transcription initiation site. This is true of the human initiator tRNA genes although the corresponding genes from *D. melanogaster* and *X. laevis* start transcription at two or even three points. Each primary transcript that has been examined contains an uncapped purine nucleoside triphosphate as the first base. These occur within 20 nucleotides, and usually within 10 nucleotides of the sequences encoding the 5′ end of the mature tRNAs. Despite this similarity in location, no obvious sequence homologies surround the initiating

TABLE 14.2. Transcription initiation sites of eukaryotic tRNA genes

Species	Gene	5′ Flanking sequences[a]	Reference
		−20 −10 −1	
Schizosaccharomyces pombe	tRNA^{Ser}–tRNA^{Met}	TTACAATGAGTGTCAGATAA	67
Saccharomyces cerevisiae	tRNA^{Tyr}	GATGGTTATCAGTTAATTGA	27
	SUP4-o tRNA^{Tyr}	CTTTCTTCAACAATTAAATA	60
Bombyx mori	tRNA$^{Ala}_2$	TTTTGCAAGCTTTTTCCGTT	37, 44
Drosophila melanogaster	tRNA$^{Arg}_2$	GTTACACTCGCACGTCAAGC	96
	tRNA$^{Lys}_2$	TATCGGCGATCTTCACAAGT	24
	tRNA$^{Lys}_2$	AGTTTTTGGCTCATCAAGTG	24
	tRNA$^{Met}_1$	ATGACGCAACTTTGTATGCA	97
	tRNA$^{Met}_1$	ATGACGCAACTTTATATGCA	97
Xenopus laevis	tRNA^{Asn}	GGTGGCCGGTGGCAAGCGTC	74
	tRNA^{Leu}	CACAACGGCTGGCTTTCACA	36
	tRNA^{Lys}	GGAGGAACGAGCGACGCAGT	69
	tRNA$^{Met}_1$	TCACCCAAAGGACGGCAATC	59
	tRNA$^{Met}_1$ Variant	TGCGCGTGCCAGCTGCGATT	50
Homo sapiens	tRNA$^{Met}_1$	GACCGTGTGCTTGGCAGAAC	118
	tRNA$^{Met}_1$ Variant	GTAGAAGCGTGTTTTCCGTT	118

[a] Sequences are numbered negatively from the 5′ nucleotide of the mature tRNAs. The non-coding DNA strand is shown in each case. Initiating nucleotides are arrowed.

nucleotides except that they are preceded by a pyrimidine. The one possible exception is a yeast tRNA^{Tyr} gene which appears to start transcription at the second of two adjacent purines when injected into *X. laevis* oocytes[27]. In this case, however, the position of the initiating nucleotide has not been unambiguously characterized.

No eukaryotic tRNA gene has yet been found to encode the characteristic 3′ CCA terminus of mature tRNAs. Instead, their 3′ flanking regions contain short clusters of T residues in the non-coding DNA strand sequence. These are thought to be transcription termination sites because the longest detectable tRNA precursors extend up to these T tracts and the transcripts contain a variable number of U residues at their 3′ ends, the number corresponding to the length of the particular T tract[37, 44, 59, 98]. In addition, mutations which create runs of five or six T residues within a yeast tRNA^{Tyr} gene result in prematurely terminated *in vitro* transcripts with up to five or six U residues at their 3′ ends[60]. This 3′ terminal heterogeneity could be explained by a rapid but imprecise nucleolytic cleavage of even longer transcripts. The number of 3′ U residues can be reduced, however, by lowering the UTP concentration of *in vitro* transcription reactions[44, 59]. This is an unexpected result for a processing event. It suggests instead that RNA polymerase III pauses within the T tracts following tRNA genes and that this pausing is an integral part of termination.

These clusters of T residues usually oocur within a few nucleotides of the 3' end of the tRNA coding region but occasionally are found much further downstream. A T_4 tract, for example, first occurs 18 bp after a *B. mori* tRNA$^{Ala}_2$ gene[37,44] and 75 bp after a *X. laevis* tRNA^{Lys} gene[69]. Transcription terminates very efficiently at these two sites, yielding primary transcripts of 98 and 157 nucleotides, respectively. Thus, although some eukaryotic tRNA precursors are potentially long enough to include two or more tRNA sequences, they may in fact contain just a single tRNA species together with an extensive 3' trailer sequence.

Transcription terminates efficiently within clusters of five or more T residues[59, 67, 98]. A sequence of the type TTNTTT can also permit termination at a T_3 tract[36]. It is perhaps surprising, then, to find T clusters within some tRNA genes. The *X. laevis* tRNA^{Lys} gene, for example, contains an internal T_4 tract which corresponds to part of the anticodon loop. More than 70% of its transcripts terminate prematurely at this intragenic site, both *in vitro* and in injected oocytes[69]. This gene thus appears to be at a transcriptional disadvantage, with only a minor fraction of initiation events giving rise to the mature product. Other cases are known in which four consecutive T residues are not enough for efficient termination, implying that this process depends on additional factors. These might include the base composition of the DNA surrounding the termination site, the conformation of the nascent transcript, and perhaps even a specific termination protein akin to the bacterial rho factor. For eukaryotic 5S RNA, another RNA polymerase III product, an efficient termination site comprises a T_4 tract flanked on either side by two or more GC base pairs[8]. A sequence with these features is found 26 bp downstream of a *X. laevis* tRNA^{Phe} gene[70]. Moreover, it is preceded by dyad symmetric DNA of high G + C content and thus resembles prokaryotic termination sites[99]. Yet *in vitro* transcripts of this tRNA^{Phe} gene terminate only weakly at this T_4 tract[69]. It also acts as a weak terminator *in vitro* when preceded by the tRNA^{Lys} gene, thereby showing that termination efficiency is independent of the particular tRNA sequence contained within the precursor. Transcription is efficiently stopped at this site, however, when the tRNA^{Phe} gene is microinjected into *X. laevis* oocytes[69]. Hence, the cell-free extract may have lost, or contain in an active form, a factor needed for efficient termination at this site.

tRNA gene promoters

The promoters for tRNA genes in prokaryotes are conserved both in sequence and position, being located just upstream of the transcription initiation points[99]. It thus seemed reasonable to expect that DNA sequence analyses of the comparable regions of eukaryotic tRNA genes would reveal the nature of their promoters. Disillusionment quickly set in, however, when the first few sequences failed to show significant homologies in these 5' flanking regions. This was reinforced by the first transcriptional analyses of manipulated tRNA genes. Progressive deletion of the DNA lying upstream of a *X. laevis* tRNA$^{Met}_1$ gene showed that the gene could still be transcribed when it contained only 22 bp of its 5' flanking sequence[104]. A *B. mori* tRNA$^{Ala}_2$ gene fortuitously cloned together with just 14 bp of its 5' flanking sequence was also transcribed well in injected oocytes[37,44]. Both results, when contrasted with the large size of RNA polymerase III, implied that the promoters are within rather than upstream of tRNA genes.

Subsequent work has confirmed this view and it has also suggested that many, and perhaps all RNA polymerase III genes possess intragenic promoters. Deletion of the DNA around and within a 120 bp *X. borealis* 5S RNA gene has shown that only nucleotides 50–83 are absolutely required for transcription[9, 89]. Moreover, the same region of the gene specifically binds a 37 K polypetide that is needed for accurate 5S RNA transcription *in vitro*[30]. A similar resection approach with the genes coding for tRNA$^{\text{Met}}_1$[51] and tRNA$^{\text{Leu}}_{\text{CUG}}$[36] of *X. laevis* and for tRNA^{Pro} of *Caenorhabditis elegans*[15] has defined the boundaries of their intragenic promoters as residues 8 and 62, using the standard tRNA gene numbering system[102]. However, not all of the DNA in between is essential for transcription. Chimaeric tRNA genes, for example, comprising the 5′ half of one gene and the 3′ half of another are transcribed well in injected oocytes[16, 36]. Transcription can also occur after the substitution of heterologous DNA for the central region of tRNA genes[16, 36].

Eukaryotic tRNA gene promoters are therefore both intragenic and discontinuous. The two regions of essential sequences, which have been termed the A and B blocks[36], encompass nucleotides 8–19 and 52–62, respectively. Their locations with respect to the tRNA cloverleaf are shown in *Figure 14.4*. A remarkable feature of these sequences is their close correlation with the most conserved nucleotides of all non-organelle tRNAs. Specifically, they encode nine of the ten nucleotides found in the D and T stem and loop regions: U8, A14, G18 and G19 by the A block, and G53, T55, C56, A58 and C61 by the B block. Substitution of a T or G for the invariant C56 within a yeast tRNA^{Tyr} gene reduces or even abolishes its transcription *in vitro*[60]. Some invariant nucleotides thus possess dual functions in eukaryotes: they are important for transcription initiation as well as for tRNA functions such as aminoacyl–tRNA synthetase interaction or ribosome binding. Indeed, it is even possible that some of these conserved nucleotides are not actually needed for tRNA functions.

The locations of these A and B blocks are notable for two other reasons. The distance between transcription initiation points and the A block sequences is usually restricted to 10–16 bp. In contrast, the distance between the A and B blocks can vary from 31 to >74 bp, the variability being due to the length of the extra arm and the presence or absence of an intervening sequence. While the DNA between the A and B blocks shows little if any sequence specificity for transcription, its length appears to be important, the optimum being 30–40 bp[16]. Hence, although intervening sequences are not essential for transcription[109], they may influence its efficiency by expanding the distance between the A and B blocks, thereby diminishing promoter strength[45].

Many of the invariant nucleotides of the A and B blocks are found within other genes transcribed by RNA polymerase III, such as those encoding the adenovirus VAI and VAII RNAs[35, 43], the Epstein-Barr virus EBER 1 and EBER 2 RNAs[87], and certain members of the Alu family of middle repetitive sequences[48, 82]. Moreover, these conserved nucleotides are contained within the region of the adenovirus VAI RNA gene that has been shown by deletion analyses to be essential for transcription[35, 43]. Whether the essential sequences are split into two blocks is not yet known, but, like tRNA gene promoters, they encompass a large region of the gene (residues 9–76) and the conserved nucleotides are found at the ends of this region. The 5S RNA gene promoter, in contrast, contains only three of

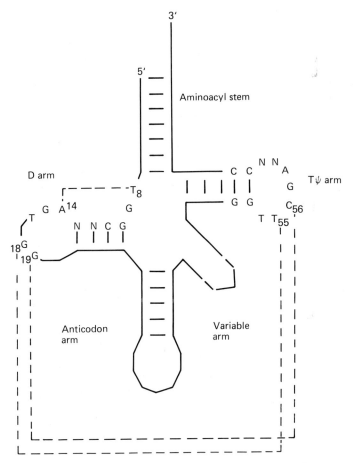

Figure 14.4. The eukaryotic tRNA gene promoter. The non-coding DNA strand is shown in cloverleaf form. The essential A block sequences extend from T8 to G19 and the B block sequences from G52 to C62. Dotted lines indicate the tertiary interactions proposed in reference 45.

the five invariant nucleotides of the B block and an even weaker homology with the A block[36]. In keeping with these sequence comparisons, tRNA genes are efficient competitors of VAI RNA transcription whereas 5S RNA genes are not [112], and the 37 K polypeptide which binds to the 5S RNA promoter is not needed for the transcription of either VAI RNA or tRNA genes[91]. Both do require, however, at least two other components in addition to RNA polymerase III for their activity. Although these have been partially purified[68, 91], no factors have yet been found that are uniquely required for VAI RNA or tRNA gene transcription.

Together, these results suggest that the A and B blocks within tRNA genes may be viewed as the paradigm of a second category of RNA polymerase III promoter. They also suggest that the A and B blocks function by interacting with one or more transcription factors, rather than with the polymerase directly. This conclusion is strengthened by the results of competition experiments. For example, while neither the 5′ nor 3′ half of a *X. laevis* tRNA$^{Met}_1$ gene can be transcribed *in vitro*, the 3′ half is able to inhibit transcription of the intact gene[62]. Sequences in the 3' half of the VAI RNA gene can similarly inhibit VAI RNA transcription *in vitro*. Moreover, this ability is lost in deletion mutants which lack the conserved nucleotides of the B block[35,43]. Hence the B blocks of both genes, but not their other essential sequences, appear to be able to sequester a limiting factor required for transcription. This factor need not be the same in both cases, although the available evidence suggests that it could well be. In any event, the primary interaction required for transcription of both tRNA and VAI RNA genes seems to be the binding of a transcription factor to their B block sequences.

What are the recognition signals that permit this binding? As with repressor binding to the *E. coli lac* operator, the interaction may simply involve direct protein binding to double helical DNA. In this case, the five invariant nucleotides within the B block (G53, T55, C56, A58 and C61) provide likely candidates for the essential contact points. An alternative model has been proposed recently[45] which ascribes different functional roles to the B block sequences. It embodies the fact that the essential A and B blocks are precisely those regions which encode single stranded loops (the D and T loops) that, in every tRNA molecule, engage in a strong mutual tertiary interaction. Specifically, as indicated by the dashed lines in *Figure 14.4*, tertiary base pairs are found between G18 and Ψ55, and between G19 and C56. It is also notable that base paired regions of the T stem are encoded by the B block whereas the A block sequences encode only the 5' portion of the D stem, together with the invariant nucleotides U8 and A14 which form an additional tertiary base pair in tRNA (*Figure 14.4*).

This 'tertiary interaction model'[45] proposes that the essential A and B block sequences promote tRNA gene transcription in the following ways. The primary event is the binding of a transcription factor to the B block region when it exists in an intra-strand stem and loop conformation. This binding then promotes tertiary interactions between sequences coding for some of the invariant nucleotides of the D and T loops. These tertiary interactions, in turn, position RNA polymerase III so that it initiates transcription at a purine 10–16 bp upstream of the A block sequences. According to this model, some of the conserved nucleotides of the A and B blocks are required for mutual tertiary interactions, some nucleotides of the B block, but not A block, are needed as inverted repeats, and relatively few nucleotides of both blocks are required as sequences *per se* that transcription factors recognize.

5' flanking sequences

Although transcription of eukaryotic tRNA genes appears to depend first on intragenic events, the DNA sequences preceding the genes can influence both the choice of initiation site and the efficiency of transcription. Moreover, these two effects are not necessarily related. Transcription of a *D. melanogaster* tRNA^{Arg} gene, for example, is much reduced when all but 8 bp of its 5' flanking sequence is replaced by plasmid and synthetic linker DNA, but initiation still occurs normally at the G at -7[96]. In contrast, deletion-substitutions in the 5' flanking sequence of a *X. laevis* tRNA$^{Met}_1$ gene can change the type of precursor made without affecting the overall transcription efficiency. This gene normally initiates with equal frequency at two positions, at a G at -7 and at an A at -4 (*Figure 14.2*). The latter is used preferentially when the normal 5' flanking sequence is replaced by foreign DNA at position -22. Further resection to position -13 leads to preferential use of the G start at -7 while removal of just one additional base pair of the 5' flanking region permits both sites to be used again with equal frequency[51]. That signals in the 5' flanking region can similarly influence the expression of other RNA polymerase III genes is shown by a natural deletion mutant of the adenovirus VAI RNA gene. This mutant, which lacks 2 bp of DNA at positions -22 and -23, is able to initiate transcription both *in vitro* and *in vivo* at only one of the two sites used by the wild type gene[106].

Sequence alterations in 5' flanking regions, if they have any quantitative effect on gene expression, usually reduce transcriptional output. The most extreme example is provided by a *B. mori* $tRNA_2^{Ala}$ gene which becomes totally inactive in homologous transcription extracts when all but 14 bp of its normal 5' flanking region is replaced by foreign DNA[101]. In such cases, the 5' flanking signals exert a positive control of RNA polymerase III activity. Two examples are known, however, of 5' flanking sequences that have a negative influence on eukaryotic tRNA gene expression. Several *D. melanogaster* $tRNA_2^{Lys}$ genes are preceded by a highly conserved but not quite identical 11 bp tract with the consensus non-coding strand sequence GGCAGTTTTTG. This tract inhibits transcription when located from positions -13 to -23 but not when moved closer to, or further away from tRNA genes[24, 25]. How these tracts inhibit transcription is far from clear but their resemblance to RNA polymerase III termination sites is very striking. The variable levels of transcription that are observed when their positions are changed may reflect an altered balance between two opposing factors – the recognition of these T-rich regions as termination sites, and the ability of such sequences to melt more easily. A second example of an inhibitory sequence is found from positions -12 to -20 in front of a variant $tRNA_1^{Met}$ gene of *X. laevis*[19]. The special feature of this sequence, TGCGCGTGC, is its potential ability to form a short stretch of left-handed or Z-form DNA[50]. Expression of this gene is also inhibited *in vitro* when it is preceded by short synthetic oligonucleotides that are able to crystallize as Z-DNA[50]. The *Drosophila* and *Xenopus* 5' flanking sequences thus appear to inhibit transcription in very different ways, but their almost identical locations emphasize the importance of the DNA lying just upstream of the transcription initiation sites.

In addition to these functional studies, recent structural analyses have demonstrated some significant sequence homologies between the 5' flanking regions of certain tRNA genes. The most intriguing example is provided by *sigma*, a 340 bp transposable-like element found 16–18 bp upstream of several yeast tRNA genes[26]. The almost identical locations of these elements could imply that tRNA genes are involved in the mechanism of *sigma* transposition. Alternatively, *sigma* could be part of a regulatory control for a subset of tRNA genes.

tRNA gene regulation

A key question, so far unresolved, is whether eukaryotic tRNA genes are really subject to transcriptional regulation or whether they are expressed constitutively. In cells engaged in making large amounts of a few proteins, such as human reticulocytes, the aminoacyl–tRNA population is very well matched to codon usage requirements[47]. Isoacceptor tRNA levels are also 'adapted'[38] to the requirements of protein synthesis in less specialized cells. In yeast, for example, serine is located at 101 positions in three abundant proteins – actin[72], glyceraldehyde-3-phosphate dehydrogenase[52] and alcohol dehydrogenase isozyme I[5] – yet none is specified by the codon UCG. Instead, 51 are specified by UCU and 45 by UCC, both of which are read by the most abundant serine tRNA isoacceptor, $tRNA_2^{Ser}$. As noted earlier, this tRNA is encoded by at least 11 genes whereas $tRNA_{UCG}^{Ser}$ is derived from a single gene. This example would thus seem to favour the idea of constitutive expression of eukaryotic tRNA genes.

The case for differential transcription of tRNA genes is more persuasive, although not yet proven, for certain specialized tissues. One example is provided by the posterior silk gland of *B. mori* which produces enormous amounts of fibroin during the fifth larval instar. At this stage, tRNAs specific for glycine, alanine and serine – the most common amino acids of fibroin – constitute about 70% of the tRNA population of the posterior silk gland. One of the tRNAs that accumulate at the time of fibroin synthesis is a novel alanine tRNA containing U (or Ψ) instead of C at a single position in the anticodon stem[100]. Differentiation of the bovine lens from epithelial to fibre cells is similarly accompanied by the appearance of a novel phenylalanine tRNA containing an A → G transition at a single position in the TΨ loop[65]. While it is conceivable that these substitutions occur post-transcriptionally, they have no known precedent. Instead, the appearance of these tissue-specific isoacceptors is more likely to be the result of the selective stabilization of the products of different tRNA genes or, even more interestingly, the selective expression of normally silent tRNA genes.

Conclusions and perspectives

One of the hallmarks of eukaryotic tRNA genes is the disparate ways they are organized. Seemingly unrelated genes can be quite distant, set close together or sometimes even juxtaposed to permit their co-transcription. A more uniform feature of these small genes is their possession of intragenic sequences that are essential for their transcription by RNA polymerase III. While this transcription can be modulated by 5' flanking sequences, no long-range sequence effects have yet been detected. Instead, tRNA gene function seems to depend on the DNA sequences within and immediately around the genes. Hence the variability of tRNA gene organization may in part reflect the fact that suitable surrounding sequences can be found at many chromosomal locations.

For the immediate future, we can be confident that many new tRNA genes will be cloned, sequenced and manipulated in order to better define the DNA sequences important for transcription. This approach has its limitations, of course, for it provides no information about the nature and modes of action of the other essential components. With the exception of RNA polymerase III, these have not yet been purified. Clearly, there is a need for further fractionation of crude transcriptional extracts but success with this approach is by no means guaranteed, particularly if some key factors are present in only one copy per gene.

An equally difficult problem is the chromatin state of tRNA genes. While tRNA genes appear to be packaged into regularly spaced nucleosomes in transcriptionally inactive erythrocytes, present evidence conflicts on whether the nucleosomes are phased with respect to tRNA genes[12, 113]. No such phasing is obvious in transcriptionally active tissues[12]. Moreover, S-100 extracts are thought to contain very little if any histones yet they transcribe a variety of RNA polymerase III genes with fidelity[111]. Recently, however, a procedure for site-directed nucleosome assembly has suggested that tRNA genes are most active *in vitro* when the nucleosome middle axis covers the essential intragenic B block sequences[114]. Perhaps the binding of a transcription factor to these sequences acts as a phase trigger for the deposition of nucleosomes, at least over a short range. Alternatively, the binding of such a factor may displace the nucleosomes from a minor fraction of the DNA and it may be this fraction that is responsible for the observed transcription. Once again, further progress will require the purification and subsequent reconstitution of the essential components.

Finally, on a more positive note, it is worth recalling the influence that suppressor tRNAs have had on the development of prokaryotic molecular biology. Their eukaryotic counterparts seem assured of an equally important future. It now appears feasible to detect and characterize naturally occurring nonsense mutations in the structural genes of higher organisms and in animal viruses by site-directed mutagenesis of cloned tRNA genes[105] and by their subsequent introduction into mammalian cells via a vector such as SV40[64]. This, in turn, offers the exciting prospect of an approach towards gene therapy.

Acknowledgements

I am most grateful to Dr R. A. Hipskind for his helpful criticisms of this review and to colleagues, too numerous to list, for communicating results prior to publication. Personal work cited was supported by grants from the Swiss National Science Foundation.

References

1. Addison, W. R., Astell, C. R., Delaney, A. D., Gillam, I. C., Hayashi, S., Miller, R. C., Rajput, B., Smith, M., Taylor, D. M. and Tener, G. M. (1982). The structures of genes hybridizing with tRNA$_4^{Val}$ from *Drosophila melanogaster*. *Journal of Biological Chemistry*, **257**, 670–673
2. Anderson, S., Bankier, A. T., Barrell, B. G., de Bruijn, M. H. L., Coulson, A. R., Drouin, J., Eperon, I. C., Nierlich, D. P., Roe, B. A., Sanger, F., Schreier, P. H., Smith, A. J. H., Staden, R. and Young, I. G. (1981). Sequence and organization of the human mitochondrial genome. *Nature, London*, **290**, 457–465
3. Baker, R. E., Eigel, A., Voegtel, D. and Feldmann, H. (1982). Nucleotide sequences of yeast genes for tRNA$_2^{Ser}$, tRNA$_2^{Arg}$ and tRNA$_1^{Val}$: homology blocks occur in the vicinity of different tRNA genes. *The EMBO Journal*, **1**, 291–295
4. Beckmann, J. S., Johnson, P. F. and Abelson, J. (1977). Cloning of yeast transfer RNA genes in *Escherichia coli*. *Science*, **196**, 205–208
5. Bennetzen, J. L. and Hall, B. D. (1982). Codon selection in yeast. *Journal of Biological Chemistry*, **257**, 3026–3031
6. Bibb, M. J., Van Etten, R. A., Wright, C. T., Walberg, M. W. and Clayton, D. A. (1981). Sequence and gene organization of mouse mitochondrial DNA. *Cell*, **26**, 167–180
7. Birnstiel, M. L., Sells, B. H. and Purdom, I. F. (1972). Kinetic complexity of RNA molecules. *Journal of Molecular Biology*, **63**, 21–39
8. Bogenhagen, D. F. and Brown, D. D. (1981). Nucleotide sequences in *Xenopus* 5S DNA required for transcription termination. *Cell*, **24**, 261–270
9. Bogenhagen, D. F., Sakonju, S. and Brown, D. D. (1980). A control region in the center of the 5S RNA gene directs specific initiation of transcription: II. The 3' border of the region. *Cell*, **19**, 27–35
10. Brenner, D. J., Fournier, M. J. and Doctor, B. P. (1970). Isolation and partial characterization of the transfer ribonucleic acid cistrons from *Escherichia coli*. *Nature, London*, **227**, 448–451
11. Broach, J. R., Friedman, L. and Sherman, F. (1981). Correspondence of yeast UAA suppressors to cloned tRNA$_{UCA}^{Ser}$ genes. *Journal of Molecular Biology*, **150**, 375–387
12. Bryan, P. N., Hofstetter, H. and Birnstiel, M. L. (1981). Nucleosome arrangement on tRNA genes of *Xenopus laevis*. *Cell*, **27**, 459–466

13. Burdon, R. H., Martin, B. T. and Lal, B. M. (1967). Synthesis of low molecular weight ribonucleic acid in tumour cells. *Journal of Molecular Biology*, **28**, 357–371
14. Carrara, G., Di Segni, G., Otsuka, A. and Tocchini-Valentini, G. P. (1981). Deletion of the 3′ half of the yeast tRNA$^{Leu}_3$ gene does not abolish promoter function *in vitro*. *Cell*, **27**, 371–379
15. Ciliberto, G., Castagnoli, L., Melton, D. A. and Cortese, R. (1982). Promoter of a eukaryotic tRNA^{Pro} gene is composed of three noncontiguous regions. *Proceedings of the National Academy of Sciences, USA*, **79**, 1195–1199
16. Ciliberto, G., Traboni, C. and Cortese, R. (1982). Relationship between the two components of the split promoter of eukaryotic tRNA genes. *Proceedings of the National Academy of Sciences, USA*, **79**, 1921–1925
17. Clarkson, S. G., Birnstiel, M. L. and Purdom, I. F. (1973). Clustering of transfer RNA genes of *Xenopus laevis*. *Journal of Molecular Biology*, **79**, 411–429
18. Clarkson, S. G., Birnstiel, M. L. and Serra, V. (1973). Reiterated transfer RNA genes of *Xenopus laevis*. *Journal of Molecular Biology*, **79**, 391–410
19. Clarkson, S. G., Koski, R. A., Corlet, J. and Hipskind, R. A. (1981). Influence of 5′ flanking sequences on tRNA transcription *in vitro*. In *Developmental Biology using Purified Genes*, ICN-UCLA Symposia on Molecular and Cellular Biology, Volume XXIII (Brown, D. D. and Fox, C. F., Eds.), pp. 463–472. New York, Academic Press
20. Clarkson, S. G. and Kurer, V. (1976). Isolation and some properties of DNA coding for tRNA$^{Met}_1$ from *Xenopus laevis*. *Cell*, **8**, 183–195
21. Clarkson, S. G., Kurer, V. and Smith, H. O. (1978). Sequence organization of a cloned tDNA$^{Met}_1$ fragment from *Xenopus laevis*. *Cell*, **14**, 713–724
22. Colby, D., Leboy, P. S. and Guthrie, C. (1981). Yeast tRNA precursor mutated at a splice junction is correctly processed *in vivo*. *Proceedings of the National Academy of Sciences, USA*, **78**, 415–419
23. Cortese, R., Melton, D., Tranquilla, T. and Smith, J. D. (1978). Cloning of nematode tRNA genes and their expression in the frog oocyte. *Nucleic Acids Research*, **5**, 4593–4611
24. DeFranco, D., Schmidt, O. and Söll, D. (1980). Two control regions for eukaryotic tRNA gene transcription. *Proceedings of the National Academy of Sciences, USA*, **77**, 3365–3368
25. DeFranco, D., Schmidt, O. and Söll, D. (1981). Identification of regulatory sequences contained in the 5′-flanking region of *Drosophila* lysine tRNA$_2$ genes. *Journal of Biological Chemistry*, **256**, 1242–12429
26. Del Rey, F. J., Donahue, T. F. and Fink, G. R. (1982). Sigma, a repetitive element found adjacent to tRNA genes of yeast.*Proceedings of the National Academy of Sciences, USA*, **79**, 4138–4142
27. DeRobertis, E. M. and Olson, M. V. (1979). Transcription and processing of cloned yeast tyrosine tRNA genes microinjected into frog oocytes. *Nature, London*, **278**, 137–143
28. Dingermann, T., Sharp, S., Appel, B., DeFranco, D., Mount, S., Heiermann, R. R., Pongs, O. and Söll, D. (1981). Transcription of cloned tRNA and 5S RNA genes in a *Drosophila* cell free extract. *Nucleic Acids Research*, **9**, 3907–3918
29. Eigel, A., Olah, J. and Feldmann, H. (1981). Structural comparison of two yeast tRNA$^{Glu}_3$ genes. *Nucleic Acids Research*, **9**, 2961–2970
30. Engelke, D. R., Ng, S.-Y., Shastry, B. S. and Roeder, R. G. (1980). Specific interaction of a purified transcription factor with an internal control region of 5S RNA genes. *Cell*, **19**, 717–728
31. Etcheverry, T., Colby, D. and Guthrie, C. (1979). A precursor to a minor species of yeast tRNA^{Ser} contains an intervening sequence. *Cell*, **18**, 11–26
32. Etcheverry, T., Salvato, M. and Guthrie, C. (1982). Recessive lethality of yeast strains carrying the *SUP*61 suppressor results from loss of a transfer RNA with a unique decoding function. *Journal of Molecular Biology*, **158**, 599–618
33. Feldmann, H., Olah, J. and Friedenreich, H. (1981). Sequence of a yeast DNA fragment containing a chromosomal replicator and a tRNA$^{Glu}_3$ gene. *Nucleic Acids Research*, **9**, 2949–2959
34. Fostel, J., Pardue, M. L. and Clarkson, S. G. Manuscript in preparation
35. Fowlkes, D. M. and Shenk, T. (1980). Transcriptional control regions of the adenovirus VAI RNA gene. *Cell*, **22**, 405–413
36. Galli, G., Hofstetter, H. and Birnstiel, M. L. (1981). Two conserved sequence blocks within eukaryotic tRNA genes are major promoter elements. *Nature, London*, **294**, 626–631
37. Garber, R. L. and Gage, L. P. (1979). Transcription of a cloned *Bombyx mori* tRNA$^{Ala}_2$ gene: Nucleotide sequence of the tRNA precursor and its processing *in vitro*. *Cell*, **18**, 817–828
38. Garel, J.-P. (1974). Functional adaptation of tRNA population. *Journal of Theoretical Biology*, **43**, 211–225
39. Gergen, J. P., Loewenberg, J. Y. and Wensink, P. C. (1981). tRNA$^{Lys}_2$ gene clusters in *Drosophila*. *Journal of Molecular Biology*, **147**, 475–499
40. Gilbert, W. (1978). Why genes in pieces? *Nature, London*, **271**, 501

41. Goodman, H. M., Olson, M. V. and Hall, B. D. (1977). Nucleotide sequence of a mutant eukaryotic gene: The yeast tyrosine-inserting *ochre* suppressor *SUP4-o*. *Proceedings of the National Academy of Sciences, USA*, **74**, 5453–5457

42. Grigliatti, T. A., White, B. N., Tener, G. M., Kaufman, T. C., Holden, J. J. and Suzuki, D. T. (1973). Studies on the transfer RNA genes of *Drosophila*. *Cold Spring Harbor Symposium on Quantitative Biology*, **38**, 461–474

43. Guilfoyle, R. and Weinmann, R. (1981). Control region for adenovirus VA RNA transcription. *Proceedings of the National Academy of Sciences, USA*, **78**, 3378–3382

44. Hagenbüchle, O., Larson, D., Hall, G. I. and Sprague, K. U. (1979). The primary transcription product of a silkworm alanine tRNA gene: Identification of *in vitro* sites of initiation, termination and processing. *Cell*, **18**, 1217–1229

45. Hall, B. D., Clarkson, S. G. and Tocchini-Valentini, G. (1982). Transcription initiation of eukaryotic rRNA genes. *Cell*, **29**, 3–5

46. Han, J. H. and Harding, J. D. (1981). Manuscript submitted

47. Hatfield, D., Varricchio, F., Rice, M. and Forget, B. G. (1982). The aminoacyl-tRNA population of human reticulocytes. *Journal of Biological Chemistry*, **257**, 3183–3188

48. Haynes, S. R. and Jelinek, W. R. (1981). Low molecular weight RNAs transcribed *in vitro* by RNA polymerase III from Alu-type dispersed repeats in Chinese hamster DNA are also found *in vivo*. *Proceedings of the National Academy of Sciences, USA*, **78**, 6130–6134

49. Hershey, N. D. and Davidson, N. (1980). Two *Drosophila melanogaster* tRNA^{Gly} genes are contained in a direct duplication at chromosomal locus 56F. *Nucleic Acids Research*, **8**, 4899–4910

50. Hipskind, R. A. and Clarkson, S. G. Manuscript submitted

51. Hofstetter, H., Kressmann, A. and Birnstiel, M. L. (1981). A split promoter for a eucaryotic tRNA gene. *Cell*, **24**, 573–585

52. Holland, J. P. and Holland, M. J. (1980). Structural comparison of two nontandemly repeated yeast glyceraldehyde-3-phosphate dehydrogenase genes. *Journal of Biological Chemistry*, **255**, 2596–2605

53. Hopper, A. K., Banks, F. and Evangelidis, V. (1978). A yeast mutant which accumulates precursor tRNAs. *Cell*, **14**, 211–219

54. Hosbach, H. A., Silberklang, M. and McCarthy, B. J. (1980). Evolution of a *D. melanogaster* glutamate tRNA gene cluster. *Cell*, **21**, 169–178

55. Hovemann, B., Sharp, S., Yamada, H. and Söll, D. (1980). Analysis of a *Drosophila* tRNA gene cluster. *Cell*, **19**, 889–895

56. Johnson, J. D., Ogden, R., Johnson, P., Abelson, J., Dembeck, P. and Itakura, K. (1980). Transcription and processing of a yeast tRNA gene containing a modified intervening sequence. *Proceedings of the National Academy of Sciences, USA*, **77**, 2564–2568

57. Knapp, G., Beckmann, J. S., Johnson, P. F., Fuhrman, S. A. and Abelson, J. (1978). Transcription and processing of intervening sequences in yeast tRNA genes. *Cell*, **14**, 221–236

58. Knapp, G., Ogden, R. C., Peebles, C. L. and Abelson, J. (1979). Splicing of yeast tRNA precursors: Structures of the reaction intermediates. *Cell*, **18**, 37–45

59. Koski, R. A. and Clarkson, S. G. (1982). Synthesis and maturation of *Xenopus laevis* methionine tRNA gene transcripts in homologous cell-free extracts. *Journal of Biological Chemistry*, **257**, 4514–4521

60. Koski, R. A., Clarkson, S. G., Kurjan, J., Hall, B. D. and Smith, M. (1980). Mutations of the yeast *SUP4* tRNA^{Tyr} locus: Transcription of the mutant genes *in vitro*. *Cell*, **22**, 415–425

61. Kressmann, A., Clarkson, S. G., Pirrotta, V. and Birnstiel, M. L. (1978). Transcription of cloned tRNA gene fragments and subfragments injected into the oocyte nucleus of *Xenopus laevis*. *Proceedings of the National Academy of Sciences, USA*, **75**, 1176–1180

62. Kressmann, A., Hofstetter, H., Di Capua, E., Grosschedl, R. and Birnstiel, M. L. (1979). A tRNA gene of *Xenopus laevis* contains at least two sites promoting transcription. *Nucleic Acids Research*, **7**, 1749–1763

63. Kubli, E. (1982). The genetics of transfer RNA in *Drosophila*. *Advances in Genetics*, **21**, 123–172

64. Laski, F. A., Alzner-DeWeerd, B., RajBhandary, U. L. and Sharp, P. A. (1982). Expression of a *X. laevis* tRNA^{Tyr} gene in mammalian cells. *Nucleic Acids Research*, **10**, 4609–4646

65. Lin, F.-K., Furr, T. D., Chang, S. H., Horwitz, J., Agris, P. F. and Ortwerth, B. J. (1980). The nucleotide sequence of two bovine lens phenylalanine tRNAs. *Journal of Biological Chemistry*, **255**, 6020–6023

66. Long, E. O. and Dawid, I. B. (1980). Repeated genes in eukaryotes. *Annual Review of Biochemistry*, **49**, 727–764

67. Mao, J., Schmidt, O. and Söll, D. (1980). Dimeric transfer RNA precursors in *S. pombe*. *Cell*, **21**, 509–516

68. Mattoccia, E., Baldi, M. I., Carrara, G., Fruscolini, P., Benedetti, P. and Tocchini-Valentini, G. P. (1979). Separation of RNA transcription and processing activities from *X. laevis* germinal vesicles. *Cell*, **18**, 643–648
69. Mazabraud, A. and Clarkson, S. G. Manuscript in preparation
70. Müller, F. and Clarkson, S. G. (1980). Nucleotide sequence of genes coding for tRNA[Phe] and tRNA[Tyr] from a repeating unit of *X. laevis* DNA. *Cell*, **19**, 345–353
71. Müller, F., Clarkson, S. G. and Galas, D. J. Manuscript in preparation
72. Ng, R. and Abelson, J. (1980). Isolation and sequence of the gene for actin in *Saccharomyces cerevisiae. Proceedings of the National Academy of Sciences, USA*, **77**, 3912–3916
73. Ng, S.-Y., Parker, C. S. and Roeder, R. G. (1979). Transcription of cloned *Xenopus* 5S RNA genes by *X. laevis* RNA polymerase III in reconstituted systems.*Proceedings of the National Academy of Sciences, USA*, **76**, 136–140
74. Niederhauser, U. and Clarkson, S. G. Unpublished observation
75. Nishikura, K., Kurjan, J., Hall, B. D. and DeRobertis, E. M. (1982). Genetic analysis of the processing of a spliced tRNA. *The EMBO Journal*, **1**, 263–268
76. O'Farrell, P. Z., Cordell, B., Valenzuela, P., Rutter, W. J. and Goodman, H. M. (1978). Structure and processing of yeast precursor tRNAs containing intervening sequences. *Nature, London*, **274**, 438–445
77. Olah, J. and Feldmann, H. (1980). Structure of a yeast non-initiating methionine–tRNA gene. *Nucleic Acids Research*, **8**, 1975–1986
78. Olson, M. V., Hall, B. D., Cameron, J. R. and Davis, R. W. (1979). Cloning of the yeast tyrosine transfer RNA genes in bacteriophage lambda. *Journal of Molecular Biology*, **127**, 285–295
79. Olson, M. V., Montgomery, D. L., Hopper, A. K., Page, G. S., Horodyski, F. and Hall, B. D. (1977). Molecular characterisation of the tyrosine tRNA genes of yeast. *Nature, London*, **267**, 639–641
80. Olson, M. V., Page, G. S., Sentenac, A., Piper, P. W., Worthington, M., Weiss, R. B. and Hall, B. D. (1981). Only one of two closely related yeast suppressor tRNA genes contains an intervening sequence. *Nature, London*, **291**, 464–469
81. Page, G. S. and Hall, B. D. (1981). Characterization of the yeast tRNA$^{Ser}_2$ gene family: genomic organization and DNA sequence. *Nucleic Acids Research*, **9**, 921–934
82. Pan, J., Elder, J. T., Duncan, C. H. and Weissman, S. M. (1981). Structural analysis of interspersed repetitive polymerase III transcription units in human DNA. *Nucleic Acids Research*, **9**, 1151–1160
83. Peebles, C. I., Ogden, R. C., Knapp, G. and Abelson, J. (1979). Splicing of yeast tRNA precursors: a two-stage reaction. *Cell*, **18**, 27–35
84. Peffley, D. M. and Sogin, M. L. (1981). A putative tRNA[Trp] gene cloned from *Dictyostelium discoideum*: Its nucleotide sequence and association with repetitive deoxyribonucleic acid. *Biochemistry*, **20**, 4015–4021
85. Robinson, R. R. and Davidson, N. (1981). Analysis of a *Drosophila* tRNA gene cluster. Two tRNA[Leu] genes contain intervening sequences. *Cell*, **23**, 251–259
86. Roeder, R. G. (1976). Eukaryotic nuclear RNA polymerases. In *RNA Polymerase* (Losick, R. and Chamberlin, M., Eds.), pp. 285–329. Cold Spring Harbor Laboratory
87. Rosa, M. D., Gottlieb, E., Lerner, M. R. and Steitz, J. A. (1981). Striking similarities are exhibited by two small Epstein-Barr virus-encoded ribonucleic acids and the adenovirus-associated ribonucleic acids VAI and VAII. *Molecular and Cellular Biology*, **1**, 785–796
88. Rosenthal, D. and Doering, J. L. (1983). The genomic organization of dispersed tRNA and 5S RNA genes in *Xenopus laevis*. *Journal of Biological Chemistry*, **258**, 7402–7410
89. Sakonju, S., Bogenhagen, D. F. and Brown, D. D. (1980). A control region in the center of the 5S border of the region. *Cell*, **19**, 13–25
90. Santos, T. and Zasloff, M. (1981). Comparative analysis of human chromosomal segments bearing nonallelic dispersed tRNA$^{Met}_1$ genes. *Cell*, **23**, 699–709
91. Segall, J., Matsui, T. and Roeder, R. G. (1980). Multiple factors are required for the accurate transcription of purified genes by RNA polymerase III. *Journal of Biological Chemistry*, **255**, 11986–11991
92. Sekiya, T., Kuchino, Y. and Nishimura, S. (1981). Mammalian tRNA genes: nucleotide sequence of rat genes for tRNA[Asp], tRNA[Gly] and tRNA[Glu]. *Nucleic Acids Research*, **9**, 2239–2250
93. Selker, E. and Yanofsky, C. (1980). A phenylalanine tRNA gene from *Neurospora crassa*: conservation of secondary structure involving an intervening sequence. *Nucleic Acids Research*, **8**, 1033–1042
94. Schmidt, O., Mao, J., Ogden, R., Beckmann, J., Sakano, H., Abelson, J. and Söll, D. (1980). Dimeric tRNA precursors in yeast. *Nature, London*, **287**, 750–752

95. Schmidt, O., Mao, J., Silverman, S., Hovemann, B. and Söll, D. (1978). Specific transcription of eukaryotic tRNA genes in *Xenopus* germinal vesicle extracts. *Proceedings of the National Academy of Sciences, USA*, **75**, 4819–4823

96. Sharp, S., DeFranco, D., Dingermann, T., Farrell, P. and Sóll, D. (1981). Internal control regions for transcription of eukaryotic tRNA genes. *Proceedings of the National Academy of Sciences, USA*, **78**, 6657–6661

97. Sharp, S., DeFranco, D., Silberklang, M., Hosbach, H. A., Schmidt, T., Kubli, E., Gergen, J. P., Wensink, P. C. and Söll, D. (1981). The initiator tRNA genes of *Drosophila melanogaster*: evidence for a tRNA pseudogene. *Nucleic Acids Research*, **9**, 5867–5882

98. Silverman, S., Schmidt, O., Söll, D. and Hovemann, B. (1979). The nucleotide sequence of a cloned *Drosophila* arginine tRNA gene and its *in vitro* transcription in *Xenopus* germinal vesicle extracts. *Journal of Biological Chemistry*, **254**, 10290–10294

99. Söll, D., Abelson, J. N. and Schimmel, P. R. (Eds.) (1980). *Transfer RNA: Biological Aspects*. Cold Spring Harbor Laboratory

100. Sprague, K. U., Hagenbüchle, O. and Zuniga, M. C. (1977). The nucleotide sequence of two silk gland alanine tRNAs: Implications for fibroin synthesis and for initiator tRNA structure. *Cell*, **11**, 561–570

101. Sprague, K. U., Larson, D. and Morton, D. (1980). 5' flanking sequence signals are required for activity of silkworm alanine tRNA genes in homologous *in vitro* transcription systems. *Cell*, **22**, 171–178

102. Sprinzl, M. and Gauss, D. H. (1982). Complication of sequences of tRNA genes. *Nucleic Acids Research*, **10**, r57–r81

103. Standring, D. N., Venegas, A. and Rutter, W. J. (1981). Yeast tRNA$^{Leu}_3$ gene transcribed and spliced in a HeLa cell extract. *Proceedings of the National Academy of Sciences, USA*, **78**, 5963–5967

104. Telford, J. L., Kressmann, A., Koski, R. A., Grosschedl, R., Müller, F., Clarkson, S. G. and Birnstiel, M. L. (1979). Delimitation of a promoter for RNA polymerase III by means of a functional test. *Proceedings of the National Academy of Sciences, USA*, **76**, 2590–2594

105. Temple, G. F., Dozy, A. M., Roy, K. L. and Kan, Y. W. (1982). Construction of a functional human suppressor tRNA gene: an approach to gene therapy for β-thalassaemia. *Nature, London*, **296**, 537–540

106. Thimmappaya, B., Jones, N. and Shenk, T. (1979). A mutation which alters initiation of transcription by RNA polymerase III on the Ad5 chromosome. *Cell*, **18**, 947–954

107. Valenzuela, P, Venegas, A., Weinberg, F., Bishop, R. and Rutter, W. J. (1978). Structure of yeast phenylalanine-tRNA genes: An intervening DNA segment within the region coding for the tRNA. *Proceedings of the National Academy of Sciences, USA*, **75**, 190–194

108. Venegas, A., Guiroga, M., Zaldivar, J., Rutter, W. J. and Valenzuela, P. (1979). Isolation of yeast tRNA^{Leu} genes. *Journal of Biological Chemistry*, **254**, 12306–12309

109. Wallace, R. B., Johnson, P. F., Tanaka, S., Schöld, M., Itakura, K. and Abelson, J. (1980). Directed deletion of a yeast transfer RNA intervening sequence. *Science*, **209**, 1396–1400

110. Weber, L. and Berger, E. (1976). Base sequence complexity of the stable RNA species of *Drosophila melanogaster*. *Biochemistry*, **15**, 5511–5519

111. Weil, P. A., Segall, J., Harris, B., Ng, S.-Y. and Roeder, R. G. (1979). Faithful transcription of eukaryotic genes by RNA polymerase III in systems reconstituted with purified DNA templates. *Journal of Biological Chemistry*, **254**, 6163–6173

112. Weinmann, R. and Guilfoyle, R. (1981). The control of adenovirus VAI RNA transcription. In *Developmental Biology using Purified Genes*. ICN-UCLA Symposia on Molecular and Cellular Biology, Volume XXIII (Brown, D. D. and Fox, C. F., Eds.). New York, Academic Press

113. Wittig, B. and Wittig, S. (1979). A phase relationship associates tRNA structural gene sequences with nucleosome cores. *Cell*, **18**, 1173–1183

114. Wittig, B. and Wittig, S. (1982). Function of a tRNA gene promoter depends on nucleosome position. *Nature, London*, **297**, 31–38

115. Yen, P. H. and Davidson, N. (1980). The gross anatomy of a tRNA gene cluster at region 42A of the *D. melanogaster* chromosome. *Cell*, **22**, 137–148

116. Yen, P. H., Sodja, A., Cohen, M., Conrad, S. E., Wu, M., Davidson, N. and Ilgen, C. (1977). Sequence arrangement of tRNA genes on a fragment of *Drosophila melanogaster* DNA cloned in *E. coli*. *Cell*, **11**, 763–777

117. Zasloff, M., Santos, T. and Hamer, D. H. (1982). tRNA precursor transcribed from a mutant human gene inserted into a SV40 vector is processed incorrectly. *Nature, London*, **295**, 533–535

118. Zasloff, M., Santos, T., Romeo, P. and Rosenberg, M. (1982). Transcription and precursor processing of normal and mutant human tRNA$^{Met}_i$ genes in a homologous cell-free system. *Journal of Biological Chemistry*, **257**, 7857–7863

Chapter 15

Mobile genetic elements in eukaryotes

A. Flavell

In all living organisms the order of genes along chromosomes is highly conserved between individual members of a species. Furthermore this arrangement seems to be very stable between related species (at least in the case of the β-globin related genes of primates[1]). However, a broad variety of DNA segments exist which are capable of movement within an organism's genome. Such sequences, termed mobile or transposable elements, appear to be ubiquitous in living systems. They have been detected either by their abnormal genetical effects (especially high rates of mutation) on adjacent genes or more recently by direct analysis of genome organization using recombinant DNA probes.

Although originally described in maize[12, 27, 28], transposable elements have been most thoroughly studied in bacteria where genetic analysis is much easier. Much of the research on eukaryotic mobile elements has been influenced by the discoveries in the prokaryotic field and several relevant features of these elements will be briefly summarized. For excellent reviews of bacterial mobile elements see references 9, 25 and 26. In bacteria many different types of 'insertion sequences' or 'transposons' have been characterized. Normally transposition of such elements involves the duplication of the element and integration at a distant chromosomal site, leaving behind the original copy which is unaffected by this process. These elements often contain genes encoding proteins that mediate or repress transposition. Bacterial mobile elements have characteristic short inverted repeat sequences at their extreme ends which are necessary for transposition. All eukaryotic mobile elements studied to date have analogous inverted termini. A key difference between the DNA sequence rearrangements associated with transposable elements and the better known homologous recombination pathways is that transposition exhibits little or no DNA sequence specificity. There is no detectable DNA sequence homology between the transposing element and either the new or old integration site, nor is there any significant homology between the two target sites for integration. Thus, while transposition target sites are not random, it appears that mobile elements have a huge number of possible locations to which they are capable of movement.

Bacterial mobile elements have several effects upon the host cell's DNA. The mobile elements often carry genes encoding antibiotic resistance between different bacteria and the DNA viruses and plasmids which occupy them. They can also act as polar insertional mutagens by integrating into operons. In some cases a mobile

263

element can activate an adjacent gene presumably by the provision of a promoter located in the element. They are also implicated in several types of chromosomal sequence rearrangement (*Figure 15.1*). Complete excision of the element or incomplete excision (leaving behind a portion of the element) is often observed. Flanking chromosomal DNA may be transposed along with a mobile element and deletions and inversions of surrounding sequences occur[13,26] (*Figure 15.1*). These rearrangements occur in both bacterial and eukaryotic systems and are believed to be connected with the mechanism of transposition of the element.

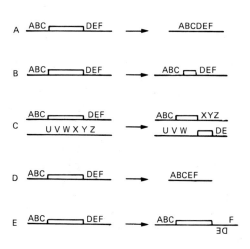

Figure 15.1. Several types of chromosomal rearrangement catalysed by transposable genetic elements. A, Perfect excision; B, imperfect excision; C, Translocation; D, Excision plus flanking deletion; E, flanking inversion. A–F, U–Z are flanking DNA sequences; □ is the mobile element. This list is not exhaustive, other types of rearrangement are known to occur.

Eukaryotic transposable elements have been best studied in plants (particularly maize), yeast and *Drosophila*, where the genetic effects associated with their activities are relatively easily monitored. The occurrence of mobile genetic elements in vertebrates is still the subject of intense research and although several lines of evidence summarized in this chapter show that certain types of mobile elements exist in this class of organism, much more work is still required to ascertain their extent and significance. The better characterized transposable elements of *Drosophila* and yeast will be dealt with first and the less thoroughly understood mobile sequences in maize and vertebrates will be covered later.

Mobile elements in *Drosophila*

Since *Drosophila* is genetically the most thoroughly studied animal it is not surprising that highly mutable alleles (a characteristic of mutations induced by mobile elements) have been observed in this organism in the past sixty years. Indeed, a fast-reverting mutant observed by Plough in 1928 was possibly the first case ever observed in eukaryotes[31]. However, we will never know, since the mutant was so unstable that it was lost before it could be studied properly!

In the past 15 years classical genetics has uncovered a number of highly mutable loci[16]. More recently recombinant DNA technology has brought to light several structurally distinct types of mobile DNA segments (see below). In the last year or two these two approaches have become fused by the molecular cloning of highly mutable loci and the identification of the mobile elements responsible for these mutations. Several examples are described below. A consequence of the molecular cloning of the bithorax region of *Drosophila* DNA has been the discovery that the majority of spontaneous mutations in this region seem to have been caused by the insertion of mobile pieces of DNA (W. Bender, personal communcation). The genetic analysis of several highly mutable alleles in *Drosophila* will be dealt with first and then the biochemical characterization of DNA elements which are presumed (and in some cases known) to cause such unstable mutations.

In 1967 Green described a mutant allele of the *white* locus termed *mutable white-crimson* (w^c). This has rapidly been followed by several other unstable mutants, mostly at the *white* locus (for reviews see references 16 and 36). The *white* locus has been used in these studies largely because of its easily scored eye colour phenotype and is not presumed to be exceptional. w^c mutates rapidly to a variety of different states, some of which are stable (such as reversion to wild type) while others are themselves highly mutable. This phenomenon has been interpreted to reflect chromosomal rearrangements associated with the integration of a transposable element at the affected locus in an analogous manner to that observed for prokaryotic mobile elements (see *Figure 15.1*). The exact molecular nature of *white* mutants is accessible to study as the cloning of DNA from the *white* locus was achieved recently by Bingham, Levis and Rubin[3]. By comparing the restriction map of the cloned *white* locus from wild type and w^c flies the w^c mutation has recently been shown to be associated with the presence of an 11 kilobase (kb) DNA insertion into the *white-ivory* (w^i) locus from which it was derived (G. Rubin, personal communication). w^i is a comparatively stable mutant and has been shown by a similar approach to contain a tandem duplication of a 3 kb segment within the *white* locus.

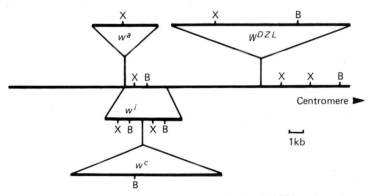

Figure 15.2. DNA insertions at the *white* locus. w^a, w^c and W^{DZL} are insertions; w^i is a duplication of a 3 kb sequence already present in the locus. X = Xba I restriction site; B = Bam HI restriction site.

Further mutable alleles at the *white* locus include W^{DZL} and w^a. W^{DZL} is associated with the insertion of a 13 kb DNA fragment at a different site in the locus to that occupied by w^c. w^a corresponds to the integration of a *copia* element at again a different site. This particular mutant is highly stable. A simplified restriction map of the *white* locus showing the insertions and duplications corresponding to the above mutants is shown in *Figure 15.2*.

Hybrid dysgenesis elements

When male *Drosophila melanogaster* recently caught from the wild are mated with female laboratory stocks, the progeny display a wide variety of abnormalities in their germ line, such as high mutability and chromosome rearrangements. This phenomenon, termed hybrid dysgenesis by Kidwell[23] does not occur in the reciprocal cross between wild females and laboratory males. Furthermore a series of genetic experiments have shown that this syndrome is connected with the interaction between a paternal (P) factor present at multiple locations in the parental genome and maternally donated 'cytotype' (M), the physical nature of which is unknown. P factors are found in many wild stocks but are absent in most laboratory stocks. Hybrid dysgenesis appears to be due to the transposition of P elements in the germ line of an M oocyte when introduced by P sperm. A further independent dysgenesis system (I-R) has been described[5].

The physical isolation of P elements has recently been achieved by the screening of dysgenesis flies for *w* mutations in the *white* locus (G. M. Rubin, personal communication). The respective *white* loci were cloned from dysgenesis-induced *white* mutants in much the same way as in the analysis of the w^c, w^i and W^{DZL} mutants. Four out of five such mutations were found to correspond to the integration of DNA fragments of between approximately 300 and 1500 nucleotides, each of which showed close sequence homology with the others (Rubin, Kidwell and Bingham, unpublished). The other insert was an unrelated *copia* element which will be discussed below. The integration sites of the five elements in the *white* locus are shown in *Figure 15.3*. The four interrelated dysgenesis elements presumably correspond to P factors.

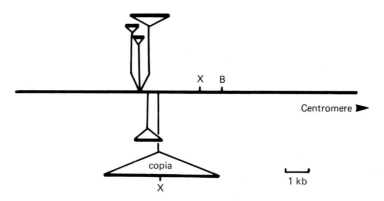

Figure 15.3. DNA insertions induced at the *white* locus by hybrid dysgenesis.
X = Xba I restriction site; B = Bam HI

It has been postulated that this phenomenon might explain how speciation occurs in *Drosophila*. It is necessary for a sub-population of any species to achieve reproductive isolation from its parent species in order to evolve independently (otherwise they would constantly drift back together by crossbreeding). Sterility between hybrids in a dysgenic cross may provide such a reproductive barrier. It is also possible that an analogous mechanism might lead to speciation in all higher eukaryotes.

Transposition of large chromosomal fragments

Several laboratories have reported the transposition in *Drosophila* of large chromosomal fragments including complete chromosome bands. Ising and Block[20] have described in detail the migrations of one element TE 1, containing the *w* and *rst* loci, to more than 150 different sites in the *D. melanogaster* genome. The size of the transferred chromosomal segment was very variable, transposition of between 1 and 8 bands were reported (one band contains very roughly 30 kb of DNA). Qualitatively similar results on the transposition of the *white* locus were obtained by Rasmuson *et al.*[33] and there appeared to be a correlation between the genetic instability of this locus and the presence of a *copia*-like element Dm225 (see below for a description of *copia*-like transposable elements). It is widely believed that these phenomena are mediated by smaller mobile elements located in the large chromosomal fragments.

Middle repetitive sequence DNA

The physical characterization of some of the DNA inserts responsible for the above mutants has resulted from a study of *Drosophila* middle repetitive DNA. In *Drosophila* three major classes of middle repetitive DNA (sequences repeated in roughly 20–200 copies/haploid genome) have been described to date, two of which have been shown to be transposable by the discovery of polymorphisms in chromosomal locations between different fly stocks[36] (*Figure 15.4*). It is presumed (but not proven) that the inserts which give rise to the various mutant alleles discussed above derive from these three classes. The first class to be considered is usually termed the *copia*-like family of transposable elements after the prototype

(a)

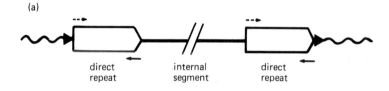

direct repeat   internal segment   direct repeat

(b)

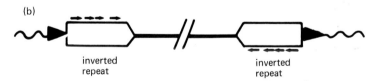

inverted repeat   inverted repeat

Figure 15.4. Drosophila middle repetitive mobile elements. (a) *Copia*-like family: An internal segment of several kilobases is bounded by a pair of identical directly repeated sequences. These repeats carry small terminal inverted repeats several base pairs long which are normally imperfect (--→ and ←). The complete element is flanked by an identical pair of host DNA sequences (▶). The flanking genomic DNA is indicated (∿∿). (b). Foldback family: A highly variable internal segment is bounded by a pair of identical (or nearly identical) inverted repeat sequences. These repeats carry small directly repeated sequences, 10–31 bp long (→) for the one element DmSFB3 whose DNA sequence is known[38]. The complete element is flanked by an identical pair of host sequences (▶; 9 bp long in the case of DmSFB3).

element *copia*[35]. *Copia* is probably the most thoroughly studied eukaryotic transposable element. There are between 15 and 50 types of *copia*-like transposable elements in the *Drosophila* genome, each of which shares little or no sequence homology with the others but the following properties are shared by all (reference 35 and references therein):

1. The elements are located at widely scattered locations in *D. melanogaster* chromosomes and are present at different locations in different fly stocks.
2. Their sequences are closely conserved and non-permuted at each genomic location; it is in these respects that they differ markedly from certain other dispersed, moderately repetitive DNA sequences found in the *D. melanogaster* genome.
3. At least five different types of element (the most thoroughly studied) possess direct repeats, several hundred base pairs long, at their ends which flank an internal segment of several kb. The direct repeats of a particular copy of an element are identical but there are often differences between different copies of a given element and the direct repeats of two unrelated elements share little or no sequence homology.
4. The direct repeats are themselves flanked by small inverted repeats several nucleotides long, in an analogous manner to prokaryotic transposable elements.
5. The complete element is flanked by an identical pair of host DNA sequences usually 4–6 bases long that are present only once at the target site for integration.
6. These five elements are each transcribed into cytoplasmic polyadenylated RNA. In the case of *copia*, this RNA can be translated *in vitro* into polypeptides of up to 51 000 daltons[15]. It is not known whether *copia*-encoded proteins exist in *Drosophila* cells.

It is thought that *copia*-like elements may account for up to 5% of *D. melanogaster* DNA. The general properties of this family are shared by the yeast transposable element Ty1 (see below) and, surprisingly, by the integrated proviruses of vertebrate retroviruses. This similarity extends to nucleotide homology at the extreme ends of the elements which are presumably important in transposition. This suggests that these elements may either share similar transposition intermediates or that they are all at least potential retroviruses. Two key experiments suggest that the latter may be the case. First, retrovirus-like entities and reverse transcriptase activity have been observed in *D. melanogaster* tissue culture cells[17]. Second, an extrachromosomal circular form of the *copia* element exists at low copy number in similar cells[14]. These circular molecules share a virtually identical structure with the circular provirus forms of retroviruses which are synthesized by reverse transcription of virion RNA. The degree of relation between *copia*-like elements and retrovirus proviruses is still under investigation.

The second family is termed the foldback family owing to the presence of inverted terminal repeats that cause foldback structures in the electron microscope whenever such DNA is denatured and reannealed[38]. Their general properties are as follows:

1. The inverted terminal repeats themselves contain many small (10–30 bp) direct repeat sequences.
2. They display considerably more structural heterogeneity than *copia*-like elements.
3. They are transposable, as judged by the same criteria ss the *copia*-like elements.

The third family of elements consists of middle repetitive sequences several hundred nucleotides long, present in clusters at scattered sites in the *D. melanogaster* chromosomes[40]. The relative order of these sequences is not fixed between clusters but appears to be scrambled. These sequences have not been tested for mobility in the *D. melanogaster* genome.

Transposable elements in yeast

The genetical behaviour of transposable elements in yeast shares close similarities with that of the maize and *Drosophila* mobile elements. A *copia*-like element, Ty1[10, 13] has been well studied both genetically and biochemically. Transposition of large DNA segments[13, 37], some containing *copia*-like elements, has also been observed. Just as the *white* locus was employed in *Drosophila* as an excellent marker for mutators, the *His* loci have been used in yeast. These loci encode enzymes necessary for histidine biosynthesis and can therefore be used as selective markers for mutants.

The Ty1 element of yeast

The prototype yeast transposable element Ty1 was found in *Saccharomyces cerevisiae* by Cameron *et al.*[8]. Extensive analysis has shown that it shares a close structural similarity (but little or no sequence homology) with the *Drosophila copia*-like elements (see above), including the presence of terminal repeats and large amounts of RNA transcripts[10, 13]. Spontaneous transposition of Ty1 into the *His*4 region has been examined in detail by Fink and co-workers. Two independent insertions of Ty1 into the presumptive control regions for the *His*4 structural genes have been studied, *His*4-912 and *His*4-917[13]. Interestingly, whereas the insert *His*4-912 shares with all known *copia*-like elements the property of identical left and right terminal direct repeats (δ sequences), there are five separate single nucleotide differences between the left and right terminal repeats of *His*4-917. Furthermore the *His*4-917 Ty1 has a 4 kb insertion/substitution in the non-δ region. The significance of these observations are still unclear. *His*4-912 partial revertants often derive from partial excision of the element by homologous recombination between δ sequences leaving a single δ behind. Others containing deletions removing part of the Ty1 element and flanking *His*4 DNA and transpositions to other chromosomes are known. Mutations in unlinked genes can affect the behaviour of the above mutants. These mutations are presumably at loci determining either enhancers or repressors of transposition or proteins affecting transcription at the affected locus. Three such unlinked mutants have been found at different sites in the yeast genome.

The effect of Ty1 on the expression of adjacent genes appears to be modulated by the mating type of the host cell (for a fuller explanation of mating type in yeast see below). Sherman and co-workers have studied Ty1 integration upstream of the gene encoding an easily assayable polypeptide, Iso-2-cyt *c*[11]). This mutation *Cyc7-h2* causes twenty-fold overproduction of the cytochrome *c* (presumably by readthrough from the Ty1 promoter) in strains which are capable of sexual reproduction. However, in heterozygous diploids, which can reproduce solely by budding, there is only a four-fold overproduction. The effect is dependent upon signals that are required for sexual conjugation. These workers have suggested that

Ty1 may normally reside upstream of mating function genes and that this modulation of Ty1-promoted transcripts in response to mating signals may reflect a role of the mobile element in coordinate regulation of genes determining sexual activity. The observation of Elder *et al.*[10] that the concentration of Ty1 RNA varies with the sexual cycle of yeast supports this hypothesis. However, this model is as yet unconfirmed and it is difficult to suggest plausible reasons for why an element which is involved in the control of expression of a set of developmentally regulated genes should be mobile.

Transposition and the mating type switch in yeast

Yeast can reproduce by sexual conjugation between haploid cells or by sexual budding from diploid cells. Sexual conjugation can occur between a pair of cells which are complementary for a pair of mating type alleles (termed *a* and *α*). The *a* and *α* mating types are controlled by the mating type locus (MAT) which contains either an *a* or *α* element. Certain strains of yeast (termed homothallic) are capable of interconverting mating type by replacing one mating type allele with the other[18].

In addition to the *a* or *α* genetic element situated at the MAT locus, one copy of each element is located at distant sites on the same chromosomes (chromosome III; see *Figure 15.5*). These two copies are 'silent', having no effect on mating type and are themselves unaffected by the mating type of the yeast cell they inhabit. Each of the three loci share some sequence homology with the others[24, 29] (*Figure 15.5*).

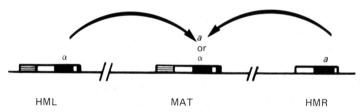

Figure 15.5. Organization of the mating type sequences on yeast chromosome III. ■: sequences specific to *a* or *α*; □: sequences common to all 3 loci; ▤: sequences present at HML and MAT only

These two 'silent' copies are responsible for the alteration in the expressed *a* or *α* 'cassette' at the MAT locus by directing the conversion of, say, an *a* element at the MAT locus, to an *α* element (the removed *a* element being destroyed in the process). This alteration is probably best considered as an example of gene conversion, that is to say 'correction' of a DNA segment by a related, but non-identical, segment at another chromosomal location. Transposition is normally considered to involve movement of a DNA segment to a site formerly unoccupied by a related element.

Mobile elements in maize

Transposition of genetic elements was first discovered over 30 years ago by a brilliant series of classical genetic experiments on maize by Barbara McClintock[12, 27, 28]. She identified mobile chromosomal elements which affected the functioning of easily recognizable genetic loci concerned with either the colour or morphology of maize kernels, Various mutant alleles were studied, the characteristic feature of each was an astonishingly rapid rate of interconversion of the mutant with an apparently wild type phenotype. This is illustrated strikingly by *Figure 15.6* which shows a corn cob, the kernels of which display visually the effects of transposable elements on the seeds. A wild type kernel whose colour is unaffected by mobile

Figure 15.6. Variegation in maize kernels induced by a mobile element. In this strain of maize a mobile element (in this case a Ds element) is inserted in the A_2 locus (one of several loci determining anthocyanin pigment production) thus inactivating it and yielding colourless kernels. The A_2 locus 'reverts' to a wild type phenotype (presumably by loss, or partial loss, of the Ds element at the locus) under the influence of an Ac element at a distant chromosomal site. Each 'revertant cell' gives rise to a clone of dark pigmented cells. Photograph by courtesy of J. Shapiro (University of Chicago) and B. Burr (Brookhaven National Laboratory).

elements, is homogeneous and dark in colour, owing to the production of the pigment anthocyanin. If the production of anthocyanin is arrested by insertion of a mobile element at one of the loci determining its synthesis then the kernels are colourless. All of the kernels in *Figure 15.6* originally possessed such an inserted element at fertilization. However, during the development of certain kernels, the affected locus 'reverts' in individual cells. The physical basis of the reversion is unknown but presumably involves loss of all or part of the mobile element from the locus. These cells and their progeny are capable of synthesizing anthocyanin and a patch of pigmented cells emerges. In this way the effects of transposition are literally visible.

Two types of transposable element system have been well characterized in maize. In the simple one-element system the mobile element affects the activity of an adjacent gene and also regulates its own transposition to other sites in the genome. In the more complex two-element system there is a 'receptor' element which resides in or close to the affected gene and a 'regulator' element which affects both its own, and the receptor element's transposition. Several two-element systems have been found in maize, each of which exhibits no cross reactivity with the functions of the others (for a review see reference 12). The Ac–Ds system is responsible for the effects shown in *Figure 15.6*. Another system termed the suppressor–mutator (Spm) or enhancer (En) system is described below. Virtually all the data on maize have been obtained using classical genetics and its interpretation in molecular terms remains largely unconfirmed.

In the Spm system either of the two mobile elements may reside in the A_1 locus which affects anthocyanin production. In both cases pigment synthesis is usually, but not always, eliminated (the exceptions presumably correspond to integration of the element in a non-essential region of the locus). The important difference between the regulator and receptor elements is that the former is capable of autonomous movement whereas the receptor element is an 'inert' sequence which can only transpose in the presence of trans-acting product(s) (presumably enzymes) specified by the regulator element. It is widely believed that the receptor element is a 'defective' version of the regulator. The regulator element possesses three independent activities: (1) a *'mutator'* activity which is responsible for transposition; (2) a *suppressor* activity which modifies the action of the target mobile element without altering its location; and (3) a *trans-activating* activity which temporarily 'repairs' another inactive regulator. The phenotypes resulting from interactions of the first two activities with either receptor or regulator elements integrated into the A_1 locus are shown in *Table 15.1*. Trans-activation will not be discussed here.

TABLE 15.1. Phenotypes of Spm mutants at the maize A_1 locus

Mutant designation	Kernel phenotype	Receptor present at A_1?	Regulator present?	Regulator activity?
A_1	●	No	No	–
a_1M-5	(image)	No	Yes, at A_1	Yes
A_1M-5	(image)	No	Yes, at A_1	No
A_1M-1	(image)	Yes, at A_1	No	No
A_1M-1	(image)	Yes	Yes	Yes
A_1M-1	(image)	Yes	No	No

The mutant a_1M-5 has a regulator element in the A_1 locus, giving a homogeneous, colourless kernel (*Table 15.1*, row 3). If mutator activity is present then frequent transposition of the regulator out of the A_1 locus gives rise to variegated kernels (row 2), if not then a homogeneous kernel results (row 3). The mutant a_1M-1 has receptor elements integrated in the A_1 locus. In this particular case anthocyanin production is reduced, not abolished, and a slightly paler kernel results (row 4). This block is enhanced by the presence of a regulator, giving rise to colourless kernels (row 4). However, the regulator can promote excision of the receptor element from the locus, leading to pigmented spots on the kernels (row 5).

In addition to mediating transposition, the mutator activity may modify the receptor element in a_1M-1 (presumably at the DNA sequence level) without excising it from the A_1 locus to yield homogeneous kernels with different levels of anthocyanin production (row 6) when the regulator is not present. Further modifications may include the development in the receptor of resistance to suppressor and/or mutator functions. Such modifications are thought to be due to the rearrangement of the receptor elements within the locus.

The activities of a regulator element are themselves modulated. Certain mutants exhibit differential regulator activity in different parts of the kernel, suggesting strongly that they correspond to developmentally controlled signals in the growing kernel. Molecular characterization of maize highly mutable loci is just beginning[6].

DNA transposition and the variation in trypanosome antigens

Trypanosomes are unicellular protozoa which are responsible for some of the most economically and socially destructive diseases facing mankind, namely sleeping sickness in humans and trypanosomiasis in cattle. These organisms occupy the bloodstream of the infected animal and escape the immunological defences of the host by producing a repertoire of distinct surface antigens, one at a time, thus sidestepping the antibody response. Borst and co-workers[4] have shown that there are over a hundred genes coding for the variants of the surface glycoproteins. By preparing cDNA clones from the RNA extracted from strains of trypanosomes expressing different antigens Borst et al.[4] showed that the expression of some of these genes is accompanied by duplication and transposition of the gene to a different chromosomal location. Further analysis by these workers[2,39] has shown that the transposed DNA fragment begins 1–2 kb upstream of the protein coding region and ends within the gene close to the 3' end (thus only a part of the gene is transposed). The 'acceptor' site for the transposed element appears to be the same in two cases studied. The 5' flanking region of each of five different genes shares a common DNA sequence (as judged by cross hybridization experiments), suggesting that this sequence might mediate the recombination between the transposed element and the target site. If this is so then the transposition of these elements may involve gene conversion in an analogous manner to that displayed by the yeast mating type switching genes (see above). It is also important to note that not all the trypanosome surface antigen genes are expressed by this mechanism. Another class of glycoprotein genes studied by Williams et al.[41] are neither duplicated nor transposed, although some form of DNA sequence rearrangement which is linked to gene expression does occur.

The occurrence of transposable elements in mammals

It is indisputable that mobile genetic elements are widespread in prokaryotes and lower eukaryotes. Are they similarly common in the higher eukaryotes such as mammals? DNA rearrangements of a highly sophisticated nature are certainly used by vertebrates in the generation of the vast assortment of antibody-producing cells (see Chapter 18). Furthermore, since mammalian genetics is so much more difficult to perform than similar studies in plants, flies and microorganisms, is it reasonable to assume that we would have detected mobile DNA elements? One study by Jenkins and co-workers suggests strongly that a mouse retrovirus can act as a mutator element[22]. These workers have shown that the coat colour mutation *dilute* (*d*) of DBA mice is closely linked with the presence of a murine leukaemia virus (MuLV) provirus integrated at the *d* locus. It is likely that the original integration of the MuLV-specific DNA occurred as a result of retroviral infection but loss of the MuLV sequence could only occur by genetic rearrangement. Owing to the way in which the *d* revertant was originally carried, it is imposible to say whether the MuLV provirus migrated to another chromosomal location or was simply destroyed.

Obviously, since repetitious DNA figures so strongly in the spectrum of transposable elements in lower eukaryotes, the analogous sequences in mammals are good candidates for mobile elements. However, the experiments to test the mobility of these sequences are only just beginning to be done. One such sequence class, the Alu family, has been extensively studied[21]. Approximately 300 000 copies of this 300 bp sequence exist in the human haploid genome.

In humans there is evidence that certain regions flanking the α-globin genes[19] and an unidentified fragment of single copy DNA[42] are highly polymorphic, suggesting that at least these regions are highly susceptible to DNA rearrangements. However, a comprehensive study of the stability of the β-globin gene family, between a wide variety of monkeys, apes and man, suggests strongly that at least in this case there is little mobility of DNA elements in an evolutionary sense[1]. The number and relative order of the five functional embryonic, fetal and adult β-related genes is identical in man, gorilla and the yellow baboon, which are separated evolutionarily by approximately 20 million years. Perhaps even more suprisingly, the restriction site analysis of the β-globin gene cluster between these species suggests that the regions between the genes have not suffered any permanent gain or loss of detectable DNA segments. Therefore transposition in this region is either virtually non-existent or it is effectively countered by selective evolutionary pressure against events which damage the gene family. Very few regions of mammalian genomes have been scrutinized at this level and much work needs to be done to describe the degree of fluidity in mammalian DNA.

The question of chromosomal lability in humans is potentially very significant in connection with the origin of some forms of cancer. There is a growing body of evidence to suggest that the chromosomes of all neoplastic cells have suffered detectable alterations in their structure[7, 32, 34]. As mobile genetic elements are demonstrably capable of promoting analogous alterations in each of the organisms in which they have been studied thoroughly it is reasonable to suppose that they may play a similar role in cancer cells.

References

1. Barrie, P. A., Jeffreys, A. J. and Scott, A. F. (1981). Evolution of the β-globin gene cluster in man and the primates. *Journal of Molecular Biology*, **149**, 319–336
2. Bernards, A., Van Der Ploeg, L. H. T., Frasch, A. C. C., Boothroyd, J., Coleman, S. and Cross, G. (1981). Activation of trypanosome surface glycoprotein genes involves a duplication-transposition leading to an altered 3' end. *Cell*, **27**, 497–505

3. Bingham, P. M., Levis, R. and Rubin, G. M. (1981). Cloning of DNA sequences from the *white* locus of *D. melanogaster* by a novel and general method. *Cell*, **25**, 693–704

4. Borst, P., Frasch, A. C. C., Bernards, A., Van Der Ploeg, L. H. T., Hoiejmakers, J. H. F., Arnberg, A. C. and Gross, G. A. M. (1981). DNA rearrangements involving the genes for variant antigens in *Trypanosoma brucei*. *Cold Spring Harbor Symposium on Quantitative Biology*, **45**, 935–944

5. Bregliano, J. C., Picard, G., Bucheton, A., Pelisson, A., Lavige, J. M. and L'Heritier, P. (1980). Hybrid dysgenesis in *Drosophila melanogaster*. *Science*, **207**, 606–611

6. Burr, B. and Burr, F. A. (1981). Controlling-elements events at the shrunken locus in maize. *Genetics*, **98**, 143–156

7. Cairn, J. (1981). On the origin of human cancers. *Nature, London*, **289**, 353

8. Cameron, J. R., Loh, E. Y. and Davis, R. W. (1979). Evidence for transposition of dispersed repetitive DNA families in yeast. *Cell*, **16**, 739–751

9. Cohen, S. N. and Shapiro, J. A. (1980). Transposable genetic elements. *Scientific American*, **242**, 36–45

10. Elder, R. T., St. John, T. P., Stinchcomb, D. T., Scherer, S. and Davis, R. W. (1981). Studies on the transposable elements Ty1 of yeast. *Cold Spring Harbor Symposium on Quantitative Biology*, **45**, 581–591

11. Errede, B., Cardillo, T. S., Wever, G. and Sherman, F. (1981). *Roam* mutations causing increased expression of yeast genes: Their activation by signals directed toward conjugation functions and their formation by insertion of Ty1 repetitive elements. *Cold Spring Harbor Symposium on Quantitative Biology*, **45**, 593–602

12. Fincham, J. R. S. and Sastry, G. R. K. (1974). Controlling elements in maize. *Annual Review of Genetics*, **8**, 15–50

13. Fink, G., Farabaugh, P., Roeder, G. and Chaleff, D. (1981). Transposable elements (Ty) in yeast. *Cold Spring Harbor Symposium on Quantitative Biology*, **45**, 575–580

14. Flavell, A. J. and Ish-Horowicz, D. (1981). Extrachromosomal circular copies of the eukaryotic transposable elements *copia* in cultured *Drosophila* cells. *Nature, London*, **292**, 591–595

15. Flavell, A. J., Ruby, S. W., Toole, J. J., Roberts, B. E. and Rubin, G. M. (1980). Translation and developmental regulation of RNA encoded by the eukaryotic transposable element *copia*. *Proceedings of the National Academy of Sciences, USA*, **77**, 7107–7111

16. Green, M. M. (1980). Transposable elements in *Drosophila* and other Diptera. *Annual Review of Genetics*, **14**, 109–120

17. Heine, C. W., Kelly, D. C. and Avery, R. J. (1980). The detection of intracellular retrovirus-like elements in *Drosophila melanogaster* cell cultures. *Journal of General Virology*, **49**, 385–395

18. Hicks, J. B., Strathern, J. N. and Herskowitz, I. (1977). The cassette model of mating-type interconversion. In *DNA Insertion Elements, Plasmids and Episomes* (Bukhari, A. *et al.*, Eds.), p. 457. New York, Cold Spring Harbor Laboratory

19. Higgs, D. R., Goodbourn, S. E. Y., Wainscoat, J. S., Clegg, J. B. and Weatherall, D. J. (1981). Highly variable regions of DNA flank the human α-globin genes. *Nucleic Acids Research*, **9**, 4213–4224

20. Isling, G. and Block, K. (1981). Derivation-dependent distribution of insertion sites for a *Drosophila* transposen. *Cold Spring Harbor Symposium on Quantitative Biology*, **45**, 527–544

21. Jagadeeswaran, P., Forget, B. and Weissman, S. (1981). Short interspersed repetitive DNA elements in eukaryotes: transposable DNA elements generated by reverse transcription of RNA Pol ll transcripts? *Cell*, **26**, 141–142

22. Jenkins, N. A., Copeland, N. G., Taylor, B. A. and Lee, B. K. (1981). *Dilute (d)* coat colour mutation of DBA/2J mice is associated with the site of integration of an ecotropic MuLV genome. *Nature, London*, **293**, 370–374

23. Kidwell, M. G., Kidwell, J. F. and Sved, J. A. (1977). Hybrid dysgenesis in *Drosophila melanogaster*: A syndrome of aberrant traits including mutation, sterility, and male recombination. *Genetics*, **86**, 813

24. Klar, A. J. S., Hicks, J. B. and Strathern, J. N. (1981). Irregular transpositions of mating type genes in yeast. *Cold Spring Harbor Symposium on Quantitative Biology*, **45**, 983–990

25. Kleckner, N. (1977). Translocatable elements in prokaryotes. *Cell*, **11**, 11–23

26. Kleckner, N. (1981). Transposable elements in prokaryotes. *Annual Review of Genetics*, **15**, 341–404

27. McClintock, B. (1951). Chromosome organization and genic expression. *Cold Spring Harbor Symposium on Quantitative Biology*, **16**, 13–47

28. McClintock, B. (1956). Controlling elements and the gene. *Cold Spring Harbor Symposium on Quantitative Biology*, **21**, 197–216

29. Naysmith, K. A., Tatchell, K., Hall, B. D., Astell, C. and Smith, M. (1981). Physical analysis of mating type loci in *Saccharomyces cerevisiae*. *Cold Spring Harbor Symposium on Quantitative Biology*, **45**, 961–981
30. Nevers, P. and Saedler, H. (1977). Transposable genetic elements as agents of gene instability and chromosomal rearrangements. *Nature, London*, **268**, 109–115
31. Plough, H. H. (1928) Black Suppressor—a sex-linked gene in *Drosophia* causing apparent anomalies in crossing over in the second chromosome. *Zeitschrift für induktive Abstammungsund VererbLehre*, Suppl. **11**, 1193–1200
32. Ponder, B. A. J. (1981). Looking for genomic changes in cancer cells. *Nature, London*, **293**, 98–99
33. Rasmuson, B., Westerberg, B. M., Rasmuson, A., Bvozdev, V. A., Belyaeva, E. S. and Ilyin, Y. V. (1981). Transpositions, mutable genes and the dispersed gene family Dm225 in *Drosophila melanogaster. Cold Spring Harbor Symposium on Quantitative Biology*, **45**, 545–551
34. Rowley, J. D. (1980). Chromosome abnormalities in human leukaemia. *Annual Review of Genetics*, **14**, 17–93
35. Rubin, G. M., Brorein, W. J., Dunsmuir, P., Flavell, A. J., Levis, R., Strobel, E., Toole, J. J. and Young, E. (1981). *Copia*-like transposable elements in the *Drosophia* genome. *Cold Spring Harbor Symposium on Quantitative Biology*, **45**, 619–628
36. Spradling, A. C. and Rubin, G. M. (1982). *Drosophila* genome organization: conserved and dynamic aspects. *Annual Review of Genetics*, 219–264
37. Stiles, J. I., Friedman, L. R. and Sherman, F. (1981). Deletions, duplications and transpositions of the *Cor* segment that encompasses the structural gene of yeast iso-l-cytochrome *c*. *Cold Spring Harbor Symposium on Quantitative Biology*, **45**, 602–607
38. Truett, M. A., Jones, R. S. and Potter, S. S. (1981). Unusual structure of the FB family of transposable elements in *Drosophila. Cell*, **24**, 753–763
39. Van Der Ploeg, L. H. T., Bernards, A., Rijsewijk, F. A. M. and Borst, P. (1982). Characterization of the DNA duplication–transposition that controls the expression of two genes for variant surface glycoproteins in *Trypanosoma brucei. Nucleic Acids Research*, **10**, 593–609
40. Wensink, P. C., Tabata, S. and Pachl, C. (1979). The clustered and scrambled arrangement of moderately repetitive elements in *Drosophila* DNA. *Cell*, **18**, 1231–1246
41. Williams, R. D., Young, J. R., Majiwa, P. A. O., Doyle, J. J. and Shapiro, S. Z. (1981). Contextual genomic rearrangements of variable-antigen genes in *Trypanosoma brucei.Cold Spring Harbor Symposium on Quantitative Biology*, **45**, 945–950
42. Wyman, A. R. and White, R. (1980). A highly polymorphic locus in human DNA. *Proceedings of the National Academy of Sciences, USA*, **77**, 6745–6758

Chapter 16

Histone genes

R. Maxson, T. Mohun and L. Kedes

Introduction

Histone proteins comprise a class of five, small, basic proteins which are ubiquitous in the eukaryotic kingdom[50]. They associate with each other and with nuclear DNA to form the basic unit of chromatin organization, the nucleosome[65]. Their structure is highly conserved between the most widely divergent species, reflecting the fundamental and unchanging functions they perform within the cell.

During the eukaryotic cell cycle, the nucleosome units of chromatin are organized in concert with a number of nucleoproteins to generate many higher orders of chromatin structure the variety of which has intrigued biologists for many decades. Yet underlying such diversity in chromosome organization is an equally remarkable constancy of chromatin composition; histone proteins are complexed with DNA in an unvarying 1:1 stoichiometric ratio[6]. Furthermore, the individual histone types that comprise the nucleosomal core (H2A, H2B, H3 and H4) are present in equimolar quantities whilst the fifth histone type, H1, is approximately half as abundant. Clearly then, precise controls must exist to enable histone:DNA ratios to be maintained by eukaryotic cells. In particular, such controls must ensure that dividing cells synthesize sufficient histones to organize newly replicated DNA into functional daughter chromosomes.

The demand for histones during the cell cycle may be periodic and light, as for example in cultured cells that exhibit relatively long doubling times; it may also be sustained and heavy, as in the case of rapidly cleaving embryos. Not surprisingly, different strategies for histone gene regulation are employed by cells in these two very different biological circumstances. In this chapter we will review our present understanding of these strategies and our more limited knowledge of the molecular mechanisms by which the stoichiometry of nucleosome components is maintained. We will not attempt an exhaustive review of the literature; in particular we will only summarize the most recent studies of histone gene structure since a comprehensive account of this subject has been published recently[43].

Histone gene regulation

Histone proteins and transcripts during the cell cycle

Over the past two decades, a considerable body of evidence has accumulated to indicate that histone biosynthesis and chromosome replication are temporally linked during the cell cycle. Thus, histone protein synthesis in HeLa cells is confined to S phase of the cell cycle[58, 94]; such synthesis takes place on a distinct class of small polysomes, utilizing slowly sedimenting (7–9S) RNA whose identity as authentic histone mRNA has been established by cell-free translation and hybridization[7, 24, 33, 75, 81, 95, 106]; isolated S phase nuclei also demonstrate histone mRNA synthesis, as do preparations of chromatin derived from S phase cells[23, 96]. In contrast, nuclei and chromatin preparations from other periods of the HeLa cell cycle show no such biosynthetic capacity. Finally, in situ hybridization methods have demonstrated a similar restriction of histone mRNA synthesis to the S phase of unsynchronized cells[105]. Similar, though much less extensive results have also been reported for a number of normal and transformed mammalian cell lines[21, 40, 53, 57, 74, 99].

Histone and DNA synthesis in these cells are both coincident and coupled, with the result that the inhibition of one results in the cessation of the other. For example, treatment of synchronized HeLa cells within inhibitors of DNA synthesis such as cytosine arabinoside, hydroxyurea or 5-fluorodeoxyuridine, results in a rapid and proportional inhibition of histone protein synthesis. Removal of the inhibitor allows the continuation of DNA replication and the restoration of histone synthesis[7, 8, 15, 31, 33, 57, 75, 80, 81].

In addition to the extensive documentation of coupled histone and DNA synthesis, numerous examples of uncoupled histone production have been presented. For example, in several early studies of synchronized HeLa cells, a basal level of histone synthesis was detected outside of S phase[32, 84]. A 5–10S cytoplasmic RNA fraction derived from HeLA G1 phase cells directed histone synthesis in a cell-free translation system[32]. In a more recent report, newly synthesized transcripts hybridizing to a sea urchin histone gene DNA probe were reportedly detected throughout the HeLa cell cycle[66]. Similarly, turnover of isotopically pre-labelled histones in synchronous Chinese Hamster Ovary cell cultures maintained at the G1/S phase boundary has been interpreted to indicate uncoupled, G1 phase histone synthesis[39]. More recently, homologous histone cDNA probes have been used in titration experiments to demonstrate a detectable level of histone mRNA in both the nuclear and cytoplasmic compartments of synchronous, G1 phase, mouse 3T6 cells[74]. Finally, evidence from studies of S49 mouse lymphoma and Chinese Hamster Ovary cell cultures has been interpreted to indicate that histone synthesis occurs throughout the cell cycle; only the incorporation of histones into chromatin is restricted to S phase[37].

Can these contradictory results be explained by differences in experimental procedure, in particular the methods used to achieve cell synchrony and the choice of hybridization probe, or do they indicate fundamental differences in histone gene regulation between HeLa and other cell lines? This question continues to provoke considerable controversy. However, a recent re-examination[112] of the various types of histones synthesized during the cell cycle may provide an explanation for at least some of the conflicting results. Wu and Bonner have documented the synthesis of a large number of histone variants throughout the HeLa cell cycle. Using the most sensitive methods for electrophoretic separation they have shown that in addition to the coupled synthesis of major histone types, the HeLa cell also synthesizes a number of specific histone forms at all stages of the cell cycle. Such basal histone synthesis may be distinguished from that restricted to S phase since the latter is abolished in the presence of DNA synthesis inhibitors. Furthermore, similar results have been found for a number of other cell types in culture. The authors suggest that basal histone variants perform some function in facilitating cell growth rather than cell division since they are present in the appropriate, nucleosomal proportions but in ratios with S phase specific histones that vary inversely with the rate of cell cycling.

What molecular mechanisms are responsible for the coupling of histone and DNA synthesis? The use of RNA and protein synthesis inhibitors has provided some indication of the types of regulation that may be involved. Foremost amongst these is the modulation of the steady state levels of histone mRNA. When synchronous HeLa cell cultures enter S phase, the quantity of histone gene transcripts (detected by hybridization to homologous probes) increases dramatically[24, 58, 80]. If DNA synthesis is briefly inhibited, histone transcript levels fall and the restoration of DNA replication may be prevented by actinomycin D[8, 33]. Little is known of the mechanisms by which such transcriptional control is effected. One approach has been to use *in vitro* models to define those components of S phase chromatin that confer transcriptional activity to histone DNA. For example, a number of studies by Stein and coworkers have suggested that the non-histone proteins of S phase chromatin will stimulate histone gene transcription in reconstituted chromatin containing G1, G2 or M phase DNA. Conversely, no inhibitory factor has been detected in G1, G2 or M phase chromatin preparations[73, 95, 96].

Post-transcriptional mechanisms

In addition to transcriptional regulation, it is generally agreed that the tight coupling between histone biosynthesis and DNA replication may involve post-transcriptional mechanism(s). Histone mRNA is relatively stable during S phase[76] with a half-life in HeLa cells approximately equal to the length of S. However, if DNA replication is blocked, histone gene transcripts are rapidly lost from polysomes. This loss appears to be the result of a destabilization of the histone mRNAs, resulting in their rapid degradation[7]. Thus, within 60 minutes of hydroxyurea treatment, cytoplasmic RNA preparations from synchronous, S phase, HeLa cells show an 80% loss of histone template activity[31]. The half life of such transcripts has been estimated as 13 minutes which corresponds to at least a tenfold decline in response to hydroxyurea treatment. Such a change in message stability is dependent on continued protein synthesis; simultaneous treatment of synchronous, S phase HeLa cultures with hydroxyurea and cycloheximide results in both the inhibition of histone synthesis and the stabilization of histone mRNA on previously active polysomes[31]. A simple model to account for such results was suggested a decade ago by Butler and Mueller. Briefly, they proposed that coupling is achieved through a negative feedback loop[15]; cells entering S phase begin DNA replication and, in doing so, facilitate the transcription of histone genes and the synthesis of histone proteins. Inhibition of DNA synthesis results in a transient accumulation of histones which are then free to bind to the transcriptionally active histone genes. This results in gene inactivation with a consequent decline in histone transcript levels. If protein synthesis is simultaneously inhibited, no free histones accumulate and histone gene transcription continues. Of course, in the light of later results, the accuracy of such a model may be questioned. No account, for example, is taken of the role of non-histone proteins in the activation of histone gene transcription of S phase chromatin. Nor can this model explain the precipitous decline in histone mRNA stability at the end of S phase. For this, the authors advanced the additional proposal that cytoplasmic histones acted to stabilize their own transcripts. However, the proposal that coupling involves a feedback loop remains attractive, providing as it does a mechanism by which transcriptional and post-transcriptional controls may be coordinated.

In recent years, molecular cloning techniques have been utilized in a re-investigation of the mechanisms of coupled histone and DNA synthesis. One organism in particular, yeast, has proved especially useful for such studies. Yeast cultures are readily synchronized by treatment with the pheromone, α-factor and show many of the properties of histone gene regulation observed in earlier studies of mammalian cell cultures[68]. In addition, cloned DNA sequences for yeast H2A and H2B genes have been thoroughly characterized[44, 104]. Using such homologous sequences as probes, Hereford and co-workers have demonstrated that transcription of H2A and H2B genes increases twenty-fold as yeast cells enter S phase[45]. Interestingly, peak levels of transcripts are detected early in S phase, when less than 25% of DNA replication has been accomplished. Employing a temperature-sensitive mutant strain defective for DNA replication at the non-permissive temperature, the authors found that inhibition of DNA synthesis resulted in a rapid and specific decay of histone mRNAs. An equally specific, though less rapid, decay of histone transcripts was detected as cells reached the end of S phase. Similar kinetics of histone mRNA accumulation and decay were found by *in vitro* translation methods for the other two nucleosomal histones, H3 and H4.

The efficacy of the post-transcriptional controls regulating yeast histone gene expression has been tested using recombinant strains bearing a duplication of a closely linked H2A/H2B gene pair[72]. Such recombinants show gene dosage compensation; that is, levels of histone protein synthesis remain unaffected by the presence of additional gene copies. In contrast, the rates of histone gene transcription increase in parallel with the increase in gene number and dosage compensation results from a similarly proportionate decrease in H2A and H2B mRNA stability. Transformants show gene dosage compensation irrespective of the site of integration of the additional H2A/H2B gene pair within the yeast genome and in each case the same post-transcriptional mechanism is utilized to maintain a fixed ratio between steady state levels of histone mRNA and the DNA replication rate. Such results indicate that coupling in yeast involves a feedback regulatory loop similar to that originally suggested by Butler and Mueller.

In summary, a major fraction of histone biosynthesis is closely coupled to DNA replication during the cell cycle. The mechanism of coupling involves both transcriptional regulation (ensuring the rapid increase in levels of histone transcripts at the start of S phase) and post-transcriptional control of histone mRNA stability (which results in the rapid and specific degradation of transcripts as DNA replication is completed). Some basal synthesis of histones may also be detected throughout the cell cycle, the proportion of which varies between cell types. Histone variants synthesized throughout the cell cycle constitute an electrophoretically distinct group whose specific cellular function is unclear.

Uncoupled histone biosynthesis

A number of examples of uncoupled histone synthesis have been found in cultured cell lines and differentiated cell types though these appear to be exceptional. For example, when a strain of Friend erythroleukemic cells is treated with the inducer, *n*-butyrate, the cells first enter a prolonged G1 phase during which significant histone synthesis may be detected[116]. Total protein synthesis declines over the same period suggesting that specific controls may be responsible for the increase in histone gene expression. Uncoupling is also shown by some neuronal cells during mammalian brain development[12]. Shortly after birth, these cells undergo a change in their chromatin structure with the result that the length of 'linker' DNA between adjacent nucleosomes is reduced from 60 to 20 nucleotides. Immediately prior to this, a transient peak of histone synthesis occurs despite the absence of any DNA replication. Lastly, a temperature-sensitive mutant has been isolated from the Balb/C-3T3 mouse fibroblast cell line that is defective in DNA replication; at the non-permissive temperature DNA synthesis is inhibited but histone gene expression continues unaffected[87]. Such mutants may prove extremely valuable in analyzing the mechanisms of coupling in the cell cycle of higher eukaryotes.

The most widespread example of uncoupled histone synthesis in eukaryotes occurs during the earliest period of their life cycle. After fertilization, the eggs of many organisms undergo a period of rapid, repeated cell division in the course of which distinct cell lineages are established. Large quantities of histones are needed to sustain the very rapid rates of chromatin replication characteristic of embryonic cleavage (see *Table 16.1*).

TABLE 16.1. Comparison of maximal DNA synthesis rates and histone gene numbers

Organism	DNA (pg/cell haploid)	Shortest known doubling (hours)	Maximal DNA synthetic rate (pg/hour)	Number of histone gene copies (haploid)[m]
Human	3[a], [b]	14[e]	0.21	10–40
Mouse	3[a], [b]	12[f]	0.25	10–40
Chicken	1.26[a]	0.35[g]	3.6	10
Xenopus	4.2[a]	0.25[h]	16.8	20–50
Urchin	0.77[c]	0.75[i]	1.0	300–600
Drosophila	0.17[a]	0.17[j]	1.0	100
Caenorhabditis elegans	0.09[d]	0.20[k]	0.45	15
Yeast	0.025[a]	1.0[l]	0.025	2

[a]   *Handbook of Biochemistry and Biology*
[b]   Hecht *et al.*, reference 41
[c]   Giudice, reference 34
[d]   H. Epstein, personal communication
[e]   Witshci, reference 109
[f]   Snell, reference 93
[g]   Eyal-Giladi and Kochav, reference 28
[h]   Nieuwkoop and Faber, reference 71a
[i]   Giudice, reference 34
[j]   Anderson and Lengyel, reference 1
[k]   Deppe *et al.*, reference 22
[l]   Pringle and Mor, reference 77
[m]   References for gene copy number are supplied in the text.

We may define embryonic uncoupling to occur when the rate of histone gene expression at the RNA or protein level diverges from the curve describing the rate of DNA synthesis. Such a situation is shown in *Figure 16.1* which depicts early development in the echinoderm (sea urchin), *S. purpuratus*. These and other rapidly cleaving embryos employ a variety of means to maintain DNA/histone stoichiometry, including the utilization of stored histone proteins and mRNA templates to supplement *de novo* transcription and translation[60, 85]. The length of

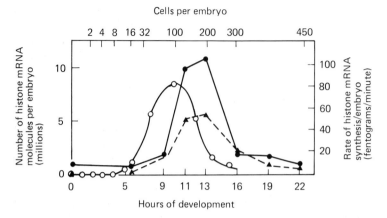

Figure 16.1. Rates of synthesis and accumulation of histone mRNAs during early sea urchin embryogenesis. The data are a composite taken from Maxson and Wilt[62] and from Mauron *et al.*[60]. Open circles (○): rate of histone mRNA synthesis; filled circles (●): number of core histone mRNA molecules/embryo; filled triangles (▲): number of H1 mRNA molecules/embryo.

time between synthesis and use of such stored products may vary widely from the few minutes separating successive G1 and S phases of the embryo cell cycle[3] to the many months that may intervene between the commencement of oogenesis and the fertilization of mature eggs[60]. In addition to these features of uncoupling, the presence of many copies of histones genes in several organisms may permit unusually rapid synthesis of large quantities of histone mRNA.

In the development of the sea urchin embryo, all three elements – storage of histone proteins, gene transcripts and transcription of highly reiterated histone genes – are coordinated in a highly complex, developmentally regulated pattern the outlines of which have emerged from the work of various laboratories. Histone demand in this organism is extremely high throughout early embryo cleavage when S phase can be as short as 10–12 minutes (*Table 16.1*)[48].

Storage of histone mRNA and protein during urchin oogenesis

The unfertilized sea urchin egg contains a significant store of histone mRNA[60] and protein[85]. Measurements of the former have been obtained using DNA excess hybridization methods with individual, homologous gene probes. Mauron *et al.*[60] estimate that the egg contains a store of approximately one million mRNA transcripts for both H2B and H3 histone genes. Egg histone protein stores have been quantitated by electrophoresis and by the ability of unfertilized eggs to reorganize the chromatin they receive after polyspermic fertilization. At minimum, eggs appear to contain sufficient histones for the assembly of approximately 12 nuclei.

Oogenesis in the sea urchin can extend for many months since oocyte growth and vitellogenesis can be delayed for as long as a year after the completion of oogonial DNA replication[20]. The difficulty encountered in fractionating oocytes at the various stages of oogenesis has hindered attempts to investigate the onset of egg histone protein and transcript accumulation. However, differences in size allow pre-vitellogenic oocytes to be separated from later vitellogenic and post-vitellogenic stages and using such populations, Mauron et al.[60] have found that histone mRNA is only detectable after the onset of yolk accumulation. Sea urchin oogenesis, therefore, represents an extreme case of uncoupling in which histone gene transcription can oocur many months after the cessation of DNA replication.

How are the histone gene transcripts and proteins maintained in the oocyte and what mechanisms regulate their storage? When unfertilized eggs are disrupted, the majority of histone mRNA is found complexed in ribonucleoprotein particles[79]. However, no kinetic measurements of oocyte histone gene transcript synthesis and turnover have been made and so the relationship between the oocyte store and egg RNPs remains unclear. Nor is much known about the nature of these particles; an unexpectedly large fraction appears to be located in the germinal vesicle (nucleus) but their structure and composition have yet to be analyzed[103].

The unfertilized egg synthesizes histone mRNA at a low rate compared with post-fertilization stages and virtually all the resulting transcripts are translated and degraded[10]. This metabolically active population of histone mRNA is not in equilibrium with the pool of previously synthesized histone mRNA present in RNP particles. Since the urchin embryo is known to utilize a number of histone variants during the course of early development[71, 85] it is of interest to know whether the compartmentalization of egg transcripts into active and inactive pools results from selective storage of particular variant mRNAs. Information on this point is fragmentary. Three sets of histone variants are found in urchin embryo nuclei; from the first cell division, cleavage stage (CS) variants are found in chromatin and their synthesis continues until about the 8-cell stage; another class, the α variants, are synthesized and used throughout early cleavage until at least the mesenchyme blastula stage after which they are largely replaced by the third, or γ, class[18, 71, 82]. Cell-free translation of total egg RNA reveals the presence of mRNAs coding for both α and at least some CS proteins[16] but no comparable evaluation of the individual transcript pools has been undertaken. Newly synthesized histone mRNA comigrates with α variant transcripts from early embryo stages but the comparable electrophoretic behavior of CS mRNAs has yet to be established[10].

The maternal stockpile of histone protein comprises both α and CS variants[85]. Curiously, these variants are not found on early embryo chromatin in the same proportion in which they are present in the egg, rather the stored CS protein variants are under-represented. One explanation for this postulates that a high concentration of CS variants is necessary to achieve exchange with sperm histones and the consequent activation of the male pronucleus[85].

Utilization of stored, maternal histone mRNA during early embryo cleavage

Histone gene expression in the sea urchin zygote can be divided into three consecutive phases distinguished by the kinetic parameters that underlie the regulation of zygotic histone mRNA levels. In the first six fours after fertilization, the egg undergoes four synchronous cleavages to form a 16-cell embryo. During this, RNA labelling studies have detected a limited amount of histone gene transcription[62] but this contributes less than 15% to the zygotic store of histone mRNA[60, 62, 108]. In contrast, the rate of histone protein synthesis, expressed as a proportion of total protein synthesis, shows a significant and steady increase throughout early cleavage[35]. Since this occurs in the absence of any net increase in histone mRNA content, it must result from regulatory mechanisms operating at the post-transcriptional and/or translational levels. Indeed, rates of polypeptide initiation and elongation have been shown to remain constant in the early urchin embryo and the increase in the rate of histone synthesis results from the mobilization of maternal RNA stores. Thus, the rate of flux of transcripts from their storage forms to the polysomal RNA pool limits the expression of histone genes during this earliest period of sea urchin embryogenesis.

Transcription of reiterated histone genes in the developing zygote

By the 16-cell stage, the rate of histone synthesis is determined solely by the level of histone gene transcripts on polysomes[79]. After the fourth cleavage, the embryo enters a period of extremely rapid, asynchronous cell division, accumulating 200 cells in only four hours[48]. The rate of DNA synthesis increases exponentially and the maintenance of an adequate histone supply requires the rapid synthesis of large quantities of histone mRNA. The rate of total embryonic RNA synthesis rises sharply at the start of this period (see *Figure 16.1*)[62, 108]. Even more dramatic is the change in the rate of histone mRNA synthesis; from the 16- to the 64-cell embryo stage (i.e. 120 minutes) the relative proportion of histone mRNA synthesis increases ten-fold from 4% to almost 40% of total synthesis and the absolute rate of histone mRNA synthesis rises five-fold[61]. A similar increase is detected in the steady state levels of histone mRNA transcripts[60], indicating that the regulatory site in histone gene expression lies in the processes comprising mRNA synthesis.

How does the embryo achieve such swift and massive increases in histone mRNA levels? More precisely, is the change in the rate of histone mRNA synthesis the result of a change in histone gene transcription rate or are other, post-transcriptional events involved? Histone genes coding for early embryonic protein forms (α) are repeated several hundred times in the sea urchin genome and if most or all copies are transcriptionally active, relatively small changes in transcription rate will be amplified into major fluctuations in the rate of transcript synthesis. Two lines of evidence suggest that an increase in histone gene transcription rate does in fact occur. Firstly, the lag time between the introduction of an isotopic label *in vivo* and the appearance of labelled histone mRNA in the cytoplasm is very short. Secondly histone mRNA is not known to be processed and a very thorough study has failed to detect high molecular weight histone mRNA transcripts in sea urchin embryos[59]. Together, these observations indicate that no significant post-transcriptional regulation of histone gene expression occurs in this phase of exponential DNA replication. Rather, the rate-limiting step in histone production resides at the level of transcription.

The switch from early to late histone synthesis

A day after fertilization, the sea urchin embryo undergoes gastrulation and its 600 cells become organized into a number of morphologically distinct tissues. A complex series of morphogenetic movements and cell differentiation events give rise to a bilaterally symmetrical (prism stage) embryo that develops into the pluteus larval form[49]. Throughout this period, DNA replication slows as the rate of cell division declines[48]. Not surprisingly then, histone gene regulation enters a new phase, characterized by a sharp decline in embryonic histone mRNA content and the appearance of a third set of histone variants (the late, or γ forms) in embryonic chromatin[18, 82]. These are encoded by a distinct group of late mRNAs which are produced by transcription of a corresponding set of late histone genes[16, 38, 46, 71]. The significance of this switch in histone type is still unclear. Recent studies have indicated that nucleosomes composed exclusively of early or late histones differ from each other in their physical properties, including the DNA repeat length they display upon digestion with DNase I. Compared with their early counterparts, twenty-fold fewer copies of late genes are present in the sea urchin embryo genome[64] so that even if histone gene transcription rates are maintained at previous levels, a switch from early to late gene transcription will result in a sharp drop in the rate of histone mRNA synthesis. Is this the means by which the quantitative regulation of histone gene expression in late embryos is achieved? Recent studies have revealed a more complicated picture. Late gene transcripts have been detected as early as ten hours after egg fertilization at which time they constitute less than 10% of the total histone mRNA pool[16]. Only five hours later, after cell cleavage has slowed, the proportions of early and late gene transcripts are reversed; greater than 90% of the mRNA pool comprises late histone templates. This switch can largely be explained by changes in the kinetics of early mRNA synthesis and turnover. The rate of early gene transcription reaches a peak value between 10–12 hours after fertilization and subsequently shows a steep decline[62]; this rate profile results in a similar, later peak in early mRNA accumulation, the decline of which is accentuated by an approximate 50% reduction in early transcript half-life (see *Figure 16.1*)[60, 62]. Thus, the transition from early to late histone mRNA synthesis need not result from coordinate transcriptional regulation of both early and late genes. Rather, it could simply result from a combination of relative early and late gene dosage, the rapid increase in cell number during early embryonic cleavage and the transcriptional inactivation of the early gene set. Quantitative measurements of rates of late histone mRNA synthesis will distinguish between these two possibilities.

Further examples of embryonic uncoupling: Drosophila *and* Xenopus

Histone gene regulation during early development has only been studied in any detail in two other organisms: the fly (*Drosophila melanogaster*) and the frog (*Xenopus laevis*). Embryonic development in both is rapid[9, 78], (Nieuwkoop and Faber[71a], DNA replication times being as brief as 10–15 minutes[1, 36]. The demand for histones in each is met by the temporal uncoupling of histone and DNA synthesis.

Little is known about histone gene expression in the earliest stages of fly development. For 90 minutes following fertilization, the *Drosophila* egg undergoes nine synchronous nuclear divisions without concomitant cell division[114]. No embryonic gene activity has been detected during this time and so the embryo must presumably rely on the utilization of stored histone mRNAs and/or proteins to meet the demands of rapid chromatin assembly[113]. The resulting, syncytial blastoderm embryo passes through four more nuclear divisions in the subsequent hour, after which membrane furrows can first be detected. Formation of cell membranes is completed in another 60–90 minutes, by which time the cellular blastoderm embryo has accumulated approximately 6000 nuclei. By gastrulation, DNA doubling time has increased to 70–80 minutes and in later stages still lower rates of DNA synthesis may be detected[2]. Nevertheless, within 14 hours of fertilization, the *Drosophila* embryo has acquired sufficient histones to organize the chromatin of 83 000 diploid nuclei. In an elegant study, Anderson and Lengyel[2] have examined the manner by which the embryo of this organism maintains appropriate histone:DNA stoichiometry. Using homologous, cloned histone DNA probes, they have measured the rates of synthesis accumulation and decay of histone mRNA at various embryo stages. Their investigations demonstrate that many features of sea urchin histone gene regulation are shared by the fly. The *Drosophila melanogaster* egg for example contains sufficient stored, maternal, histone mRNA to meet the demands of pre-blastoderm development. During blastoderm and gastrula stages, transcription of a repetitive class of histone genes, containing several hundred members, results in the rapid accumulation of histone mRNA. The rate of histone mRNA synthesis increases to a peak six hours after oviposition and coincides with the peak in the rate curve for DNA replication. As the rate of DNA synthesis declines, both the histone mRNA synthesis and the total mass of histone transcripts decline sharply. By combining the results of quantitative and kinetic studies, Anderson and Lengyel have shown that a dramatic change in histone gene transcript stability accompanies this drop in the rate of histone mRNA synthesis; in the few hours from syncytial blastoderm stage to the end of gastrulation, histone mRNA half-life falls seven-fold from 2.3 hours to around 20 minutes. Such a change in mRNA stability can be explained in several ways. It could result from a developmentally regulated change in histone mRNA stability that affects the entire steady state histone mRNA pool; alternatively it could arise from the changing proportions of maternal and embryonic transcripts in the histone mRNA pool. If maternal histone mRNA is significantly more stable than its embryonic counterpart then the burst of histone gene transcription that begins in the syncytial blastoderm will alter the composition of the histone mRNA pool and the more rapidly degraded transcripts will soon predominate.

How do the programs of embryonic histone gene expression in the fly and the sea urchin compare? In both organisms we can establish three distinct regulatory phases. During the initial nuclear divisions, histone synthesis relies on the mobilization of stored, maternal mRNA. In the second (accumulation) phase rapid transcription of reiterated genes provides the templates for histone synthesis. Finally, in the third (decay) phase, as DNA replication rates fall, histone supply is coordinately reduced by a simultaneous decline in synthesis and increase in turnover of histone gene transcripts.

Studies of early development in the anuran amphibian, *Xenopus laevis*, demonstrate that this program of histone gene expression is not shared by all rapidly cleaving embryos. The frog egg begins cleavage about 75 minutes after fertilization (Nieuwkoop and Faber[71a]). Over the next five hours, DNA replication rate increases exponentially and the cell cycle is reduced to a brief, 12 minute S phase[36]. Since histone genes are repeated only about 40 times in the *Xenopus* genome[51, 101] and cell division in the early embryo is so rapid, it is not surprising that the major source of histones for the developing embryo is provided by maternally derived stores of histone mRNA and protein. Indeed, Woodland[110] has calculated that only by the 1000–2000 cell stage does the zygote possess sufficient numbers of histone genes to synthesize histones at a rate adequate to meet the demands of DNA replication.

The size and origin of these maternal stores have been the subject of a number of studies by Woodland and co-workers[110]. They have found that the unfertilized egg contains sufficient histones (excluding H1) to organize over 20 000 diploid nuclei. These proteins are synthesized throughout oogenesis and retained largely within the germinal vesicle (nucleus) of the oocyte. During the hormonally-induced maturation of full grown oocytes, the germinal vesicle disintegrates and the histone stores are released into the egg cytoplasm. At the same time, the rate of core histone synthesis increases fifty-fold. By the completion of maturation (only a few hours later) the full complement of maternal histone stores has been accumulated. This change in histone synthetic rate at the onset of maturation is unaffected by either physical or chemical enucleation of the oocyte and must therefore be the result of post-transcriptional or translational events. Certainly, it coincides with a change in the structure of egg histone mRNA. Prior to maturation, a significant proportion of histone gene transcripts are polyadenylated (in contrast to those of other organisms); in the mature egg, $poly(A)^+$ histone mRNA is no longer detectable although the total quantity of histone mRNA is unchanged. However, no evidence has yet been presented to demonstrate a role for deadenylation in translational regulation of histone synthesis and so the significance of these structural changes is unclear.

In contrast to the protein stores, oocyte histone mRNA pools appear to remain unchanged during maturation. Indeed, at least for those encoding histones of the nucleosomal core, steady state levels are established early in vitellogenesis (prior to the lampbrush chromosome stage)[102]. No kinetic data are available to assess how this stockpile of histone mRNA is maintained through the months of oogenesis. However, together with stored histone proteins, translation of these mRNAs supports embryo chromatin formation during the earliest period of rapid cleavage.

Because of the low repetition frequency of *Xenopus* histone genes, embryonic transcription cannot make a significant contribution to the embryo pool of histone mRNA until quite late in development. Using interspecies hybrids with electrophoretically distinct histones, Woodland and co-workers have demonstrated that the embryonic proteins are first detected at the 1000–2000 cell blastula stage and become predominant only in the early gastrula embryo[111]. Similarly, androgenetic haploid embryos can maintain normal patterns of H1 histone synthesis until the early blastula stage, suggesting that maternal stores support chromatin formation for at least the first seven hours of development.

From late blastula stage, DNA replication rate declines and the demand for histones is met by the utilization of embryonic transcripts. Within 24 hours, the embryo consists of many differentiating cell types that exhibit widely differing cell division rates. Average cell cycle times are then sufficiently long to permit the re-establishment of coupled histone and DNA biosynthesis.

The achievement of histone stoichiometry

We have seen that histone gene expression is regulated temporally and quantitatively to provide sufficient histones for chromatin assembly. Such regulation must include mechanisms to ensure that the appropriate relative proportions of all five histone types are incorporated into newly synthesized nucleosomes. One simple way of achieving this would be through the self assembly of chromatin from appropriately sized free histone pools. If histones are synthesized in stoichiometric quantities then self assembly will guarantee that they are incorporated into newly formed chromatin in similar ratios. Alternatively, nucleosome histone stoichiometry may result from specific regulatory controls exerted during chromatin assembly.

In the sea urchin, histone mRNAs are synthesized in amounts that reflect the molar ratio of histones in chromatin. Two independent lines of evidence support this conclusion. Using homologous, gene coding region probes for titration of individual transcript levels, Mauron et al.[60] have found that H1, H2B and H3 mRNAs are present in an approximate 0.5:1:1 ratio at the mid-blastula stage of embryonic development. Similar results have been obtained by an in vivo labelling procedure in which the accumulation of newly synthesized, embryonic transcripts was determined by gel electrophoresis[63]. Both approaches indicate that the levels of each of the mRNAs encoding core histones are comparable and remain so throughout early development. By contrast, H1 mRNA is approximately half as prevalent as the other histone gene transcripts (though whether this results from a lower rate of synthesis or a higher rate of mRNA decay is presently unknown). Since individual histone mRNA types appear to be translated at equal rates, and since H1 protein is present in a half molar amount relative to the histones of the nucleosome core particle, we may conclude that histone proteins are synthesized with the appropriate stoichiometry for chromatin assembly.

An additional finding to emerge from these studies is the apparent under-representation of H1 mRNA in the unfertilized sea urchin egg. At this stage, H1 transcripts are present at a ten-fold lower concentration that those from H2B and H3 genes[60]. No measurements have been made of egg chromatin histone content so the significance of reduced egg H1 mRNA levels is unclear. In another organism, Xenopus, the relative rate of H1 histone synthesis is known to be reduced in early embryogenesis and Woodland has suggested that embryonic chromatin may be depleted in H1 protein[110].

Synthesis of stoichiometric quantities of the various histone mRNAs does not alone explain how chromatin assembly is achieved during embryogenesis. The oocytes and eggs of organisms such as the frog contain such large maternal stockpiles of histone proteins so that further mechanisms must exist to regulate their orderly use. For example, Xenopus eggs maintain a concentration of histones that will precipitate in vitro under physiological conditions. In a series of studies, Laskey and co-workers[54, 55, 67] have identified a protein factor from extracts of oocytes and eggs that prevents non-specific histone association in vivo and organizes chromatin formation in vitro. This factor, present in the germinal vesicle during oogenesis and released into the cytoplasm during egg maturation, consists of several, tightly bound, 29 000 dalton subunits. The protein complex is highly acidic and appears to function by binding free histones. Chromatin formation involves an interaction of the histone/factor complex with DNA and the activity of a nicking/closing enzyme to permit DNA supercoiling. The 29 000 dalton subunit is itself present in large quantities in the oocyte, comprising 7–10% of total nuclear protein. An analogous chromatin assembly activity has been detected in extracts from early Drosophila embryos[70] though the factor(s) responsible in this organism remain to be purified.

Histone gene structure

During the past decade, recombinant DNA techniques have made possible a detailed examination of histone gene organization and structure in many diverse organisms. In almost every case, histone genes are closely associated within the genome but the nature and extent of such clustering varies widely. Indeed, copy number and topology differ so markedly between even closely related organisms that it has proved impossible to detect any phylogenetic trend in histone gene organization. This suggests that in contrast to the exceptional conservation of polypeptide sequence, histone gene organization shows a remarkably rapid rate of evolutionary change.

Primitive organisms

Table 16.1 summarizes histone gene organization in various organisms, illustrating the evolutionary plasticity of this gene family. In yeast, a representative of the ancestral kingdom Protista, only eight histone gene sequences have been demonstrated. Two copies each of the H2A and H2B genes are arranged in heterotypic, divergently transcribed pairs. Two copies each of the H3 and H4 genes are similarly arranged. All four pairs are located remote from one another[44, 104]. (The lack of a detectable H1 gene is consistent with the absence of any H1 protein in yeast chromatin.) The functional significance of the duplicate H2A/H2B gene pairs is not obvious; the proteins encoded by the H2B genes differ in amino acid sequence but mutant yeast strains lacking one or other functional gene show no impairment of growth[83]. From comparison of the extent of amino acid sequence divergence with the known rate of H2B protein sequence evolution, Grunstein and co-workers have estimated that the H2B genes resulted from duplication of an ancestral gene 190 million years ago. In another protist, the ciliated protozoan, *Stylonichia mytilus* histone genes show no apparent clustering. Preliminary results from Southern blot analysis (using individual heterologous, sea urchin histone gene probes), suggest that this organism contains several copies of each histone gene. The members of each major class (H1, H2A, etc.) are found on a unique set of genomic DNA restriction fragments indicating a dispersed manner of organization[25]. In the nematode, *Caenorhabdits elegans*, histone genes show an irregular, clustered arrangement (Campanelli, Emmons and Childs, personal communication). This simple, acoelomate metazoan possesses approximately 15 histone gene clusters which are distinct in size after digestion of genomic DNA with a restriction enzyme and appear to contain sequences encoding most or all of the five major histones. Thus, regular tandem repetition of clusters is not evident.

Higher eukaryotes

The histone genes of higher eukaryotes are also generally clustered and they can be divided into two categories: those that have an ordered array of gene sequences tandemly repeated many times throughout the genome and those that do not.

The histone genes of the chicken are arranged in random clusters with apparently no one cluster sharing structural homologies with any other[26]. Similarly, mammalian histone genes to date do not show any ordered polarity or uniformity of repetition (mouse[90]; human[42, 88]). On the other hand, the genomes of the fly (*Drosophila melanogaster*), several species of sea urchin and the newt (*Notophthalamus viridescens*) all contain hundreds of tandemly arranged histone gene sets, each tandem unit containing genes for the five histone proteins and adjacent spacer DNA (reviewed in reference 43). Within each species, the repeats are generally uniform, indicating the existence of extremely efficient correction mechanisms that maintain their sequence homogeneity. Curiously, in the newt the tandem repeat units are separated from each other by large (50–60 kb) and variable lengths of satellite DNA. The length of spacer, relative gene order and polarity of transcription differ between these organisms suggesting that each orginated by the evolutionary amplification of a unique histone gene cluster. Alternatively, a single ancestral cluster may have undergone various rearrangements and sequence changes to produce the diversity of fine structures in the histone gene clusters found in different organisms. For example, the differences between the sea urchin and *Drosophila* major repeat units can be explained by a relatively small number of gene rearrangements[56]. However, such an explanation becomes unwieldy when it is applied to the differences between clusters in some other organisms. Thus, the evolution of a single ancestral arrangement to form the very different gene orders and polarities in the sea urchin and the newt would require many rearrangement events.

Evolutionary amplification occurs in a phylogenetically arbitrary manner, since histone gene numbers and organizations may differ drastically in closely related organisms possessing similar life cycles. Consider, for example, the differences in histone gene arrangement amongst the Amphibia. A general trend of increasing copy number with increasing haploid DNA content has been observed[47] but the two are not directly proportional. Other factors are presumably involved. The anuran, *Xenopus laevis*, possesses approximately 40 histone genes organized in several types of dispersed, irregular repeats[69, 115]. In contrast, the histone genes of the urodele, *Notophthalamus viridescens*, are repeated 600 times and organized in regular tandem units[97]. Equally as striking is the divergence between the two closely related species of frog, *Xenopus laevis* and *Xenopus borealis*; most (70%) of the histone gene clusters in the latter are members of a single, highly homogeneous group (P. Turner and H. R. Woodland, personal communication). Clearly, major differences in histone gene number and topology may be established over brief periods of evolutionary time.

What genetic mechanisms account for the rapid evolution of histone gene organization? Comparison of the organization and nucleotide sequence of histone repeats found in two species of sea urchin may provide a clue. Birnstiel and co-workers have examined the gene coding sequences for both a major and minor repeat unit in the species *Psammechinus miliaris*[13]. The predominant repeat, (clone h22) differs from the minor variant (clone h19) by 12.4%. In contrast, the sequence divergence between h19 and the major repeat of another species, *Strongylocentrotus purpuratus*, is as little as 1.74%. Perhaps then, h19 of *P. miliaris* and the major repeat of *S. purpuratus* are derived from a common ancestral histone gene cluster that predated the divergence of the two species. During evolution, amplified copies of this ancestral cluster became stabilized within the genome of one species (*S. purpuratus*), while a different ancestral cluster became amplified in the other (*P. miliaris*). A similar explanation may account for the coexistence of two major repeat structures in another sea urchin species, *L. pictus*[19]; in this case, the repeats differ only in sequence of spacer DNA that surrounds the histone genes and may have arisen from the evolutionary amplification of two different ancestral clusters. A simple mechanism for the expansion and contraction of multigene families through evolution need involve only the processes of gene duplication and unequal crossover[91]. Moreover, such a mechanism may account for the emergence of structurally distinct histone gene sets that acquire their own patterns of expression.

An alternative explanation for the origin of the minor h19 variant of *P. miliaris* has recently been proposed by Busslinger *et al.*[13]. It is their opinion that the ancestral h19 cluster may have been derived from the distantly related *S.*

dröbachiensis which inhabits the same waters as *P. miliaris*. Horizontal gene transfer by some as yet unknown mechanism could have resulted in the incorporation of one or more of the *S. dröbachiensis* histone clusters into the *P. miliaris* genome. Certainly the degree of sequence homology between the H2B/H3 spacer regions of these two clusters is very striking (98.7%) especially considering that these two families diverged 130 million years ago.

Multiple histone gene families: early and late genes in the sea urchin embryo

An example of the differential expression of structurally distinct histone gene sets may be provided by the early and late histone genes in the sea urchin zygote. Indeed, the recent isolation and structural analysis of several DNA clones containing late histone gene sequences has demonstrated the extent to which these two gene sets differ in both sequence organization and genetic fine structure[64] (G. Childs, personal communication). Unlike the tandemly organized early genes, late genes are loosely clustered and heterogeneously arranged. Thus, two of four clones from the *S. purpuratus* genome contain isolated, single genes surrounded by long stretches of non-histone DNA; the remaining two contain individual, heterotypic pairs of late genes (H2B/H2A and H2B/H4). Genomic Southern transfer experiments in which late H2A and H2B histone gene coding regions are used to probe restriction enzyme digests of *S. purpuratus* genomic DNA have confirmed this view of late gene organization. They have also indicated that only about 20 copies of the late H2A and H2B genes are present within the genome. The arrangement and number of late histone genes in the *L. pictus* genome appear to be similar to the case of *S. purpuratus* (G. Childs, personal communication).

The late gene family is heterogeneous not only in overall organization, but also in genetic fine structure. Restriction enzyme mapping and DNA sequence analysis have revealed few similarities between late gene clones outside of gene coding regions[64] (Gormezano *et al.*, unpublished observations). This is true even for those regions of DNA that immediately flank analogous late genes. The coding regions themselves show considerable variation. For example, both the 5′ leader and (to a lesser extent) protein coding regions of three late H2B genes are substantially different in nucleotide sequence.

When during the course of echinoderm evolution did early and late histone gene families diverge? From both DNA and protein sequence comparisions[13, 98] (Gormezano, Mohun, Maxson and Kedes, submitted for publication) we have estimated that the bifurcation of the two families occurred perhaps as early as the Precambrian, prior to the branching of the Echinodermata from the line leading to vertebrate phyla (see Appendix). If this estimate proves accurate, we may expect to detect derivatives from each of the two ancestral histone gene sets in organisms from other lineages. Of course, such derivatives need not all be functional, nor need they show distinct patterns of expression. Differential regulation of early and late genes in sea urchins may be a recently evolved feature common only to echinoderms. In this light, the failure to detect stage-specific synthesis of histone variants in other phyla may not be surprising.

Orphons

An alternative mechanism for the evolution of the histone multigene gene family is suggested by the recent discovery of 'orphons' – single, isolated genes derived from tandem repeats[17]. These genes are presumed to result from a recombination event between the repeat unit and a remote genomic locus. Once removed from the constraints that retain homogeneity amongst the co-evolving, tandem repeats, orphons would be free to diverge in both sequence and function. If, for example, chromosomal location affects gene regulation, then the creation of orphons might result in histone gene derivatives adopting a new spatial or temporal pattern of expression as they become subject to new regulatory controls. Orphon generation and selective gene amplification are not of course mutually exclusive. Both may have contributed to the rapid evolution of histone gene topologies.

It was amongst the orphon population that the first histone genes to contain intervening sequences were discovered. One cloned sea urchin H2B orphon was found to contain a transposon-like element within the coding sequence (Weinthal, J., Childs, G., Maxson, R., Liebermann, D., Kedes, L., manuscript in preparation). Recently, a cloned chicken H3 gene has also been described that contains two introns within its coding sequence[27]. Both introns are delimited by the dinucleotides common to all other intron splice junctions, namely 5'-GT and 3'-AG[11]. Another intervening sequence may be present in the 5'-untranslated leader since there are no 'CCAAT' or 'ATA' consensus promoter sequences[11] within the 873 nucleotides immediately 5' to the initiation codon. Alternatively, this orphon may be a 'pseudogene'. However, the strong hybridization of these sequences to chick poly(A)$^+$ RNA and the absence of internal stop codons within the three exons might suggest that this sequence is capable of being both transcribed and translated.

The primary structure of histone genes and the identification of putatitive regulatory signals is discussed at greater length in Chapter 9.

Function of tandem repeats

The existence of tandem repetition in several species has provoked considerable speculation about the functional significance of the various histone gene topologies. Tandem organization, for example, could solve the problem of coordinate expression for the five histone genes through mechanisms such as polycistronic transcription[52]. Location of the five genes together in a repeat unit could prevent any deleterious effects on histone gene stoichiometry arising from unequal crossover events[30]. Finally, it has also been suggested that tandem repeats facilitate rapid transcription at times of high histone demand[5]. In the light of recent studies none of these earlier suggestions seem tenable.

Polycistronic transcription alone could not account for coordinate expression of the five histone genes since only half as much H1 is synthesized compared with each of the other four proteins. The tandem repeats of both *Drosophila* and *Notophthalamus* contain divergently oriented genes[56, 97], so for these two species, coordinate synthesis of histone mRNAs requires coordinate transcription of both DNA strands. In *Xenopus laevis*, a repeat unit has been isolated that contains two H4 genes; participation of repeats bearing such a disproportionate arrangement of genes in polycistronic transcription would result in a similar imbalance in the stoichiometry of mRNA synthesis. Only in the case of sea urchins are the histone genes arranged in a common orientation and, with the exception of H1, in an appropriate ratio for coordinate transcription[43]. Yet in a careful study, Mauron *et al.*[59] have demonstrated that, at least for the early genes, monocistronic transcription is the physiological mechanism of mRNA synthesis in this organism.

It is attractive to speculate that tandem organization enables histone gene stoichiometry to be maintained through crossover events. However, this proposal could only apply to those organisms with highly conserved repeat units that maintain a uniform and unitary ratio of the various histone genes. Furthermore, gene stoichiometry can obviously be ensured by other means since many organisms show a dispersed arrangement of their histone genes.

Does tandem repetition facilitate rapid histone gene transcription? The measured rate of sea urchin histone gene transcription is approximately one mRNA molecule/gene/minute[62]. Hentschel and Birnstiel[43] have pointed out that this value falls well below the theoretical maximum transcription rate, the rapid accumulation of histone gene transcripts during early development resulting instead from the high repetition frequency of the genes.

We are left therefore with the remaining speculation that tandem organization reflects a form of gene regulation used by histone genes. If so, such a regulatory mechanism is neither necessary nor sufficient to provide the developmentally modulated patterns of histone gene expression described in earlier sections. Not all organisms possess tandem repeats and at least in the case of the sea urchin, these coexist with a dispersed set of late genes that exhibit their own coordinate program of expression[64]. Similar arguments apply when we consider the variation in histone gene copy numbers amongst diverse organisms. It has previously been suggested that high repetition frequencies are associated with rapid rates of embryogenesis. *Table 16.1* indicates that this is untrue. The rate of development in the frog is comparable with that of the sea urchin yet the former contains ten-fold fewer histone genes than the latter. Equally, early development in the chick is rapid, yet only about 20 histone genes are present in its genome. Repetition frequency is probably best viewed as only one element in a mosaic of functionally interdependent and co-evolving factors that includes embryonic cell cleavage rates, the stockpiling of histone mRNA and/or proteins in the egg along with various transcriptional and post-transcriptional regulatory mechanisms; together these factors give rise to the quantitative program of histone gene expression through the life cycle.

Genomic arrangements of histone genes, then, can diverge without corresponding changes in the program of histone gene expression. How are we to interpret this paradoxical result? It is possible that tandem organization of histone genes has no functional significance whatsoever and is merely the consequence of the recombinational forces consequent upon high gene number. Once the number of tandem units reaches a threshold value, unequal crossover between repeats during meiosis would maintain sequence homogeneity, giving rise to a regular and stable, tandem organization[97]. Such a theory could account for the absence of tandem arrangement among organisms that contain relatively small numbers (i.e. tens rather than hundreds) of histone genes. Perhaps the genomic organization of histone genes in these organisms is also without functional significance. On the other hand, the variety of histone gene topologies so far described may equally well testify to the multiplicity of control mechanisms by which histone gene regulation in different organisms is achieved.

Appendix

Calculation of the evolutionary time of divergence of early and late sea urchin H2B histone genes

The recent discovery that mutations accumulate in protein sequences at rates which are nearly constant for any given protein[107] provides a basis for estimating the time at which early and late sea urchin histone genes diverged from an ancestral sequence. Two elements of information are needed to make this calculation: first, the rate of evolution of histone protein sequences and, second, the extent of the differences in sequence between early and late histone genes.

It has been estimated that the H2B protein accumulates 1% of change in amino acid sequence in approximately 60 000 000 years of evolution[107]. This value is based on a comparison of the protein sequences in mammals and bony fish. New information on the rate of H2B sequence evolution within the echinoid lineage agrees closely with this figure. A comparison of the H2B sequence in the *P. miliaris* clone H19 and the *S. purpuratus* clone pSp102 reveals an 0.83% difference in amino acid sequence (one change in 121 amino acids)[13, 98] (M. Strickland, unpublished). These two proteins are functional analogs in the two species. Thus, even though the genes coding for these proteins are repeated in the hundreds, their nucleotide sequences are probably orthologously related and can therefore be used to estimate a rate of sequence evolution. The reader is referred to the article by Wilson *et al.*[107] for a discussion of methods of estimating rates of sequence evolution and Busslinger *et al.*[13] for a description of the relationship between these two gene sequences. Since *S. purpuratus* and *P. miliaris* diverged approximately 65 000 000 years ago[29] (J. Wyatt Durham, personal communication) we conclude that a 1% difference in nucleotide sequence accumulates in 80 000 000 years, an interval which, because of the uncertainty in the echinoderm fossil data, cannot be considered to differ significantly from that calculated by Wilson from mammal and teleost sequence data.

Complete nucleotide sequences of two late H2B histone genes from *S. purpuratus* are now available. These sequences differ from the early H2B of *S. purpuratus* in 14 amino acids out of a total of 121. Thus, if we assume that early and late H2B histone sequences have been evolving at equal rates since they diverged, then the time elapsed since their split is approximately 700 000 000 years (11.6% × 60 000 000 years/1% difference).

References

1. Anderson, K. V. and Lengyel, J. A. (1980). Changing rates of histone mRNA synthesis and turn-over in *Drosophila* embryos. *Cell*, **21**, 717–727
2. Anderson, K. V. and Lengyel, J. A. (1981). Changing rates of DNA and RNA synthesis in *Drosophila* embryos. *Developmental Biology*, **82**, 127–138
3. Arceci, R. and Gross, P. R. (1977). Noncoincidence of histone and DNA synthesis in cleavage cycles of early development. *Proceedings of the National Academy of Sciences, USA*, **74**, 5016–5020
4. Arceci, R. and Gross, P. R. (1980). Histone variants and chromatin structure during sea urchin development. *Developmental Biology*, **80**, 186–209
5. Birnstiel, M., Portmann, R., Busslinger, M., Schaffner, W., Probst, E. and Kressman, A. (1979). Functional organization of the histone genes in the sea urchin *Psammechinus*: A progress report. *Alfred Benson Symposium*, **13**, 117–129

6. Bonner, J., Dahmus, M. E., Fainbrough, D., Huang, R. C., Marushige, K. and Tuan, D. Y. H. (1968). The biology of isolated chromatin. *Science*, **159**, 47–56
7. Borun, T. W., Scharff, M. D. and Robbins, E. (1967). Rapidly labelled polyribosome-associated RNA having the properties of histone messenger. *Proceedings of the National Academy of Sciences, USA*, **58**, 1977–1983
8. Borun, T. W., Gabrielli, F., Ajiro, K., Zweidler, A. and Baglioni, C. (1975). Further evidence of transcriptional and translational control of histone messenger RNA during the HeLa S3 cycle. *Cell*, **4**, 59–67
9. Bownes, M. A. (1975). Photographic study of development in the living embryo of *Drosophila melanogaster*. *Journal of Embryology and Experimental Morphology*, **33**, 789–801
10. Brandhorst, B. P. (1980). Simultaneous synthesis translation and storage of mRNA including histone mRNA in sea urchin eggs. *Developmental Biology*, **79**, 139–149
11. Breathnach, R. and Chambon, P. (1981). Organization and expression of eucaryotic split genes coding for proteins. *Annual Review of Biochemistry*, **50**, 349–383
12. Brown, I. R. (1980). Histone synthesis in isolated neuronal perikaryon relative to the post-natal appearance of a short DNA repeat length. *Developmental Biology*, **80**, 248–252
13. Busslinger, M., Portmann, R., Irminger, J. and Birnstiel, M. (1980). Ubiquitous and gene-specific regulatory 5′ sequences in a sea urchin histone DNA clone coding for histone protein variants. *Nucleic Acids Research*, **8**, 957–977
14. Busslinger, M., Rusconi, S. and Birnstiel, M. L. (1982). An unusual evolutionary behaviour of a sea urchin histone gene cluster. *EMBO Journal*, **1**, 27–33
15. Butler, W. B. and Mueller, G. C. (1973). Control of histone synthesis in HeLa cells. *Biochimica et Biophysica Acta*, **294**, 481–496
16. Childs, G., Maxson, R. and Kedes, L. H. (1979). Histone gene expression during sea urchin embryogenesis: Isolation and characterization of early and late messenger RNAs of *Strongylocentrotus purpuratus* by gene specific hybridization and template activity. *Developmental Biology*, **73**, 153–173
17. Childs, G., Maxson, R., Cohn, R. and Kedes, L. H. (1981). Orphons: Dispersed genetic elements derived from tandem repetitive genes of eukaryotes. *Cell*, **23**, 651–663
18. Cohen, L. H., Newrock, K. M. and Zweidler, (1975). Stage specific switches in histone synthesis during embryogenesis of the sea urchin. *Science*, **190**, 994–997
19. Cohn, R. H. and Kedes, L. H. (1979). Nonallelic histone gene clusters of individual sea urchins (*Lytechinus pictus*): Mapping of homologies in coding an spacer DNA. *Cell*, **18**, 855–864
20. Davidson, E. H. (1968). *Gene Activity in Early Development*. New York, Academic Press
21. Delegeane, A. M. and Lee, A. S. (1981). Coupling of histone and DNA synthesis in the somatic cell cycle. *Science*, **215**, 79–81
22. Deppe, U., Schierenberg, E., Cole, T., Krieg, C., Schmitt, D., Yoder, B., von Ehrenstein, G. (1978). Cell lineages of the embryo of the nematode *Caenorhabditis elegans*. *Proceedings of the National Academy of Sciences, USA*, **75**, 376–380
23. Detke, S., Stein, J. L. and Stein, G. S. (1978). Synthesis of histone mRNAs by RNA polymerase II in nuclei from S phase HeLa S3 cells. *Nucleic Acids Research*, **5**, 1515–1528
24. Detke, S., Lichtler, A., Phillips, I., Stein, J. and Stein, G. (1979). Reassessment of histone gene expression during cell cycle in human cells by using homologous H4 histone DNA. *Proceedings of the National Academy of Sciences, USA*, **76**, 4995–4999
25. Elsevier, S. M., Lipps, H. J. and Steinbruck, G. (1978). Histone genes in macronuclear DNA of the ciliate *Stylonichia mytilus*. *Chromosoma*, **69**, 291–306
26. Engel, J. D. and Dodgson, J. B. (1981). Histone genes are clustered but not tandemly repeated in the chicken genome. *Proceedings of the National Academy of Sciences, USA*, **78**, 2856–2860
27. Engel, J. D., Sugarman, B. J. and Dodgson, J. B. (1982). A chick histone H3 gene contains intervening sequences. *Nature, London*, **297**, 434–436
28. Eyal-Giladi, H. and Kochav, S. (1976). From cleavage to primative streak formation: a complementary normal table and a new look at the first stages of the development of the chick. I. Normal development. *Developmental Biology*, **49**, 321–337
29. Fell, H. B. (1966). *Echinodermata*, pages U374. Kansas, Lawrence
30. Finnegan, D., Rubin, G., Young, M. and Hogness, D. (1977). Repeated gene families in *Drosophila melanogaster*. *Cold Spring Harbor Symposium on Quantitative Biology*, **42**
31. Gallwitz, D. (1975). Kinetics of inactivation of histone mRNA in the cytoplasm after inhibition of DNA replication in synchronized HeLa cells. *Nature, London*, **257**, 247–248
32. Gallwitz, D. and Brendl, M. (1972). Synthesis of histones in a rabbit reticulocyte cell-free system directed by a polyribosomal RNA fraction from synchronized HeLa cells. *Biochemical and Biophysical Reasearch Communications*, **47**, 1106–1111

33. Gallwitz, D. and Mueller, G. (1969). Histone synthesis *in vitro* on HeLa cell microsomes. *Journal of Biological Chemistry*, **244**, 5947–5952
34. Giudice, G. (1973). *Developmental Biology of the Sea Urchin*. New York, Academic Press
35. Goustin, A. S. (1981). Two temporal phases for the control of histone gene activity in cleaving sea urchin embryos *S. purpuratus*. *Developmental Biology*, **87**, 163–175
36. Graham, C. F. and Morgan, R. W. (1966). Changes in the cell cycle during early amphibian development. *Developmental Biology*, **14**, 439–460
37. Groppi, V. E. Jr. and Coffino, P. (1980). G1 and S phase mammalian cells synthesize histones at equivalent rates. *Cell*, **21**, 195–204
38. Grunstein, M. (1978). Hatching in the sea urchin *Lytechinus pictus* is accompanied by a shift in H4 gene activity. *Proceedings of the National Academy of Sciences, USA*, **75**, 4135–4139
39. Gurley, L. R. and Hardin, J. M. (1969). The metabolism of histone fractions II: Conservation and turnover of histone fractions in mammalian cells. *Archives of Biochemistry and Biophysics*, **130**, 1–6
40. Gurley, L. R., Walters, R. A. and Tobey, R. (1972). The metabolism of histone fractions IV: Synthesis of histones during the G1 phase of the mammalian life cycle. *Archives of Biochemistry and Biophysics*, **148**, 633–641
41. Hecht, R. M., Schomer, D. F., Oroja, J., Bartel, A. H. and Hungerford, E. D. (1981). Simple adaptations to extend the range of flow cytometry 5 orders of magnitude of DNA analysis of uni- and multi-cellular systems. *Journal of Histology and Cytology*, **29**, 771–774
42. Heintz, N., Zernik, M. and Roeder, R. G. (1981). The structure of the human histone genes: clustered but not tandemly repeated. *Cell*, **24**, 661–668
43. Hentschel, C. and Birnstiel, M. (1981). The organization and expression of histone gene families. *Cell*, **25**, 301–313
44. Hereford, L., Fahrner, K., Woolford, Jr. J., Rosbash, M. and Kaback, D. B. (1979). Isolation of yeast histone genes H2A and H2B. *Cell*, **18**, 1261–1271
45. Hereford, L. M., Osley, M. A., Ludwig, J. R. and McLaughlin, C. S. (1981). Cell cycle regulation of yeast histone mRNA. *Cell*, **24**, 367–376
46. Hieter, P. A., Hendricks, M., Hemminki, K. and Weinberg, E. S. (1979). Histone gene switch in the sea urchin embryo. Identification of late embryonic histone mRNAs and the controls of their synthesis. *Biochemistry*, **18**, 2707–2715
47. Hilder, V. A., Livesey, R. N., Turner, P. C. and Vlad, M. T. (1981). Histone gene number in relation to C-value in amphibians. *Nucleic Acids Research*, **9**, 5737–5746
48. Hinegardner, R. T. (1967). Echinoderms. In *Methods in Developmental Biology*, pp. 139–155. New York, Crowell
49. Horstadius, S. (1973). *Experimental Embryology of Echinoderms*. Oxford, Clarendon Press
50. Isenberg, I. (1979). Histones. *Annual Review of Biochemistry*, **48**, 159–191
51. Jacob, E. (1976). Histone gene reiteration in the genome of the mouse. *European Journal of Biochemistry*, **65**, 275–284
52. Kedes, L. H. (1975). Histone messengers and histone genes. *Cell*, **8**, 321–331
53. Kemper, B. and Rich, A. (1973). Identification of presumptive histone mRNA in rapidly labelled polysomal RNA of mouse myelomas. *Biochimica et Biophysica Acta*, **319**, 364–372
54. Laskey, R. A., Mills, A. D. and Morris, N. R. (1977). Assembly of SV40 chromatin in a cell free system from *Xenopus* eggs. *Cell*, **10**, 237–243
55. Laskey, R. A., Honda, B. M., Mills, A. D. and Finch, J. T. (1978). Nucleosomes are assembled by an acidic protein which binds histones and transfers them to DNA. *Nature, London*, **275**, 416–420
56. Lifton, R. P., Goldberg, M. L., Karp, R. W. and Hogness, D. S. (1977). The organization of the histone genes in *Drosophila melanogaster*: Functions and evolutionary implication. *Cold Spring Harbor Symposium on Quantitative Biology*, **42**, 1047–1051
57. Littlefield, J. N. and Jabos, P. S. (1965). The relationship between DNA and protein synthesis in mouse fibroblasts. *Biochimica et Biophysica Acta*, **108**, 652–658
58. Marashi, F., Baumbach, L., Rickles, R., Sierra, F., Stein, J. L. and Stein, G. S. (1982). Histone proteins in HeLa S3 cells are synthesized in a cell-cycle stage specific manner. *Science*, **215**, 683–685
59. Mauron, A., Levy, S., Childs, G. and Kedes, L. H. (1981). Monocistronic transcription is the physiological mechanism of sea urchin embryonic histone gene expression. *Molecular Cellular Biology*, **1**, 661–671
60. Mauron, A., Kedes, L., Hough-Evans, B. and Davidson, E. H. (1982). Accumulation of individual histone mRNAs during embryogenesis of the sea urchin *Strongylocentrotus purpuratus*. *Developmental Biology*, **94**, 435–440
61. Maxson, R. E. (1978). PhD Thesis, University of California, Berkeley

62. Maxson, R. E. and Wilt, F. H. (1981). The rate of synthesis of histone mRNA during the development of sea urchin embryos *Strongylocentrotus purpuratus*. *Developmental Biology*, **83**, 380–386

63. Maxson, R. and Wilt, F. H. (1982). Accumulation of the early histone mRNAs during the development of *S. purpuratus*. *Developmental Biology*, **94**, 425–434

64. Maxson, R., Mohun, T., Gormezano, G., Childs, G. and Kedes, L. H. (1983). Sea urchin early and late histone genes are differently arranged. *Nature*, **301**, 120–125

65. McGhee, J. and Felsenfeld, G. (1980). Nucleosome structure. *Annual Review of Biochemistry*, **49**, 1115–1156

66. Melli, M., Spinelli, G. and Arnold, E. (1977). Synthesis of histone messenger RNA of HeLa cells during the cell cycle. *Cell*, **12**, 167–174

67. Mills, A. D., Laskey, R. H., Black, P. and DeRobertis, E. M. (1980). An acidic protein which assembles nucleosomes *in vitro* is the most abundant protein in *Xenopus* oocyte nuclei. *Journal of Molecular Biology*, **139**, 561–567

68. Moll, R. and Wintersberger, E. (1976). Synthesis of yeast histones in the cell cycle. *Proceedings of the National Academy of Sciences, USA*, **73**, 1863–1867

69. Moorman, A. F. M., DeLaaf, R. T. M., Destree, O. H. J., Telford, J. and Birnstiel, M. L. (1980). Histone genes from *Xenopus laevis*: molecular cloning and initial characterization. *Gene*, **10**, 185–193

70. Nelson, T., Hsieh, T. S. and Brutlag, D. (1979). Extracts of *Drosophila* embryos mediate chromatin assembly *in vitro*. *Proceedings of the National Academy of Sciences, USA*, **76**, 5510–5514

71. Newrock, K., Cohen, L., Hendricks, M. Donnelly, R. and Weinberg, E. (1978). Stage-specific mRNAs coding for subtypes of H2A and H2B histones in the sea urchin embryo. *Cell*, **14**, 327–336

71a. Nieuwkoop and Faber (1967). A normal table of *Xenopus laevis*. *Daudin*. Amsterdam, North Holland

72. Osley, M. A. and Hereford, L. M. (1981). Yeast histone genes show dosage compensation. *Cell*, **24**, 377–384

73. Park, W. D., Stein, J. L. and Stein, G. S. (1976). Activation of *in vitro* histone gene transcription from HeLa S3 chromatin by S-phase nonhistone chromosomal proteins. *Biochemistry*, **15**, 3296–3300

74. Parker, M. L. and Fitschen, W. (1980). Histone mRNA metabolism during the mouse fibroblast cell cycle. *Cell Differentiation*, **9**, 23–30

75. Pederson, T. and Robbins, E. (1970). Absence of translational control of histone synthesis during the HeLa cell life cycle. *Journal of Cell Biology*, **45**, 509–513

76. Perry, R. P. and Kelley, D. E. (1973). Messenger RNA turn-over in mouse L-cells. *Journal of Molecular Biology*, **79**, 681–696

77. Pringle, J. R. and Mor, J. (1975). Methods for monitoring the growth of yeast cultures and for dealing with the clumping problem. *Methods in Cell Biology*, (Prescott, D. M. Ed.), pp. 131–168. New York, Academic Press

78. Rabinowitz, M. (1941). Studies on the cytology and early embryology of the eggs of *Drosophila melanogaster*. *Journal of Morphology*, **69**, 1–51

79. Raff, R. (1980). In *Cell Biology*, (Goldstein, L., Prescott, D. M., Eds), New York, Academic Press

80. Rickles, R., Marashi, F., Sierra, F., Clark, S., Wells, J., Stein, J. and Stein, G. (1982). Analysis of histone gene expression during the cell cycle in HeLa cells using cloned human histone genes. *Proceedings of the National Academy of Sciences, USA*, **79**, 749–753

81. Robbins, E. and Borun, T. W. (1967). The cytoplasmic synthesis of histones in HeLa cells and its temporal relationship to DNA replication. *Proceedings of the National Academy of Sciences, USA*, **57**, 409–416

82. Ruderman, J. V. and Gross, P. R. (1974). Histones and histone synthesis in sea urchin development. *Development Bioilogy*, **36**, 286–298

83. Rykowski, M., Wallis, J., Choe, J. and Grunstein, M. (1981). Histone H2B subtypes are dispensible during the yeast cell cycle. *Cell*, **25**, 477–487

84. Sadgopal, A. and Bonner, J. (1969). The relationship between histone and DNA synthesis in HeLa cells. *Biochimica et Biophysica Acta*, **186**, 349–357

85. Salik, J., Herlands, L., Hoffman, H. P. and Piccia, D. (1981). Electrophoretic analysis of the stored histone pool in unfertilized sea urchin eggs: Quantification and identification by antibody binding. *Journal of Cell Biology*, **90**, 385–393

86. Sharp, P. A. (1981). Speculations on RNA splicing. *Cell*, **23**, 643–646

87. Sheinan, R. and Lewis, P. N. (1980). DNA and histone synthesis in mouse cells which exhibit temperature-sensitive DNA synthesis. *Somatic Cell Genetics*, **6**, 225–239

88. Sierra, F., Lichtler, A., Marashi, F., Rickles, R., Van Dyke, T., Clark, S., Wells, J., Stein, G. and Stein, J. (1982). Organization of human histone genes. *Proceedings of the National Academy of Sciences, USA*, **79**, 1795–1799

89. Simpson, R. (1981). Modulation of nucleosome structure by histone subtypes in sea urchin embryos. *Proceedings of the National Academy of Sciences, USA*, **78**, 6803–6807

90. Sittman, D. B., Chiv, I.-M., Pan, C.-J., Cohn, R. H., Kedes, L. H. and Marzluff, W. F. (1981). Isolation of two clusters of mouse histone genes. *Proceedings of the National Academy of Sciences, USA*, **78**, 4078–4082

91. Smith, G. (1976). Evolution of repeated DNA sequences by unequal crossover. *Science*, **191**, 528–535

92. Smith, H. O. and Birnstiel, M. L. (1976). A simple method for DNA restriction site mapping. *Nucleic Acids Research*, **3**, 2387–2397

93. Snell, G. D. (1941). *Biology of the Laboratory Mouse*. New York, Dover

94. Spalding, J., Kaijwara, K. and Mueller, G. C. (1966). An extracted basic protein isolated from HeLa nuclei and resolved by electrophoresis. *Proceedings of the National Academy of Sciences, USA*, **56**, 1535–1542

95. Stein, J. L., Thrall, C. L., Park, W. D., Mans, R. J. and Stein, G. S. (1975). Hybridization analysis of histone mRNA: Association with polyribosomes during the cell cycle. *Science*, **189**, 557–558

96. Stein, G. S., Park, W. D., Thrall, C. L., Mans, R. J. and Stein, J. L. (1975). Regulation of cell cycle stage-specific transcription of histone genes from chromatin by non-histone chromosomal proteins. *Nature, London*, **257**, 764–767

97. Stephenson, E., Erba, H. and Gall, J. (1981). Histone gene clusters of the newt *Notophthalamus* are separated by long tracts of satellite DNA. *Cell*, **24**, 639–647

98. Sures, I., Lowry, J. and Kedes, L. H. (1978). The DNA sequence of sea urchin (*S. purpuratus*) H2A, H2B and H3 histone coding and spacer regions. *Cell*, **15**, 1033–1044

99. Tarnowska, M. A., Baglioni, C. and Basilico, C. (1978). Synthesis of H1 histones by BHK cells in G1. *Cell*, **15**, 163–171

100. Tobin, A. (1979). Evaluating the contribution of post-transcriptional processing to differential gene expression. *Developmental Biology*, **68**, 47–58

101. Van Dongen, W., De Laaf, L., Moorman, A. and Destree, O. (1981). The organization of histone genes in the genome of *Xenopus laevis*. *Nucleic Acids Research*, **9**, 2297–2311

102. Van Dongen, W., Zaal, R., Moorman, A. and Destree, O. (1981). Quantitation of the accumulation of histone messenger RNA during oogenesis in *Xenopus laevis*. *Developmental Biology*, **86**, 303–314

103. Venezky, D., Angerer, L. and Angerer, R. (1981). Accumulation of histone repeat transcripts in the sea urchin egg pronucleus. *Cell*, **24**, 385–391

104. Wallis, J., Hereford, L. and Grunstein, M. (1980). Histone H2B genes of yeast encode two different proteins. *Cell*, **22**, 799–805

105. Wilkes, P. R., Birnie, G. D. and Old, R. W. (1978). Histone gene expression during the cell cycle studied by *in situ* hybridization. *Experimental Cell Research*, **115**, 441–444

106. Wilson, C. M., Melli, M. and Birnstiel, M. (1974). Reiterated frequency of histone coding sequences in man. *Biochemical and Biophysical Research Communications*, **61**, 404–409

107. Wilson, A. C., Carlson, S. S. and White, T. J. (1977). Biochemical Evolution. *Annual Review of Biochemistry*, **46**, 573–640. Palo Alto CA

108. Wilt, F. H. (1970). The acceleration of RNA synthesis in cleaving sea urchin embryos. *Developmental Biology*, **23**, 444–455

109. Witschi, E. (1956). *Development of Vertebrates*. Philadelphia, W. B. Saunders

110. Woodland, H. R. (1980). Histone synthesis during the development of *Xenopus*. *FEBS Letters*, **121**, 1–7

111. Woodland, H. R., Flynn, J. M. and Wyllie, A. J. (1979). Utilization of stored mRNA in *Xenopus* embryos and its replacement by newly synthesized transcripts: Histone H1 synthesis using interspecies hybrids. *Cell*, **18**, 165–171

112. Wu, R. S. and Bonner, W. M. (1981). Separation of basal histone synthesis from S-phase histone synthesis in dividing cells. *Cell*, **27**, 321–330

113. Zalokar, M. (1976). Autoradiographic study of protein and RNA formation during early development of *Drosophila* eggs. *Developmental Biology*, **49**, 425–437

114. Zalokar, M. and Erk, I. (1976). Division and migration of nuclei during early embryogenesis in *Drosophila melanogaster*. *Journal de Microscopie Biologie Cellulaire*, **25**, 97–106

115. Zernik, M., Heintz, N., Boime, I. and Roeder, R. (1980). *Xenopus laevis* histone genes: Variant H1 genes are present in different clusters. *Cell*, **22**, 807–815

116. Zlatanova, J. and Swetly, P. (1978). Uncoupled synthesis of histones and DNA during Friend cell differentiation. *Nature, London*, **276**, 276–277

Chapter 17

Globin genes: their structure and expression

R. A. Flavell and F. G. Grosveld

Introduction

Of all the gene systems that have served as models for the study of eukaryotic gene structure and expression, surely the globin genes have been among the most useful. This is explained by a number of facets of the globin system:

1. The globins are major gene products in animals; they are also accessible to analysis by their ready presentation in blood.
2. As a result of this, the biology and biochemistry of the globin system has advanced rapidly: by the mid-seventies the primary structure of many globin proteins and their mRNAs was already known.
3. The globin genes exhibit many of the aspects of eukaryotic gene expression that interest us: their expression is tissue-specific and developmental stage-specific. Moreover, the expression of globin genes must be coordinated. At any one time, the level of α-like and β-like globin must be equal otherwise the excess free α- or β-chains aggregate and cause early red cell death[52].
4. Finally, inherited diseases which result from defects in β-globin gene expression are common in the human population. The frequency of these cases is the consequence of the fact that a reduction in the level of α- or β-globin relative to its complementary polypeptide chain gives rise to red cells which are more resistant to malaria. The study of these defective globin genes is of value both as a model for the molecular basis of genetic defects and for the study, and hopefully in the long term, diagnosis and therapy of these specific diseases.

Haemoglobin biosynthesis

The accumulation of haemoglobin accompanies the differentiation of erythroid precursor cells from the pluripotent stem cell (capable of generating lymphocytes, granulocytes and macrophages, megakaryocytes and erythrocytes) to the erythrocyte. Around the proerythroblast stage the globin genes are switched on and globin mRNA accumulates; by the reticulocyte stage most of the poly(A)$^+$ mRNA in the cells is globin mRNA; and in the final product, the erythrocyte, haemoglobin is by far the major protein.

The site of erythropoiesis changes during development. In the early mammalian embryo erythropoiesis occurs in the yolk sac while in the mid-gestation embryo erythropoiesis switches to the fetal liver and gradually to the bone marrow. In the normal adult mammal, erythropoiesis is restricted to the bone marrow. The type of haemoglobin chains produced depends on the developmental stage of the organism. In practically every higher organism, early embryos express so-called embryonic haemoglobins. In the case of man, ζ is the embryonic α-globin and ε is the embryonic β-globin. In some species, notably the primates, so-called fetal globins are expressed in the latter stages of embryonic life. Thus, man expresses a fetal γ-globin chain, the fetal β-like globin, during mid to late fetal life. Other animals, however, switch directly from embryonic to adult β-globins; and in the α-globin gene family this, to our knowledge, always occurs.

Globin gene structure

The essence of the structure of the globin gene was elucidated by two groups independently in 1977[26, 45]. As has now transpired for most cellular genes, globin genes are split, in this specific case by the presence of (usually) two introns which are found at a constant position with respect to the coding sequence (see, for

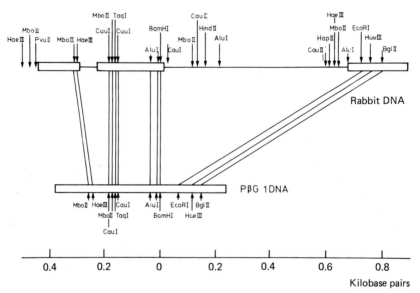

Figure 17.1. A schematic drawing of the structure of a typical animal globin gene, in this case, the rabbit β-globin gene. The genomic DNA is compared with the cDNA copy of rabbit β-globin mRNA. The exons are shown as blocks and the introns as connecting lines.

example, *Figure 17.1*). In the case of leghaemoglobin, however, a third intron is present[36]. As in other systems, the position of introns correlates with the discontinuities in the protein molecule that mark the boundaries of the domains of the protein[8]. This topic is given ample coverage and will not therefore be discussed here. DNA sequence comparisons have shown that the introns have accumulated significantly more base substitutions than coding sequences[11, 49]. It is therefore believed that most of the intronic DNA does not play a major role in gene expression. This is substantiated by the fact that much intronic DNA can be deleted without affecting RNA splicing.

Globin gene linkage

Direct evidence for linkage of globin genes was first published in 1978[13, 30]. We now know that the α- and β-related globin genes generally form two groups of closely linked genes. In most animals these are found on separate chromosomes (for example, in man the α-genes are on chromosome 16 while the β-genes are found on chromosome 11. In *Xenopus*, however, the α- and β-globin genes are closely linked[27]). The order of genes on the linkage map generally reflects the order of expression during development. Thus, in man the β-related globin genes are 5′ ε-GγAγ-δ-β 3′ and all genes are transcribed from the same DNA strand (*Figure 17.2*). This is not so in chicken[10] and mouse[12].

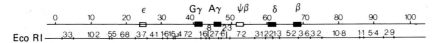

Figure 17.2. The structure of the human β-globin locus. Each of the genes is transcribed from left to right. The bottom line represents the Eco RI restriction map; introns are not indicated.

Intergenic distances are large compared with the size of genes themselves. For example, in man the distances are ε, 13.5 kb; Gγ, 3.5 kb; Aγ, 13.5 kb; δ, 5,5 kb β?. Each globin gene, including introns, is about 1.5 kb in length.

Repetitive DNA within clusters

Much of the intergenic DNA is repetitive. These structures have been partially mapped in the case of the rabbit[23, 41], mouse[22] and human β-globin loci (J. Groffen, F. Grosveld, A. de Kleine and R. A. Flavell, in preparation). Interestingly, A. de Kleine and F. Grosveld in our laboratory have recently shown that the position of some of the repeated sequences is conserved in evolution while others are specific to only one of the given species (unpublished data). Of the repeated sequences in the human genome, the so-called Alu family has received most attention. About 300 000 of these sequences, each approximately 200 bp in length, are present in the human genome and they have a set of intriguing properties:

1. They are homologous to an abundant class of small nuclear RNAs and thus at least some of the repeats are transcribed. In fact, *in vitro* many of these repeats can be transcribed by RNA polymerase III.
2. They also contain a sequence that is homologous to the origin of replication region of papova virus genomes. Attempts to show that Alu sequences can serve as origins of replication have, however, failed.

Recently, much attention has been given to the notion that Alu sequences are eukaryotic transposons; this is, however, beyond the scope of this article (see discussion in Chapter 15).

Pseudogenes

The availability of large cloned fragments from a given gene region also makes it possible to ask whether other mRNA coding sequences are linked to the genes under analysis. When short DNA fragments from the human β-globin gene regions were hybridized with probes for the globin gene coding sequences, additional fragments were detected as well as the known globin gene fragments. Since all the human β-related globins were apparently accounted for, the possibility arose that these globin gene-related sequences were defective genes which are not expressed and that had hitherto remained undetected. These genes have therefore been called pseudogenes in accordance with the terminology first used for a defective *Xenopus* 5S RNA gene which lacked its 3′ terminal sequences. This view has been confirmed in a number of cases by detailed DNA sequence analysis. Thus, there are two human α-related pseudogenes that are defective. The first ψα$_1$ is 70–80% homologous to the adult α-globin gene throughout its sequence, but it has a deletion of 20 nucleotides in the middle of the gene that puts the coding sequence out of phase[40]. The second, ψζ is homologous to the embryonic ζ-gene but it has a termination codon at amino acid 6 (N. Proudfoot, personal communication).

Several other pseudogenes have been characterized. Except for the fact that they are defective, they have few common structural similarities. Thus, the entire DNA sequence has been determined for a mouse pseudo α-globin gene[35, 51]. This gene lacks both introns and has several additional short deletions and insertions which serve to add stop codons in all reading frames. Pseudogenes have also been described in the β-related globin genes and at this stage it looks as if they will be a structural feature of all globin gene clusters and by extrapolation, all gene families. If we assume that the proportion of pseudogenes to active genes in other gene systems is similar to that for the globin genes, then it would follow that there are approximately equal numbers of pseudogenes and structural genes present in higher organisms.

Surprisingly, the mouse α-pseudogene of Nishioka *et al.*[35] and Vanin *et al.*[51] is localized on a different chromosome to that of the mouse α-globin gene locus[31]. This raises the interesting possibility that this gene was transposed to its new position. The proximity of retrovirus-related 'A-type particle' genomes linked to this α-globin pseudogene is suggestive evidence for this idea. Specifically, Leder *et al.*[31] suggest that the retroviral genome was inserted 5′ to the productive α-globin gene. This viral promoter then initiated the synthesis of RNA which was processed by removal of the introns; the viral reverse transcriptase could convert the RNA to cDNA and the cDNA could be inserted back into the genome. This event would need to occur (ultimately) in the germ line of the mouse to ensure the inheritance of the pseudogene in the mouse population.

Why are there pseudogenes? One view is that they represent the defective products of gene duplication and that they have no function. There is little evidence against such an idea. If this were so, then we would expect pseudogenes to be relatively recent in evolutionary terms; after longer times they would be lost by deletion. Alternatively, it could be argued that pseudogenes serve an important function. In this case, it is reasonable to expect them to be conserved throughout evolution. A detailed comparison of pseudogenes from related organisms should provide a solution to this problem.

Globin gene transcription – the promoter

A variety of transcription mapping experiments have shown that the globin genes are transcribed by the initiation of RNA synthesis at the nucleotide corresponding to the 5′ end of mature β-globin mRNA[14, 19, 53]. The question of the termination of RNA synthesis is more controversial. In adenovirus and papova viruses it is clear that transcription can and does proceed beyond a given poly(A) addition site[34]. Some data in favour of this are available for the mouse β-major-globin gene. Sequences complementary to the 3′ extragenic regions of the β-major-globin gene have been detected on pulse-labelled RNAs of a size expected for β-globin pre-mRNA[24].

The comparison of DNA sequences in the 5′ regions of a number of genes has shown the presence of two conserved sequences, the ATA box ($^{T}_{C}$ATA$^{T}_{A}$AAG)[17, 39] and the CAAT box (GG$^{T}_{C}$CAATCT)[3, 11].

From the analysis of a number of deletion mutants the following can be deduced:

1. The ATA box is required for efficient transcription *in vivo* and *in vitro*[2, 18, 19, 20]. Deletion of the ATA box causes a loss of transcriptional specificity *in vivo*; initiation of RNA synthesis occurs at a large number of sites instead of the natural 'cap' site preferred with the unmutated gene.
2. Mutations in and around the CAAT box cause a decrease in the efficiency of transcription.

Deletion of the region which contains the CAAT box causes a decrease in the steady state level of rabbit β-globin mRNA[9, 29]. Grosveld *et al.*[20, 21] studied the effect of short internal deletions of DNA in the CAAT box region using the HeLa cell SV40 system of Banerji *et al.*[1]. In this system a given gene, in this case a rabbit β-globin gene, is coupled to a plasmid which contains a segment of SV40 that carries the 72 base pair repeat region located close to the origin of replication. Under these conditions, the globin gene is expressed from its own promoter. A summary of the deletions examined and the transcription efficiencies of the respective mutants is shown in *Figure 17.3*. Deletions in the conserved CAAT box cause a reduction of five-fold to twelve-fold in transcription efficiency. Interestingly, deletion of the two conserved G residues at the 5′ end of the CAAT box causes a strong reduction in the level of β-globin transcripts[21]. Deletions to the 5′ side of this structure show an even more pronounced effect on transcription levels with down mutations of the order of fifty-fold.

P. Diercks, C. Weissmann and colleagues (1982, unpublished) have also carried out an extensive study of both point mutations and deletions in the rabbit β-globin promoter using a similar expression system based upon polyoma virus. These results also establish a role of the ATA and CAAT sequences in the promotion of RNA synthesis. Moreover, their data point to the presence of three C-rich elements which have an important role in the promoter centered around −100 (*Figure 17.3*). Sequential deletion of these structures causes a stepwise progressive decrease in transcriptional efficiency[9].

The picture that emerges for the β-globin promoter is that of a tripartite structure which consists of a C-rich region around −100 to −85, the CAAT box at −75 and the ATA box at −30. It is, of course, not clear how the RNA polymerase and its accessory factors interact with these structures; for example, whether it sees a single 'structure' or interacts with each separate sequence sequentially. Another fundamental point remains to be answered. Transcription of the globin gene in

Sequence	Deletion range	Transcription level
ACAGGGGTGCTGTCATCACCCAGACCTCACCCTGCAGAGCCACACCCTGGTGTTGGCCAATCTACACACGGGGTAGGGATT		1
ACAGGGGTGCTGTCATCACCCAGACCTCACCCTGCAGAGCCACACCCTGGTGTT · CCAATCTACACACGGGGTAGGGATT	−76 to −77	0.12
ACAGGGGTGCTGTCATCACCCAGACCTCACCCTGCAGAGCCACACCCTGG · AATCTACACACGGGGTAGGGATT	−74 to −81	0.2
ACAGGGGTGCTGTCATCACCCAGACCTCACCCTGCAGAGCCACACCCTGGTGTTGGCC · ACACACGGGGTAGGGATT	−69 to −73	0.2
ACAGGGGTGCTGTCATCACCCAGACCTCACCCTGCAGAGCCACACCCTGGTG · TCTACACACGGGGTAGGGATT	−72 to −79	0.08
ACAGGGGTGCTGTCATCACCCAGACCTCACCCTGCAGAGCCACACCCT · GTAGGGATT	−60 to −83	0.25
ACAGGGGTGCTGTCATCACCCAGACCTCACCCTGCAGAGCCACACCC · GGTAGGGATT	−61 to −84	0.15
ACAGGGGTGCTGTCATCACCCAGACCTCACCCTGCAG · GCCAATCTACACACGGGGTAGGGATT	−76 to −93	0.02
ACAGGGGT · CGGGGTAGGGATT	−64 to −123	0.005
ACAGGGGTGCTGTCATCAC · CCCTGGTGTTGGCCAATCTACACACGGGGTAGGGATT	−88 to −122	0.2
ACAGGGGTGCTGTCATCACCCAGACCT · AGCCACACCCTGGTGTTGGCC · ACACACGGGGTAGGGATT	−94 to −104	0.1
ACAGGGGTGCTGTCATCAC · AGAGCCACACCCTGGTGTTGGCC · ACACACGGGGTAGGGATT	−97 to −112	0.1
ACAGGGGTGCTGTCATCACCCAGACCTCACCCTGCAGAGCCACACCCTGGTGTTGGCCAATCTACACACGGGGTAGGGATT		0.4
GCTGCAGC (insertion)		0.7
CCAAGCTTGG (insertion)		0.25
CCAAGCTTGGCCAAGCTTGG (insertion)		0.25
CCAAGCTTGGCCAAGCTTGGCCAAGCTTGGCCAAGCTTGG (insertion)		0.2

Position scale: −130   −120   −110   −100   −90   −80   −70   −60   −50

Figure 17.3. The transcription efficiencies of deletion and insertion mutants in the −80 region of the rabbit β-globin gene. The sequence in the −80 region of each mutant template is presented, and the gaps show the DNA residues deleted. Transcription level was determined by scanning the autoradiographs of the S_1 mapping experiments and correcting the signal for the deletion mutants by reference to the internal human β-globin gene control.

these cell systems requires the presence of the SV40 or polyoma enhancer sequence. To date, however, nobody has discovered the cellular counterpart of this DNA structure which would (hypothetically) potentiate globin gene expression in the red cell.

Sequences required for splicing

The DNA sequences at the 5' and 3' splice junctions show sequence conservation which reflects the requirements for recognition, cleavage and ligation of the pre-mRNA by the splicing enzymes. The comparison of numerous splice junctions from many different genes has enabled the derivation of consensus sequences of $^C_AAG/GT^G_AAGT$ and $(^C_T)_nN^T_CAG/G$ (where $n \geqslant 11$) for the 5' and 3' splice junctions respectively. Alteration of the consensus sequences of the rabbit β-globin gene by deletion or site-directed mutagenesis has shown the following[56]:

1. Deletion of the 5' splice site in IVS2 eliminates the normal splicing at this position. Instead, a cryptic splice site within exon 2 is used in these mutants.
2. Alteration of the consensus donor GT → AT eliminates splicing at this position and the generation of three novel mRNAs. Again these result from splicing to three different cryptic splice sites within exon 2. Two of these contain the

Consensus	$^C_AAG/GT^A_GAGT$
Normal IVS2	AGG/GTGAGT
Cryptic 1	TGA/GTTTGG
Cryptic 2	TAA/GCTGAG
Cryptic 3	AAG/GTGAAG

Figure 17.4. DNA sequences at the splice sites used in normal and mutant rabbit β-globin genes. See text for details.

expected GT but one of these contains a GC dinucleotide. In the latter case, Wieringa *et al.*[56] suggest that the C may be looped out to give a reasonable splice consensus sequence G^CUGAGU. The cryptic sites present in the exons show a relatively poor match with the consensus sequences (*Figure 17.4*). This presumably reflects the fact that these are poor splice sites, which would be advantageous to the organism since a strong splice site within an exon would cause wastage of mRNA by aberrant splicing.

Genetic defects in globin gene expression: the thalassaemias

We can divide genetic lesions in human globin gene expression into two groups. The first group of mutations lie within the structural gene and cause an altered globin protein to be produced. Numerous examples of this type of mutation have been described, the best known of which is probably sickle cell haemoglobin which results from a mutation in the codon for the sixth amino acid of the β-globin chain. The second group of mutations in the diseases collectively called the thalassaemias do not affect the structure of the globin protein itself; rather, they cause the level of otherwise normal globin to be reduced, in some cases, to undetectable levels.

Naturally, thalassaemic lesions occur in both the α- and β-globin genes. Since a large number of lesions occur and since there are in general similarities between the lesions in the α- and β-thalassaemias, we shall restrict ourselves to the β-thalassaemias. Several good reviews of the α-thalassaemias have been published recently[52].

$(\delta\beta)^0$ thalassaemia and HPFH

A heterogeneous collection of diseases grouped under the heading $(\delta\beta)$ thalassaemia and Hereditary Persistence of Fetal Haemoglobin (HPFH) have a common and interesting phenotype. In both cases the fetal globin genes are expressed at elevated levels with HPFH showing the highest levels of haemoglobin F. The nature of the genetic lesions in these disorders has been determined in many cases by Southern blotting. A summary of these results is shown in *Figure 17.5*. The

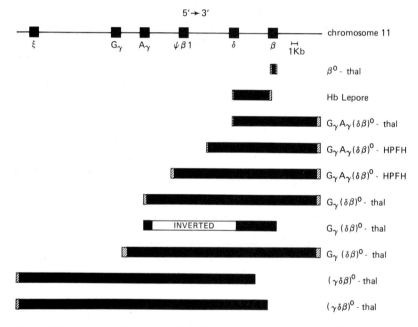

Figure 17.5. Structure of the human β-globin locus in a series of deletion thalassaemias. The deletions are shown as black blocks, and the hatched areas represent the uncertainties in the measurement of the extent of deletion.

most common form of GγAγ$(\delta\beta)^0$ thalassaemia is found in Southern Italy and results from a deletion with its 5′ break point in the δ-globin gene and its 3′ break point past the β-globin gene on the 3′ side[5, 16]. A number of other forms of this disease exists. Thus, in a Spanish form of GγAγ$(\delta\beta)^0$ thalassaemia, a similar deletion has occurred but the break point is some 5 kb further upstream than in the Southern Italian form of the disease[37]. In two rare forms of $(\delta\beta)^0$ thalassaemia, the Gγ-globin gene, but not the Aγ-globin gene, is expressed. Two of these result from a single deletion which removes part or all of the Aγ-globin gene in addition to the δ- and β-globin genes[16, 28, 35a]. The third is a complex rearrangement which includes an inversion in the region from the Aγ-globin gene to the δ-globin gene[28].

HPFH is another complex series of genetic disorders. Two molecular forms of the GγAγ HPFH are found in blacks of African descent. Both are deletions similar to that of the $(\delta\beta)^0$ thalassaemia but in both cases the deletion extends further away from the 5' side of the δ-globin gene than in the $\delta\beta^0$ cases[4, 16, 48]; however, the distance is only about 1 kb or less between the 5' breakpoints in the Spanish $(\delta\beta)^0$ thalassaemia and one form of GγAγ HPFH.

In addition, there are several forms of HPFH in which no deletion can be detected by blotting[4, 29, 48]; the lesion must be either a small deletion/insertion or some base substitution. The best known of these is the Aγ HPFH found in Greeks; as the name suggests, only the Aγ-globin gene is expressed to a high level, although a low level of expression of the Gγ-globin gene can be detected. Several other similar non-deletion HPFH cases have been characterized recently.

The main feature of interest in $(\delta\beta)^0$ thalassaemia and HPFH is the mechanism of elevation of haemoglobin F. If we assume, as is reasonable, that the deletions themselves cause the effects, then inspection of the deletions in *Figure 17.5* suggests that changes in DNA sequences far from the γ-globin gene causes a change in their expression levels, and, as we shall consider briefly, sequences to the 5' side of the δ-globin gene and the 3' side of the β-globin gene are implicated. How these effects are transmitted is not known. Several models have been proposed[4, 28, 42], all of which consider in one way or another the organization of actively expressed globin genes in chromosomal domains.

The comparison of the 5' break point of the deletions in $\delta\beta^0$ thalassaemia and HPFH shows that a short region (perhaps less than 1 kb) defines the difference between the two different diseases. It is of interest that this region contains two Alu repeat sequences; in fact, one of these is completely deleted[4, 48] and the other partially deleted in HPFH[25] whereas both are certainly present in the Italian form of GγAγ$(\delta\beta)^0$ thalassaemia and one but not the other is present in Spanish $\delta\beta^0$ thalassaemia (S. Ottolenghi, personal communication). This has been taken to suggest an involvement of these sequences with the partial suppression of γ-globin gene expression in $(\delta\beta)^0$ thalassaemia. The situation is, however, probably more complex. $(\delta\beta)^0$ thalassaemias show a high level of haemoglobin F expression (usually about 10% in heterozygotes) while haemoglobin Lepore, which is also a deletion, has a low fetal haemoglobin level. Inspection of *Figure 17.5* shows that both Lepore and GγAγ$(\delta\beta)^0$ thalassaemia are deletions starting in the δ-globin gene; in the former case the 3' break point is in the β-globin gene whereas it is well beyond the 3' end of the β-globin gene in $\delta\beta^0$ thalassaemia. This suggests that the sequences to the 3' side of the β-globin gene may also play a role in the suppression of γ-globin gene expression. It should also be remembered that the 3' break points of the HPFH and $\delta\beta^0$ thalassaemia deletions will also probably be different and this has been shown to be true for a number of cases (B. Forget, personal communication). It is therefore possible that the different extent of deletion of DNA to the 3' side of the β-globin locus is responsible for the difference in phenotype. It is equally possible that the phenotypic effects seen are the result of disruption of this region of the chromosome by the loss of different amounts of DNA rather than the loss of a specific given DNA sequence which has a function in the switching process. This type of problem has been discussed previously and of course must apply to any model which attempts to explain these effects[5, 16, 32, 50].

Specifically, it has been suggested:

1. The deletions affect the boundaries of these hypothetical chromosomal domains and therefore alter their ability to present the γ-globin genes in active form[4].

2. It has been shown that the γ-genes are expressed early in erythropoiesis in adults. A second model seeks to explain the enhancement of γ-globin gene expression in adults with a block in the progression of globin gene expression during normal erythropoiesis (ε? → γ → β) caused by the given mutation[28].
3. Finally, a proposal has been made that the ability of a chromosomal region to be expressed depends on how it is replicated. According to this model, regulation would be achieved by replication from different origins[42].

In its active state, a given globin gene is replicated from an upstream origin, whereas in its inactive state the downstream origin is used. It is further postulated (as in model 1) that the active γ state and the active β state are alternatives; when the β-globin gene is expressed replication is proposed to proceed from an origin between the Aγ- and δ-globin gene. Bi-directional replication from this site causes an inactive γ-globin gene (downstream origin) and an active β-globin gene (upstream origin). Smithies suggests that deletion of this origin occurs in GγAγ HPFH but not in GγAγ(δβ)⁰ thalassaemia and suggests further that the Alu repeats in this region may well be implicated in this[42].

How could the use of different replication origins affect the active state of a gene? Weintraub and his colleagues[54] have shown that during DNA replication nucleosomes segregate conservatively, that is, they are retained on a single branch of the replication fork. The second branch of the fork must receive new nucleosomes which can be of the 'active' (e.g. able to carry HMG14/17 proteins) or inactive types. It is obvious that if replication occurs from a different direction, then it is the other fork which retains the old nucleosomes in this case. The crux of the matter is, however, how the decision might be made to add an active rather than an inactive nucleosome. To us, one attractive possibility is that DNA methylation determines this. Specifically, a number of experiments suggest (but certainly do not prove) that DNA methylase is a processive enzyme; that is, it binds to the DNA and then proceeds to methylate suitable sites along that molecule. We suggest that the binding of methylase occurs at the origin of replication (so-called upstream origins in the discussion above) and that origins which generate active genes, one way or another, cannot bind methylase. It would follow that replication from this site generates unmethylated DNA strands after two rounds of replication. We suggest that the unmethylated DNA can bind 'active' nucleosomes while methylated DNA cannot. Conversely, origins which can bind methylase (the so-called downstream origins in this discussion) generate methylated DNA and maintain the status quo.

γδβ thalassaemia

Another rare thalassaemia which results from a deletion is γδβ thalassaemia (see, for example, reference 50). The disease manifests itself as a 'γ thalassaemia' around birth followed by a β thalassaemia in adult life.

In one form of γδβ thalassaemia a large deletion has occurred which eliminates the GγAγ and δ-globin genes but which leaves the β-globin gene, together with some 2.5 kb of 5' upstream sequences intact[50]. Nonetheless, the β-globin gene functions poorly or not at all since heterozygous adults with this condition are β thalassaemic. The sequence of the β-globin gene (including the promoter) from these individuals is, however, normal (D. Kioussis and F. Grosveld, unpublished). The thalassaemic phenotype therefore results from the deletion of the distal DNA

TABLE 17.1 Mutations in β-thalassemia

Mutation class	Type (β⁰ or β⁺)	Ethnic origin	References
Nonsense mutation			
codon 17	β^0	Chinese	Chang and Kan[7]
codon 39	β^0	Mediterranean	Orkin and Goff[35c]
			Trecartin et al.[46]
			Moschonas et al.[33]
Frameshift			
−2 codon 8	β^0	Turkish	Orkin and Goff[35c]
−1 codon 16	β^0	Indian	S. H. Orkin and H. H. Kazazian (pers. comm.)
−1 codon 44	β^0	Kurdish	Kinniburgh et al.[29a]
−4 codons 41/42	β^0	Indian	S. H. Orkin and H. H. Kazazian (pers. comm.)
+1 codon 8/9	β^0	Indian	S. H. Orkin and H. H. Kazazian (pers. comm.)
Splice junction substitution			
IVS1-position 1 G→A	β^0	Mediterranean	Orkin et al.[35d]
IVS2-position 1 G→A	β^0	Mediterranean	Treisman et al.[47]
Consensus substitution			
IVS1-position 5 G→C	β^+	Indian	S. H. Orkin and H. H. Kazazian (pers. comm.)
IVS1-position 6 T→C	β^+	Mediterranean	Orkin et al.[35d]
Internal IVS substitution			
IVS1-position 110	β^+	Mediterranean	Spritz et al.[44]
			Westaway and Williamson[55]
IVS2-position 705	β^+	?Mediterranean	Spence et al.[43a]
IVS2-position 745	β^+	Mediterranean	Orkin et al.[35d]
Transcriptional mutant			
−87 C→G	β^+ (probable)	Mediterranean	Orkin et al.[35d]
−28	?	Kurdish	Poncz et al.[38]
Coding region substitution affecting RNA processing			
codon 26 G→A	β^E	Southeast Asian	Orkin et al.[35e]
codon 24 T→A	β^+	Black	A. Nienhuis (pers. comm.)

sequences as discussed previously[50]. Thus, in γδβ thalassaemia, just as in HPFH and γδ[0] thalassaemia deletion of sequences far from a given gene influences its activity.

β[+] and β[0] thalassaemia

These are the most common forms of thalassaemia. Molecular analysis of cloned genes from a large number of cases has shown that the following types of lesions occur (*Table 17.1*). The abundance of different forms of the disease reflect the strong selective advantage of the heterozygous thalassaemic phenotype on the individual.

Stop codons (β[0])

Two types are known with terminators at position 17[7] and 39[33, 46] in Chinese and Mediterranean patients, respectively.

Frameshifts (β[0])

Five forms have been found at codons located within the first exon[35c]; (see *Table 17.1*).

Altered splice junctions (β[0] and β[+])

Forms have been found with GT → AT changes at the 5' splice sites of both IVS1 and IVS2 in Mediterranean β[0] thalassaemics[47, 35d]. The absence of the 5' splice site (GT) at IVS2 causes the use of a cryptic splice site in IVS2 to generate an mRNA which contains an additional 47 nucleotides of IVS2 between exons 2 and 3[47]. In addition, in two β[+] thalassaemic types, the consensus donor sequence for splicing at IVS1 is altered at positions 5 and 6 downstream from the doner G residue, respectively (*Table 17.1*).

Finally, three cases of the introduction of splice sites within introns have occurred. In the first, an AG acceptor is generated 21 nucleotides from the 3' end of IVS1 in a Cypriot form of β[+] thalassaemia[44, 55] which results in aberrant splicing to this site[6]. More recently, two similar lesions have been found in IVS2 generating AG acceptors at positions 705 and 745, respectively. Again, these cause abnormal splicing to these acceptors and the concomitant generation of non-functional β-globin mRNA[47].

Splice sites are generated within exons by point mutations which in two out of three cases simultaneously generate amino acid substitutions. In haemoglobin E codon 26 of the β-globin gene is altered to produce Lys instead of Glu. This gives the sequence AAG which contains an AG which can function as a splice acceptor. A similar phenomenon occurs in haemoglobin Knossos at codon 27 (Ala (GCC) → Ser (TCC)).

Promoter mutations

Recently, promoter mutations have been found in two β thalassaemias; a C → G at −87 which probably alters the C-rich sequence in the upstream part of the promoter and an ATA box mutation which converts CATAAAA → CATACAA.

Presumably, both these mutations cause a decrease in the efficiency of transcription although this has not yet been tested experimentally.

Partial deletions

In an Asian form of β^0 thalassaemia, a 0.6 kb deletion has occurred which eliminates the 3' terminal exon of the β-globin gene[15, 35b].

Deletion – thalassaemias suggest a novel mechanism of deletion in eukaryotic chromosomes

Southern blotting has been used to compare the DNA sequences at the break points of the deletions in a number of HPFH cases. Used as a probe was a DNA fragment derived from the sequences that were brought to the 3' side of the Aγ-globin gene by the deletion event. Interestingly, this DNA segment is present close to the break point in three other HPFH cases, the form of GγAγ HPFH with the larger deletion (see *Figure 17.5*), a Gγ HPFH (with a break point in the Aγ-globin gene (see *Figure 17.5*) and an unpublished Chinese HPFH which also has a break point in the Aγ-globin gene (P. Henthorn, personal communication). In contrast, the DNA segments transposed in a γδβ thalassaemia showed no such homology (Dixie Mager, personal communication). A similar result is found when the two different forms of γδβ thalassaemia were compared (*Figure 17.5*). In this instance the DNA sequences transposed to the 5' side of the β-globin genes were found to be the same (E. Vanin and F. Grosveld, unpublished). Together these data indicate that the generation of such deletions follow a similar mechanism involving the same DNA region for a particular phenotype. The precise end points of the deletions are however different by several kilobases in different cases. This suggests that hot spots exist for the origin of deletion events which may reflect labile structures in eukaryotic chromosomes.

Specific expression of globin genes in erythroid cells

It has been of paramount importance to the field of eukaryotic gene expression that genes such as globin genes are expressed when introduced into a number of cells such as fibroblasts (L cells) or HeLa cell. It is somewhat surprising, however, since the endogeneous globin genes in these cells are not expressed. While this phenomenon is of great use for the study of promoters, splicing, etc., it is not likely to be of value for the study of regulated β-globin expression.

An experimental model for the latter was provided by the observation that the introduction of chromosomes carrying globin genes into murine erythroleukaemia (MEL) cells permits the specific induction of α- and β-globin mRNA synthesis from this exogenous chromosome upon the induction of erythroid differentiation in these cells[57].

Several groups have therefore introduced cloned globin genes into MEL cells by calcium phosphate-mediated DNA transfer[43] (P. Mellon, M. Chan, P. Charnay, R. Axel and T. Maniatis, unpublished; S. Wright, F. Grosveld, E. de Boer and R. A. Flavell, unpublished). Under these conditions, the human and mouse β-globin promoters show inducible synthesis of β-globin mRNA which suggests that DNA present in MEL cells responds to *trans* acting factors present in these cells. The

DNA sequences required for this response appear to lie to the 3' side of -100 in the mouse β-globin gene, that is, either in the promoter or in the gene (P. Mellon, M. Chan, P. Charnay, R. Axel and T. Maniatis, personal communication).

Acknowledgements

We are grateful to our many colleagues who provided us with access to their unpublished data, in particular, S. Orkin, E. Vanin, B. Forget and C. Weissmann. Work from our laboratory was supported by the British Medical Research Council. We are grateful to Madlyn Nathanson and Cora O'Carroll for typing the manuscript.

References

1. Banerji, J., Rusconi, S. and Schaffner, W. (1981). Expression of a genomic segment of Simian virus 40 DNA. *Cell*, **27**, 299–304
2. Benoist, C. and Chambon, P. (1981). The ovalbumin gene sequence of putative control elements. *Nature, London*, **290**, 304–310
3. Benoist, C., O'Hare, K., Breathnach, R. and Chambon, P. (1980). The ovalbumin gene sequence of putative control regions. *Nucleic Acids Research*, **8**, 127–142
4. Bernards, R. and Flavell, R. A. (1980). Physical mapping of the globin gene deletion in hereditary persistance of foetal haemoglobin (HPFH). *Nucleic Acids Research*, **8**, 1521–1534
5. Bernards, R. Kooter, J. M. and Flavell, R. A. (1979). Physical mapping of the globin gene deletion in (δβ)^0 thalassaemia. *Gene*, **6**, 265–280
6. Busslinger, M., Moschonas, N. and Flavell, R. A. (1981). β$^+$ thalassaemia: Aberrant splicing results from a single point mutation in an intron. *Cell*, **27**, 289–298
7. Chang, J. C. and Kan, Y. W. (1979). β^0 thalassemia, a nonsense mutation in man. *Proceedings of the National Academy of Sciences, USA*, **76**, 2886–2889
8. Craik, C. S., Buchman, S. R. and Beycho, K. (1980). Characterization of globin domains: theme binding to the central exon product. *Proceedings of the National Academy of Sciences, USA*, **77**, 1384–1388
9. Dierks, P., Van Ooyen, A., Mantei, N. and Weissmann, C. (1981). DNA sequences preceding the rabbit β-globin gene are required for formation in mouse L cells of β-globin RNA with correct 5' terminus. *Proceedings of the National Academy of Sciences, USA*, **78**, 1411–1415
10. Dolan, M., Sugarman, B. J., Dodgson, J. B. and Engel, J. D. (1981). Chromosomal arrangement of the chicken β-type globin genes. *Cell*, **24**, 669–677
11. Efstratiadis, A., Posakony, J. W., Maniatis, T., Lawn, R. M. O'Connell, C., Spritz, R. A., DeRiel, J. K., Forget, B., Weissman, S. M., Slightom, J. L., Blechl, A. E., Smithies, O., Baralle, F. E., Shoulders, C. C. and Proudfoot, N. J. (1980). The structure and evolution of the human β-globin gene family. *Cell*, **21**, 653–668
12. Faraca, L., Brown, B., Hardy, S., Hutchinson III, C. and Edgell, M. (1983). In preparation
13. Flavell, R. A., Kooter, J. M., DeBoer, E., Little, P. F. R. and Williamson, R. (1978). Analysis of the β–δ globin gene loci in normal human DNA: Direct determination of gene linkage and intergene distance. *Cell*, **15**, 25–41
14. Flavell, R. A., Grosveld, G. C., Grosveld, F. G., Bernards, R., Kooter, J. M., DeBoer, E. and Little, P. F. R. (1979). The structure and expression of normal and abnormal globin genes. In *From Gene to Protein*, (Russell, T. R., Brew, K., Faber, H. and Schultz, J., Eds.), Volume 16, pp. 149–165. London, Academic Press
15. Flavell, R. A., Bernards, R., Kooter, J. M., DeBoer, E., Little, P. F. R., Annison, G. and Williamson, R. (1979). The structure of the human β-globin gene in thalassaemia. *Nucleic Acids Research*, **6**, 2749–2760
16. Fritsch, E. F., Lawn, R. M. and Maniatis, T. (1979). Characterization of deletions which affect the expression of fetal globin genes in man. *Nature, London*, **279**, 598–603
17. Goldberg, M. (1979). Sequence analysis of *Drosophila* histone genes. PhD Thesis, Stanford University, Palo Alto, California

18. Grosschedl, R. and Birnstiel, M. L. (1980). Identification of regulatory sequences in the prelude sequences of an H$_2$A histone gene by the study of specific deletion mutants *in vivo. Proceedings of the National Academy of Sciences, USA*, **77**, 1432–1436

19. Grosveld, G. C., Koster, A. and Flavell, R. A. (1981). A transcription map for the rabbit β-globin gene. *Cell*, **23**, 573–584

20. Grosveld, G. C., Shewmaker, C. K., Jat, P. and Flavell, R. A. (1982). Localization of DNA sequences necessary for transcription of the rabbit β-globin gene *in vitro. Cell*, **25**, 215–226

21. Grosveld, G. C., DeBoer, E., Shewmaker, C. K. and Flavell, R. A. (1982). DNA sequences necessary for the transcription of the rabbit β-globin gene *in vivo. Nature, London*, **295**, 120–126

22. Haigwood, N. L., Jahn, C. L., Hutchinson III, C. A. and Edgell, M. H. (1981). Location of three repetitive sequence families found in BALB/c adult β-globin clones. *Nucleic Acids Research*, **9**, 1133–1150

23. Hoeijmakers-van Dommelen, H. A. M., Grosveld, G. C., DeBoer, E., Flavell, R. A., Varley, J. M. and Jeffreys, A. J. (1980). The localization of repetitive and unique DNA sequences neighbouring the rabbit β-globin gene. *Journal of Molecular Biology*, **140**, 531–547

24. Hofer, E. and Darnell, Jr, J. E. (1981). The primary transcription unit of the mouse β-major globin gene. *Cell*, **23**, 585–593

25. Jagadeeswaran, P., Tuan, D., Forget, B. G. and Weissman, S. M. (1982). A gene deletion ending at the midpoint of a repetitive DNA sequence in one form of hereditary persistence of fetal haemoglobin. *Nature, London*, **296**, 469–470

26. Jeffreys, A. J. and Flavell, R. A. (1977). The rabbit β-globin gene contains a large insert in the coding sequence. *Cell*, **12**, 1097–1108

27. Jeffreys, A. J., Wilson, V., Wood, D., Simons, J. P., Kay, R. M. and Williams, J. G. (1980). Linkage of adult α- and β-globin gene in *X. laevis* and gene duplication by tetraploidization. *Cell*, **21**, 555–564

28. Jones, R. W., Old, J. M., Trent, R. J., Clegg, J. B. and Weatherall, D. J. (1981). Restriction mapping of a new deletion responsible for Gγ(δβ)^0 thalassaemia. *Nucleic Acids Research*, **9**, 6813–6825

29. Jones, R. W., Old, J. M., Wood, W. G., Clegg, J. B. and Weatherall, D. J. (1982). Restriction endonuclease maps of the β-like globin gene cluster in the British and Greek forms of HPFH and for one example of Gγβ HPFH. *British Journal of Haematology*, **50**, 415–422

30. Lawn, R. M., Fritsch, E. F., Parker, R. C., Blake, G. and Maniatis, T. (1978). The isolation and characterization of linked δ- and β-globin genes from a cloned library of human DNA. *Cell*, **15**, 1157–1174

31. Leder, A., Swan, D., Ruddle, F., D'Eustachio, P. and Leder, P. (1981). Dispersion of β-like globin genes of the mouse to three different chromosomes. *Nature, London*, **293**, 196–200

32. Maniatis, T., Fritsch, E. F., Lauer, J. and Lawn, R. M. (1980). Molecular genetics of human hemoglobins. *Annual Review of Genetics*, **14**, 145–178

33. Moschonas, N., DeBoer, E., Grosveld, F. G., Dahl, H. H. M., Wright, S., Shewmaker, C. K. and Flavell, R. A. (1981). Structure and expression of a cloned β^0 thalassaemic globin gene. *Nucleic Acids Research*, **9**, 4391–4401

34. Nevins, J. R. and Darnell, Jr. J. E. (1978). Steps in the processing of Ad2 mRNA: Poly(A) nuclear sequences and conserved and poly(A) addition precedes splicing. *Cell*, **15**, 1477–1493

35. Nishioka, Y., Leder, A. and Leder, P. (1980). Unusual α-globin-like gene that has cleanly lost both globin intervening sequences. *Proceedings of the National Academy of Sciences, USA*, **77**, 2806–2809

35a. Orkin, S. H., Old, J. M., Weatherall, D. J. and Nathan, D. B. (1979). Partial deletion of β-globin gene DNA in certain patients with β^0 thalassaemia. *Proceedings of the National Academy of Sciences, USA*, **76**, 2400–2404

35b. Orkin, S. H., Kolodner, R., Michelson, A. and Husson, R. (1980). Cloning and direct examination of a structurally abnormal human β^0 thalassemia globin gene. *Proceedings of the National Academy of Sciences, USA*, **77**, 3558–3562

35c. Orkin, S. H. and Goff, S. C. (1981). Nonsense and frameshift mutations in β^0 thalassemia detected in cloned β-globin genes. *Journal of Biological Chemistry*, **256**, 9782–9784

35d. Orkin, S. H., Kazazian, H. H., Antonarakis, S. E., Goff, S. C., Boehm, C. D., Sexton, J. P., Waber, P. G. and Giardina, P. J. V. (1982). Linkage of β-thalassaemia mutations and β-globin gene polymorphisms with DNA polymorphisms in human β-globin gene cluster. *Nature, London*, **296**, 627–631

35e. Orkin, S. H., Kazazian, H. H., Antonarakis, S. E., Ostrer, H., Goff, S. C. and Sexton, J. P. (1983). *Nature, London* (in press)

36. Østergaard Jensen, E., Paludan, K., Hyldig-Nielsen, J. J., Jørgensen, P. and Marcker, K. A. (1981). The structure of a chromosomal leghaemoglobin gene from soybean. *Nature, London*, **291**, 677–679

37. Ottolenghi, S., Giglioni, B., Taramaelli, R., Comi, P., Mazza, U., Saglio, G., Camaschella, C., Izzo, P., Cao, A., Galanello, R., Gimferrer, E., Baiget, M. and Gianni, A. M. (1982). Molecular comparison of δβ-thalassaemia and hereditary persistence of fetal hemoglobin DNAs: Evidence of a regulatory area? *Proceedings of the National Academy of Sciences, USA*, **79**, 2347–2351

38. Poncz, M., Ballantine, M., Solowiejczyk, D., Barak, I., Schwartze, E. and Surrey, S., (1982). β-thalassemia ina Kurdish Jew. Single base change in the T-A-T-A box. *Journal of Biological Chemistry*, **257**, 5994–5996

39. Proudfoot, N. J. (1979). Eukaryotic promotors? *Nature, London*, **279**, 376

40. Proudfoot, N. J. and Maniatis, T. (1980). The structure of a human α-globin pseudogene and its relationship to α-globin gene duplication. *Cell*, **21**, 537–544

41. Shen, C.-K. J. and Maniatis, T. (1980). The organization of repetitive sequences in a cluster of rabbit β-like globin genes. *Cell*, **19**, 379–392

42. Smithies, O. (1982). The control of globin and other eukaryotic genes. *Journal of Cell Physiology, Supplement*, **1**, 137–143

43. Spandidos, D. and Paul, J. (1982). Transfer of human globin genes to erythroleukemic mouse cells. *EMBO Journal*, **1**, 15–20

43a. Spence, S. E., Pergolizzi, R. G., Donovan-Peluse, M., Kosche, K. A., Dobkin, C. S. and Bank, A. (1982). Five nucleolide changes in the large intervening sequence of a β-globin gene in β⁺ thalassemia. *Nucleic Acids Research*, **10**, 1283–1294

44. Spritz, R. A., Jagadeeswaran, P., Choudary, P. V., Biro, P. A., Elder, J. T., DeRiel, J. K., Manley, J. L. Gefter, M. L., Forget, B., G. and Weissman, S. M. (1981). Base substitution in an intervening sequence of a β⁺ thalassemic human globin gene. *Proceedings of the National Academy of Sciences, USA*, **78**, 2455–2459

45. Tilghman, S. M., Curtis, P. J., Tiemeier, D. C., Leder, P. and Weissmann, C. (1978). The intervening sequence ot a mouse β-globin gene is transcribed within the 15S-globin mRNA precursor. *Proceedings of the National Academy of Sciences, USA*, **75**, 1309–1313

46. Trecartin, R. F., Liebhaber, S. A., Chang, J. C., Lee, K. Y., Kan, Y. W., Furbetta, M., Angius, A. and Cao, A. (1981). β⁰ thalassemia in Sardinia is caused by a nonsense mutation. *Journal of Clinical Investigation*, **68**, 1012–1017

47. Treisman, R., Proudfoot, N. J., Shander, M. and Maniatis, T. (1982). A single-base change at a splice site in a β⁰ thalassemic gene causes abnormal RNA splicing. *Cell*, **29**, 903–911

48. Tuan, D., Murnane, M. J., DeRiel, J. K. and Forget, B. G. (1980). Heterogeneity in the molecular basis of hereditary persistence of foetal haemoglobin. *Nature, London*, **285**, 335–336

49. Van Den Berg, J., Van Ooyen, A., Mantei, N., Schambóck, A., Grosveld, G. C., Flavell, R. A. and Weissmann, C. (1978). Comparison of cloned rabbit and mouse β-globin genes showing strong evolutionary divergence of two homologous pairs of introns. *Nature, London*, **275**, 37–44

50. Van Den Ploeg, L. H. T., Konings, A., Oort, M., Roos, D., Bernini, L. F. and Flavell, R. A. (1980). γβ thalassaemia: deletion of the γ- and δ-genes influences β-globin gene expression in man. *Nature, London*, **283**, 637–642

51. Vanin, E. F., Goldberg, G. I., Tucker, P. W. and Smithies, O. (1980). A mouse β-globin-related pseudogene lacking intervening sequences. *Nature, London*, **286**, 222–226

52. Weatherall, D. J. and Clegg, J. (1979). *The Thalassaemia Syndromes*, 3rd Edition. Oxford, Blackwell Scientific Publications

53. Weaver, R. F. and Weissmann, C. (1979). Mapping of RNA by a modification of the Berk-Sharp procedure: The 5′ termini of 15S β-globin mRNA precursor and mature 10S β-globin mRNA have identical map coordinates. *Nucleic Acids Research*, **6**, 1175–1193

54. Weintraub, H., Fling, S. J., Leffak, I. M., Groudine, M. and Grainger, R. M. (1977). The generation and propagation of variegated chromosome structures. *Cold Spring Harbor Symposium on Quantitative Biology*, **42**, 401–407

55. Westaway, D. and Williamson, R. (1981). An intron nucleotide sequence variant in a cloned β⁺ thalassaemia globin gene. *Nucleic Acids Research*, **9**, 1777–1788

56. Wieringa, B., Meyer, F., Reiser, J. and Weissmann, C. (1982). Cryptic splice sites in the rabbit β-globin gene are revealed following inactivation of an authentic 5′ splice site by mutagenesis *in vitro*. In *Gene Regulation* (O'Malley, B. W. and Fox, C. F., Eds.), pp. 65–85. Academic Press

57. Willing, M. C., Nienhuis, A. W. and Anderson, W. F. (1979). Selective activation of human β- but not γ-globin gene in human fibroblast × mouse erythroleukaemia cell hybrids. *Nature, London*, **277**, 534–538

Chapter 18

The egg white protein genes

A. E. Sippel

Introduction

The genes for the major egg white proteins ovalbumin, ovotransferrin (=
conalbumin), ovomucoid and lysozyme are a group of genes defined by their
common tissue-specific and steroid regulated expression. They are not a family of
evolutionarily related genes nor are their chromosomal loci linked to each other in
the chicken genome[21].

This chapter is restricted to the chicken genes for the abovementioned four egg
white proteins and to two additional members of the ovalbumin-like gene family[66].
Data on the genes for minor egg white proteins, such as avidin and avian
ovoinhibitor and for any egg white proteins from other species, are not available at
the present time.

In the tubular gland cells of the mature chicken oviduct, the synthesis of the
major egg white proteins comprises over 70% of the total translational activity[55].
This synthesis is dependent on the continued presence of gonadal steroid
hormones[53]. The high level of expression of a few proteins and the fact that their
synthesis can be readily controlled by extracellular effector molecules is the basis
for the interest in the oviduct system. Steriod hormones induce specific
accumulation of the egg white protein mRNAs by an increase in the rate of
transcription[43, 49, 74] as well as by an increase in mRNA stability[12, 23, 57].
Knowledge of the structure of the genes is a prerequisite for understanding the
level of differential gene activation. Therefore data on the organization of the
genes will first be summarized and thereafter some of the problems which have to
be solved before the processes involved in regulation can be explained on the
molecular level will be discussed.

The ovalbumin gene family

Ovalbumin and its mRNA

Ovalbumin is the major nutritional protein in the avian egg white (~65% $^W\!/w$). It is
a phosphoprotein consisting of 386 amino acid residues (molecular weight ~43 kd)
and related to human antithrombin III[29]. Its synthesis comprises 55–60% of total
protein synthesis in the tubular gland cells of the oviduct of laying hens. The

315

synthesis of ovalbumin became one of the most favourable model systems for studies on regulation of gene expression in higher cells. The high relative rate of synthesis stimulated early attempts to isolate ovalbumin mRNA (reviewed in reference 54) and to measure the parameters of mRNA metabolism[56]. Measurements of ovalbumin mRNA concentrations and the kinetics of its changes upon hormonal manipulations in the immature chick were initially done by cell-free translation assays and later by cDNA titration in many groups. It could be shown that oestrogens and progesterone cause a rise in the steady state level of ovalbumin mRNA from a few molecules per cell in the hormone withdrawn organ to about 50 000–100 000 molecules/cell in the hormonally stimulated organ[19, 44, 69].

Cloning of a full sized cDNA and its sequence analysis[45, 52] determined the organization of the ovalbumin mRNA and provided for the first time the full amino acid sequence of ovalbumin. The mRNA has an abnormally long 3' non-coding part of 637 (650) residues, and the 5' end of the mRNA is able to form a terminal hairpin structure[45]. Direct determination of the cap oligonucleotides of purified ovalbumin mRNA from laying hen oviducts revealed a microheterogeneity, resulting in mRNAs with a 5' non-coding region of 62 (3%), 64 (61%) and 66 (36%) residues in length[46].

According to sequence analysis of the protein[58] and of the cDNA, ovalbumin seemed to be a first example of a secreted protein synthesized without a typical hydrophobic signal peptide. It was later suggested, however, that ovalbumin might have an internal correlate from amino acid 234 to 253, a part of the protein encoded by the end of exon 7[42].

The ovalbumin gene

The ovalbumin gene was one of the first genes in which the fragmented structure of eukaryotic genes became apparent[6, 13, 30, 79]. The split organization, initially demonstrated by genomic DNA blotting, was established rapidly upon subsequent cloning of the gene sequences (for reference see reference 28). Once all genomic sequences coding for the mRNA had been found[14, 16], the following final sequence

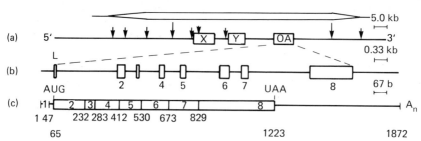

Figure 18.1. Schematic diagram of the cluster of (a) the chicken ovalbumin-like genes, (b) the ovalbumin gene, and (c) the ovalbumin mRNA. (a) Data are derived from Lawson *et al.*[36]. Arrows indicate the midpoint of restriction fragments containing middle repetitive DNA; larger arrows indicate more highly reiterated sequences. The open bar on top represents the domain of increased DNase I sensitivity of DNA in oviduct chromatin. Regions of transition between sensitive and resistant chromatin are indicated by the tips of the bar. (b) Data are derived from Dugaiczyk *et al.*[14]. Exons are numbered in 5' to 3' direction; the first exon, coding only for part of the 5' non-coding region of the mRNA, is called a 'leader' (= L)[6]. (c) Data are derived from Gannon *et al.* The coding region of the mRNA is represented by the open bar (nucleotide 65 to 1223). Numbers below give the last nucleotide encoded in each exonic region.

organization could be established (*Figure 18.1*). The ovalbumin gene consists of eight exons. Six introns interrupt the coding region for ovalbumin at the codons for amino acids no. 56, 73, 117, 156, 203 and 255. The first intron interrupts the 5′ non-coding region of the mRNA. It separates a 47 base pair gene leader region about 1.5 kb from the 17 base pairs corresponding to the second part of the 5′ non-coding region. The last exon of the gene is one of the longest exonic regions known and contains more than half of the region coding for the mRNA (1043 bp). Upon sequence analysis of the exon–intron junctions it was noticed that the boundaries share common features. In all cases sequences could be aligned so that introns would start with GT and end with AG. This observation led to the general 5′-GT–AG-3′ rule for intron structure in eukaryotic mRNA producing genes[6a].

The ovalbumin gene was mapped in genomic DNA from germline cells and actively transcribing and non-transcribing tissue[6, 79]. No gross differences in restriction maps could be detected between the genes in sperm, oviduct and erythrocytes. Analysis of cloned DNA from oviduct and erythrocytes showed no alterations at the nucleotide level of the 5′ terminal part of the gene, assumed to be important for the initiation of transcription[15]. Mapping in genomic DNA and cloned DNA indicated, however, frequent differences in intron-specific restriction enzyme cleavage sites. Since these gene variants could be detected homozygously and heterozygously in individual chickens, they were taken to be the result of genetic allelism rather than indicative for the presence of more than one ovalbumin gene/haploid genome[31].

In genomic DNA, sequences corresponding to the ovalbumin mRNA are contained in a gene region of roughly four times the size of mRNA. Analysis of oviduct nuclear RNA with exon and intron-specific gene DNA probes[63, 64, 77] revealed many of the steps involved in ovalbumin mRNA synthesis. The primary transcript of roughly 7.8 kb in length, is processed in a preferred but not obligatory order. Spliced intervening RNA sequences are preferentially degraded and comprise less than 10% of the steady state level of nuclear mRNA specific sequences. No indications were detected for steroid regulation of nuclear RNA processing. The ends of the primary transcript were coincident with the beginning and end of the structural sequences in the gene. Potentially regulatory signal sequences for initiation of transcription, splicing and polyadenylation are discussed in detail in Benoist *et al.*[5]. No significant sequence homology beside correlates of the TATA motif could be detected in the first 250 bp upstream from the cap site between the ovalbumin gene and other steroid-regulated major egg white protein genes[18]. Studies of the functional aspects of DNA signals for the expression and regulation of the ovalbumin gene by cell-free transcription[76, 78] and by DNA-mediated gene transfer[7, 33] are initiated but are not yet developed to a degree to which clues for the mechanism of gene activation could be extracted.

The X–Y–ovalbumin gene cluster

Cosmid cloning of sequences upstream from the ovalbumin gene led to the discovery of two related genes, called X and Y[66]. When increasing DNA regions were cloned by chromosomal walking, it became evident that the X, the Y and the ovalbumin gene were a small multigene family contained in about 40 kb of DNA from the beginning of the X gene to the end of the ovalbumin gene[10, 20]. The three genes are transcribed in the same direction. They have a comparable exon–intron pattern and show a certain degree of sequence homology, indicating that they have evolved by duplications of an already split ancestor gene. Comparison of selected regions of these genes[20], shows that the locations and sequences of the splice junctions have been strongly conserved, and that intron sequences have diverged far more than the exon regions. The only striking difference in the structural parts of the genes is the variability in length and sequence of the 3' non-coding region on the last exon. From the drift in silent sites in the amino acid coding part, a period of around 55×10^6 years was estimated for the divergence of the X and the ovalbumin gene.

All three genes are expressed in the oviduct under the influence of steroid hormones. However, the relative rates of accumulation of nuclear transcripts in the oestrogen-induced organ differ greatly (X:Y: ovalbumin = 1:10:100)[10]. The sizes of the largest nuclear mRNA precursors correspond to the respective gene sizes. Within the limitations of the sensitivity of detection, this excludes the possibility of a common primary transcript. There are indications for a non-coordinated response to steroids other than oestrogens[39]. Progesterone, when compared with oestradiol, has a much weaker effect on X mRNA accumulation than on the accumulation of Y and of ovalbumin mRNA. The biological role of the ovalbumin related protein products of the X and Y gene is unknown at the present.

Efforts were made to characterize the ovalbumin gene cluster further, in view of the additional aspects involved in regulation of a multigene family as compared to the regulation of a single gene. In an analysis of DNA methylation sites by 5mC-sensitive restriction endonucleases, overall undermethylation could be seen in and around the ovalbumin gene in the oviduct, as compared to the gene in non-active tissue. However, no uninterrupted unmethylated compartment covering the active ovalbumin gene region was found[47]. A different picture emerged from studies in which DNase I sensitivity of chromatin was tested[4, 35]. DNase sensitivity in oviduct nuclei appeared to extend homogeneously over the ovalbumin gene and its flanking sequences and extended also over the X and Y gene region. Recently the precise dimensions of the DNAase sensitive domain in oviduct chromatin could be determined[36]. Regions of transition from the sensitive to the resistant chromatin configuration were found 20–30 kb upstream from the X gene and 15–25 kb downstream from the ovalbumin gene, delineating a size of approximately 85 kb for the structural chromatin domain of the X–Y–ovalbumin gene cluster (*Figure 18.1*).

There is a notable uneven distribution of dispersed middle repetitive sequence elements throughout the multigene domain. Regions of repetitive sequences lie within the first and the third intron of gene X and two others lie in the intergenic region between X and Y[20]. None of these elements is harboured by the other gene or intergene regions. Recently a set of related middle repetitive elements were identified which seem only to occur close to the upstream and downstream transition regions for chromatin DNase I sensitivity[36, 73].

The ovotransferrin gene

Ovotransferrin and its mRNA

Avian transferrin synthesis, as model for tissue-specific expression, is of interest because the protein is expressed under different control in liver and oviduct (see p. 322). Chicken ovotransferrin, the second most abundant egg white protein (10–15%) is identical in primary structure to serum transferrin, secreted from chicken liver[38]. The 77 kd proteins differ from each other only in the carbohydrate groups which are attached to the same site on the protein molecule[80]. The structure, the evolutionary relationship and the iron-binding property of the transferrins were reviewed recently[1].

The primary structure of ovotransferrin, often called by its older name conalbumin, was obtained recently by protein sequencing[81] and via cloned cDNA[25]. Its mRNA contains 2376 nucleotides, including 76 nucleotides of 5' non-coding region and 182 nucleotides of 3' non-coding region. The RNA codes for a 705 amino acid preprotein, including a 19 amino acid N-terminal signal peptide. The mRNA sequence confirms earlier suggestions that transferrin, which contains two iron-binding sites, has evolved by an intragenic duplication[1]. More insight into the process of evolution of the protein is expected from the elucidation of the exon organization of the chromosomal gene.

The gene for ovotransferrin

From genomic DNA blotting it was inferred that there is only one transferrin gene present/haploid chicken genome and that it is identical in oviduct and liver DNA[38]. The gene has been cloned and shown to consist of at least 17 exons, distributed over a DNA region of 10.3 kb[9, 62]. No sequence data have been published at the present time other than a 460 bp stretch encompassing the first exon of the gene. The first intron interrupts the structural sequences within the coding region for the signal peptide. The ovotransferrin gene promoter has in common with the Adeno 2 major late promoter a sequence of 12 bp (-32 to -21), including the TATA box, and a second homology region at -83 to -90. Both promoters show comparable efficiency in a cell-free transcription system and their *in vitro* sequence requirements are very similar[11].

Knowledge of the DNA signal requirements for regulation would be of great value, since transferrin mRNA synthesis is strongly induced by oestrogens in the oviduct and is insignificantly induced in the liver[23, 37]. The implications of this tissue-specific difference in regulation for the architecture of the gene in the two cell types are discussed later (p. 322).

No significant sequence homology beside the TATA box could be found in the immediate promoter upstream region between the transferrin and the ovalbumin gene. Several middle repetitive elements of differing genomic abundance were, however, detected upstream of the gene and in the second and third intron[9].

The ovomucoid gene

Ovomucoid and its mRNA

Avian ovomucoids are a family of proteinase inhibitors (reviewed in reference 34). Chicken ovomucoid, a glycoprotein with trypsin inhibitory activity, accounts for about 10% of the protein content of the egg white. The primary structure, functional aspects and the disulphide bond pattern of the protein[27] led to the conclusion that ovomucoid is a polypeptide divided into three homologous functional domains, each having its own trypsin inhibitory site.

Complementary DNA corresponding to purified ovomucoid mRNA was used to titrate the content of mRNA specific sequences in the oviduct tissue at various endocrine states[22, 23]. The level of ovomucoid mRNA was consistently regulated parallel to the level of ovalbumin mRNA. The sequence organization of ovomucoid mRNA was determined from cloned cDNA[8]. A 630 nucleotide coding region includes 24 codons for a hydrophobic signal peptide and 186 codons for mature ovomucoid, framed by 53 and 138 residues of the 5' and 3' non-coding region.

The gene for ovomucoid

The gene for ovomucoid convincingly illustrates how knowledge of eukaryotic gene structure can be used to aid the understanding of the evolutionary history of a protein. The structural sequence of the ovomucoid gene was shown to be divided into eight exons[32, 40]. Exons 3 + 4, 5 + 6, and 7 + 8 exactly specify the three protein domains of structural homology, providing the strongest argument for the evolution of the protein by two intragenic duplications of an ancestral split gene[72]. The family of avian proteinase inhibitors presents more evidence for exon duplication as one of the driving forces for gene evolution. On the basis of protein sequence studies, birds code for at least ten paralogous functional domains for proteinase inhibition – six in ovoinhibitor, a minor egg white protein, three in ovomucoid and one in avian pancreatic secretory inhibtor[34]. Sequence data of the non-coding DNA of these genes might pinpoint the DNA sequences involved in ancient recombinational events. Thus far, DNA sequences encompassing only the transcriptional initiation site and the poly(A) addition point of the ovomucoid gene are available[32]. The genomic distance between the two points is consistent with the presence of a 5.5 kb nuclear primary transcript. The seven regions of intervening sequence RNA are spliced from the primary transcript in a preferred but not necessarily obligatory order[77].

The lysozyme gene

Lysozyme and its mRNA

Lysozyme is the least abundant of the four major chicken egg white proteins (~2%) and the only one with a known enzymatic function. Its antibacterial activity is most likely the sole physiological purpose for its presence in the avian egg white.

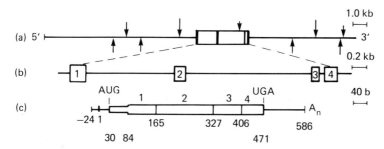

Figure 18.2. Schematic diagram of (a) the chicken lysozyme gene region, (b) the lysozyme gene and (c) the lysozyme mRNA. (a) Vertical bars represent the exons of the gene. Downward arrows indicate the midpoint of restriction fragments containing middle repetitive DNA; larger arrows indicate more highly reiterated sequences (Sippel *et al.*, unpublished data). Upward arrows indicate the midpoint of restriction fragments containing specific high affinity binding sites for nuclear protein components[51]. (b) Data are derived from Jung *et al.*[26]. The exons are numbered in 5′ to 3′ direction. (c) Data are derived from Jung *et al.*[26] and Grez *et al.*[18]. The coding region of the mRNA is represented by the open bar (nucleotide 30 to 83, protein signal sequences; 84 to 470, mature lysozyme). Numbers below give the last nucleotide encoded in each exonic region. Vertical lines indicate the 5′ ends (1, −2 and −24) of the three lysozyme mRNAs found in the oviduct.

Early sequence analysis and early X-ray diffraction analysis made lysozyme a model for the general understanding of enzymatic catalysis[24].

Using complementary DNA probes from purified mRNA, it was established that the level of lysozyme-specific mRNA in the cell is coordinately regulated by steroid hormones with the level of ovalbumin mRNA[22,23]. The sequence organization of the mRNA (*Figure 18.2*) was determined via analysis of cloned cDNA[26,71]. The lysozyme mRNA contains a coding region for the 129 amino acids of mature lysozyme and for the 18 residues of the N-terminal signal peptide and includes a 116 nucleotide 3′ non-coding region (*Figure 18.2*). By S1 mapping and via cDNA sequence analysis[18] evidence was presented that the 5′ non-coding region of the lysozyme mRNA is heterogeneous in length. The 5′ termini map 29, 31 and 53 nucleotides upstream from the common translation initiation codon. The sequence at the 5′ end of the three different oviduct mRNAs is organized in such a way that only the longest mRNA is able to form a terminal hairpin structure. The relevance of the mRNA heterogeneity for the tissue- and hormone-specific expression of lysozyme is unknown at the present.

The gene for lysozyme

In total genomic DNA[50] and in cloned gene DNA[2,41] it was established that the lysozyme gene is split into four exons, distributed over a total DNA stretch of 3.8 kb (*Figure 18.2*). All introns interrupt the coding region for mature lysozyme[26]. The protein product encoded by exon 2 + 3 contains the complete substrate binding site of lysozyme. Exon 2, encoding the catalytically active centre of the enzyme (amino acid 28–82), can be considered a primitive β, 1–4 glycosidase minigene. The product encoded on exon 3 (amino acid 82–108) forms the so called specificity site of lysozyme. This second substrate binding region is responsible for the pattern of cleavage during hydrolysis of the N-acetylmuramic acid:N-acetyl glucosamine copolymer in bacterial cell walls. The products encoded by exon 1 and exon 4 do not participate directly in the catalytic function of the protein.

The immediate 5′ flanking sequence of exon 1 shows a surprising repetition of sequence elements similar to those found in the promoter region of other genes[18]. Two overlapping sets of motifs exist, each consisting of a TATA-like sequence and a CAAT-like sequence. Both sets of motifs are most likely used for initiation of transcription. They reside at equivalent positions upstream from the start positions at −2/+1 and −24, found in S1 mapping experiments with hnRNA and mRNA.

The analysis of the DNA region encompassing the lysozyme gene revealed repeated sequences[3] which belong to different families of interspersed middle repetitive elements of the chicken genome (Sippel *et al.*, unpublished result). Some of these elements (*Figure 18.2*), several kilobases away from the gene on both sides, are in the vicinity of a number of new DNA elements which are highly specific for the binding of a nuclear protein component[51].

Control by steroid hormones

Expression of the egg white protein genes

In sexually immature female chicks, differentiation of the oviduct is induced and maintained by the administration of steroid hormones. Oestrogens are required for the primary stimulation, in which egg white protein-producing tubular gland cells become the majority of the oviduct cells. Discontinuation of oestrogen stimulation (withdrawal) results in a rapid decline of egg white protein synthesis and in the establishment of a withdrawal state of the oviduct for about three days before the oviduct regresses. Egg white protein synthesis can be rapidly restimulated in the remaining tubular gland cells (secondary stimulation) by oestrogens and non-oestrogenic steroid hormones. Considerable work has been done to show that increasing or decreasing rates of egg white protein synthesis are closely parallel to changing levels of the respective mRNAs (for early reviews see references 65 and 67; for more recent contributions see references 59 and 69). Changes in mRNA abundance have been shown to be due to changes in rate of transcription of the respective genes[43, 49, 74], although mRNA stability also appears to be affected by steroid hormones[12, 23, 43, 57]. Measurements of the kinetics of specific mRNA accumulation[23, 60] and of transcriptional activation[43, 49] upon secondary steroid induction have shown that the egg white protein genes are not activated in a strictly coordinate fashion. Following steroid application, the synthesis of ovotransferrin mRNA sequences directly increases with maximal rate, starting out from a significant base level in the withdrawn state. However, transcription of the ovalbumin gene (and, most likely, transcription of the ovomucoid and the lysozyme gene) starts out from an undetectable low level in the withdrawn state. Transcriptional rates, even though rising measurably within the first hour, reach constant high level only after a lag period of roughly three hours.

The mechanisms by which steroid hormones control mRNA transcription (and stability) are unknown. It is generally assumed that steroids act via a common mechanism in which transcriptional regulation is mediated by steroid receptors that bind to specific sites on DNA, possibly near the genes they regulate. When specific gene transcription was correlated to nuclear oestrogen receptor levels[61], it became apparent that only ovotransferrin gene transcription initiates as rapidly as oestrogen receptors reach the nucleus and with rates proportional to nuclear receptor levels. In contrast, ovalbumin gene transcription is activated considerably more slowly and is related to receptor levels in a way that suggests cooperative interactions among receptors. A model that accounts for the various aspects of hormonal regulation of the ovalbumin and the transferrin gene in the oviduct has been presented by Palmiter and associates[61, 75]. Basically, the model predicts that a single receptor binding site is involved in the regulation of the ovotransferrin gene, whereas there are several such sites involved in ovalbumin gene regulation. A considerable strengthening of this multiple site hypothesis for receptor interaction with the ovalbumin gene may come from the recent finding of several high affinity binding regions for the chicken oviduct progesterone receptor in cloned ovalbumin gene DNA[48]. Studies on steroid receptor interaction with gene DNA can be expected to elucidate the mechanisms by which the same hormone is able to regulate different genes in the same cell differently.

Expression of the transferrin gene in oviduct and liver

By comparing transferrin gene activity in oviduct and liver the problem was studied as to how the same hormone can regulate the same gene in different cells differently[67,68]. The transferrin gene is active in both liver and oviduct, but the response to steroid hormones is qualitatively and quantitatively different[23, 37]. The insignificant change of the ovotransferrin mRNA level in chicken liver upon oestrogen administration cannot be explained by a lack of functional oestrogen receptor, since hepatocytes are synthesizing egg yolk proteins in a strictly oestrogen-dependent manner. This means that the gene must be organized in an active but different state in both organs, assuming that there are no differences between steroid receptor proteins and in the transferrin gene DNA and flanking sequences[38] in liver and oviduct. At present we do not know which structural features of DNA and/or chromatin might explain the differential programming of this gene in the two cell types.

Transcription and differentiation

When nucleases were used to probe for chromatin structure, it was found that the ovalbumin gene was degraded selectively in hen oviduct nuclei[17]. DNase sensitivity persists in primary induced chick oviducts even after oestrogens are withdrawn or when tamoxifen, an oestrogen antagonist, is applied, leaving the gene transcriptionally inactive[60, 70]. Furthermore, in the ovalbumin gene region nuclease sensitive chromatin extends far beyond the transcription unit[4, 35, 36]. Thus the sensitive state of chromatin does not reflect transcriptional activity *per se*, but rather appears to be a general prerequisite for the expression of the gene. Secondary induction of gene activity by steroids, in interaction with their specific receptor proteins, therefore must be a different process from the one which leads to changes in overall structure of the chromatin domain. However, oestrogens also trigger tubular gland cell differentiation and the cell-specific commitment of the egg white protein genes during primary induction, a process most likely intimately connected with the transition of the chromatin domains of these genes from the 'closed' to the 'open' state.

Now that the DNA sequence organization of the egg white protein genes has been largely elucidated, it will be necessary to define the proteins and the DNA signal sequences which are functionally responsible for the various regulatory processes. Promoter regions, on the one hand, and the boundaries between the nuclease sensitive and resistant chromatin, on the other hand, might be the respective focal points for regulation of transcriptional activity and gene commitment.

References

1. Aisen, P. and Listowsky, I. (1980). Iron transport and storage proteins. *Annual Review of Biochemistry*, **49**, 357–393
2. Baldacci, P., Royal, A., Cami, B., Perrin, F., Krust, A., Garapin, A. and Kourilsky, O. (1979). Isolation of the lysozyme gene of chicken. *Nucleic Acids Research*, **4**, 2667–2681
3. Baldacci, P., Royal, A., Brëgëgère, F., Abastado, J. P., Cami, B., Daniel, F. and Kourilsky, P. (1981). DNA organization in the chicken lysozyme gene region. *Nucleic Acids Research*, **9**, 3575–3588
4. Bellard, M., Kuo, M. T., Dretzen, G. and Chambon, P. (1980). Differential nuclease sensitivity of the ovalbumin and β-globin chromatin regions in erythrocytes and oviduct cells of laying hen. *Nucleic Acids Research*, **8**, 2737–2750
5. Benoist, C., O'Hare, K. O., Breathnach, R. and Chambon P. (1980). The ovalbumin gene-sequence of putative control regions. *Nucleic Acids Research*, **8**, 127–142
6. Breathnach, R., Mandel, J. L. and Chambon, P. (1977). Ovalbumin gene is split in chicken DNA. *Nature, London*, **270**, 314–319
6a. Breathnach, R., Benoist, C., O'Hara, K., Gannon, F. and Chambon, P. (1978). *Proceedings of the National Academy of Sciences, USA*, **75**, 4853–4857

7. Breathnach, R., Mantai, N. and Chambon, P. (1980). Correct splicing of a chicken ovalbumin gene transcript in mouse L cells. *Proceedings of the National Academy of Sciences, USA*, **77**, 740–744
8. Catterall, J. F., Stein, J. P., Kristo, P., Means, A. R. and O'Malley, B. W. (1980). Primary sequence of ovomucoid messenger RNA as determined from cloned complementary DNA. *Journal of Cell Biology*, **87**, 480–487
9. Cochet, M., Gannon, F., Hen, R., Maroteaux, L., Perrin, F., and Chambon, P. (1979). Organisation and sequence studies of the 17-piece chicken conalbumin gene. *Nature, London*, **282**, 567–574
10. Colbert, D. A., Knoll, B. J., Woo, S. L. C., Mace, M. L., Tsai, M. J. and O'Malley, B. W. (1980). Differential hormonal responsiveness of the ovalbumin gene and its pseudogenes in the chick oviduct. *Biochemistry*, **19**, 5586–5592
11. Cordon, J., Wasylyk, B., Buchwalder, A., Sassone-Corsi, P., Kedinger, C. and Chambon, P. (1980). Promoter sequences of eukaryotic protein-coding genes. *Science*, **209**, 1406–1414
12. Cox, R. F. (1977). Estrogen withdrawal in chick oviduct. Selective loss of high abundance classes of polyadenylated messenger RNA. *Biochemistry*, **16**, 3433–3443
13. Doel, M. T., Houghton, M., Cook, E. A. and Carey, N. H. (1977). The presence of ovalbumin mRNA coding sequences in multiple restriction fragments of chicken DNA. *Nucleic Acids Research*, **4**, 3701–3713
14. Dugaiczyk, A., Woo, S. L. C., Colbert, D. A., Lai, E. C., Mace, M. L. and O'Malley, and O'Malley, B. W. (1979). The ovalbumin gene: cloning and molecular organization of the entire natural gene. *Proceedings of the National Academy of Sciences, USA*, **76**, 2253–2257
15. Gannon, F., Jeltsch, J. M. and Perrin, F. (1980). A detailed comparison of the 5'-end of the ovalbumin gene cloned from chicken oviduct and erythrocyte DNA. *Nucleic Acids Research*, **8**, 4405–4421
16. Gannon, F., O'Hare, K., Perrin, F., LePennec, J. P., Benoist, C., Cochet, M., Breathnach, R., Royal, A., Garapin, A., Cami, B. and Chambon, P. (1979). Organisation and sequences at the 5' end of a cloned complete ovalbumin gene. *Nature, London*, **278**, 428–434
17. Garel, A. and Axel, R. (1976). Selective digestion of transcriptionally active ovalbumin genes from oviduct nuclei. *Proceedings of the National Academy of Sciences, USA*, **73**, 3966–3970
18. Grez, M., Land, H., Giesecke, K., Schütz, G., Jung, A. and Sippel, A. E. (1981). Multiple mRNAs are generated from the chicken lysozyme gene. *Cell*, **25**, 743–752
19. Harris, S. E., Rosen, J. M., Means, A. R. and O'Malley, B. W. (1975). Use of a specific probe for ovalbumin messenger RNA to quantitate estrogen-induced gene transcripts. *Biochemistry*, **14**, 2072–2080
20. Heilig, R., Perrin, F., Gannon, F., Mandel, J. L. and Chambon, P. (1980). The ovalbumin gene family: structure of the X gene and evolution of duplicated split genes. *Cell*, **20**, 625–637
21. Hughes, S. H., Stubblefield, E., Payvar, F., Engel, J. D., Dodgson, J. B., Spector, D., Cordell, B., Schimke, R. and Varmus, H. E. (1979). Gene localization by chromosome fractionation: Globin genes are on at least two chromosomes and three estrogen-inducible genes are on three chromosomes. *Proceedings of the National Academy of Sciences, USA*, **76**, 1348–1352
22. Hynes, N. E., Groner, B., Sippel, A. E., Nguyen-Huu, M. C. and Schütz, G. (1977). mRNA complexity and egg white protein mRNA content in mature and hormone-withdrawn oviduct. *Cell*, **11**, 923–932
23. Hynes, N. E., Groner, B., Sippel, A. E., Jeep, S., Wurtz, T., Nguyen-Huu, M. C., Giesecke, K. and Schütz, G. (1979). Control of cellular content of chicken egg white protein specific RNA during estrogen administration and withdrawal. *Biochemistry*, **18**, 616–624
24. Imoto, T., Johnson, L. N., North, A. C. T., Phillips, D. C. and Rupley, J. A. (1972). Vertebrate lysozymes. In *The Enzymes*, Volume 1, (Boyer, P. D., Ed.) pp. 665–868. New York, Academic
25. Jeltsch, J. M. and Chambon, P. (1982). The complete nucleotide sequence of the chicken ovotransferrin mRNA. *European Journal of Biochemistry*, **122**, 291–295
26. Jung, A., Sippel, A. E., Grez, M., Schütz, G. (1980). Exons encode functional and structural units of chicken lysozyme. *Proceedings of the National Academy of Sciences, USA*, **77**, 5759–5763
27. Kato, I., Kohr, W. J. and Laskowski, M. Jr. (1978). Evolution of avian ovomucoids. *Federation of European Biochemistry Society Proceedings of 11th Meeting, Copenhagen*, **47**, 197–206
28. Kourilsky, P. and Chambon, P. (1978). The ovalbumin gene: an amazing gene in eight pieces. *Trends in Biochemical Science*, **3**, 244–247
29. Kurachi, K., Chandra, T., Friezner Degen, S. J., White, T. T., Marchioro, T. L., Woo, S. L. C. and Davie, E. W. (1981). Cloning and sequence of cDNA coding for α_1-antitrypsin. *Proceedings of the National Academy of Sciences, USA*, **78**, 6826–6830
30. Lai, E. C., Woo, S. L. C., Dugaiczyk, A., Catterall, J. F. and O'Malley, B. W. (1978). The ovalbumin gene: structural sequences in native chicken DNA are not contiguous. *Proceedings of the National Academy of Sciences, USA*, **75**, 2205–2209

31. Lai, E. C., Woo, S. L. C., Dugaiczyk, A. and O'Malley, B. W. (1979). The ovalbumin gene: Alleles created by mutations in the intervening sequences of the natural gene. *Cell*, **16**, 201–211

32. Lai, E. C., Stein, J. P., Catterall, J. F., Woo, S. L. C., Mace, M. L., Means, A. R. and O'Malley, B. W. (1979). Molecular structure and flanking nucleotide sequences of the natural chicken ovomucoid gene. *Cell*, **18**, 829–842

33. Lai, E. C., Woo, S. L. C., Bordelon-Riser, M., Fraser, T. H. and O'Malley, B. W. (1980). Ovalbumin is synthesized in mouse cells transformed with the natural chicken ovalbumin gene. *Proceedings of the National Academy of Sciences, USA*, **77**, 244–248

34. Laskowski, M. Jr. and Kato, I. (1980). Protein inhibitors of proteinases. *Annual Review of Biochemistry*, **49**, 593–626

35. Lawson, G. M., Tsai, M. J. and O'Malley, B. W. (1980). Desoxyribonuclease I sensitivity of the nontranscribed sequences flanking the 5' and 3' ends of the ovomucoid gene and the ovalbumin and its related X and Y genes in hen oviduct nuclei. *Biochemistry*, **19**, 4403–4411

36. Lawson, G. M., Knoll, B. J., March, C. J., Woo, S. L. C., Tsai, M. J. and O'Malley, B. W. (1982). Definition of 5' and 3' structural boundaries of the chromatin domain containing the ovalbumin multigene family. *Journal of Biological Chemistry*, **257**, 1501–1507

37. Lee, D. C., McKnight, G. S. and Palmiter, R. D. (1978). The action of estrogen and progesteron on the expression of the transferrin gene. A comparison of the response in chicken liver and oviduct. *Journal of Biological Chemistry*, **253**, 3494–3503

38. Lee, D. C., McKnight, S. G. and Palmiter, R. D. (1980). The chicken transferrin gene. Restriction endonuclease analysis of gene sequences in liver and oviduct DNA. *Journal of Biological Chemistry*, **255**, 1442–1450

39. LeMeur, M., Glanville, N., Mandel, J. L., Gerlinger, R., Palmiter, R. and Chambon, P. (1981). The ovalbumin Gene family: Hormonal control of X and Y gene transcription and mRNA accumulation. *Cell*, **23**, 561–571

40. Lindenmaier, W., Nguyen-Huu, M. C., Lurz, R., Blin, N., Stratmann, M., Land, H., Jeep, S., Sippel, A. E. and Schütz, G. (1979). Isolation and characterization of the chicken ovomucoid gene. *Nucleic Acids Research*, **7**, 1221–1232

41. Lindenmaier, W., Nguyen-Huu, M. C., Lurz, R., Stratmann, M., Blin, N., Wurtz, T., Hauser, H. J., Sippel, A. E. and Schütz, G. (1979). Arrangement of coding and intervening sequences of chicken lysozyme gene. *Proceedings of the National Academy of Sciences, USA*, **76**, 6196–6200

42. Lingappa, V. R., Lingappa, J. R. and Blobel, G. (1979). Chicken ovalbumin contains an internal signal sequence. *Nature, London*, **281**, 117–121

43. McKnight, G. S. and Palmiter, R. D. (1979). Transcriptional regulation of the ovalbumin and conalbumin genes by steroid hormones in chick oviduct. *Journal of Biological Chemistry*, **254**, 9050–9058

44. McKnight, G. S., Pennequin, P. and Schimke, R. T. (1975). Induction of ovalbumin mRNA sequences by estrogen and progesteron in chick oviduct as measured by hybridization to complementary DNA. *Journal of Biological Chemistry*, **250**, 8105–8110

45. McReynolds, L. A., O'Malley, B. W., Misbet, A. D., Fothergill, J. E., Givole, D., Fields, S., Robertson, M. and Brownlee, G. G. (1978). Sequence of chicken ovalbumin mRNA. *Nature, London*, **273**, 723–728

46. Malek, L. T., Eschenfeldt, W. H., Munns, T. W. and Rhoads, R. E. (1981). Heterogeneity of the 5' terminus of hen ovalbumin messenger ribonucleic acid. *Nucleic Acids Research*, **9**, 1657–1673

47. Mandel, J. L. and Chambon, P. (1979). DNA methylation: organ specific variations in the methylation pattern within and around ovalbumin and other chicken genes. *Nucleic Acids Research*, **7**, 2081–2103

48. Mulvihill, E. R., LePennec, J. P. and Chambon, P. (1982). Chicken oviduct progesterone receptor: Location of specific regions of high-affinity binding in cloned DNA fragments of hormone-responsive genes. *Cell*, **24**, 621–632

49. Nguyen-Huu, M. C., Barrett, K. J., Giesecke, K., Wurtz, T., Sippel, A. E. and Schütz, G. (1978). Transcription of the chicken ovalbumin and conalbumin gene during early secondary induction with estrogens. *Hoppe-Seyler's Zeitschrift für Physiologische Chemie*, **359**, 1307–1313

50. Nguyen-Huu, M. C., Stratmann, M., Groner, B., Wurtz, T., Land, H., Giesecke, K., Sippel, A. E. and Schütz, G. (1979). Chicken lysozyme gene contains several intervening sequences. *Proceedings of the National Academy of Sciences, USA*, **76**, 76–80

51. Nowock, J. and Sippel, A. E. (1982). Specific protein–DNA interaction at four sites flanking the chicken lysozyme gene. *Cell*, **30**, 607–615

52. O'Hare, K., Breathnach, R., Benoist, C. and Chambon, P. (1979). No more than seven interruptions in the ovalbumin gene: comparison of genomic and doublestranded cDNA. *Nucleic Acids Research*, **7**, 321–334

53. O'Malley, B. W., McGuire, W. L., Kohler, P. O. and Koreman, S. G. (1969). Studies on the mechanism of steroid hormone regulation of synthesis of specific proteins. *Recent Progress in Hormone Research*, **25**, 105–160

54. O'Malley, B. W., Roop, D. R., Lai, E. C., Nordstrom, J. L., Catterall, J. F., Swaneck, G. E., Colbert, D. A., Tsai, M. J., Dugaiczyk, A. and Woo, S. L. C. (1979). The ovalbumin gene: Organization, structure, transcription and regulation. *Recent Progress in Hormone Research*, **35**, 1–46

55. Palmiter, R. D. (1972). Regulation of protein synthesis in chick oviduct. I. Independent regulation of ovalbumin, conalbumin, ovomucoid and lysozyme induction. *Journal of Biological Chemistry*, **247**, 6450–6461

56. Palmiter, R. D. (1975). Quantitation of parameters that determine the rate of ovalbumin synthesis. *Cell*, **4**, 189–197

57. Palmiter, R. D. and Carey, N. H. (1974). Rapid inactivation of ovalbumin messenger ribonucleic acid after acute withdrawal of estrogen. *Proceedings of the National Academy of Sciences, USA*, **71**, 2357–2361

58. Palmiter, R. D., Gagnon, J. and Walsh, K. A. (1978). Ovalbumin: A secreted protein without a transient hydrophobic leader sequence. *Proceedings of the National Academy of Sciences, USA*, **75**, 94–98

59. Palmiter, R. D., Moore P. D., Mulvihill, E. R. and Emtage, S. (1976). A significant lag in the induction of ovalbumin messenger RNA by steroid hormones: a receptor translocation hypothesis. *Cell*, **8**, 557–572

60. Palmiter, R. D., Mulvihill, E. R., McKnight, G. S. and Senear, A. W., (1978). Regulation of gene expression in the chick oviduct by steroid hormones. *Cold Spring Harbor Symposium on Quantitative Biology*, **42**, 639–647

61. Palmiter, R. D., Mulvihill, E. R., Shepard, J. H. and McKnight, G. S. (1981). Steroid hormone regulation of ovalbumin and conalbumin gene transcription: A model based upon multiple regulatory sites and intermediary proteins. *Journal of Biological Chemistry*, **256**, 7910–7916

62. Perrin, F., Cochet, M., Gerlinger, P., Cami, B., LePennec, J. P. and Chambon, P. (1979). The chicken conalbumin gene: studies of the organization of cloned DNAs. *Nucleic Acids Research*, **6**, 2731–2748

63. Roop, D. R., Tsai, M. J. and O'Malley, B. W. (1980). Definition of the 5' and 3' ends of transcripts of the ovalbumin gene. *Cell*, **19**, 63–68

64. Roop, D. R., Nordstrom, J. L., Tsai, S. Y., Tsai, M. J. and O'Malley, B. W. (1978). Transcription of structural and intervening sequences in the ovalbumin gene and identification of potential ovalbumin mRNA precursors. *Cell*, **15**, 671–685

65. Rosen, J. M. and O'Malley, B. W. (1975). Hormonal regulation of specific and gene expression in the chick oviduct. In *Biochemical Actions of Hormones* (Litwack, G., Ed.), Volume 3, pp. 271–315. New York, Academic Press

66. Royal, A., Garapin, A., Cami, B., Perrin, F., Mandel, J. L., LeMeur, M., Brégégère, F., Gannon, F., LePennec, J. P., Chambon, P. and Kourilsky, P. (1979). The ovalbumin gene region: Common features in the organisation of three genes expressed in chicken oviduct under hormonal control. *Nature, London*, **279**, 125–132

67. Schimke, R. T., McKnight, G. S. and Shapiro, D. J. (1975). Nucleic acid probes and analysis of hormone action in oviduct. In *Biochemical Actions of Hormones* (Litwack, G., Ed.) Volume 3, pp. 245–269. New York, Academic Press

68. Schimke, R. T., McKnight, G. S., Payvar, F. and Pennequin, P. (1975). Hormone control of protein synthesis in hen oviduct. In *Developmental Biology. Pattern formation. Gene Regulation* (McMahon, D. and Fox, C. F., Eds.) pp. 384–394. Mento Park, Benjamin

69. Schütz, G., Nguyen-Huu, M. C., Giesecke, K., Hynes, N. E., Groner, B., Wurtz, T. and Sippel, A. E. (1978). Hormonal control of egg white protein messenger RNA synthesis in the chicken oviduct. *Cold Spring Harbor Symposium on Quantitative Biology*, **42**, 617–624

70. Shepard, J. H., Mulvihill, E. R., Thomas, P. S. and Palmiter, R. D. (1980). Commitment of chick oviduct tubular gland cells to produce ovalbumin mRNA during hormonal withdrawal and restimulation. *Journal of Cell Biology*, **87**, 142–151

71. Sippel, A. E., Land, H., Lindenmaier, W., Nguyen-Huu, M. C., Wurtz, T., Timmis, K. N., Giesecke, K. and Schütz, G. (1978). Cloning of chicken lysozyme structural gene sequences synthesized *in vitro*. *Nucleic Acids Research*, **5**, 3275–3294

72. Stein, J. P., Catterall, J. F., Kristo, P., Means, A. R. and O'Malley, B. W. (1980). Ovomucoid intervening sequences specify functional domains and generate protein polymorphism. *Cell*, **21**, 681–687

73. Stumph, W. E., Kristo, P., Tsai, M. J. and O'Malley, B. W. (1981). A chicken middle-repetitive DNA sequence which shares homology with mammalian ubiquitous repeats. *Nucleic Acids Research*, **9**, 5383–5397
74. Swaneck, G. E., Nordstrom, J. L., Kreuzaler, F., Tsai, M. J. and O'Malley, B. W. (1979). Effect of estrogen on gene expression in chicken oviduct: Evidence for transcriptional control of ovalbumin gene. *Proceedings of the National Academy of Sciences, USA*, **76**, 1049–1053
75. Thomas, P. S. and Teller, D. C. (1981). A multiple site hypothesis for the interaction of steroid hormone receptors with the ovalbumin gene. *Journal of Molecular Biology*, **150**, 160–166
76. Tsai, S. Y., Tsai, M. J. and O'Malley, B. W. (1981). Specific 5' flanking sequences are required for faithful initiation of *in vitro* transcription of the ovalbumin gene. *Proceedings of the National Academy of Sciences, USA*, **78**, 879–883
77. Tsai, M. J., Ting, A. C., Nordstrom, J. L., Zimmer, W. and O'Malley, B. W. (1980). Processing of high molecular weight ovalbumin and ovomucoid precursor RNAs to messenger RNA. *Cell*, **22**, 219–230
78. Wasylyk, B., Kédinger, C., Cordon, J., Brison, O. and Chambon, P. (1980). Specific *in vitro* initiation of transcription on conalbumin and ovalbumin genes and comparison with adenovirus-2 early and late genes. *Nature, London*, **285**, 367–373
79. Weinstock, R., Sweet, R., Weiss, M., Cedar, H. and Axel, R. (1978). Intragenic DNA spacers interrupt the ovalbumin gene. *Proceedings of the National Academy of Sciences, USA*, **75**, 1299–1303
80. Williams, J. (1968). A comparison of glycopeptides from the ovotransferrin and serum transferrin of the hen. *Biochemical Journal*, **108**, 57–67
81. Williams, J., Elleman, T. C., Kingston, I. B., Wilkins, A. G. and Kuhn, K. A. (1982). The primary structure of hen ovotransferrin. *European Journal of Biochemistry*, **122**, 297–303

Chapter 19

Regulation and structure of the vitellogenin genes

G. U. Ryffel and W. Wahli

Vitellogenins are female-specific lipoglycoproteins synthesized outside of the ovary and subsequently taken up by the growing oocyte to become the major constituent of the yolk, the vitellins. The vitellogenins are found in all animal species producing yolky eggs and their synthesis is tightly regulated by hormones.

The hormonal regulation of vitellogenin synthesis offers a challenging system to analyze the molecular mechanism of hormone activated expression of specific genes. Since vitellogenins are produced in large amounts during oogenesis, they can be isolated and characterized quite easily. Their high levels of synthesis allow the purification of the corresponding mRNA, a prerequisite to isolation and characterization of the gene itself. Furthermore, vitellogenins undergo protein modification, secretion, uptake and cleavage all of which may be analyzed in detail in this system.

In this chapter we will discuss data obtained in several animal species, including examples from insects as well as from vertebrates. We concentrate on experimental systems, where recent experiments have succeeded in extending the analysis from the protein to the gene level. Data mainly confined to the protein level have been reviewed a few years ago by Hagedorn and Kunkel[24] for insects and by Clemens[13] for vertebrates.

Vitellogenesis in insects

Large and small yolk precursor proteins

In female insects vitellogenin is synthesized mainly in the fat body and subsequently secreted into the hemolymph to be taken up by the growing oocyte. We can distinguish species such as the African migratory locust (*Locusta migratoria*) where vitellogenins of high molecular weight, up to 140 000 d, are found in the hemolymph from species where the vitellogenins have relatively low molecular weights (~45 000 d) as, for example, in the fruit fly *Drosophila melanogaster*. In *Locusta* the primary translation products are cleaved into smaller proteins with molecular weights of 50 000 to 140 000 d, while in *Drosophila* only small changes in the molecular weight of the primary translational products and the deposited yolk proteins can be observed[57]. Since no obvious change in circulatory and deposited proteins are evident in *Drosophila*, the proteins are already designated yolk proteins when they are present in the hemolymph. These proteins have, however, typical characteristics of vitellogenins, i.e. they are synthesized at least partially outside of the ovary and they undergo some physical changes after uptake into the oocyte prior to their incorporation into the yolk granules. In *Drosophila*, the three different yolk proteins YP1, YP2 and YP3 with molecular weights of 47 000, 46 000 and 45 000 d respectively, have been identified[56].

In vitro translation experiments with isolated mRNAs clearly established that these three proteins are derived from three different mRNAs of correspondingly low molecular weight and that some post-translational modifications occur after cleavage of a short signal peptide[8,57].

In *Locusta*, *in vitro* incubation of isolated fat bodies revealed two very large vitellogenin primary translational products of 250 000 and 265 000 d. These large protein sizes were verified by the isolation and characterization of a large vitellogenin mRNA of about 6300 nucleotides[1, 10, 12]. Incubation of isolated fat bodies demonstrated that the two primary translational products were converted into at least eight different polypeptides ranging from 50 000 to 140 000 d which are found as main constituents of the yolk. Most of these proteins are extensively modified by the addition of carbohydrates and lipids[11]. Although the experiments did not rigorously exclude artificial degradation during preparation of the yolk proteins, it seems clear that extensive post-translational cleavages occur intracellularly in the fat body. This contrasts with the situation in vertebrates (see below) where cleavage of precursor proteins into the main yolk protein takes place after uptake into the oocyte.

Regulation of vitellogenin synthesis

Vitellogenin synthesis is regulated in most insects by juvenile hormone but in a few cases by ecdysone[24]. Juvenile hormone is produced in all stages of insect development in both sexes. Vitellogenin, however, is usually produced only in adult females (for exceptions see reference 17).

Though juvenile hormone induces polyploidy in male *Locusta* fat body cells, a process normally occurring in females, the hormone does not stimulate vitellogenin synthesis[37]. This restricted responsiveness of male fat body cells is an intrinsic character of this tissue, since a female fat body transplanted into a male can be induced to vitellogenin synthesis, while the male fat body remains inactive.

In the mosquito *Aedes aegypti* there is good evidence from *in vitro* organ cultures that ecdysone controls vitellogenin synthesis[25], whereas in the fruit fly *Drosophila melanogaster* both ecdysone and juvenile hormone regulate yolk protein synthesis. It seems clear that in *Drosophila* ecdysone induces yolk protein synthesis only in the fat body, whereas juvenile hormone stimulates vitellogenin synthesis in fat body and in the ovary[39]. It is not clear, however, whether both these hormones act directly on the target tissues, since experiments were successful only on isolated abdomens and not on isolated fat bodies or ovaries.

Quantification of vitellogenin mRNA levels in *Drosophila*[5] and *Locusta*[1, 12] revealed that induced vitellogenin synthesis is due to the accumulation of the corresponding mRNAs. These observations suggest that the hormone regulates vitellogenin gene expression on a transcriptional level.

In *Drosophila* several mutations which alter or even suppress vitellogenin synthesis have been described[39]. So far, one mutant, *fs* (1) 1163 has been identified which produces a structurally abnormal primary translation product of the YP1 gene, leading to a reduction in levels of YP1. Another mutation, *RI*, maps very close to the YP3 structural gene and prevents the synthesis of YP3. In this case, the gene and the corresponding mRNA are detectable in normal amounts but YP3 mRNA cannot be translated. Both these mutants are of great interest in understanding the requirements for correct translation of the vitellogenin mRNAs and for post-translational modifications of the newly synthesized vitellogenin. Other mutants may provide additional features more closely related to the genetic elements responsible for hormonal control.

Gene structure and chromosomal localization of the vitellogenin genes

All three yolk protein genes of *Drosophila* have been isolated by cloning techniques[6, 27, 40]. A detailed analysis of the isolated genes, mainly performed in Wensink's group, revealed their structure as summarized in *Figure 19.1*.

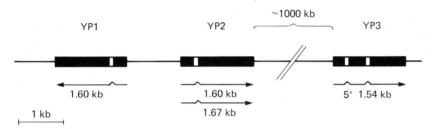

Figure 19.1. Structure of the yolk protein genes in *Drosophila*. The exons (black boxes) and the introns (white boxes) of the three yolk protein genes are given. The known transcripts are indicated by arrows and their size given in kb. Drawn from data of Barnett and Wensink[5] and Hung *et al.*[30].

S1 digestion of hybrids between yolk protein mRNAs and the corresponding genomic clone showed one intron each in YP1 and YP2 and two introns in YP3. Sequencing of the genomic DNA of the YP1 gene revealed the amino acid sequence of the signal peptide and the mature YP1 protein[27, 30].

Formation of heteroduplexes between the three protein genes demonstrated that YP1 and YP2 are more closely related to one another than to YP3. *In situ* hybridization of cloned fragments of the yolk protein genes to polytene chromosomes revealed that all these genes are localized on the X chromosome. Whereas YP1 and YP2 are closely linked (in fact cloning experiments yielded clones containing both genes), YP3 is about 1000 kb away. The same loci had been determined previously by genetic analysis[39]. The fact that the two linked genes are closely related in sequence and have a similar intron–exon pattern may suggest that these genes have arisen by duplication quite recently, after the separation from the YP3 gene. The closely linked YP1 and YP2 genes are transcribed from opposite strands and therefore represent different transcriptional units. Obviously, some mechanism must ensure coordinate control of these three genes which each have their own promotor for RNA transcription.

Analyzing the size of the RNA coding for the three yolk proteins, Hung *et al.*[30] observed that the YP2 gene codes for two mRNA species which may arise from alternative sites of transcription termination. Since these two YP2 transcripts can be isolated from polysomal RNA of fat bodies, it seems likely that both are translated. The significance of this heterogeneity at the 3' end is unknown but has also been observed for transcripts of other genes[18, 45, 47].

In *Locusta* Southern blot analysis of restricted genomic DNA with cloned vitellogenin cDNA probes indicated the existence of multiple, partially homologous sequences in the genome[64]. Since the genome of *Locusta* is about forty-fold larger than that of *Drosophila*, isolation of the vitellogenin genes of *Locusta* is correspondingly more difficult. Nevertheless, fragments of one vitellogenin gene have been isolated[64].

Vitellogenesis in vertebrates

In fish, amphibians, reptiles and birds, the yolk proteins deposited in the growing oocytes are derived from a precursor protein, vitellogenin, which is synthesized under oestrogen control in the liver of mature females, but can also be induced artificially in males by hormone injection.

At this time, the amphibian and avian systems are better characterized than are the fish and reptile systems. Since information for both gene structure and gene expression are available only for amphibians and birds, we shall focus on these two systems.

Avian and amphibian vitellogenins

Vitellogenin has a monomer molecular weight of 210 000–230 000 d in chicken and 190 000–210 000 d in *Xenopus* and is highly modified in both species. Glycosylation, phosphorylation and lipidation occur prior to secretion from hepatocytes[52]. Two distinct vitellogenins have been found in the serum of oestrogen-stimulated roosters suggesting the possible presence of two genes within a haploid chromosome set[53]. Multiple vitellogenins have also been reported in *Xenopus*[60] for which there is direct evidence demonstrating vitellogenin gene repetition (see below). In *Xenopus*, vitellogenin is taken up by the oocyte about 50 times faster than are other serum proteins, by receptor-mediated endocytosis. Newly formed endosomes fuse to form transitional yolk bodies in which proteolytic processing of vitellogenin into the yolk proteins takes place[38]. The cleavage products subsequently deposited in the yolk granules are as heterogeneous as the vitellogenins from which they are derived. They are grouped into lipovitellins with molecular weights from 30 000–121 000 d carrying non-covalently bound lipids and into phosvitins which are highly phosphorylated proteins of molecular weights from 13 000–34 000 d[61].

Control of vitellogenin synthesis

The specificity of the oestrogen response is very high. Other steroids, such as testosterone, progesterone, cortisone and dexamethasone do not trigger synthesis of vitellogenin as demonstrated in *Xenopus* males[52]. The possibility of induced synthesis in male hepatocytes has proved very useful in studying the cell and molecular biology of the response to oestrogen. Through this induction, massive production and secretion of vitellogenin by males occurs in fully differentiated hepatocytes. Profound cytological changes are observed in these cells, including formation of Golgi elements and rough endoplasmatic reticulum, loss of glycogen and modification of nuclear structures[7]. Induction of vitellogenin synthesis can be achieved in organ cultures as well as in primary cultures of hepatocytes from amphibians, demonstrating that oestrogen acts directly on the hepatocytes. The *in vitro* studies have also shown that induction is not dependent upon DNA synthesis in the parenchymal cells and that the cell population is responding homogeneously to the hormone stimulus[20, 23, 44, 46, 54, 55].

The high concentration and large size of the vitellogenin mRNA facilitated its isolation from oestrogen-induced animals. Purified *Xenopus* 6.3 kb and chicken 6.9 kb mRNA which accumulate in the cytoplasm of hepatocytes after estrogen induction were shown to code for vitellogenin by *in vitro* translation followed by immunoprecipitation of the products with antibodies against vitellogenin. The accumulation of vitellogenin mRNA during hormonal stimulation has been demonstrated by titration with complementary DNA probes. Prior to induction there is no detectable vitellogenin mRNA in either male *Xenopus* or roosters. After hormone administration massive accumulation of vitellogenin mRNA has been observed. In roosters, the accumulation is more rapid than in *Xenopus*. However, the accumulation level reached is higher in the latter. The mRNA appears less than one hour after induction in roosters, compared with 4.5 hours in *Xenopus*. The accumulation peak level is reached after 2–3 days with 5000–10 000 mRNA molecules/cell in rooster and after 7–12 days with 30 000–50 000 mRNA molecules/hepatocyte in *Xenopus*. In both species, the time and level of maximal response is dose dependent. After reaching a maximum, the levels of mRNA decline and finally are indistinguishable from those of unstimulated males. Restimulation (secondary stimulation) at this time leads to a faster and more efficient response. In both rooster and *Xenopus*, the lag phase is shorter, the accumulation is faster and the levels reached are 10 000–15 000 mRNA molecules/cell in roosters and 80 000–100 000 mRNA molecules/hepatocyte in *Xenopus*[3, 4, 9, 15, 16, 32, 41]. The molecular basis for the faster secondary response is not yet elucidated but some possible explanations will be discussed below.

These observations together with others which have demonstrated that no vitellogenin precursor RNA sequences are present in nuclear RNA before stimulation[49] indicate that in the vitellogenin system, oestrogen regulates gene expression primarily at the transcriptional level. In addition, the kinetics of accumulation of vitellogenin mRNA are characteristic for molecules with a long half life. In fact, good evidence has been presented recently that the half life of avian vitellogenin mRNA is 7–8 times longer in the presence of oestrogen than in its absence[63].

At all stages of induction it is observed that the amount of vitellogenin mRNA in the hepatocytes is roughly proportional to the rate of vitellogenin synthesis and no substantial translational control has been reported.

Oestrogen receptors and ontogenic appearance of oestrogen-dependent inducibility of vitellogenin synthesis

In the classical model for steroid hormone action on gene expression, oestrogen first binds to cytoplasmic receptors present in high concentration in target cells (about 10 000 sites). The oestrogen–receptor complex then translocates to the nucleus where it binds to chromatin and regulates transcription of specific genes. In *Xenopus* the receptor system appears to deviate in several points from the model. A protein that binds oestrogen with high affinity ($K_d = 0.5 \times 10^{-9}$M) has been found in hepatocytes, but only at a frequency of 100–150 binding sites/cell in the cytoplasm and 100–500 sites/nucleus. After oestrogen stimulation the level of nuclear receptor reaches 1000–2000 binding sites within 12 hours, while there is no significant increase in the cytoplasm[26, 58, 59]. About 30 days after induction, when vitellogenin mRNA has disappeared completely from the cell, the nuclear receptor level is still three times higher than in cells of non-treated males. After 60–70 days of withdrawal the nuclear receptor level has returned to basal level (uninduced male), but over 1000 sites were measured in the cytoplasm at this time. It can be concluded that oestrogen enhances the level of its own receptor and an elevated nuclear receptor level might be a prerequisite for vitellogenin induction (see below). Compared to uninduced animals, withdrawn frogs have a higher hepatocyte receptor level which could be responsible for the shorter lag period and faster accumulation during secondary stimulation. In avian hepatocytes, interestingly, the level of oestrogen receptor is also low. Only 150 nuclear and 300 cytoplasmic high affinity sites have been found in cockerel hepatocytes. Following oestrogen stimulation the level of nuclear binding sites rises to about 2500 and the level of cytoplasmic sites remains at 250[34]. These values compare well with those found in *Xenopus* and probably reflect strong similarities in the induction mechanisms.

In *Xenopus*, vitellogenin synthesis becomes inducible by oestrogen when the developing tadpole reaches stage 62, that is during metamorphic climax[28, 35]. Low oestrogen receptor levels of 50–100 sites/nucleus are present well before this stage. However, oestrogen is able to elevate its receptor level only from stage 62 onwards, when vitellogenin becomes inducible[36]. This coincidence between inducibility and ability to enhance the receptor level again suggests that an elevated receptor level is necessary for vitellogenin induction. Synthesis of several cellular proteins is modulated by oestrogen before stage 62, though the receptor level is still very low. In chicken, the acquisition of the hepatic oestrogen receptor system occurs on days 10–12 of development, whereas vitellogenin synthesis becomes inducible between days 13–15. The inducibility of vitellogenin is delayed compared with another oestrogen-stimulated protein, i.e. apoprotein B of VLDL which can be stimulated from day 11 onwards[33]. This suggests that in both *Xenopus* and chicken different levels of receptors are needed for the regulation of specific genes, possibly revealing different affinities of the controlling regions for the hormone–receptor complex.

Multiplicity and structure of the vitellogenin gene

The analysis of cloned complementary DNA prepared from *Xenopus* vitellogenin mRNA has revealed that vitellogenin is encoded in at least four genes, all expressed coordinately in hepatocytes of individual, stimulated animals[51]. The four mRNAs were designated A1, A2 and B1, B2. A and B sequences differ in about 20% of their nucleotides, whereas A1 and A2 or B1 and B2 differ only in 5% of their nucleotides. The four mRNAs have similar lengths and direct synthesis, *in vitro*, of four 200000 d polypeptides with differences in their primary structure proportional to the divergence between the mRNAs[20, 31]. Genomic sequences of the four genes have been isolated from *Xenopus* DNA libraries[48]. Linkage between different genes has been observed but the organization of the whole family is not yet elucidated (Wahli *et al.*, unpublished). The two A genes (A1 and A2) have been extensively studied and their structural organization is summarized in *Figure 19.2*.

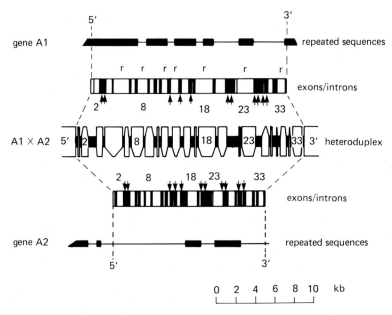

Figure 19.2. Structure of the A1 and A2 vitellogenin genes of *Xenopus*. The exons (black boxes) and the 33 introns (white boxes or arrows for small introns) are given for the A1 and A2 vitellogenin genes. The map derived from heteroduplex analysis shows pairing of homologous exons and heterology loops of unpaired introns. Regions where repetitive DNA was found are indicated in the top and bottom maps as heavy bars. The letter 'r' indicates repetitive DNA elements which have been mapped to specific introns of the A1 gene. Drawn from data of Wahli *et al.*[49] and Ryffel *et al.*[43].

Gene A1 has a length of 21 kb and gene A2 of 16 kb. Since both genes produce a 6.3 kb mRNA, differences in their intron structure must account for the length divergence. In both genes, the coding regions are interrupted 33 times by introns[50]. Thus, the coding region of amphibian genes may be interrupted as frequently as avian or mammalian genes as opposed to insect genes where longer uninterrupted sequences are described. Heteroduplexes between the A1 and A2 genes have shown that the introns interrupt the two genes at homologous positions. This structural organization suggests that the two genes arose by the duplication of an ancestral gene with a similar structure. After the duplication event, coding regions were well conserved (5% divergence) while, as revealed by electron microscopy, analogous introns changed rapidly in sequence and length. Only events such as deletions, insertions and duplications, in addition to point mutations, can explain the rapid evolution of the introns.

Recently the chicken vitellogenin gene has been isolated and characterized[2, 62]. Its length is 22 kb, similar to the A1 *Xenopus* vitellogenin gene (21 kb) and its coding region is interrupted 33 times as are the two characterized *Xenopus* genes[2]. As shown in *Figure 19.3*, there is a significant resemblance between the exon length of the chicken and *Xenopus* A2 vitellogenin gene if the comparison begins with *Xenopus* exon 1 and chicken exon 2 (correlation coefficient $p = 0.89$ and even 0.97 if only the first 24 exon pairs are considered). The shift could be explained by assuming that *Xenopus* and chicken have lost one exon at the proximity of their

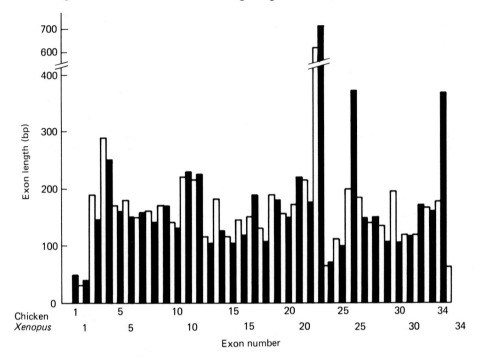

Figure 19.3. Comparison of the exon lengths between the chicken and the A2 *Xenopus* vitellogenin genes. Chicken exons are represented by black bars and *Xenopus* exons by open bars. Figure taken from Arnberg *et al.*[2] and kindly provided by Dr. G. Ab.

opposite end or that small introns have been missed at those positions due to limited resolution with the electron microscope. A region of homology between the chicken and the *Xenopus* A2 vitellogenin genes has been found close to exon 23 revealing a potentially important functional domain[2]. Although there is evidence in chicken for vitellogenin heterogeneity at the mRNA and protein level[14,53], no evidence has yet been presented from genomic DNA clones for more than one vitellogenin gene. Expression of the vitellogenin genes as well as other genes involves the transcription of the introns. In this system, the high number of introns reveals the extreme precision of the splicing mechanism which assures the connection of the large number of exons. The introns of the *Xenopus* A1 vitellogenin pre-mRNA, are not spliced out in a single unique order[42].

Closer analysis of the *Xenopus* A1 and A2 vitellogenin genes has revealed that middle repetitive DNA is present not only in the flanking regions of the genes but also within several introns[49]. At least seven different repeats are found in different introns of the A1 gene and at least two in introns of gene A2. The repeats in gene A1 differ from the ones in gene A2[43]. If the two genes arose by the duplication of an ancestral gene, repeats entered or left the genes after the duplication event, resulting in a unique distribution for each gene. Therefore, the repeats show characteristics of genetic mobile elements and they contribute significantly to the sequence and length difference found in analogous introns. Experiments with the repeats of gene A1 have demonstrated that they are also present in transcriptional units, other than the vitellogenin gene, active in the absence of oestrogen[43]. This suggests that the repeats are not involved in the regulation of oestrogen-inducible gene expression.

Oestrogen-induced changes in the chromatin conformation of the vitellogenin genes

Transcriptionally active chromatin is characterized by a higher sensitivity to nucleases such as DNase I or micrococcal nuclease. *Figure 19.4* shows that the A1 and A2 vitellogenin genes are approximately twice as sensitive to DNase I in hepatocyte chromatin of oestrogen-stimulated *Xenopus* compared with chromatin of untreated males. In fact, the DNase I sensitivity in hepatocyte chromatin of

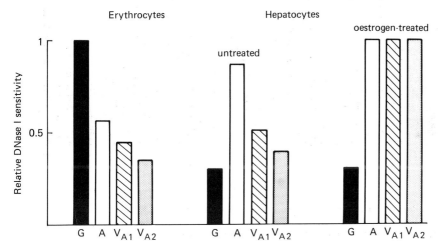

Figure 19.4. Oestrogen-induced changes in the chromatin conformation specific for the vitellogenin genes. The relative DNase I sensitivities for the β_1-globin (G), the 68 kd albumin (A), the A1 (V_{A1}) and A2 (V_{A2}) vitellogenin gene are given as measured in *Xenopus* chromatin of erythrocytes as well as of hepatocytes before and after oestrogen treatment. The values were calculated using the data published by Felber *et al.*[21] and Gerber-Huber *et al.*[22]. In the case of the A1 and A2 vitellogenin genes, the mean of the values for the different gene fragments analyzed was used.

untreated males is identical to the sensitivity found in non-vitellogenic cells such as the erythrocytes. The oestrogen-induced change of chromatin conformation is tissue-specific since no alteration in nuclease sensitivity of these genes was observed in erythrocytes from induced animals compared to non-induced. In hepatocytes, the conformational change is gene-specific, since the β_1-globin gene, a silent gene, and the albumin gene, an active gene, have unchanged sensitivity during oestrogen stimulation[21, 22]. It remains to be analyzed whether the oestrogen-induced change in chromatin conformation of the vitellogenin genes is permanent or disappears in withdrawn hepatocytes.

Concluding remarks

Taking into account that the inducing hormones for vitellogenin synthesis differ in insects and vertebrates and that the vitellogenin genes analyzed in these two systems have no common structural features, one is tempted to conclude that vitellogenins in insects and vertebrates have no common evolutionary origin.

Vitellogenesis in insects and vertebrates is a very attractive system to analyze hormone-induced activation of defined genes in a tissue-specific manner. The successful isolation and characterization of vitellogenin genes achieved in several species will now allow a careful functional analysis of these genes using *in vitro* transcription systems to define the factors responsible for tissue-specific activation by the hormone. The possibility of introducing the isolated genes into cell cultures or even into the whole organism with techniques established in the last few years offers a handle to study the molecular mechanism involved in hormone-activated gene expression. One major goal will be to prove that the oestrogen receptors we can describe at the present time are indeed the factors mediating the hormone effect.

Acknowledgment

We are grateful to our numerous colleagues who provided data not yet published and to Marianne and Thomas Brown, Susan Gerber-Huber and Philippe Walker for helpful comments on the manuscript.

References

1. Applebaum, S. W., James, T. C., Wreschner, D. H. and Tata, J. R. (1981). The preparation and characterization of locust vitellogenin mRNA and the synthesis of its complementary DNA. *Biochemical Journal*, **193**, 209–216
2. Arnberg, A. C., Meijlink, F. C. P. W., Mulder, J. Van Bruggen, E. F. J., Gruber, M. and Ab, G. (1981). Isolation and characterization of genomic clones covering the chicken vitellogenin gene. *Nucleic Acids Research*, **9**, 3271–3286
3. Baker, H. J. and Shapiro, D. J. (1977). Kinetics of estrogen induction of *Xenopus laevis* messenger RNA as measured by hybridization to complementary DNA. *Journal of Biological Chemistry*, **252**, 8428–8434
4. Baker, H. J. and Shapiro, D. J. (1978). Rapid accumulation of vitellogenin messenger RNA during secondary estrogen stimulation of *Xenopus laevis*. *Journal of Biological Chemistry*, **253**, 4521–4524
5. Barnett, T. and Wensink, P. C. (1981). Transcription and translation of yolk protein mRNA in the fat bodies of *Drosophila*. *The ICN/UCLA Winter Symposium*, **23**, 97–106
6. Barnett, T., Pachl, C., Gergen, J. P. and Wensink, P. C. (1980). The isolation and characterization of *Drosophila* yolk protein genes. *Cell*, **21**, 729–738
7. Bergink, E. W., Tseng, M. T. and Wittliff, J. L. (1977). Sequential changes in the structure and function of hepatocytes in estrogen-treated *Xenopus laevis* males. *European Journal of Cell Biology*, **14**, 362–377
8. Brennan, M. D., Warren, T. G. and Mahowald, A. P. (1980). Signal peptides and signal peptidase in *Drosophila melanogaster*. *Journal of Cell Biology*, **87**, 516–520
9. Burns, A. T. H., Deeley, R. G., Gordon, J. I., Udell, D. S., Mullinix, K. P. and Goldberger, R. F. (1978). Primary induction of vitellogenin mRNA in the rooster by 17 β-estradiol. *Proceedings of the National Academy of Sciences, USA*, **75**, 1815–1819
10. Chen, T. T. (1980). Vitellogenin in *Locusta* (*Locusta migratoria*). Translation of vitellogenin mRNA in *Xenopus* oocytes and analysis of the polypeptide products. *Archives of Biochemistry and Biophysics*, **201**, 266–276
11. Chen, T. T., Strahlendorf, P. W. and Wyatt, G. R. (1978). Vitellin and vitellogenin from locusts (*Locusta migratoria*). *Journal of Biological Chemistry*, **253**, 5325–5331
12. Chinzei, Y., White, B. N. and Wyatt, G. R. (1982). Vitellogenin mRNA in locust fat body. Identification, isolation and quantitative changes induced by juvenile hormone. *Canadian Journal of Biochemistry*, **60**, 243–251
13. Clemens, M. J. (1974). The regulation of egg yolk protein synthesis by steroid hormones. *Progress in Biophysics and Molecular Biology*, **28**, 71–108

14. Cozens, P. J., Cato, A. C. B. and Jost, J.-P. (1980). Characterization of cloned complementary DNA covering more than 6000 nucleotides (97%) of avian vitellogenin mRNA. *European Journal of Biology*, **112**, 443–450
15. Deeley, R. G., Gordon, J. I., Burns, A. T. H., Mullinix, K. P., Bina-Stein, M. and Goldberger, R. F. (1977). Primary activation of the vitellogenin gene in the rooster. *Journal of Biological Chemistry*, **252**, 8310–8319
16. Deeley, R. G., Udell, D. S., Burns, A. T. H., Gordon, J. I. and Goldberger, R. F. (1977). Kinetics of avian vitellogenin messenger RNA induction. *Journal of Biological Chemistry*, **252**, 7913–7915
17. Dhadialla, T. S. and Wyatt, G. R. (1981). Competence for juvenile hormone-stimulated vitellogenin synthesis in the fat body of female and male *Locusta migratoria*. In *Juvenile Hormone Biochemistry*, (Pratt, G. E. and Brooks, G. T., Eds.), pp. 257–262. Amsterdam, Elsevier/North Holland
18. Early, P., Rogers, J., Davis, M., Calame, K., Bond, M., Wall, R. and Hood, L. (1980). Two mRNAs can be produced from a single immunoglobulin μ gene by alternative RNA processing pathways. *Cell*, **20**, 313–319
19. Felber, B. K., Ryffel, G. U. and Weber, R. (1978). Estradiol induced accumulation of vitellogenin mRNA and secretion of vitellogenin in liver cultures of *Xenopus*. *Molecular and Cellular Endocrinology*, **12**, 151–166
20. Felber, B. K., Maurhofer, S., Jaggi, R. B., Wyler, T., Wahli, W., Ryffel, G. U. and Weber, R. (1980). Isolation and translation *in vitro* of four related vitellogenin mRNAs of estrogen stimulated *Xenopus laevis*. *European Journal of Biochemistry*, **105**, 17–24
21. Felber, B. K., Gerber-Huber, S. Meier, C., May, F. E. B., Westley, B., Weber, R. and Ryffel, G. U. (1981). Quantitation of DNase I sensitivity in *Xenopus* chromatin containing active and inactive globin, albumin and vitellogenin genes. *Nucleic Acids Research*, **9**, 2455–2474
22. Gerber-Huber, S., Felber, B. K., Weber, R. and Ryffel, G. U. (1981). Estrogen induces specific changes in the chromatin conformation of the vitellogenin genes in *Xenopus*. *Nucleic Acids Research*, **9**, 2475–2494
23. Green, C. D. and Tata, J. R. (1976). Direct induction of estradiol of vitellogenin synthesis in organ cultures of male *Xenopus laevis*. *Cell*, **7**, 131–139
24. Hagedorn, H. H. and Kunkel, J. G. (1979). Vitellogenin and vitellin in insects. *Annual Review in Entomology*, **24**, 475–505
25. Hagedorn, H. H., O'Connor, J. D., Fuchs, M. S., Sage, B., Schlaeger, D. A. and Bohm, M. K. (1975). The ovary as a source of α-ecdysone in an adult mosquito. *Proceedings of the National Academy of Sciences, USA*, **72**, 3255–3259
26. Hayward, M. A., Mitchel, T. A. and Shapiro, D. J. (1980). Induction of estrogen receptor and reversal of the nuclear–cytoplasmic receptor ratio during vitellogenin synthesis and withdrawal in *Xenopus laevis*. *Journal of Biological Chemistry*, **255**, 11308–11313
27. Hovemann, B., Gallar, R., Walldorf, U., Küpper, H. and Bautz, E. K. F. (1981). Vitellogenin in *Drosophila melanogaster*: sequence of the yolk protein 1 gene and its flanking regions. *Nucleic Acids Research*, **9**, 4721–4734
28. Huber, S., Ryffel, G. U. and Weber, R. (1979). Thyroid hormone induces competence for estrogen dependent vitellogenin synthesis in developing *Xenopus laevis* liver. *Nature, London*, **278**, 65–67
29. Hung, M. C. and Wensink, P. C. (1981). The sequence of the *Drosophila melanogaster* gene for yolk protein 1. *Nucleic Acids Research*, **9**, 6407–6419
30. Hung, M. C., Barnett, T., Woolford, C. and Wensink, P. C. (1982). Transcript maps of *Drosophila* yolk protein genes. *Journal of Molecular Biology*, **154**, 581–602
31. Jaggi, R. B., Felber, B. K., Maurhofer, S., Weber, R. and Ryffel, G. U. (1980). Localization of phosvitin within vitellogenin and characterization of different vitellogenin proteins in *Xenopus laevis*. *European Journal of Biochemistry*, **109**, 343–347
32. Jost, J.-P. Ohno, T., Panyim, S., Schuerch, A. R. (1978). Appearance of vitellogenin mRNA sequences in chicken liver following primary and secondary stimulation by 17β-estradiol. *European Journal of Biochemistry*, **84**, 355–361
33. Lazier, C. B. (1978). Ontogeny of the vitellogenin response to estradiol and of soluble nuclear estrogen receptor in embryonic chick liver. *Biochemical Journal*, **172**, 143–152
34. Lazier, C. B. and Haggarty, A. J. (1979). A high-affinity estrogen-binding protein in cockerel liver cytosol. *Biochemical Journal*, **180**, 347–353
35. May, F. E. B. and Knowland, J. (1980). The role of thyroxine in the transition of vitellogenin synthesis from non-inducibility to inducibility during metamorphosis in *Xenopus laevis*. *Developmental Biology*, **77**, 419–430
36. May, F. E. B. and Knowland, J. (1981). Developmental study of the relationship between nuclear estrogen receptor levels and vitellogenin synthesis in *Xenopus laevis* liver. *Nature, London*, **292**, 853–855

37. Nair, K. K., Chen, T. T. and Wyatt, G. R. (1981). Juvenile hormone-stimulated polyploidy in adult locust fat body. *Developmental Biology*, **81**, 356–360
38. Opresko, L., Wiley, H. S. and Wallace, R. A. (1980). Differential postendocytotic compartmentation in *Xenopus* oocytes is mediated by a specifically bound ligand. *Cell*, **22**, 47–57
39. Postlethwait, J. H. and Shirk, P. D. (1981). Genetic and endocrine regulation of vitellogenesis in *Drosophila. American Zoologist*, **21**, 687–700
40. Riddel, D. C., Higgins, M. J., McMillan, B. J. and White, B. N. (1981). Structural analysis of the three vitellogenin genes in *Drosophila melanogaster. Nucleic Acids Research*, **9**, 1323–1338
41. Ryffel, G. U., Wahli, W. and Weber, R. (1977). Quantitation of vitellogenin messenger RNA in the liver of male *Xenopus* toads during primary and secondary stimulation by estrogen. *Cell*, **11**, 213–221
42. Ryffel, G. U., Wyler, T., Muellener, D. B. and Weber, R. (1980). Identification, organization and processing intermediates of the putative precursors of *Xenopus* vitellogenin messenger RNA. *Cell*, **19**, 53–61
43. Ryffel, G. U., Muellener, D. B., Wyler, T., Wahli, W. and Weber, R. (1981). Transcription of single-copy vitellogenin gene of *Xenopus* involves expression of middle repetitive DNA. *Nature, London*, **291**, 429–431
44. Searl, P. F. and Tata, J. R. (1981). Vitellogenin gene expression in male *Xenopus* hepatocytes during primary and secondary stimulation with estrogen in cell cultures. *Cell*, **23**, 741–746
45. Setzer, D. R., McGrogan, M., Nunberg, J. H. and Schimke, R. T. (1980). Size heterogeneity in the 3'end of dihydrofolate reductase messenger RNAs in mouse cells. *Cell*, **22**, 361–370
46. Stanchfield, J. E. and Yager, J. D. (1978). An estrogen responsive primary amphibian liver cell culture system. *Experimental Cell Research*, **116**, 239–252
47. Tosi, M., Young, R. A., Hagenbüchle, O. and Schibler, U. (1981). Multiple polyadenylation sites in a mouse α-amylase gene. *Nucleic Acids Research*, **9**, 2313–2323
48. Wahli, W. and Dawid, I. B. (1980). Isolation of two closely related vitellogenin genes, including their flanking regions from a *Xenopus laevis* gene library. *Proceedings of the National Academy of Sciences, USA*, **77**, 1437–1441
49. Wahli, W., Dawid, I. B., Ryffel, G. U. and Weber, R. (1981). Vitellogenesis and the vitellogenin gene family. *Science*, **212**, 298–304
50. Wahli, W., Dawid, I. B., Wyler, T., Weber, R. and Ryffel, G. U. (1980). Comparative analysis of the structural organization of two closely related vitellogenin genes in *Xenopus laevis. Cell*, **20**, 107–117
51. Wahli, W., Dawid, I. B., Wyler, T., Jaggi, R. B., Weber, R. and Ryffel, G. U. (1979). Vitellogenin in *Xenopus laevis* is encoded in a small family of genes. *Cell*, **16**, 535–549
52. Wallace, R. A. (1978). Oocyte growth in non-mammalian vertebrates. In *Vertebrate Ovary* (Jones, R. E., Ed.), pp. 469–502. New York, Plenum
53. Wang, S.-Y. and Williams, D. L. (1980). Identification, purification and characterization of two distinct avian vitellogenins. *Biochemistry*, **19**, 1557–1563
54. Wangh, L. J. and Knowland, J. (1975). Synthesis of vitellogenin in cultures of male and female frog liver regulated by estradiol treatment *in vitro. Proceedings of the National Academy of Sciences, USA*, **72**, 3172–3176
55. Wangh, L. J., Osborner, J. A., Hentschel, C. L. and Tilly, R. (1979). Parenchymal cells purified from *Xenopus* liver and maintained in primary culture synthesize vitellogenin in response to 17β-estradiol and serum albumin in response to dexamethasone. *Developmental Biology*, **70**, 479–499
56. Warren, T. G. and Mahowald, A. P. (1979). Isolation and partial chemical characterization of the three major yolk polypeptides from *Drosophila melanogaster. Developmental Biology*, **68**, 130–139
57. Warren, T. G., Brennan, M. D. and Mahowald, A. P. (1979). Two processing steps in maturation of vitellogenin polypeptides in *Drosophila melanogaster. Proceedings of the National Academy of Sciences, USA*, **76**, 2848–2852
58. Westley, B. and Knowland, J. (1978). An estrogen receptor from *Xenopus laevis* liver possibly connected with vitellogenin synthesis. *Cell*, **15**, 367–374
59. Westley, B. and Knowland, J. (1979). Estrogen causes a rapid, large and prolonged rise in the level of nuclear estrogen receptor in *Xenopus laevis* liver. *Biochemical and Biophysical Research Communications*, **88**, 1167–1172
60. Wiley, H. S. and Wallace, R. A. (1978). Three different molecular weight forms of the vitellogenin peptide from *Xenopus laevis. Biochemical and Biophysical Research Communications*, **85**, 153–159
61. Wiley, H. S. and Wallace, R. A. (1981). The structure of vitellogenin. Multiple vitellogenins in *Xenopus laevis* give rise to multiple forms of the yolk proteins. *Journal of Biological Chemistry*, **256**, 8626–8634

62. Wilks, A., Cato, A. C. B., Cozens, P. J., Mattaj, I. and Jost, J.-P. (1981). Isolation and structural organization of an avian vitellogenin gene coding for a major estrogen inducible mRNA. *Gene*, **16**, 249–259

63. Wiskocil, R., Bensky, P., Dower, W., Goldberger, R. F., Gordon, J. I. and Deeley, R. G. (1980). Coordinate regulation of two estrogen-dependent genes in avian liver. *Proceedings of the National Academy of Sciences, USA*, **77**, 4474–4478

64. Wyatt, G. R., Locke, J., Bradfield, J. Y., White, B. N. and Deeley, R. G. (1981). Molecular cloning of vitellogenin gene sequence from *Locusta migratoria*. In *Juvenile Hormone Biochemistry* (Pratt, G. E. and Brooks, G. T., Eds.), Amsterdam, Elsevier/North Holland

Chapter 20

Immunoglobulin genes

J. M. Adams and S. Cory

Introduction

Overview

It is now clear that in the immunoglobulin (Ig) gene system, alterations of DNA sequence and arrangement, and control of RNA processing, play vital roles in the production of specific antibody. The ability of the system to generate a vast number of different antibodies involves not only several hundred variable region (V) genes in the germ line but also such phenomena as somatic mutation and recombination of distinct DNA elements. The events assembling a complete V gene in a lymphocyte are a prerequisite for productive transcription and play a major role in restricting Ig expression to one of the two relevant alleles ('allelic exclusion'). By subsequent 'switch recombination' events a specific lymphocyte clone can alter the type (and hence the effector function) of its Ig without changing antigen specificity. Alternative pathways of RNA processing are also implicated in determining whether an antibody molecule is secreted or bound to the cell membrane as antigen receptor.

The immune system

Specific immunity is mediated by two distinct types of cells, known respectively as B and T lymphocytes. Each B cell bears receptor Ig with a single antigenic specificity; this specificity is also faithfully maintained in the antibody secreted by its progeny. T cells, which mature in the thymus, mediate cellular immune responses, such as cytotoxic lysis, and interact with B cells. Although T cells display antigen specificity, little is known about their antigen receptors. Since the presence of specific antigen triggers expansion of specific clones in both the T and B lineage, their receptor repertoires must be acquired largely prior to antigen contact.

An Ig molecule comprises a light (L) chain, either κ or λ, and one of eight types of heavy (H) chain (μ, δ, γ3, γ1, γ2a, γ2b, ε or α), the heavy chain being responsible for defining the Ig class (IgM, IgD, IgG$_3$, etc.). The major receptors of immature B cells are IgM and IgD, whereas secreted antibody is predominately IgM following the primary exposure to an antigen (primary response), and IgG in the secondary. The basic unit of receptor and secreted immunoglobulin is a tetrameric molecule with two light and two heavy chains, [H$_2$L$_2$], and with

343

disulphide links between the H chains and between L and H. IgM is secreted as a pentamer $[H_2L_2]_5$, in association with a J (joining) chain.

The N-terminal ~110 amino acids in each chain (V region) vary in different molecules and $V_L + V_H$ fold to form the antigen binding site. Variability occurs most frequently in three 'hypervariable' (HV) or 'complementarity-determining' regions, lying between three 'framework regions' (FR) in each chain[135]. HV regions are thought to comprise the antigen binding cavity, whereas FR regions determine the molecule's shape and the ability of L and H to associate.

The COOH-terminal constant (C) region has a fixed sequence within a species except for allelic ('allotypic') differences, which may be few, or numerous as in mouse $C_{\gamma 2a}$ chains[105]. The various C_H regions determine different effector functions, such as ability to fix complement; each occurs in secretory and membrane forms which differ only at their COOH termini. The receptor form has a 26 residue transmembrane segment which anchors the Ig on the cell surface.

The three immunoglobulin gene loci

Immunoglobulins are encoded at independent loci for λ, for κ and for H chains. In the mouse, the H locus lies on chromosome 12, the κ on 6 and the λ on 16[32, 47]. The human H, κ and λ loci are located on chromosomes 14, 2 and 22 respectively[28, 40, 69, 72]. *Figure 20.1* compares the murine germ line organization with that in lymphocytes.

Separate germ line V and C Genes

A striking feature is the separation of germ line V and C genes, proposed by Dreyer and Bennett[34] and first demonstrated by Hozumi and Tonegawa[52]. Moreover, a V

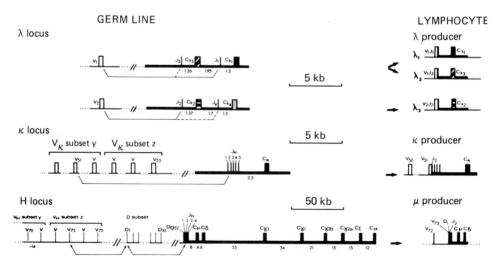

Figure 20.1. Organization of the three immunoglobulin loci before and after somatic rearrangement. Distances are in kilobase pairs (kb), the scale for the H locus differing from the κ and λ. Several aspects of λ organization are not established[14, 77]: $V_{\lambda 1}$ and $V_{\lambda 2}$ could be together; $J_3C_{\lambda 3}$–$J_1C_{\lambda 1}$ may lie either upstream or downstream of $J_2C_{\lambda 2}$–$J_4C_{\lambda 4}$; and $J_{\lambda 4}$–$C_{\lambda 4}$ is not known to be expressed. Linkage within the C_H locus is derived from a number of studies, including μ–δ[68]; γ_3–γ_{2a}[97]; γ_1–ϵ[114] and ϵ–α[83].

region is itself specified by distinct, widely separated genetic elements, as first shown for λ_1[13], where a V gene proper encodes the first 97 amino acids and a separate J (joining) element located near $C_{\lambda 1}$ encodes the last 13. Thus expression in a λ_1-producing lymphocyte requires somatic recombination of $V_{\lambda 1}$ and $J_{\lambda 1}$ and the $J_{\lambda 1}$–$C_{\lambda 1}$ intervening sequence is subsequently removed by RNA splicing (*Figure 20.1*). It is now realized that the λ locus contains four different C_λ genes (λ_1–λ_4), each having a separate J_λ[14, 77]. $C_{\lambda 4}$ is non-functional due to a faulty $J_{\lambda 4}$ splice site[78]. Since λ_1 chains predominate over other λ chains, $V_{\lambda 1}$–$J_{\lambda 1}$ joining may occur preferentially.

In contrast to the single mouse $V_{\lambda 1}$ and $V_{\lambda 2}$ genes, the κ locus has multiple V_κ genes. There are five J_κ genes clustered near the single C_κ gene, although $J_{\kappa 3}$ appears to be non-functional due to a defective splice site[75, 101]. Why two different L chains have evolved is unclear. In the mouse, κ occurs in 95% of serum Ig, but certain immune responses are restricted to λ_1. Curiously, some mammals, e.g. the horse, have more λ than κ, suggesting that the V_κ and V_λ loci undergo rapid evolutionary expansion and contraction.

The H chain locus comprises a common pool of V_H genes for the eight C_H genes and a single cluster of four J_H genes located near the C_μ gene[11, 37, 44, 81, 102], the first C_H gene used in B cell ontogeny. V_H assembly also requires a D (diversity) element[37, 65, 103, 104], which represents most of HV3 and encodes 2–15 residues. Thus both a V_H–D and a D–J_H join occur (*Figure 20.1*). The two D sets identified include a single D lying only 700 bp 5′ to J_{H1}, and a set of ~10 members, spread over some 50 kb, most likely lying somewhere between the V_H and J_H loci[65].

The mouse C_H locus, which covers ~180 kb (*Figure 20.1*), has been spanned with cloned sequences, many obtained by Honjo and colleagues[113]. The C_H spacers, which range up to 55 kb between C_δ and $C_{\gamma 3}$, partially comprise repeating units implicated in C_H switching (see p. 353) and other repetitive sequences occur[8, 125]. Whereas C_λ and C_κ are single exons, each C_H gene is divided by introns into segments matching the ~110 amino acid domains in the polypeptides. This suggests that introns were present within the ancestral (probably C_μ-like) C_H gene. The C_H locus presumably evolved largely by successive duplications, but multiple crossovers may have occurred[79, 86]. Gene conversion has also been implicated[9, 105].

Clusters of germ line V genes

V genes are split by a small intron within the leader (L) region encoding the precursor peptide required for transport into the endoplasmic reticulum. Each V_κ or V_H gene is closely homologous to 1–15 other V genes. Cloning studies show that a number of genes within each such 'family' are neighbours[15, 27, 43, 57, 61]. Although V family clusters predominate, the most homologous V genes are not always adjacent[61] and some interspersion of families is not ruled out. Based on deletions in plasmacytomas, the relative order of three V_H families[61] and five V_κ families[111] has been determined, but deletions do not always give a unique order[111].

A typical spacing between members of a V_H family is 8–16 kb. V genes of one V_κ family exhibited extensive homology of 5′ flanking sequences, prompting the suggestion that this large target for homologous recombination stimulates formation of new V genes within a family[109]. However, only one of two V_H families exhibited such homology, members of the other having essentially unique flanking sequences[61]. Poor conservation argues that most of the spacer lacks a sequence-specific function. Spacer regions may reflect evolutionary age, the most ancient families having least conservation[61].

Several V gene families exhibit greater divergence in HV than FR[15, 43], suggesting that the locus is under pressure to create new specificities. New V genes could be generated by duplication followed by mutation, but gene conversion and recombination between V genes may also contribute[9, 19, 33]. So far 6 of 18 V genes examined (one-third) are non-functional (pseudo) genes[10, 15, 27, 43]. Presumably they are continuously created by mutation and removed by processes such as non-homologous recombination, which may be facilitated by the 0.3–1.5 kb partially homologous repeats scattered within V_H spacers[61]. That some alterations within V family clusters have occurred within recent evolutionary time is indicated by differences in Southern patterns with various mouse strains[25, 56].

Mechanism of V region assembly

Signal sequences for joining

How does the joining machinery fuse into a single coding region two non-homologous sequences, probably often more than 1000 kb apart? An important clue is the conservation of two short 'signal sequences' 3′ to a germ line V gene and also, in inverted form, 5′ to J[75, 101]. *Figure 20.2(a)* shows that a V_κ gene is followed by the heptamer CACAGTG separated by 11 or 12 bp from the

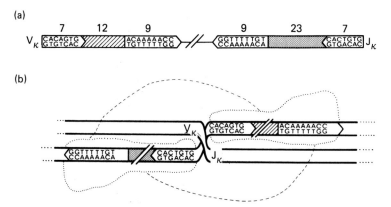

Figure 20.2. Joining of V_κ and J_κ. (a) shows the postulated heptamer and nanomer signal sequences 3′ to a V_κ gene 5′ to a J_κ (see text). (b) illustrates one possible structure of the recombination complex. The recombinase is imagined here to have closely related subunits, one of which binds to V_κ signal sequences and the other to J_κ signals[37]; other hypothetical subunits might bind the first subunits together and/or bind to sequences within V_κ and J_κ.

nanomer ACATAAACC, while a J_κ is preceded by the complementary sequences separated by 23 or 24 bp. The inverted relationship means that paired V and J flanking sequences could form a cruciform structure[75, 101]. Early *et al.*[37] emphasized instead that the 11–12 bp and 22–23 bp spacings were close to the 10.4 and 20.8 bp in one and two turns of the DNA helix respectively. It was proposed that recombination is promoted by the binding of closely related proteins, one to the

347

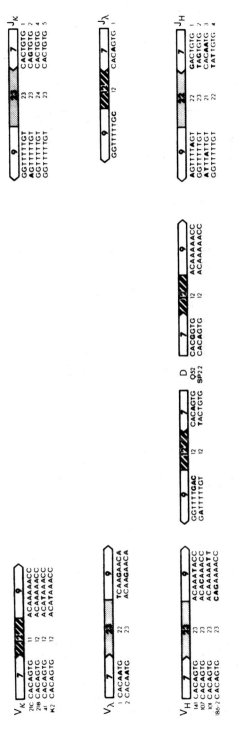

Figure 20.3. Conservation of signal sequences for joining in the three immunoglobulin loci. Since only one strand is shown, the signals 3' to V are the complement of those 5' to J (or D) (see *Figure 20.2*). Differences from the consensus sequences are indicated by bold letters. References to sequences not given in the text or in Tonegawa *et al.*[123] include V_H 101[56] and V_H 186–2[15].

signal with a 11–12 bp spacer and the other to that with a 22–24 bp spacer, thus enabling recombination only for Ig elements with spacers of different length[37, 102]. *Figure 20.2(b)* depicts one such hypothetical recombination complex.

The 12–23 bp rule for joining is now well supported (*Figure 20.3*). All D elements examined have 12 bp spacers[65], and a V_H gene (23 bp spacer) pairs only with a D and not directly with a J_H (23 bp). Whether the normal intermediate is V_H–D or D–J_H is not known; D–J joins found in some lines[65] could be intermediates or abortive.

The virtual identity of 'signal sequences' for the λ, κ and H chain (*Figure 20.3*) suggests that the same, or closely related, joining enzymes operate at the three loci[37, 102]. Moreover, the same signals flank human V genes[10]. Even a 'wrong site' V_κ join involves sequences related to J_κ signals[107]. Nevertheless, other factors may influence joining, such as conserved sequences within the V or J gene.

Cis versus trans joining events

Models for V–J joining in which a copy of the V gene, or an excised V gene, is inserted at J have been excluded because sequences upstream from the recombined V gene appear unchanged. That L chain joining involves deletions was revealed by the absence of germ line sequences 3' to V_λ and 5' to J_λ from one λ-producing line[101], and of those 3' to a V_κ and 5' to J_κ from a κ-producer[110]. Analogous V_H and J_H deletions occur in many lines[22, 23].

Figure 20.4 shows that deletions could reflect looping-out of sequences on the same chromosome (*cis* recombination), or *trans* events, as in sister chromatid exchange[129]. In the looping-out model, sequences excised between V and J would presumably be lost on cell division, whereas on the sister chromatid model they remain in the lymphocyte population, albeit segregated into different daughter cells. Contrary to the looping-out model, rearranged sequences 5' to J_κ persist in about half the plasmacytomas examined, as well as in a B cell population[129]. Moreover, sequences 5' to a J_κ were shown to be linked to those 3' to a V_κ in three tumours, in each case unrelated to the V_κ–J_κ joins in the same tumour[49, 119]. This fits well with the exchange model, which predicts that such sequences (arrows in *Figure 20.4*) arise from a V_κ–J_κ join on the other allele in an earlier cell. Thus sister chromatid exchanges, which occur only in dividing cells, may predominate. The reciprocal product would also be maintained, however, if some germ line V_κ genes were inverted with respect to J_κ, so that joining gave inversions rather than deletions[67].

Regulation of V gene joining

Is the V gene used in a given cell determined stochastically or programmed? Evidence that certain specificities arise at fixed times during ontogeny favours programming[117] but other data favours a largely random process. One pre-B cell line generates multiple subclones with different V_κ–J_κ joins[96] and unrelated V–J rearrangements are generally found on separate alleles in plasmacytomas for both κ[6, 12, 76, 107, 130] and H chains[23, 26]. The process may not be altogether random, however, because one pre-B line has the same V_H gene on both alleles, joined to different D and J elements (S. Gerondakis and O. Bernard, unpublished). Perhaps random choice is made from a restricted number of V_H genes programmed for rearrangement in each cell.

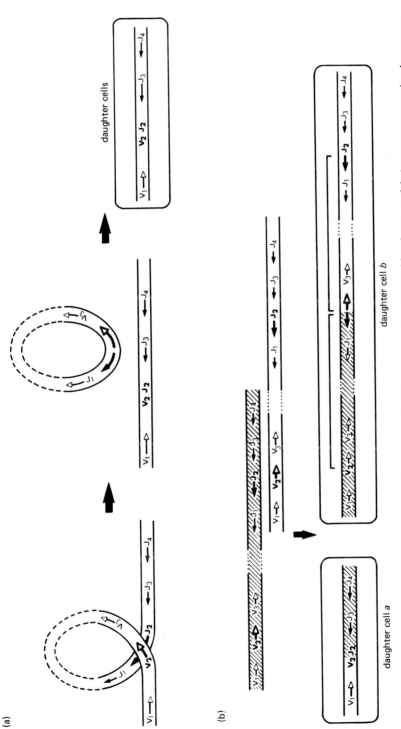

Figure 20.4. Comparison of (a) the looping-out model for joining with (b) an unequal sister chromatid exchange model. Arrows represent signal sequences 3' to V and 5' to J (see *Figure 20.2*). Joining of V_2 and J_2 is illustrated here and those genes and their signal sequences are shown bold. Only one allele is shown in each model. In the looping-out model, the 'episome' generated by joining, which bears the flanking signal sequences, would be lost on cell proliferation. In the diagrammed chromatid exchange model (based on Van Ness *et al.*[129]), the two chromatids are distinguished to illustrate the fate of sequences in the separate daughter cells. Daughter cell *a* has suffered a deletion of sequences between V_2 and J_2, whereas daughter *b* has a duplication of the region from just 3' to V_2 to just 5' to J_2, as indicated by the bars.

Somatic diversification

Extent of the germ line V repertoire

Comparison of V region amino acid sequences made a good case for multiplicity of germ line V genes (e.g. Potter[93]). Direct V gene counting is complicated by varying degrees of homology. Conclusions from initial kinetic hybridization studies were limited by impurities in uncloned probes and the uncertain effect of mismatch. With Southern band counting, several studies (e.g. Seidman et al.[109]) showed that 1–12 mouse DNA fragments hybridize with a given V_κ sequence. A study using nine unselected V_κ probes[25] revealed repeats of certain patterns and provided estimates of 90–300 V_κ genes. Moreover, in man at least 75% of κ chains are apparently encoded by no more than 20 V_κ genes[10]. Less V_H data is available but there appear to be of the order of 160 mouse V_H genes[61]. These estimates favour substantial somatic diversification, but they could be low if myeloma proteins are not representative of V sequences. The proportion of total spleen κ mRNA molecules represented by a single V_κ group suggested that there might be ~2000 germ line V_κ genes[141]. Studies with more κ probes might clarify the discrepancy with the Southern estimate.

Somatic mutation

Many V sequences are expressed unchanged[42, 61, 108, 132]. Initial evidence for somatic variation came from mouse $V_{\lambda 1}$ amino acid sequences[132], which all derive from a single gene[13]. Similarly, the 22 distinct $V_{\kappa 21}$ amino acid sequences reported[133] derive from at most 8–12 germ line genes[25, 128]. In phosphorylcholine (PC) antibodies, all the V_H amino acid sequences, which have up to eight substitutions from the prototype TEPC 15[42], may derive from a single T15 gene[27, 62] (but see Clarke et al.[19]).

The positions where rearranged V_H and V_κ genes differ from their apparent germ line analogues (*Figure 20.5*) reveal important features of somatic variation.

1. Variants typically exhibit multiple differences, not restricted to the V gene proper. V_H M167, for example, differs at 46 positions scattered up to at least 50 bp 5' to the leader (L) region, as well as within the J cluster 3' to the gene, but not in regions several kb from the V gene[62]. Thus mutations have peppered a 1–2 kb region spanning a V_H–D–J_H gene. Whereas $V_{\lambda 1}$ chains differ by only one or two amino acids[13, 132], variant V_H and V_κ sequences so far average 6.9 base changes (4.6 amino acid changes) from the germ line within the V + J coding segment.
2. In contrast to $V_{\lambda 1}$, changes occur in FR as well as HV and at a similar frequency: 63 base changes overall in FR and 33 in HV. In V_H, 21 out of 46 coding region base changes occur in HV2, which itself represents less than 20% of V_H. This could reflect a preferred target region for mutation or antigenic selection. It might be argued that FR changes are passively acquired during antigenically selected changes in HV. However, as shown in *Figure 20.5*, mutations occur in the rearranged but unexpressed V_κ T2 gene[87] and amino acid replacements occur only in FR for V_κ M167 and V_κ M511[112].
3. All reported changes are single base substitutions, except for single base deletions and additions in V_H M167. *Table 20.1* shows that all 12 possible base changes occur, although there is a bias against T to G, G to T, and possibly C to

351

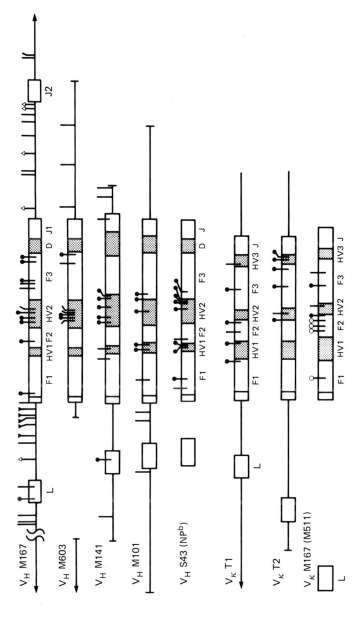

Figure 20.5. Somatic mutation around V_H and V_κ genes. The positions at which rearranged V genes of lymphoid tumours differ from their apparent germ line analogue are indicated by vertical lines for silent base changes, filled circles for amino acid changes, and (in V_H M167) open triangles for one base deletions and filled triangles for additions. For V_κ M167[112], open circles indicate the amino acid substitutions in M511, for which the nucleotide sequence has not been reported. L indicates the polypeptide leader (residues −4 to −19) and I the intervening sequence within L; HV regions are shaded. V_κ T1 is the expressed allele in plasmacytoma T and T2 the unexpressed[87]. V_H sequences were reported by Kim *et al.*[62] for the TEPC 15 derivatives V_H M167 and V_H M603; by Sakano *et al.*[102] for V_H M141, by Kataoka *et al.*[56] for V_H MC101; and by Bothwell *et al.*[15] for the nitrophenyl (NP^b) gene V_H S43. For S43 and V_κ M167, a cDNA sequence was compared with the germ line gene.

TABLE 20.1. Somatic base changes in V genes[a]

Germ line	Somatic variant				
	A	G	T	C	Total
A	–	12	8	7	27
G	15	–	2	10	27
T	10	1	–	13	24
C	3	6	8	–	17

[a] Data are based on V genes cited in the legend to *Figure 20.5* and in Bernard *et al.*[13] for $V_{\lambda 1}$, and in Ollo *et al.*[86] for V_{H441}.

A. No pattern is evident in the sequences surrounding the base change. Few, if any, of these changes can be accounted for by somatic recombination of V genes, or gene conversion[15, 27, 112], but those mechanisms are implicated for two other V_H genes[19, 33].

A special mutator mechanism

It is unlikely that the basal mutation rate accounts for V region mutation. First, mutation is focused around recombined V genes: none have been found in C regions[7, 13, 45, 46, 62], or in an unrearranged V_κ M167 gene in two plasmacytomas[112]. Hence the mechanism might be linked to joining[112], or have a special target near a recombined VJ. Secondly, mutation correlates strongly with C_H switching (see p. 353): 11 IgM PC antibodies had germ line V_H, while 6 of 9 IgGs were variants[42]. Similarly, an IgM nitrophenyl antibody has germ line V_H, while an IgG (S43) has multiple changes[15] (*Figure 20.5*) as do the IgGs from MC101[56] and M141[103]. Since V_κ in IgG antibodies has also mutated much more than in IgM, the mutator may operate in *trans*[42].

The picture that emerges is that a special mechanism operating within a substantial fraction of B cells generates a burst of random point mutations around recombined V genes. Since this appears to occur predominantly late in B cell ontogeny, probably most frequently after antigen contact, its major effect may be improved fit during the secondary response. The stability of V regions in some plasmacytomas over hundreds of generations[106] suggests that the mechanism only operates for a short period.

Contribution of D and J to diversity

Diversity is amplified by assortment of V, (D) and J elements. Even without somatic mutation, independent assortment of say 200 V_H, 20 D_H and 4 J_H genes would give 16 000 V_H regions, and assortment of these with a few hundred V_κ–J_κ regions would give a few times 10^6 distinct ($V_H + V_\kappa$) Ig molecules. What fraction would represent antigenically-distinct combining sites is however uncertain. Nor is it clear whether the genes assort entirely randomly, or what fraction of H and L chains actually form stable Ig molecules.

The recombination points between a V, (D) and J element can vary over a few base pairs, and this imprecision adds to diversity by forming hybrid codons at the junctions[75, 101, 134]. For instance, joining of a 3' V$_\kappa$ CCC (Pro) to a 5' J$_\kappa$ TGG (Trp) with crossing over after the first or second base yields CGG (Arg) and CCG (Pro) respectively. This junctional variation may allow a D$_H$ gene to be used in all three reading frames[65], expanding diversity in HV3.

Class (C$_H$) switching

Class (C$_H$) switches enable a lymphocyte clone to alter effector properties of the immunoglobulin while retaining the antigen specificity of the V$_H$–D–J$_H$. As indicated in *Figure 20.6*, recombination links a region between J$_H$ and C$_\mu$, designated S$_\mu$, with a switch region 5' to another C$_H$ gene (here S$_\mu$), so that the V$_H$–D–J$_H$ and a 5' portion of J$_H$–C$_\mu$ are retained[31, 55, 102]. Since intervening rather than coding sequences are linked, switching need not be precise and, judging by known switch points (*Figure 20.6*), recombination probably can occur over a region of a few kb. That intervening C$_H$ genes are deleted on switching (*Figure 20.6*) was first indicated by kinetic hybridization analysis of C$_H$ gene content in plasmacytomas[50]. Southern blots provided stronger evidence[20, 22, 24, 94, 138] and showed that deletions include C$_H$ flanking regions, extend close to active C$_H$ genes, and often occur on both alleles[24]. However, one cell line apparently switches without deletion of C$_\mu$[5], perhaps by processing a large transcript, as in dual μ–δ expression (p. 357), and B cells may also switch without deletion[140].

The sequences surrounding switch sites exhibit few nucleotides in common, although secondary deletions may have obscured some switch sites. However, the nearby regions comprise a tandem array of partially repetitious sequences (see reviews by Davis *et al.*[29] and Honjo *et al.*[51]). The S$_\gamma$ regions have 49 bp repeats which are 50–70% homologous to one another[54], but have little homology with S$_\mu$ or the S$_\alpha$ unit of 30 bp[30]. S$_\mu$ contains a highly repetitious region (inner box at top of *Figure 20.6*) comprised of (GAGCT)$_n$GGGGT repeats[35, 82]. YAGGTTG, which is rare in that portion of S$_\mu$, frequently occurs 5' to switch donor or acceptor sites, and may be another recognition element[14]. The possibility that sub-regions of S$_\mu$ are specific for other C$_H$ genes is made unlikely by the varied location of S$_\mu$ sites that recombine with say S$_\alpha$[26, 30] and by the nearly uniform nature of the majority of S$_\mu$[82].

One possible mechanism for switching is that homology of S regions facilitates unequal crossing-over[35, 51, 125]. While S$_\mu$ lacks extensive homology to most S regions, the multiple short common runs may aid transient synapsis[82]. This model must be reconciled with the clustering of known S$_\mu$ sites 5' to the most repetitious portion of S$_\mu$ (top of *Figure 20.6*). A second model[29] postulates different proteins which bind to the S$_\mu$ repeat and to that of, say, S$_\alpha$; their association would then promote recombination. This scheme accommodates regulated switching, whereas homologous recombination fits better with stochastic events, but both schemes may contribute.

Switching between homologous chromosomes is unlikely from genetic data and from nucleic acid results on one F$_1$ cell line[139]. However, like V–J joining, switching may be mediated by looping out (*cis*) or by sister chromatid exchange (*trans*)[51]. *Cis* deletion cannot easily account for a 490 bp S$_\alpha$ segment between S$_\mu$ and S$_{\gamma1}$ in a γ1 gene[30, 85] or for 'reverse switches' from γ2b to γ1[95]. The sister chromatid exchange model predicts duplications and occasional scrambling of C$_H$ gene order[51].

354

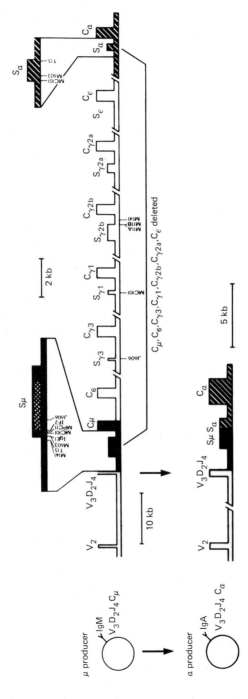

Figure 20.6. Mechanism of switching. A C_μ to C_α (IgM to IgA) switch is illustrated, the final active α gene (bottom) being a mosaic of separate germ line V_H, D_H, J_H-S_μ, and $S_\alpha-C_\alpha$ regions. Boxes denote the portion of each S region so far shown to be repetitive by sequencing; actual S regions may well be much larger[125]. Blow ups of the S_μ and S_α region are at the top; particular S sites in tumours are indicated, largely taken from Honjo *et al.*[51] and, for S_α, from Davis *et al.*[30]. The MPC 11 switch sites are from Lang *et al.*[66a]. The cross-hatched inner box for S_μ denotes the highly repetitious region (see text), which varies in length in different mouse strains[73].

Switch machinery must be induced in B cells at a late stage, because switching has never been observed to precede V assembly. Switching may require demethylation[98] and might be determined stochastically, be preprogrammed or be regulated by extrinsic factors (microenvironment, T cells, etc.). That preprogramming can occur is indicated by two cell lines that make specific switches *in vitro*[5, 16, 118].

Expression of Ig genes

Ig gene activation and B cell development

An Ig gene is activated in stages. (a) The *quiescent state* in non-lymphocytes probably reflects condensed chromatin, because the C_κ gene is nuclease-insensitive in liver but sensitive in a plasmacytoma[92,121]. (b) *C gene transcriptional competence* precedes V–J joining because an unrearranged C_κ gene in a κ-expressing lymphocyte yields an 8.4 kb transcript[89,91] and several C_μ transcripts can be made with or without rearrangement[59] (see below). The C_κ transcript starts 5′ to $J_{\kappa 1}$, while certain C_μ transcripts include 5′ segments from within $J_H–C_\mu$, implicating new splice sites[58]. (c) *Proper mRNA production* requires V–J recombination but other factor(s) may be involved because Ig transcription in some pre-B cell lines is markedly stimulated by lipopolysaccharide or altered growth conditions[88,115]. Promoter elements, such as the 'TATA' box, occur upstream from the V leader[56]. Since no transcription of germ line V genes has been observed[89,91], presumably the V promoter is active only when moved to the C region zone of transcriptional competence. (d) *Amplification of transcription* occurs during lymphocyte maturation: up to ten-fold for μ in moving from pre-B to B, and another 100-fold for both H and L in progressing to the plasma cell[59,90]. Coordinated H and L expression implicates *trans*-acting factors, as does the fact that B cell–plasmacytoma hybrids express B cell Ig mRNAs at plasmacytoma levels[90].

The earliest characterized stage of B cell development is the pre-B cell, lacking surface IgM but having cytoplasmic μ. Such cells contain membrane μ mRNA but not κ mRNA[116] because $V_\kappa–J_\kappa$ joining has not occurred[70,90]. After $V_L–J_L$ fusion, L chain expression leads to deposition of surface IgM. On antigen contact, the transition to IgM secretion requires elevated secretory μ production and expression of the J chain gene, which involves its demethylation without rearrangement[136]. C_H switching is thought to occur in mature B cells, although one Abelson pre-B line undergoes switch events[16].

J–C splicing involves V

A striking feature of the processing of Ig pre-mRNAs is that J–C splicing appears to require the V region[89,91]. Cells competent to make a $V_\kappa–J_\kappa$ to C_κ splice, for instance, do not splice the 8.4 kb C_κ transcript, which spans the J_κ and C_κ loci. Moreover, the size of κ precursor RNAs suggests that only the J_κ linked to V_κ serves as a splice donor. Such results suggest that the J–C splice requires a special recognition mechanism.

Alternative modes of RNA processing

Figure 20.7 indicates that alternative modes of RNA processing account for two major puzzles: generation of distinct secretory and membrane H chains from a single gene, and the dual production of μ and δ receptors by most immature B cells.

Membrane vs. secreted H chain

Distinct secretory and membrane mRNAs were first discovered for μ chains[38, 100]. *Figure 20.7* shows that the 2.4 kb μ secretory (μ_s) mRNA has a 3' end derived from a gene segment (S) contiguous to $C_{\mu 4}$, whereas the 3' end of the 2.7 kb μ membrane (μ_m) mRNA comes from two exons (M1 + M2) 1.85 kb downstream. Similarly

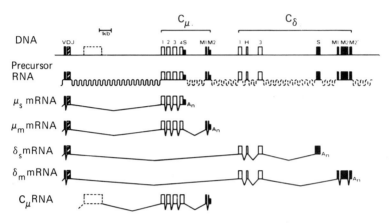

Figure 20.7. Generation of functionally distinct mRNA species for the μ-δ locus by alternative modes of RNA processing. Boxes denote gene segments (exons), the shorter portion being untranslated. Bent lines on the RNA molecules indicate splices. The broken line in the precursor RNA beyond μ_s indicates that certain RNA molecules (e.g. μ_s mRNA) may derive either from a short precursor or by scission of a larger precursor (see text). Only the two most prominent of the five δ transcripts are shown[18, 71], that for the major δ_s mRNA of 1.75 kb and the major δ_m of 2.9 kb. The dotted box indicates a gene segment from which sequences (of unknown length) are derived in two of the larger C_μ RNAs[58]; the origin of their 5' ends is not known but they have either μ_s or μ_m 3' termini.

located S, M1 and M2 exons 3' to each C_γ gene specify 3' ends of the 1.9 kb γ_s and 3.9 kb γ_m mRNAs[99, 126, 127, 137]. For δ, however, the S exon lies 4.7 kb downstream from $C_{\delta 3}$ and M exons 6.5 kb downstream[18, 68, 71, 80, 124]. At least five δ transcripts exist, some of which appear to reflect alternative splices in the 3' untranslated region.

The transmembrane segments in γ_m and μ_m are surprisingly homologous, perhaps indicative of a binding site for another membrane protein[99, 127, 137]. However, the marked difference between the 28-residue intracellular domain of γ_m and the two residues in μ_m may mean that the signal transmitted to the cytoplasm on antigen binding differs for IgG on memory cells, compared with IgM on virgin cells[127].

Two models could account for regulation of membrane vs. secreted H chains (*Figure 20.7*)[38, 100]. (a) The *two transcript* model postulates regulated termination of transcription, either between S and M1, or after M2, the longer transcript giving membrane mRNA. (b) The *single transcript* model assumes a common precursor spanning all exons which is subsequently either spliced from C4 (for μ) to M1, or cut between S and M1, perhaps at the S poly-A addition site. The regulatory mechanism must be competititve rather than absolute, because secretory and membrane mRNAs are found in different ratios in various μ- and also γ-producing cell lines[38, 60, 99, 100, 126, 127]. Since fusion of a B cell and plasmacytoma elevates the ratio of μ_s to μ_m mRNA as well as the total amount of μ mRNA[100], the processing must be regulated in a sophisticated manner, perhaps by a system specific for Ig genes.

μ vs. δ production

Dual production of μ and δ also involves alternative modes of RNA processing[63, 68, 71, 80, 124] (*Figure 20.7*), presumably analogous to one of the two models above. Both models assume that splicing from V_HDJ_H to $C_{\delta 1}$ competes effectively with V_HDJ_H to $C_{\mu 1}$. Since C_μ gene is deleted in certain IgD-only lines, IgD-expressing plasma cells may be produced by switch recombination[71, 80]. The function of the dual IgM and IgD receptors on immature B cells remains a mystery.

Allelic exclusion

Each Ig chain expressed in a given B cell derives only from a single chromosome. A related phenomenon, 'isotype exclusion', is the restriction to a single L chain type (κ, $\lambda 1$, $\lambda 2$, etc.). Both phenomena reflect events at several levels of Ig gene expression, the primary being V–J joining:

1. *Absence of rearrangement* on one J_κ allele accounts for allelic exclusion in about half of κ-expressing plasmacytomas[91] and in the majority of B cells[21, 53]. Moreover, λ rearrangement is rare in κ-expressors of mice[21] or man[48]. J_H rearrangement does occur on both alleles in nearly all plasmacytomas[26] and in at least 80% of B cells[21, 84], but some joins are incomplete (DJ only)[36].
2. *Errors in recombination* inactivating an allele include recombination at incorrect sites, deletions, out-of-phase joins, and joins to defective elements. In MPC 11 a V_κ gene has recombined between J_κ and C_κ[107] and unidentified sequences have recombined between J_H and C_μ in several tumours[23, 26]. Deletion of C_κ is common in human λ-expressors[48] and J_H deletion occurs in some mouse lines[26]. Out-of-phase V_κ–J_κ joins exist on one allele in several plasmacytomas[6, 12, 76, 130] and a pre-B line has a V_HDJ_H with D and J out of phase (Gerondakis and Bernard, unpublished). Non-functional joins presumbly also include the pseudo V genes, or pseudo J(D) elements, such as $J_{\kappa 3}$ and a pseudo J_H[44].
3. *Transcription blocks*, resulting from, say, a defective promoter, may contribute to allelic exclusion. A plasmacytoma survey[89] suggested that usually only one rearranged allele was transcribed. However, since κ mRNA from a second allele occurs in several randomly chosen plasmacytomas[12, 107, 130] selective transcription of one allele cannot be the predominant mechanism.
4. *A post-translational mechanism* may account for the apparent allelic exclusion of κ in M173 and S107, both of which contain two properly joined κ mRNAs[12, 66]. Poor assembly with the H chain, due to the absence of an 'invariant' tryptophan in M173 and of the first two J_κ amino acids in S107, may exclude those κ chains from the secreted immunoglobulins.

Models for allelic exclusion

In a stochastic model[1, 3, 12, 90, 91] each allele behaves independently and allelic exclusion results from the low frequency of productive events (designated for κ as κ^+) vs. non-productive ones (κ^-). As this model relies on random events, it predicts that a minor fraction of lymphocytes will be 'double expressors' (e.g. $\kappa^+\kappa^+$), as in M173 and S107[12, 66], and that a large fraction will be inactive (e.g. $\kappa^-\kappa^-$). The high error frequency required[12] is not implausible. If we assume that joining occurs independently of phase, that say 40% of V genes are pseudo, and that perhaps 30% of joins are to wrong sites, the frequency of productive joins would be $\frac{1}{3} \times 0.6 \times 0.7 = {\sim}14\%$, which would give 7.5% of κ^+ cells as $\kappa^+\kappa^+$ and leave 74% of B cells inactive $(\kappa^-\kappa^-)$[12].

Some programming probably is superimposed on a stochastic process, since μ rearrangement occurs before κ[70, 90] and κ probably before λ[48]. Alt et al.[4] proposed a feedback scheme for L loci: the transition from germ line $\kappa^0\kappa^0$ to $\kappa^+\kappa^0$ would give surface IgM and its appearance would trigger differentiation events that shut off further L chain rearrangement. An initial $\kappa^-\kappa^0$ step however could be followed by $\kappa^-\kappa^+$ or $\kappa^-\kappa^-$, and the latter cell might later activate λ. The presence of both $\kappa^+\kappa^0$ and $\kappa^+\kappa^-$ B cells favours this scheme over the strict stochastic model, but the latter remains viable, particularly for the H chain locus, since a feedback signal from cytoplasmic μ chain would presumably be more complex.

The continuing puzzle of the T cell receptor

Whether the T cell receptor involves any Ig gene product is unclear. No κ or λ chain or conventional C_H region is detectable. In certain immune responses, however, B and T cells apparently display the same idiotype (an antigenic determinant involving the combining site) and certain suppressor T cells are reported to secrete a factor which binds to an anti-idiotype serum as well as to antigen[39]. Such results are sometimes interpreted to mean that B and T cells use the same V_H genes.

Nucleic acid studies (reviewed by Kronenberg et al.[64]) raise serious doubts about conventional Ig gene involvement. 'Activation' of the J_H–C_μ region does occur in certain T cells, as reflected in nuclease sensitivity of thymocyte chromatin[120] and C_μ transcripts are produced[58, 60] but not μ mRNAs[58, 131]. Since a number of functional helper and cytotoxic T cell lines have all J_H genes in germ line context[17, 64, 65], their receptor cannot include a conventional V_H–D_H–J_H region. Moreover, joining of a V_H gene to a putative J_T (C_T) region lying 3' to J_H might delete the J_H locus from one allele but that has not been observed[17]. J_H rearrangements are seen in some T cells[23, 41, 65], but those cloned to date are 'DJ-only'[65] (Gerondakis and Bernard, unpublished). These rearrangements may be relics of events which occurred in a precursor cell before divergence of the T and B lineages, or they may indicate that T cells utilize a similar recombination machinery to join a V_T and J_T.

Oncogene translocation to immunoglobulin loci

Two B lymphoid malignancies, murine plasmacytomas and Burkitt's lymphomas of man, display specific chromosome translocations involving band regions bearing immunoglobulin loci. Significantly, recent work (reviewed by Klein, 1983) indicates that these translocations shift the cellular myc (c-myc) oncogene to an immunoglobulin locus, most often to an S_H region[2, 122]. The plasmacytoma translocations, perhaps mediated by the switch recombination machinery, disrupt the normal c-myc transcriptional unit and trigger distinct c-myc mRNA species proposed to function in oncogenesis[2]. Thus the long suspected link between chromosome aberrations and oncogene activation has been forged.

References

1. Adams, J. M. (1980). Organization and expression of immunoglobulin genes. *Immunology Today*, **1**, 10–17
2. Adams, J. M., Gerondakis, S., Webb, E., Corcoran, L. and Cory, S. (1983). Cellular *myc* oncogene is altered by chromosome translocation to an immunoglobulin locus in murine plasmacytomas and rearranged similarly in human Burkitt lymphomas. *Proceedings of the National Academy of Sciences, USA*, **80**, 1982–1986
3. Adams, J., Kemp, D., Bernard, O., Gough, N., Webb, E., Tyler, B., Gerondakis, S. and Cory, S. (1981). Organization and expression of murine immunoglobulin genes. *Immunological Reviews*, **59**, 5–32
4. Alt, F., Enea, V., Bothwell, A. and Baltimore, D. (1980). Activity of multiple light chain genes in murine myeloma cells producing a single, functional, light chain. *Cell*, **21**, 1–12
5. Alt, F., Rosenberg, N., Lewis, S., Casanova, R. and Baltimore, D. (1981). Variations in immunologlobulin heavy *and* light chain gene expression in a cloned Abelson murine leukemia virus-transformed cell line. In *B Lymphocytes in the Immune Response* (Klinman, N. *et al.*, Eds.), pp. 33–41. Amsterdam, Elsevier/North Holland
6. Altenburger, W., Steinmetz, M. and Zachau, H. (1980). Functional and non-functional joining in immunologlobuin light chain genes of a mouse myeloma. *Nature, London*, **287**, 603–607
7. Altenburger, W., Neumaier, P., Steinmetz, M. and Zachau, H. (1981). DNA sequence of the constant gene region of the mouse immunoglobulin kappa chain. *Nucleic Acids Research*, **9**, 971–981
8. Arnheim, N., Seperack, P., Banerji, J., Lang, R., Miesfeld, R. and Marcu, K. B. (1981). Mouse rDNA nontranscribed spacer sequences are found flanking immunoglobulin C_H genes and elsewhere throughout the genome. *Cell*, **22**, 179–185
9. Baltimore, D. (1981). Gene conversion: some implications for immunoglobulin genes. *Cell*, **24**, 592–594
10. Bentley, D. and Rabbitts, T. (1980). Human immunoglobulin variable region genes: DNA of two V_κ genes and a pseudogene. *Nature, London*, **288**, 730–733
11. Bernard, O. and Gough, N. (1980). Nucleotide sequence of immunoglobulin heavy chain joining segments between translocated V_H and μ constant region genes. *Proceedings of the National Academy of Sciences, USA*, **77**, 3630–3634
12. Bernard, O., Gough, N. and Adams, J. (1981). Plasmacytomas with more than one immunoglobulin κ mRNA: implications for allelic exclusion. *Proceedings of the National Academy of Sciences, USA*, **78**, 5812–5816
13. Bernard, O., Hozumi, N. and Tonegawa, S. (1978). Sequences of mouse immunoglobulin light chain genes before and after somatic changes. *Cell*, **15**, 1133–1144
14. Blomberg, B., Traunecker, A., Eisen, H. and Tonegawa, S. (1981). Organization of four mouse λ light chain immunoglobulin genes. *Proceedings of the National Academy of Sciences, USA*, **78**, 3765–3769
15. Bothwell, A., Paskind, M., Reth, M., Imanishi-Kari, T., Rajewsky, K. and Baltimore, D. (1981). Heavy chain variable region contribution to the NP^b family of antibodies; somatic mutation evident in a γ2a variable region. *Cell*, **24**, 625–637
16. Burrows, P., Beck, G. and Wabl, M. (1981). Expression of μ and γ immunoglobulin heavy chains in different cells of a cloned mouse lymphoid line. *Proceedings of the National Academy of Sciences, USA*, **78**, 564–568
17. Cayre, Y., Pallidino, M., Marcu, K. and Stavnezer, J. (1981). Expression of an antigen receptor on T cells does not require recombination at the immunoglobulin J_H–C_μ locus. *Proceedings of the National Academy of Sciences, USA*, **78**, 3814–3818
18. Cheng, H.-L., Blattner, F. R., Fitzmaurice, L., Mushinski, J. F. and Tucker, P. W. (1982). Structure of genes for membrane and secreted murine IgD heavy chains. *Nature, London*, **296**, 410–414
19. Clarke, S. H., Claflin, J. L. and Rudikoff, S. (1982). Polymorphisms in immunoglobulin heavy chains suggesting gene conversion. *Proceedings of the National Academy of Sciences, USA*, **79**, 3280–3284
20. Coleclough, C., Cooper, D. and Perry, R. (1980). Rearrangement of immunoglobulin heavy chain genes during B lymphocyte development as revealed by studies of mouse plasmacytoma cells. *Proceedings of the National Academy of Sciences, USA*, **77**, 1422–1426
21. Coleclough, C., Perry, R., Karjalainen, K. and Weigert, M. (1981). Aberrant rearrangements contribute significantly to the allelic exclusion of immunoglobulin gene expression. *Nature, London*, **290**, 372–377
22. Cory, S. and Adams, J. (1980). Deletions are associated with somatic rearrangement of immunoglobulin heavy chain genes. *Cell*, **19**, 37–51

23. Cory, S., Adams, J. and Kemp, D. (1980). Somatic rearrangements forming active immunoglobulin μ genes in B and T lymphoid cell lines. *Proceedings of the National Academy of Sciences, USA*, **77**, 4943–4947

24. Cory, S., Jackson, J. and Adams, J. (1980). Deletions in the constant region locus can account for switches in immunoglobulin heavy chain expression. *Nature, London*, **285**, 450–455

25. Cory, S., Tyler, B. and J. Adams (1981). Sets of immunoglobulin V genes homologous to ten cloned V_κ sequences: implications for the number of germline V_κ genes. *Journal of Molecular and Applied Genetics*, **1**, 103–116

26. Cory, S., Webb, E., Gough, J. and Adams, J. (1981). Recombination events near the immunoglobulin C_μ gene join variable and constant region genes, switch heavy chain expression, or inactivate the locus. *Biochemistry*, **20**, 2662–2670

27. Crews, S., Griffin, J., Huang, H., Calame, K. and Hood, L. (1981). A single V_H gene segment encodes the immune response to phosphorylcholine: somatic mutation is correlated with the class of the antibody. *Cell*, **25**, 59–66

28. Croce, C. M., Shander, M., Martinis, J., Cicurel, L., D'Amona, G. G., Dolby, T. W. and Koprowski, H. (1979). Chromosomal location of the genes for human immunoglobulin heavy chains. *Proceedings of the National Academy of Sciences, USA*, **76**, 3416–3419

29. Davis, M., Kim, S. and Hood, L. (1980). Immunoglobulin class switching: developmentally regulated DNA rearrangements during differentiation. *Cell*, **22**, 1–2

30. Davis, M., Kim, S. and Hood, L. (1980). DNA sequences mediating class switching in α-immunoglobulins. *Science*, **209**, 1360–1366

31. Davis, M., Calame, K., Early, P., Livant, D., Joho, R., Weissman, I. and Hood, L. (1980). An immunoglobulin heavy chain gene is formed by at least two recombinatorial events. *Nature, London*, **283**, 733–739

32. D'Eustachio, P., Bothwell, A., Takaro, T., Baltimore, D. and Ruddle, F. (1981). Chromosomal locations of structural genes encoding murine immunoglobulin λ light chains: genetics of murine λ light chains. *Journal of Experimental Medicine*, **153**, 793–800

33. Dildrop, R. M., Brüggeman, M., Radbruch, A., Rajewsky, K. and Bayreuther, D. (1982). II. Recombination between V genes. *EMBO Journal*, **1**, 635–640

34. Dreyer, W. and Bennett, J. (1965). The molecular basis of antibody formation: a paradox. *Proceedings of the National Academy of Sciences, USA*, **54**, 864–869

35. Dunnick, W., Rabbitts, T. and Milstein, C. (1980). An immunoglobulin deletion mutant with implications for the heavy-chain switch and RNA splicing. *Nature, London*, **286**, 669–675

36. Early, P. and Hood, L. (1981). Allelic exclusion and nonproductive immunoglobulin gene rearrangements. *Cell*, **24**, 1–3

37. Early, P., Huang, H., Davis, M., Calame, K. and Hood, L. (1980). An immunoglobulin heavy chain variable region gene is generated from three segments of DNA: V_H, D and J_H. *Cell*, **19**, 981–992

38. Early, P., Rogers, J., Davis, M., Calame, K., Bond, M., Wall, R. and Hood, L. (1980). Two mRNAs produced from a single immunoglobulin μ gene by alternative RNA processing pathways. *Cell*, **20**, 313–319

39. Eichmann, K. (1978). Expression and function of idiotypes on lymphocytes. *Advances in Immunology*, **26**, 195–254

40. Erikson, J., Martinis, J. and Croce, C. M. (1981). Assignment of the genes for the human λ immunoglobulin chains to chromosome 22. *Nature, London*, **294**, 173–175

41. Forster, A., Hobart, M., Hengartner, H. and Rabbitts, T. H. (1980). An immunoglobulin heavy-chain gene is altered in two T-cell lines. *Nature, London*, **286**, 897–899

42. Gearhart, P., Johnson, N., Douglas, R. and Hood, L. (1981) IgG antibodies to phosphorylcholine exhibit more diversity than their IgM counterparts. *Nature, London*, **291**, 29–34

43. Givol, D., Zahut, R., Eftan, K., Rechavi, G., Ram, D. and Cohen, J. (1981). The diversity of germline immunoglobulin V_H genes. *Nature, London*, **292**, 426–430

44. Gough, N. and Bernard, O. (1981). Sequences of the joining region genes for immunoglobulin heavy chains and their role in generation of antibody diversity. *Proceedings of the National Academy of Sciences, USA*, **78**, 509–513

45. Gough, N., Cory, S. and Adams, J. (1979). Identical 3′ non-coding sequences in five mouse immunoglobulin kappa chain mRNAs favour a unique C_κ gene. *Nature, London*, **281**, 394–396

46. Hamlyn, P., Gillam, S., Smith, M. and Milstein, C. (1977). Sequence analysis of the 3′ non-coding region of mouse immunoglobulin light chain messenger RNA. *Nucleic Acids Research*, **4**, 1123–1134

47. Hengartner, H., Meo, T. and Müller, E. (1978). Assignment of genes for immunoglobulin κ and heavy chains to chromosomes 6 and 12 in the mouse. *Proceedings of the National Academy of Sciences, USA*, **75**, 4494–4498

48. Hieter, P., Korsmeyer, S., Waldmann, T. and Leder, P. (1981). Human immunoglobulin κ light-chain genes are deleted or rearranged in λ-producing B cells. *Nature, London*, **290**, 368–372
49. Höchtl, J., Müller, C. and Zachau, H. (1982). Recombined flanks of the variable and joining segments of immunoglobulin genes. *Proceedings of the National Academy of Sciences, USA*, **79**, 1383–1387
50. Honjo, T. and Kataoka, T. (1978). Organization of immunoglobulin heavy chain genes and allelic deletion model. *Proceedings of the National Academy of Sciences, USA*, **75**, 2140–2144
51. Honjo, T., Kataoka, T., Yaoita, Y., Shimizu, A., Takahashi, N., Yamawaki-Kataoka, Y., Nikaido, T., Nakai, S., Obata, M., Kawakami, T. and Nishida, Y. (1980). Organization and reorganization of immunoglobulin heavy chain genes. *Cold Spring Harbor Symposium on Quantitative Biology*, **XLV**, 913–923
52. Hozumi, N. and S. Tonegawa (1976). Evidence for somatic rearrangement of immunoglobulin genes coding for variable and constant regions. *Proceedings of the National Academy of Sciences, USA*, **73**, 3628–3632
53. Joho, R. and Weissman, I. (1980). V–J joining of immunoglobulin κ genes only occurs on one homologous chromosome. *Nature, London*, **284**, 179–181
54. Kataoka, T., Miyata, T. and Honjo, T. (1980). Repetitive sequences in class switch recombination regions of immunoglobulin heavy chain genes. *Cell*, **23**, 357–368
55. Kataoka, T., Kawakami, T., Takahashi, N. and Honjo, T. (1980). Rearrangement of immunoglobulin γ1-chain gene and mechanism for heavy-chain class switch. *Proceedings of the National Academy of Sciences, USA*, **77**, 919–923
56. Kataoka, T., Nikaido, T., Miyata, T., Moriwaki, K. and Honjo, T. (1982). The nucleotide sequences of rearranged and germline immunoglobulin V_H genes of a mouse myeloma MC101 and evolution of V_H genes in mouse. *Journal of Biological Chemistry*, **257**, 277–285
57. Kemp, D., Cory, S. and Adams, J. (1979). Cloned pairs of variable region genes for immunoglobulin heavy chains isolated from a clone library of the entire mouse genome. *Proceedings of the National Academy of Sciences, USA*, **76**, 4627–4631
58. Kemp, D., Harris, A. and Adams, J. (1980). Transcripts of the immunoglobulin C_μ gene vary in structure and splicing during lymphoid development. *Proceedings of the National Academy of Sciences, USA*, **77**, 7400–7404
59. Kemp, D., Harris, A., Cory, S. and Adams, J. (1980). Expression of the immunoglobulin C_μ gene in mouse T and B lymphoid and myeloid cell lines. *Proceedings of the National Academy of Sciences, USA*, **77**, 2876–2880
60. Kemp, D., Wilson, A., Harris, A. and Shortman, K. (1980). The immunoglobulin μ constant region gene is expressed in mouse thymocytes. *Nature, London*, **286**, 168–170
61. Kemp, D., Tyler, B., Bernard, O., Gough, N., Gerondakis, S., Adams, J. and Cory, S. (1981). Organization of genes and spacers within the mouse immunoglobulin V_H locus. *Journal of Molecular and Applied Genetics*, **1**, 245–261
62. Kim, S., Davis, M., Sinn, E., Patten, P. and Hood, L. (1981). Antibody diversity somatic hypermutation of rearranged V_H genes. *Cell*, **27**, 573–581
62a. Klein, G. (1983). Specific chromosomal translocations and the genesis of B-cell-derived tumors in mice and men. *Cell*, **32**, 311–315
63. Knapp, M. R., Liu, C.-P., Newell, N., Wara, R. B., Tucker, P. W., Strober, S. and Blattner, F. (1982). Simultaneous expression of immunoglobulin μ and δ heavy chains by a cloned B-cell lymphoma: A single copy of the V_H gene is shared by two adjacent C_H genes. *Proceedings of the National Academy of Sciences, USA*, **79**, 2996–3000
64. Kronenberg, M., Kraig, E., Horvath, S. and Hood, L. (1981). Cloned T cells as a tool for molecular geneticists: approaches to cloning genes which encode T-cell antigen receptors. In *Isolation, Characterization and Utilization of T Lymphocyte Clones* (Fathman, C. G. and Fitch, F., Eds.). Academic Press
65. Kurosawa, Y., Von Boehmer, H., Haas, W., Sakano, H., Traunecker, A. and Tonegawa, S. (1981). Identification of D segments of immunoglobulin heavy-chain genes and their rearrangement in T lymphocytes. *Nature, London*, **290**, 565–570
66. Kwan, S.-P., Max, E., Seidman, J., Leder, P. and Scharff, M. (1981). Two kappa immunoglobulin genes are expressed in the myeloma S107. *Cell*, **26**, 57–66
66a. Lang, R. B., Stanton, L. W. and Marcu, K. B. (1982). On immunoglobulin heavy chain gene switching: two γ2b genes are rearranged via switch sequences in MPC-11 cells but only one is expressed. *Nucleic Acids Research*, **10**, 611–630
67. Lewis, S., Rosenberg, N., Alt, F. and Baltimore, D. (1982). Continuing kappa-gene rearrangement in a cell line transformed by Abelson murine leukemia virus. *Cell*, **30**, 807–816
68. Liu, C.-P., Tucker, P., Mushinski, J. and Blattner, F. (1980). Mapping of heavy chain genes for mouse immunoglobulins M and D. *Science*, **209**, 1348–1353

69. McBride, O. W., Hieter, P. A., Hollis, G. F., Swan, D., Otey, M. D. and Leder, P. (1982). Chromosomal location of human kappa and lambda immunoglobuin light chain constant region genes. *Journal of Experimental Medicine*, **155**, 1480–1490
70. Maki, R., Kearney, J. Paige, C. and Tonegawa, S. (1980). Immunoglobulin gene rearrangement in immature B cells. *Science*, **209**, 1366–1369
71. Maki, R., Roeder, W., Traunecker, A., Sidman, C., Wabe, M., Raschke, W. and Tonegawa, S. (1981). The role of DNA rearrangement and alternative RNA processing in the expression of immunoglobulin delta genes. *Cell*, **24**, 353–365
72. Malcolm, S., Barton, P., Murphy, C., Ferguson-Smith, M. A., Bentley, D. L. and Rabbitts, T. H. (1982). Localisation of human kappa light chain variable region genes to the short arm of chromosome 2 by *in situ* hybridization. *Proceedings of the National Academy of Sciences, USA*, **79**, 4957–4961
73. Marcu, K., Banerji, J., Penncavage, N., Lang, R. and Arnheim, N. (1980). 5' flanking region of immunoglobulin heavy chain constant region genes displays length heterogeneity in germlines of inbred mouse strains. *Cell*, **22**, 187–196
74. Marcu, K. B., Lang, R. B., Stanton, L. W. and Harris, J. J. (1982). A model for the molecular requirements of immunoglobulin heavy chain class switching. *Nature, London*, **298**, 87–89
75. Max, E., Seidman, J. and Leder, P. (1979). Sequences of five potential recombination sites encoded close to an immunoglobulin R constant region gene. *Proceedings of the National Academy of Sciences, USA*, **76**, 3450–3454
76. Max, E., Seidman, J., Miller, H. and Leder, P. (1980). Variation in the crossover point of kappa immunoglobulin gene V–J recombination: evidence from a cryptic gene. *Cell*, **21**, 793–799
77. Miller, J., Bothwell, A. and Storb, U. (1981). Physical linkage of the constant region genes for immunoglobulins λ_I and λ_{III}. *Proceedings of the National Academy of Sciences, USA*, **78**, 3829–3833
78. Miller, J., Selsing, E. and Storb, U. (1982). Structural alterations in J regions of mouse immunoglobulin λ genes are associated with differential gene expression. *Nature, London*, **295**, 428–430
79. Miyata, T., Yasunaga, T., Yamawaki-Kataoka, Y., Obata, M. and Honjo, T. (1980). Nucleotide sequence divergence of mouse immunoglobulin γ_1 and γ_{2b} chain genes and the hypothesis of intervening sequence-mediated domain transfer. *Proceedings of the National Academy of Sciences, USA*, **77**, 2143–2147
80. Moore, K., Rogers, J., Hunkapiller, T., Early, P., Nottenburg, C., Weissman, I., Bazin, H., Wall, R. and Hood, L. (1981). Expression of IgD may use both DNA rearrangement and RNA splicing mechanisms. *Proceedings of the National Academy of Sciences, USA*, **78**, 1800–1804
81. Newell, N., Richards, J., Tucker, P. and Blattner, F. (1980). J genes for heavy chain immunoglobulins of mouse. *Science*, **209**, 1128–1132
82. Nikaido, T., Nakai, S. and Honjo, T. (1981). Switch region of immunoglobulin C_μ gene is composed of simple tandem repetitive sequences. *Nature, London*, **292**, 845–848
83. Nishida, Y., Kataoka, T., Ishida, N., Nakai, S., Kishimoto, T., Böttcher, I. and Honjo, T. (1981). Cloning of mouse immunogloublin ϵ gene and its location within the heavy chain gene cluster. *Proceedings of the National Academy of Sciences, USA*, **78**, 1581–1585
84. Nottenberg, C. and Weissman, I. L. (1981). C_μ gene rearrangement of mouse immunoglobulin genes in normal B cells occurs on both the expressed and non-expressed chromosomes. *Proceedings of the National Academy of Sciences, USA*, **78**, 484–488
85. Obata, M., Kataoka, T., Nakai, S., Yamagishi, H., Takahashi, N., Yamawaki-Kataoka, Y., Nikaido, T., Shimizu, A. and Honjo, T. (1981). Structure of rearranged γ1 chain gene and its implication to immunoglobulin class-switch mechanism. *Proceedings of the National Academy of Sciences, USA*, **78**, 2437–2441
86. Ollo, R., Auffray, C., Sikorav, J.-L. and Rougeon, F. (1981). Mouse heavy chain variable regions: nucleotide sequence of germ-line V_H gene segment. *Nucleic Acids Research*, **9**, 4099–4109
87. Pech, M., Höchtl, J., Schnell, H. and Zachau, H. (1981). Differences between germ-line and rearranged immunoglobulin V_κ coding sequences suggest a localized mutation mechanism. *Nature, London*, **291**, 668–670
88. Perry, R. and Kelley, D. E. (1979). Immunoglobulin messenger RNAs in murine cell lines have characteristics of immature B lymphocytes. *Cell*, **18**, 1333–1339
89. Perry, R., Coleclough, C. and Weigert, M. (1980). Reorganization and expression of immunoglobulin genes: the status of allelic elements. *Cold Spring Harbor Laboratory of Quantitative Biology*, **XLV**, 925–933
90. Perry, R., Kelley, D., Coleclough, C. and Kearney, J. (1981). Organization and expression of immunoglobulin genes in fetal liver hybridomas. *Proceedings of the National Academy of Sciences, USA*, **78**, 247–251
91. Perry, R., Kelley, D., Coleclough, C., Seidman, J. Leder, P., Tonegawa, S., Matthyssens, G. and

Weigert, M. (1980). Transcription of mouse κ chain genes: implications for allelic exclusion. *Proceedings of the National Academy of Sciences, USA*, **77**, 1937–1941

92. Pfeiffer, W. and Zachau, H. (1980). Accessibility of expressed and non-expressed genes to a restriction nuclease. *Nucleic Acids Research*, **8**, 4621–4638

93. Potter, M. (1977). Antigen-binding myeloma proteins of mice. *Advances in Immunology*, **25**, 141–212

94. Rabbitts, T., Forster, A., Dunnick, W. and Bentley, D. (1980). The role of gene deletion in the immunogloublin heavy chain switch. *Nature, London*, **283**, 351–356

95. Radbruch, A., Liesegang, B. and Rajewsky, K. (1980). Isolation of variants of mouse myeloma X63 that expressed changed immunoglobulin. *Proceedings of the National Academy of Sciences, USA*, **77**, 2909–2913

96. Riley, S., Brock, E. and Kuehl, W. (1981). Induction of light chain expression in a pre-B cell line by fusion to myeloma cells. *Nature, London*, **289**, 804–806

97. Roeder, W., Maki, R., Traunecker, A. and Tonegawa. S. (1981). Linkage of the four γ subclass heavy chain genes. *Proceedings of the National Academy of Sciences, USA*, **78**, 474–478

98. Rogers, J. and Wall, R. (1981). Immunoglobulin heavy chain genes: demethylation accompanies class switching. *Proceedings of the National Academy of Sciences, USA*, **78**, 7497–7501

99. Rogers, J., Choi, E., Souza, L., Carter, C., Word, C., Kuehl, M., Eisenberg, D. and Wall, R. (1981). Gene segments encoding transmembrane carboxyl termini of immunoglobulin γ chains. *Cell*, **26**, 19–27

100. Rogers, J., Early, P., Carter, C., Calame, K., Bond, M., Hood, L. and Wall, R. (1980). Two mRNAs with different 3′ ends encode membrane-bound and secreted forms of immunoglobulin μ chain. *Cell*, **20**, 303–312

101. Sakano, H., Huppi, K., Heinrich, G. and Tonegawa, S. (1979). Sequences at the somatic recombination sites of immunoglobulin light chain genes. *Nature, London*, **280**, 288–294

102. Sakano, H., Maki, R., Kurosawa, Y., Roeder, W. and Tonegawa, S. (1980). Two types of somatic recombination are necessary for the generation of complete immunoglobulin heavy chain genes. *Nature, London*, **286**, 676–683

103. Sakano, H., Kurosawa, Y., Weigert, M. and Tonegawa, S. (1981). Identification and nucleotide sequence of a diversity DNA segment (D) of immunoglobulin heavy-chain genes. *Nature, London*, **290**, 562–565

104. Schilling, J., Clevinger, B., Davie, J. and Hood, L. (1980). Amino acid sequence of homogeneous antibodies to dextran and DNA rearrangements in heavy chain V region gene segments. *Nature, London*, **283**, 35–40

105. Schreier, P., Bothwell, A., Mueller-Hill, B. and Baltimore, D. (1981). Multiple differences between the nucleic acid sequences of the IgG2a[a] and IgG2a[b] alleles of the mouse. *Proceedings of the National Academy of Sciences, USA*, **78**, 4495–4499

106. Secher, D., Cotton, R., Cowan, N. and Milstein, C. (1974). Spontaneous mutation in immunoglobulin genes. In *The Immune System: Genes, Receptors, Signals* (Sercarz, E., Williamson, A. and Fox, C., Eds.), pp. 353–356. London and New York, Academic Press

107. Seidman, J. and Leder, P. (1980). A mutant immunoglobulin light chain is formed by aberrant DNA- and RNA-splicing events. *Nature, London* , **286**, 779–783

108. Seidman, J., Max, E. and Leder, P. (1979). A κ-immunoglobulin gene is formed by site-specific recombination without further somatic mutation. *Nature, London*, **280**, 370–375

109. Seidman, J., Leder, A., Nau, M., Norman, B. and Leder, P. (1978). Antibody diversity. *Science*, **202**, 11–17

110. Seidman, J., Nau, M., Norman, B., Kwan, S.-P., Scharff, M. and Leder, P. (1980). Immunoglobulin V/J recombination is accompanied by deletion of joining site and variable region segments. *Proceedings of the National Academy of Sciences, USA*, **77**, 6022–6026

111. Selsing, E. and Storb, U. (1981). Mapping of immunoglobulin variable region genes: relationship to the 'deletion' map of immunoglobulin rearrangement. *Nucleic Acids Research*, **9**, 7525–7535

112. Selsing, E. and Storb, U. (1981). Somatic mutation of immunoglobulin light-chain variable-region genes. *Cell*, **25**, 47–58

113. Shimizu, A., Takahashi, N., Yaoita, Y. and Honja, T. (1982). Organisation of the constant region gene family of the mouse immunoglobulin heavy chain. *Cell*, **28**, 499–506

114. Shimizu, A., Takahashi, N., Yamawaki-Kataoka, Y., Nishida, Y., Kataoka, T. and Honjo, T. (1981). Ordering of mouse immunoglobulin heavy chains by molecular cloning. *Nature, London*, **289**, 149–153

115. Siden, E., Baltimore, D., Clark, D. and Rosenberg, N. (1979). Immunoglobulin synthesis by lymphoid cells transformed *in vitro* by Abelson murine leukemia virus. *Cell*, **16**, 389–396

116. Siden, E., Alt, F., Shinefield, L., Sato, V. and Baltimore, D. (1981). Synthesis of immunoglobulin μ chain gene products precedes synthesis of light chains during B-lymphocyte development. *Proceedings of the National Academy of Sciences, USA*, **78**, 1823–1827

117. Sigal, N., Pickard, A., Metcalf, E., Gearhart, P. and Klinman, N. (1977). Expression of phosphorylcholine-specific B cells during murine development. *Journal of Experimental Medicine*, **146**, 933–948

118. Sitia, R., Rubartelli, A. and Hämmerling, U. (1981). Expression of 2 immunoglobulin isotypes, IgM and IgA, with identical idiotype in the B cell lymphoma 1.29. *Journal of Immunology*, **127**, 1388–1394

119. Steinmetz, M., Altenburger, W. and Zachau, H. (1980). A rearranged DNA sequence related to the translocation of immunoglobulin gene segments. *Nucleic Acids Research*, **8**, 1709–1720

120. Storb, U., Arp, B. and Wilson, R. (1981). The switch region associated with immunoglobulin C_μ genes is DNase I hypersensitive in T lymphocytes. *Nature, London*, **294**, 90–92

121. Storb, U., Wilson, R., Selsing, E. and Walfield, A. (1981). Rearranged and germline immunoglobulin κ genes: different states of DNase I sensitivity of constant κ genes in immunocompetent and nonimmune cells. *Biochemistry*, **20**, 990–996

122. Taub, R., Kirsch, I., Morton, C., Lenoir, G., Swan, D., Tronick, S., Aaronson, S. and Leder, P. (1982). Translocation of the c-*myc* gene into the immunoglobulin heavy chain locus in human Burkitt's lymphoma and murine plasmacytoma cells. *Proceedings of the National Academy of Sciences, USA*, **78**, 7837–7841

123. Tonegawa, S., Sakano, H., Maki, R., Traunecker, A., Heinrich, G., Roeder, W. and Kurosawa, Y. (1980). Somatic reorganization of immunoglobulin genes during lymphocyte differentiation. *Cold Spring Harbor Symposium on Quantitative Biology*, **45**, 839–858

124. Tucker, P., Liu, C., Mushinski, J. and Blattner, F. (1980). Mouse immunoglobulin D: messenger RNA and genomic DNA sequences. *Science*, **209**, 1353–1360

125. Tyler, B. and Adams, J. (1980). Organization of the sequences flanking immunoglobulin heavy chain genes and their role in class switching. *Nucleic Acids Research*, **8**, 5579–5598

126. Tyler, B., Cowman, A., Adams, J. and Harris, A. (1981). Generation of long mRNA for membrane immunoglobulin γ_{2a} chains by differential splicing. *Nature, London*, **293**, 406–408

127. Tyler, B., Cowman, A., Gerondakis, S., Adams, J. and Bernard, O. (1982). mRNA for surface immunoglobulin γ chains encodes a highly conserved transmembrane sequence and a 28-residue intracellular domain. *Proceedings of the National Academy of Sciences, USA*, **79**, 2008–2012

128. Valbuena, O., Marcu, K., Weigert, M. and Perry, R. (1978). Multiplicity of germline genes specifying a group of related mouse κ chains with implications for the generation of immunoglobulin diversity. *Nature, London*, **276**, 780–784

129. Van Ness, B., Coleclough, C., Perry, R. and Weigert, M. (1982). DNA between variable and joining gene segments of immunoglobulin κ light chain is frequently retained in cells that rearrange the κ locus. *Proceedings of the National Academy of Sciences, USA*, **79**, 1

130. Walfield, A., Selsing, E., Arp, B. and Storb, U. (1981). Misalignment of V and J gene segments resulting in a nonfunctional immunoglobulin gene. *Nucleic Acids Research*, **9**, 1101–1109

131. Walker, I. and Harris, A. (1980). Immunoglobulin C_μ RNA in T lymphoma cells is not translated. *Nature, London*, **288**, 290–293

132. Weigert, M., Cesair, I., Yonkovich, S. and Cohn, M. (1970). Variability in the lambda light chain sequences of mouse antibody. *Nature, London*, **228**, 1045–1047

133. Weigert, M., Gatmaitan, L., Loh, E., Schilling, J. and Hood, L. (1978). Rearrangement of genetic information may produce immunoglobulin diversity. *Nature, London*, **276**, 785–790

134. Weigert, M., Perry, R., Kelley, D., Hunkapiller, T., Schilling, J. and Hood, L. (1980). The joining of V and J gene segments creates antibody diversity. *Nature, London*, **283**, 497–499

135. Wu, T. and Kabat, E. (1970). An analysis of the sequences of the variable regions of Bence Jones proteins and myeloma light chains and their implications for antibody complementarity. *Journal of Experimental Medicine*, **132**, 211–250

136. Yagi, M. and Koshland, M. E. (1981). Expression of the J chain gene during B cell differentiation is inversely correlated with DNA methylation. *Proceedings of the National Academy of Sciences, USA*, **78**, 4907–4911

137. Yamawaki-Kataoka, Y., Nakai, S., Miyata, T. and Honjo, T. (1982). Nucleotide sequences of gene segments encoding membrane domains of immunoglobulin γ chains. *Proceedings of the National Academy of Sciences, USA*, **79**, 2623–2627

138. Yaoita, Y. and Honjo, T. (1980). Deletion of immunoglobulin heavy chain genes accompanies the class switch rearrangement. *Biomedical Research*, **1**, 164–175

139. Yaoita, Y. and Honjo, T. (1980). Deletion of immunoglobulin heavy chain genes from expressed allelic chromosome. *Nature, London*, **286**, 850–853

140. Yaoita, Y., Kumagai, Y., Okumura, K. and Honjo, T. (1982). Expression of lymphocyte surface IgE does not require switch recombination. *Nature, London*, **297**, 697–699

141. Zeelon, E., Bothwell, A., Kantor, F. and Schechter, I. (1981). An experimental approach to enumerate the genes coding for immunoglobulin variable regions. *Nucleic Acids Research*, **9**, 3809–3820

Chapter 21

Contractile protein genes

M. E. Buckingham and A. J. Minty

It has been evident for a number of years that contractile proteins exist as multiple isoforms encoded by multigene families. The advent of recombinant DNA technology has made it possible to look directly at the structure of the contractile protein genes, at their organization in the genome, and at their transcriptional products in different tissues. In this review, we shall confine our discussion to the

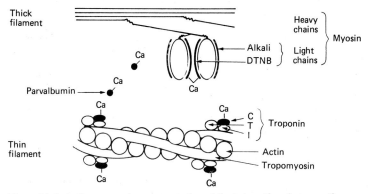

Figure 21.1. A diagrammatic representation of the interactions between the major contractile proteins of skeletal muscle. Parts of the thin and thick filaments are shown, including one crossbridge. The alkali light chains correspond to LC1 and LC3, the DTNB light chain to the phosphorylatable light chain, LC2. Adapted from reference 5.

major components of muscle and of the microfilaments of the cytoskeleton, namely actin, myosin, tropomyosin and the muscle troponins. The interaction of these proteins in skeletal muscle is illustrated in *Figure 21.1*.

Contractile protein isoforms in different muscle and non-muscle tissues

Contractile protein isoforms are expressed in a large number of different muscle types, which can be classified as smooth muscles (e.g. stomach, intestine, uterus,

365

TABLE 21.1. Contractile protein isoforms in different muscle and non-muscle tissues[a]

Protein family	Skeletal muscle					
	Fast			Slow		
	Fetal	Newborn	Adult	Fetal	Newborn	Adult
Myosin heavy chains: (MHC)	$MHC_{emb}^{165,198}$	$MHC_{NB}^{80,199}$	$2MHC_F^{166,212}$ MHC_{SF}^{148}	MHC_{emb}^{198}	MHC_{NB}^{198} MHC_F^{198}	MHC_S^{190}
Myosin light chains: non-phosphorylatable (LC_1,LC_3)	LC_{1emb}^{195} (LC_{1F})	$(LC_{1emb})^{195}$ LC_{1F}^{56} LC_{3F}^{56}	LC_{1F}^{56} LC_{3F}^{56}	LC_{1emb}^{195} (LC_{1F})	(LC_{1emb}) LC_{1S} LC_{1S}	$LC_{1S}^{188,195}$ $LC'_{1S}^{188,195}$
phosphorylatable (LC_2)	$LC_{2F}^{27,188}$	$LC_{2F}^{27,188}$	$LC_{2F}^{27,188}$	LC_{2F}^{195}	LC_{2F}^{195} LC_{2S}^{195}	$LC_{2S}^{188,195,191}$ $LC'_{2S}^{88,195,191}$
Troponins: TnC	TnC_F^{202}		TnC_F^{29}			TnC_S^{183}
TnI	TnI_F^{202}		TnI_F^{202}			TnI_S^{202}
TnT	TnT_F^2		$2TnT_F^{201}$			TnT_S^{26}
Tropomyosins (Tm_α,Tm_β)	Tm_α $Tm_\beta^{2,149}$		$?2Tm_{\alpha F}^{11,170}$ $?2Tm_{\beta F}^{11,110}$			$Tm_{\alpha S}^{11,128,168,170}$ $Tm_{\beta S}^{110,128,168,170}$
Actins (α,β,γ)			α_{Sk}^{28}			α_{Sk}^{178}

[a] The following abbreviations are used: emb, fetal; NB, newborn; F, fast; SF, superfast; S. slow; A, atrial; V, ventricular; SM, smooth muscle; NM, non-muscle; C, cardiac; Sk, skeletal.

arteries, etc.), cardiac muscles (atrium, ventricle, Purkinje fibres) and a variety of skeletal muscles (classed as slow or fast, depending on the speed of fibre contraction). The complexity of skeletal muscle development is indicated by the following considerations: (i) Different adult fast muscles may show different patterns of isoform expression[16]; (ii) Individual skeletal muscles contain several distinct fibre types (type I: slow twitch, and types II, A, B, C: fast twitch) which may express different contractile protein isoforms[12]; (iii) Developing skeletal muscles may express further fetal or new-born isoforms[199]. In *Table 21.1* a summary is presented of the currently available protein data bearing on the contractile protein isoforms expressed in different tissues. In general, the data presented are for mammals, since this is the most extensive.

The questions which are commented on in the subsequent sections are as follows: (i) What is the minimum number of isoforms (and thus genes) in each contractile protein family? (ii) Are individual isoforms expressed in a single or in several muscle types? (iii) Are there developmental changes in isoform expression? We also discuss the homology between different contractile protein isoforms, since this is relevant to the evolution of the different multigene families.

Cardiac muscle						Smooth muscle	Non-muscle	Gene No.
Atria			Ventricles					
Fetal	*Newborn*	*Adult*	*Fetal*	*Newborn*	*Adult*			
		MHC_A[153]	MHC_{V3}[102]	MHC_{V1}[102]	MHC_{V3}[76,81,152]	MHC_{SM}[19]	$2MHC_{NM}$[19]	11–13
				(MHC_{V3})[102]	(MHC_{V1})[76,81,152]			
		$?{\equiv}MHC_{V1}$[23]						
LC_{1emb}[197,200]	LC_{1emb}[197,200]	LC_{1emb}[197,200]	LC_{1emb}[193]		LC_{1S}	LC_{1NM}[114]	LC_{1NM}[114]	6
LC_{2A}		LC_{2A}[197]	LC'_{2S}[191]	LC'_{2S}[191]	LC_{2S}[97,114,191]	LC_{2NM}[18,84]	LC_{2NM}[18,84]	5
			(LC_{2S})[191]	LC_{2S}[191]	LC'_{2S}[97,114,191]			
					TnC_S[183]			2
					TnI_C[202]			3
					TnT_C[26]			3–4
					$?Tm_{\alpha C}$[79,,128]	$Tm_{\alpha SM}$[32,53,128]	$?2Tm_{NM}$[30,53,64]	5–11
						$Tm_{\beta SM}$[32,128]		
					α_C[178]	α_{SM},γ_{SM}[178,180]	β_{NM},γ_{NM}[46,176]	6

Myosins

The myosin molecule consists of two heavy chains (MHC, ~200 000 d), two phosphorylatable or regulatory light chains (LC_2, 17 000–20 000 d) and two non-phosphorylatable or alkali light chains (LC_1, 19 000–22 000 d† and in fast skeletal muscle, LC_3, 16 000 d).

Myosin heavy chains

The myosin heavy chain (MHC) proteins clearly demonstrate the presence of tissue-specific isoforms and, indeed, there may even be more than one distinct MHC in any one tissue type. In adult fast skeletal muscle, for instance, there are two MHC molecules apparently expressed in fast type IIA and fast type IIB fibres respectively[10, 12]. Two non-muscle myosin heavy chains have been described which also show tissue specificity[19].

Developmental stage-specific isoforms have been identified in striated muscle. Transitions from fetal (at 0–5 days after birth), to newborn (3–14 days), to adult

† On SDS acrylamide gels, these light chains migrate anomalously as 25 000–27 000 dalton peptides.

MHC occur sequentially in rat skeletal muscle[199]. In cardiac muscle, a different ventricular form (MHCV$_1$) appears at birth in some species and may persist (e.g in rodents). In other species, the fetal form (MHCV$_3$) persists in the adult[102].

Only very fragmentary amino acid sequence information is available for the myosin heavy chain proteins. However, from peptide mapping and immunological data it is clear that the proteins have considerable homology (see also p. 370).

Myosin light chains

In the case of the myosin light chains, fairly extensive sequence data are available for the adult striated muscle forms. It is not clear whether the slow and cardiac ventricular light chains are identical, since their complete sequences are not known for the same species. The second slow light chain, LC$_{1S}'$, is found only in mammals. This is probably true also for LC$_{1emb}$, the light chain found in fetal skeletal muscle and heart[33, 193] and in adult heart atria[35].

The myosin light chains within any one group show some sequence homology. The most striking example is that of LC$_{3F}$ which shows perfect homology with LC$_{1F}$ for the last 141 $-COOH$ terminal amino acids. It has an additional eight amino acids at the $-NH_2$ terminal which are distinct from the 49 amino acids of the $-NH_2$ terminal sequence of LC$_1$[56, 113] (see p. 373 on MLC genes). Skeletal muscle and cardiac LC$_1$ have 31.5% homology in the $-NH_2$ region, compared with about 70% in the rest of the molecule[108]. No sequence information exists for LC$_{1emb}$, and its identification is based on two-dimensional gel electrophoresis[195]. Partial sequence comparisons of the phosphorylatable or regulatory light chains (LC$_2$) from rabbit skeletal muscle, bovine heart and chicken gizzard show approximately 70% homology between these proteins[84, 97], and about 15–25% homology with Ca^{2+} binding proteins such as troponin C, parvalbumin and calmodulin[27, 187]. This should be compared with about 25% homology between LC$_2$ and LC$_1$[27].

Troponins

In vertebrate striated muscle, Ca^{2+} activation of contraction is mediated by the troponins. There are three types of troponin: TnC ($\sim$19 000 d), TnI ($\sim$24 000 d) and TnT (37 000 d). TnI inhibits the actomyosin ATPase by binding to actin; this inhibition is enhanced by tropomyosin to which troponin T binds. The inhibition is reversed by TnC binding to TnI, which is effective when the Ca^{2+} $-$TnC complex is formed. Ca^{2+} sensitivity requires TnT (see *Figure 21.1*).

Troponin C is present in at least two isoforms, one in fast skeletal, and the other in slow skeletal and cardiac muscle. The amino acid sequence of TnC$_S$ is about 63% homologous with that of TnC$_F$ and there is about 40% homology between these troponins and calmodulin, another Ca^{2+} binding protein[187]. Three distinct isoforms of troponin I have been identified in cardiac, and in fast and slow skeletal muscle. There is about 60% overall homology between these sequences[202]. This is strongest in the region which binds to actin and less in the $-NH_2$ terminal part of the molecule which interacts with TnC. Cardiac TnI has an additional 26 amino acids at the $-NH_2$ terminal compared with the other TnI isoforms. Isoforms of TnT have also been characterized in different striated muscles, although detailed sequence information is not available. Two fast isoforms, with different $-NH_2$ termini, have been identified in chick breast and leg muscle respectively. It has been suggested that this may be a result of post-translational cleavage[115].

Tropomyosins

The tropomyosins are another class of regulatory proteins, which are associated with actin in muscle and non-muscle cells. Two tropomyosins are easily distinguished by gel electrophoresis as α and β, although in fact they have the same size (284 amino acids). Sequencing of α[170] and β[110] tropomyosins of rabbit skeletal muscle (mainly fast) indicates 39 amino acid differences, mostly in the −COOH terminal. Sequence data for β-tropomyosins show heterogeneity at 11 positions; microheterogeneity in the α-tropomyosin sequence has also been reported. Different isoforms of α- and β-tropomyosin are found in slow and fast muscle fibres[11] and probably also in the two fast fibre types, IIA and IIB. The relative proportions of β to α in different fibre types may vary[16, 40]. In developing muscles, β-tropomyosin is the predominant isoform[129, 149].

In chick, it has been shown that there is a cardiac α-tropomyosin isoform distinct from that in fast or in slow muscle[128], although in rabbit, the major cardiac tropomyosin is identical in sequence to the major skeletal muscle α-tropomyosin[109].

Actins

Actin, in its polymerized form, is a major structural component of contractile systems. It is the principal protein of the microfilaments of non-muscle cells and of the muscle sarcomere where it is present in the molar ratio actin:myosin:tropomyosin:troponin, 7:1:1:1[143]. At present, six different primary sequences have been described for the actins of different mammalian tissues[178]. Their nomenclature is based on isoelectric focussing differences[65, 194]. There is an α-actin sequence in skeletal muscle and, at least in mammals, a different α-actin in the heart (see also section on p. 371). In smooth muscle, two actins have been identified whose relative proportions vary in different tissues[180]. Two non-muscle actins, β and γ, have been found in all non-muscle cells examined, and it has been shown that they are both expressed in a single cell[13].

Analyses of actin N-terminal peptides[180] have failed to reveal isoforms other than the six discussed above, although additional isoforms have been postulated[13, 15, 94, 111]. In the case of the transformed fibroblast cell line studied by Leavitt and coworkers[70, 94, 182], a mutant β-actin sequence is expressed in addition to normal β-actin. A subclone of this line exhibits a β-actin with a second site mutation. Tumorigenicity in these lines can be correlated with the expression of mutant β-actin[95]. It seems likely as reported for a variant tropomyosin[69] that one of the two β-actin alleles has mutated in these cells.

A striking feature of the actins is their very high sequence conservation. Only four out of 375 amino acids differ between cardiac and skeletal muscle actins, six or eight between smooth and skeletal muscle actins, 24 or 25 between striated muscle and non-muscle actins, and 22 or 23 between smooth muscle and non-muscle actins[178]. High sequence conservation is also seen between species, even those as different as *Dictyostelium* and mammals[179].

Characterization of different contractile protein mRNAs: cDNA clones

A large number of different cDNA recombinant plasmids coding for vertebrate contractile proteins have been characterized. In many cases, the cDNA clones have provided new insights into the numbers of isoforms, their expression in different tissues, their coding sequences and degree of sequence homology.

Myosin heavy chain cDNA clones

The partial amino acid sequence in the −COOH terminus of a myosin heavy chain isoform from adult rabbit skeletal muscle[45] has served to confirm the identification of MHC clones from many different sources, illustrating the suspected amino acid sequence conservation. Further DNA sequencing has provided original information on the amino acid sequences in this −COOH terminal rod portion of the molecule. The most detailed sequence comparisons have been made over more than a thousand nucleotides, between two different adult cardiac MHC cDNA clones[107]. The two sequences have 97% homology in amino acid residues (95% in nucleotides) in their common coding region. Analysis of the amino acid sequence shows a number of features characteristic of the α-helical nature of the rod part of MHC, such as the regular 3,4,3,4 spacing of non-polar hydrophobic residues.

Comparison of these cardiac sequences with that for a skeletal muscle isoform, MHC$_{emb}$[107, 119], indicates regions of homology interspersed with less homologous sections. Individual 'hot spots' for amino acid substitution between isoforms are also evident[192]. It is particularly striking that the last 10 or so amino acids at the −COOH termini of MHC from different tissues and species are highly diverged[107, 174, 192], compared with c. 80% homology elsewhere. Observations of this kind indicate that different evolutionary pressures are acting on different parts of the molecule, presumably reflecting areas of functional importance. As far as the coding regions of the mRNAs are concerned, nucleotide divergence is only slightly higher than that for amino acids. In contrast, the 3′ non-coding sequences of the mRNAs show no significant sequence homology and furthermore this region differs in length (50–120 nucleotides[107, 174, 192]). The position of the polyadenylation signal also varies, and indeed the site of addition of poly A varies slightly between different DNA recombinants for the same MHC sequence[107].

The characterization of MHC cDNA clones has provided new information on MHC isoform expression in different tissues (e.g. reference 162) and demonstrated the presence of additional isoforms, such as two adult V$_1$ isoforms in heart ventricles[107] and developmental MHC isoforms in the breast and leg muscles of the chicken[174].

Myosin light chain cDNA clones

The recombinant cDNA plasmid p161 contains a non-coding sequence which appears to be common to the two myosin alkali light chains MLC$_{1F}$ and MLC$_{3F}$[147]. This is an atypical situation, since 3′ non-coding sequences, even between closely related mRNAs (e.g. the actins), are usually highly diverged. The question this raises about the nature of the corresponding genes is discussed in a later section on p. 373. The cloned non-coding sequence cross-hybridizes weakly with the mRNA for MLC$_{1emb}$, which clearly has a different size[147]. This provides direct evidence that MLC$_{1emb}$ derives from a different gene than MLC$_{1F}$, while suggesting its

Contractile protein genes: their structure and organization   371

relatively recent evolutionary origin from the fast alkali light chains. Isolation of recombinant cDNA plasmids containing myosin LC_2 light chain sequences has also been reported[3, 63, 72, 90].

Tropomyosin and troponin cDNA clones

The number of different tropomyosin isoforms in muscle and non-muscle tissues is unclear. However, the isolation of a second cDNA sequence coding for a distinct α-tropomyosin, from the leg muscle of embryonic chick, suggests that a second tropomyosin isoform is present in this tissue, probably a slow skeletal muscle type[105]. The tropomyosin coding sequences have apparently diverged more than the MHCs or actins. Only faint cross-hybridization is detectable with other α-tropomyosins, and very little with β-tropomyosin[104]. Troponin C[63,72], troponin T[63] and troponin I[72] sequences have also been cloned.

Actin cDNA clones

Actin cDNA clones have been obtained from a number of avian and mammalian skeletal muscle and non-muscle tissues. Whereas those clones containing actin coding sequences cross-hybridize with other actin mRNAs (although in some cases, only rather poorly[123]), those clones containing 3′ non-coding sequences hybridize specifically to their homologous mRNA[24, 124, 160] (see also references 155 and 211). The hybridization of a cDNA plasmid corresponding to the 3′ non-coding sequence of rat skeletal muscle actin mRNA, with rat heart RNA[160] may reflect a low level of expression of this mRNA in the heart.

A cDNA plasmid corresponding to a fetal mouse skeletal muscle actin mRNA has been identified, and this mRNA appears identical to the major actin mRNA expressed in both fetal and adult heart[123]. A single gene thus appears to be expressed as a fetal isoform in one tissue (skeletal muscle) and as a constitutive tissue-specific isoform in a second tissue (heart).

Contractile protein genes: their structure and organization

Either by direct characterization of genomic fragments cloned in bacteriophage λ, or by the intermediary of cDNA clones, a number of contractile protein genes have been isolated. Structural studies using these cloned genes are beginning to provide information on the contractile protein multigene families. Essentially, the questions which are now accessible concern: (a) the internal exon–intron structure and the flanking regions, particularly the potential regulatory sequences at the 5′ end of individual genes; (b) the number of genomic sequences coding for contractile proteins and the number of these that are active genes; and (c) the relative organization of different genes within a contractile protein family and between families in terms of chromosomal localization and possible linkage.

Most of the available data concern the actin genes in a wide range of species. This is partly because actin is a highly conserved sequence, present as a major protein in eukaryotic cells, and encoded by abundant mRNAs. Actin probes, cloned for example from *Dictyostelium*[7], have been used to detect actin genes in other species, such as *Drosophila*[57]. It is not yet clear to what extent the coding sequences for other muscle contractile proteins will crossreact with their non-muscle counterparts, nor what the degree of species specificity is likely to be. In the following sections, myosin and tropomyosin genes are discussed briefly before the more extensive information on actin genes is reviewed.

Myosin heavy chain genes

Structural studies

Myosin heavy chain genomic sequences have been isolated from the rat[131, 135, 136], from dystrophic chicken[146], from *Drosophila*[9, 150] and from a nematode[106]. Structural studies on parts of rat skeletal muscle MHC genes suggest that they are interrupted by many large intervening sequences[131, 135, 136] and cover up to 30 kb of DNA (the mRNAs are about 7 kb long with about 6 kb required to code for MHC).

The most detailed analysis of an MHC gene is that carried out on the *unc* 54 myosin heavy chain gene of the nematode, *Caenorhabditis elegans*[106]. This gene has been characterized by a genetic approach based on the mapping of the nematode genome with respect to mutations affecting movement. In parallel with these genetic studies, the biochemistry of the major contractile proteins of nematode muscles is being examined[50, 51, 103]. One of the mutations (E-675) in the *unc* 54 gene which affects movement, is an internal deletion of 100 amino acids near the −COOH terminus of the protein which results in a shortened polypeptide of the major MHC of the nematode body wall[106].

A large number of different mutations have been identified, including null mutations which in the homozygote lead to the absence of the body wall MHC, internal deletion mutants, resulting in defective MHC protein, and putative mis-sense mutations which result in a normal sized MHC but abnormal myofibrillar assembly or abnormal contractility of the assembled sarcomere[125, 186]. So far, all the mutations which have been mapped are in the structural sequence of the gene (coding or intron). The possibility of isolating further mutations in the potential regulatory sequence at the 5′ end of the gene is of major interest. Another locus, *unc* 52, appears to modulate the accumulation of *unc* 54 MHC during development. However, the nature of this modulator gene is not yet understood[207].

Numbers of MHC genes

There is not yet enough information available on cloned MHC genes to draw any conclusion on gene number from this approach. However, from Southern blotting with skeletal muscle probes under non-stringent conditions, about 5–8 different genomic sequences have been detected in rat[134, 135], in mouse (P. Daubas, personal communication) and in chick[68, 146]. These numbers are within the estimated number of MHC genes (*c*. 12, see *Table 21.1*), and presumably correspond to the more homologous sequences. Non-muscle and smooth muscle MHC sequences are probably not detected with these probes, although this may depend on mRNA abundance, exposure times and stringency conditions[134, 146]. In the case of *Drosophila*, one MHC gene has been reported, using a quail skeletal muscle MHC probe[9]. This maps at the locus of a flight mutation and probably corresponds to a muscle MHC (see discussion on chromosomal localization on p. 382).

Myosin light chain genes

The phosphorylatable myosin light chain MLC$_2$

A rat genomic DNA clone has been isolated, containing approximately 80% of the MLC$_{2F}$ gene. Five introns have been located within this part of the coding sequence[136]. Under apparently non-stringent Southern blot conditions, the cDNA probe p103 hybridized to a single genomic fragment[21].

The alkali myosin light chains, MLC$_1$, MLC$_3$

The two fast alkali light chains MLC$_{1F}$, MLC$_{3F}$, of adult skeletal muscle have a common −COOH terminal amino acid sequence[56]. Any amino acid substitutions between species are identical for both proteins in this region[113]. It has therefore been proposed from the protein data that these sequences are transcribed from a common genomic fragment. The finding that the 3′ non-coding region is apparently identical also is consistent with this view[147]. Two different cloned genomic fragments from *Mus musculus* hybridizing to p161 have been isolated and one sequenced over the common −COOH terminal region. (B. Robert, personal communication). On Southern blots of restricted *Mus musculus* mouse DNA, the −COOH coding genomic sequence (and p161) hybridizes to two major bands, whereas in the DNA of *Mus spretus* and rat, only one band is seen. It would thus seem that LC$_{1F}$ and LC$_{3F}$ are encoded by the same gene, and that in *Mus musculus* there is a recently evolved intronless pseudogene corresponding to the common −COOH region.

Tropomyosin genes

Tropomyosin gene sequences have been isolated from a cloned genomic bank of *Drosophila* DNA[6]. Two contiguous tropomyosin sequences coding for muscle-type tropomyosins, tropomyosin I (34 kd) and tropomyosin II (35 kd), are within 3 kd of each other on the same genomic fragment. A third tropomyosin gene, probably coding for a non-muscle tropomyosin (30 kd) lies 1.3 kd upstream from tropomyosin II (R. Storti, personal communication). Southern blotting experiments reveal no further tropomyosin genes and therefore suggest that all the homologous tropomyosin genes of *Drosophila* lie within a 22 kb region of the DNA. The tropomyosin mRNA sizes from TmI, TmII and Tm$_{NM}$ are 1.8 kb (also minor species 1.4–1.5 kb), 1.5 kb and 2.8 kb respectively, about twice the size necessary to encode the protein. It may be functionally significant that this non-muscle tropomyosin mRNA, like the non-muscle actin mRNA of vertebrates, has considerably more non-coding sequence. The genomic regions coding for TmI and TmII cover 4.9 kb and 2.1 kb respectively, suggesting the presence of introns in these genes, and possibly of two coding sequences in the case of TmI. The demonstration that the tropomyosin genes are linked in *Drosophila* is of great interest, given the potential evolutionary and functional implications.

Actin genes

Actin gene number

The estimated sizes of the actin multigene families in different organisms are presented in *Table 21.2*. With the exception of yeast, which has only a single actin

TABLE 21.2. Actin gene number and organization

	Organism	Actin gene number			Actin gene organization			
		Gene[a] number	No. of proteins detected	No. of genes expressed	Pseudogenes	Organization	Type of experiment	Refs.
Protista	Yeast	1	1	1	–	–		133, 60
	Acanthamoeba	>3						132
	Oxytricha (ciliated protozoan)	2–3						88
	Physarum	>4	1 major (separate amoebal form?)			3 at separate loci 4th locus: several genes	Genetic	154, 177
	Dictyostelium	17	1 major (>2 minor)	>8	+	2 linked: 1 gene + 1 pseudogene; 2 others possibly linked	Structural	92, 116, 118
Protostoma	Nematode	4				3 genes linked (opposite orientations)	Structural	10
	Drosophila	6		6	–	All 6 genes dispersed	Chromosomal	173, 57, 59
Deuterostoma	Sea urchin	>11		>3		2 × 2 linked 1 × 3 linked	Structural	43, 155, 37
	Chicken	>7	4			>3 dispersed (β & γ not linked)	Chromosomal	157, 24, 25,
	Mouse Rat Human	>20	≥6		+	At least partially dispersed. 2 striated muscle actin genes not linked in mouse	Genetic Chromosomal	48, 78, 122, 178, 164, 41
	Hamster	>5						41

[a] Estimated from Cot analysis (refs. 92, 43, 157, 78), Southern blotting (refs. 60, 132, 88, 154, 92, 173, 57, 155, 37, 24, 48, 78, 122, 41) or gene isolation (refs. 155, 48, 78, 122).

gene[60, 133], these eukaryotes have actin multigene families of varying sizes. There is a general increase in actin gene number with genome size and evolutionary complexity, although several exceptions are noted, e.g. *Dictyostelium* has more actin genes than the chicken (*Table 21.2*).

An indication that the very large number (>20) of actin-related sequences found in certain mammals such as mice and men[48, 78, 122] may not have a functional significance comes from the observations of Dodemont *et al.*[41] who detected, using a hamster actin cDNA probe, a much more limited number of hybridizing bands (c. 5) on Southern blots of restricted hamster DNA.

To date, only six actin isoforms have been clearly defined in mammals (see p. 369). One explanation for the much higher number of actin-related sequences found in certain mammalian genomes might be the presence of multiple genes encoding individual isoforms. *Dictyostelium* provides an example in which multiple actin genes appear to code for the same actin isoform[116], though none of these genes has been completely sequenced. However, this may not simply reflect a requirement for multiple gene copies in order to produce sufficient quantities of the actin protein, since these genes are differentially regulated during the *Dictyostelium* life cycle[116]. In contrast, yeast functions with only one actin gene[60, 113]. In *Drosophila*, at least five of the six genes encode different actin polypeptides[58, 59].

The use of probes from the 3' non-coding regions of chicken β- and γ-actin mRNAs has shown that there are one β- and three γ-like actin genes in the chicken[24]. However, it remains to be proven that the γ-like genes do indeed code for γ-actin. In the mouse, the striated muscle actins each appear to be encoded by single genes[122], but three genomic fragments highly homologous to a non-muscle (β or γ) actin probe are seen on Southern blots[122]. Similarly, in humans, multiple actin genomic sequences homologous to β- and γ-actin mRNAs have been isolated[49]. Thus it is possible that mammalian non-muscle actins may be encoded by several genes, or that the mammalian cytoplasmic actins may be more diverse than has been thought (as found for amphibian non-muscle actins[181]). However, the coding capacity of the putative non-muscle actin genes remains to be established.

In most species it is not clear what fraction of the detected actin genes is expressed (see *Table 21.2*). The presence of a transcription product is most effectively determined by using specific probes from the 3' non-coding regions of cDNAs to hybridize to blots of cellular RNA[155, 163, 211] or by using 5' genomic sequences in S_1 mapping experiments[116]. Such experiments show that all the *Drosophila* actin genes[59, 163, 211] and at least eight of the *Dictyostelium* actin genes[116, 117] are expressed. A negative result in such an experiment may conceivably result from a low level of transcription, or the specific expression of the gene in another cell type, and cannot thus rule out the possibility that the gene is expressed.

In many gene families, genomic DNA sequences have been described which show considerable homology to (and are presumably derived from) functional genes, but which no longer give rise to functional RNA transcripts; these are called 'pseudogenes'[83, 100, 145]. Two types of pseudogenes have been encountered – 'processed' pseudogenes, deriving from the reintegration of a cDNA transcript of an RNA, possibly involving retrovirus-like sequences[77], and 'duplicated' pseudogenes, deriving from the duplication and subsequent mutation of a functional gene[100]. *Dictyostelium* contains at least one potential pseudogene which lacks a TATA box (see below), shows multiple amino acid substitutions and is not expressed (<1% of actin RNA)[116]. This pseudogene is linked to another, functional, gene[117] from which it presumably arose by duplication. Another potential pseudogene, whose expression is not detectable, is also linked to a second, undefined, actin gene[118]. Several processed 'intron-less' non-muscle actin pseudogenes have been isolated from human genomic libraries[130] (L. Kedes, personal communication; S. Alonso and S. Humphries, personal communication), and a large family of actin genomic sequences showing only poor homology to actin cDNA probes, probably representing duplicated pseudogenes, has been detected in the mouse[122]. The large size of the actin gene family in certain mammals might thus partly result from the presence of a large population of pseudogenes. The presence of multiple processed pseudogenes might reflect a high level of actin mRNA transcription in germ line cells.

Actin gene organization

Two extreme patterns of multigene family organization have been established, with the genes being either clustered or scattered in different chromosomal regions. It is now clear that such patterns are not static, and that clustered gene families may give rise to individual scattered members ('orphons')[22, 96]. In the case of the clustered gene families where sub-families can be distinguished (α_1, α_2..., β_1, β_2...), two gene arrangements have been found, which can be defined as $\alpha_1\alpha_2/\beta_1\beta_2$ (e.g. α- and β-globins[44]) and $\alpha_1\beta_1/\alpha_2\beta_2$ (e.g. A and B chorion proteins[86] and histone genes in certain organisms[73]).

The close physical linkage of certain actin genes in *Dictyostelium*[118], of three actin genes in the nematode[52] and several actin genes in the sea urchin[37, 138, 155, 156] has been established by the isolation of recombinant phage-containing overlapping genomic DNA regions. The sea urchin actin genes can be classed into sub-groups according to the homology of their 3' non-coding regions, which may reflect their evolutionary derivation or may be related to different modes of regulation[37]. Linkage occurs both of genes in the same sub-group, i.e. an $\alpha_1\alpha_2$ type organization, and of genes in different sub-groups, i.e. an $\alpha_1\beta_1$ type organization[37]. The *Dictyostelium* actin genes can be similarly grouped according to their 3' non-coding regions, although it is not clear (with the exception of Act sub-1 and sub-2) whether they have remained linked[118].

In contrast, *in situ* hybridization experiments have shown that the *Drosophila* actin genes are at separate chromosomal locations[57, 58, 173]. In mammals as well, *in situ* hybridization has indicated a high degree of scattering[164]. This might result from a prevalence of dispersed pseudogenes in mammalian actin gene families. However, observations on the segregation of mouse actin genes in recombinant inbred mouse strains has indicated that several functional actin genes are not closely linked[122]. A similar approach has shown that the *Physarum* actin multigene family consists of three unlinked actin genes and a fourth locus containing several linked actin genes[154]. It is thus clear that different taxonomic groups have evolved different patterns of actin gene organization.

The scattering of actin genes may allow rapid divergence of the duplicated gene copies by preventing the operation of unequal crossing-over events[210] or other gene correction mechanisms[85] which would otherwise tend to keep the genes homologous. In addition, it may serve to place the actin genes in different 'regulatory modules'[37].

Actin gene structure

Putative regulatory signals
Extensive DNA sequencing has now revealed features characteristic of a large number of animal genes[14, 121]. These can be summarized as follows:

1. A region approximately 70–80 nucleotides from the cap site, consensus sequence $GG_T^CCAATCT$ (the CAAT box);
2. A region approximately 30 nucleotides from the cap site, consensus sequence $TATA_T^{AA}A_T^A$ (the TATA box);
3. The presence of intron sequences bounded by GT and AG; and
4. A polyadenylation site AATAA shortly before the end of the 3′ non-coding region[14, 44].

Sequencing of actin genes in organisms widely separated in evolution from the animal kingdom (e.g. the actin genes of protists and plants) allows the universality of these regions to be examined (see *Table 21.3*).

The position of the cap site in the genomic DNA (see *Table 21.3*) has been determined by S1 mapping experiments, hybridizing mRNA to restriction fragments spanning the putative 5′ end of the gene. In the case of several of the *Dictyostelium* genes[116], several sizes of S1-protected fragments are seen. This may reflect either multiple initiation sites for transcription, or 'nibbling' by S1 nuclease in the highly AT-rich regions[116]. The yeast actin gene has no TATA sequence at the position 30 nucleotides upstream from the cap site, characteristic of most genes, although two possible TATA boxes are found further upstream[62] (see *Table 21.3*). *Oxytricha* and *Acanthamoeba* actin genes have potential TATA boxes, and all three protists have potential CAAT boxes, although the fit to the consensus sequence is not always very good (*Table 21.3*). The same is true of some of the potential *Dictyostelium* actin gene TATA boxes and those of soybean. These sequences have yet to be shown by functional tests to act as promotor sequences[185].

At the 3′ ends of the *Oxytricha*, yeast and *Dictyostelium* genes, two polyadenylation signal sequences are found. The two sequences in yeast are probably used to generate two different mRNA molecules. In general, it would seem that the various consensus sequences detected in higher vertebrate genes represent signals which have been relatively well conserved, over enormous evolutionary distances.

Intron position in actin genes
A similar evolutionary comparison can be made for the positions of introns in actin genes. All actin genes except those of *Dictyostelium* and *Oxytricha* have so far been found to contain introns (*Table 21.4*). Whilst a large amount of experimental evidence strongly suggests that *Dictyostelium* actin genes do not contain introns[117], none of these has been completely sequenced. Similarly, the sequenced actin gene in *Oxytricha*[89] was isolated from macronuclear DNA, and introns may have been eliminated during the development of the macronucleus from the micronucleus.

Intron position within multigene families has, in general, been shown to be conserved, e.g. the globin family[142], ovalbumin and related genes[74], vitellogenin genes[184] and immunoglobulin genes[205]. In contrast, intron positions in actin genes are highly variable both within and between species (see *Table 21.4*). This variation in intron position may, in part, reflect the much larger evolutionary distance spanned in the case of actin genes. Although recent reports on the leghemoglobin

TABLE 21.3. Potential regulatory sequences in protist and plant actin genes

Gene	Gene[a] expressed (5')	CAAT box[b] (^T_CCAATCT)	TATA box[b] (TATA^T_AA^A_T)	Cap[b] site	GT…intron…AG	Polyadenylation[c] signal (3')	Reference No.
Oxytricha	?	^G_TCAAT (−82, −54)	TATAATA −46 / TATAAAT −29	N.D.[d]	no introns	AATAA (+152, +164)	89
Yeast	+	CCAATCT −223	TATATA −194 / TATATA −206	−141	+	AATAA (+71, +90)	62
Acanthamoeba	+	CCAAGCG −115	CATAAA −78	−46	+	AATAC (+44)	132
Dictyostelium							116, 117
M6	+		TTTAAAAA −72	−39 / −42			
2-sub-1	+		TATAAAT −65	-		AATAAA (+54, +67)	
7	+		TTTAAAAT −72	-	no introns		
5	+		TATAAATA −71	−30 / −42			
6	+		TATAAAAT −70	−34 / −39		AATAAA (+69, +80)	
8	+		AATAAATA −77	−28 / −42		AATAAA (+8, +14)	
Soybean 1	?		TGTAAATG −98	−71? potential sequence	+		159

(a) See section on p. 375.
(b) Distances are given in nucleotides from the start of the actin coding sequence. The distance of the cap site from the start of the actin coding sequence has been determined from S1 mapping experiments.
(c) Distances are given in nucleotides from the end of the actin coding sequence.
(d) N.D. = not done

TABLE 21.4. Intron positions in actin genes

Position of intron in actin coding sequence (amino acid number)[a]

Group		5'NT	4	13	18/19	41/42	63	105/106	121/122	150	204	267	307	327/328	353/354	Refs
Deuterostoma	Human cardiac muscle	●●	–	–	–	–	–	–	–	●	●	●	–	●	–	71
	Rat skeletal muscle	–	–	–	–	●	–	–	–	●	●	●	–	●	–	37, 206
	Rat β	–	–	–	–	●	–	–	●	–	–	●	–	●	–	37, 206
	Chicken skeletal muscle	●	–	–	–	●	–	–	–	●	●	●	–	●	–	55[b][c]
	Chicken β	–	–	–	–	●	–	–	●	–	–	●	–	●	–	–
	Sea urchin C (+ probably several others)	–	–	–	–	–	–	–	●	–	●	–	–	–	–	31, 37[d]
	Sea urchin J	●	–	–	–	●	–	–	●	–	●	–	–	●	–	–
Protostoma	Drosophila 5C	–	–	–	–	–	–	–	–	–	–	–	–	–	–	58
	Drosophila 57A	–	–	●	–	–	–	–	–	–	–	–	–	–	–	58
	Drosophila 79B	–	–	–	–	–	–	–	–	–	–	–	●	–	●	58, 151
	Drosophila 88F	–	–	–	–	–	–	–	–	–	–	–	●	–	●	151
	Nematode I/II/III	–	–	–	–	–	●	–	–	–	–	–	–	–	–	52
	Nematode IV	–	–	–	●	–	–	–	–	–	–	–	–	●	–	52
Protista	Oxytricha	–	–	–	–	–	–	●	–	–	–	–	–	–	–	89
	Acanthamoeba	–	●	–	–	–	–	–	–	–	–	–	–	–	–	132
	Yeast	–	–	–	–	–	–	–	–	–	–	–	–	–	–	61, 133
	Dictyostelium	–	–	–	–	–	–	–	–	–	–	–	–	–	–	117
Plantae	Soybean	–	–	–	●	–	–	–	–	●	–	–	–	–	●	159[e]
	Corn	–	–	–	●	–	–	–	–	●	–	–	–	–	●	[e]

(a) The intron position is given using the numbering system originally used for rabbit skeletal muscle actin[28]. Since the N-terminal sequences of actin proteins show different numbers of amino acids, and internal additions and deletions may occur in the actin protein sequence[159], this position is not necessarily equivalent to the actual numbering system for the actin gene in question. ● indicates the presence of an intron. – the absence of an intron, and a blank indicates that no information was available to us as at the time of compiling this table.
(b) T. Kost and S. Hughes (personal communication). ⎫ quoted in reference 55.
(c) B. Paterson (personal communication). ⎬
(d) A. Cooper and W. Crain (personal communication).
(e) D. Shah (personal communication).

genes of soybean[82] show similar intron positions to globin genes of mammals and *Xenopus*[142], the evolutionary origin of these genes is not clear[82].

Two extreme models can be envisaged to explain the variability in intron positions: firstly, that the original actin gene contained many introns which have since been gradually eliminated[58]; or, secondly, that the original actin gene contained no introns, and that these have been introduced during evolution[37, 118]. Intron elimination has been proposed for other genes[82, 101] and would seem to explain very satisfactorily the evolution of the deuterostome actin genes from a primordial gene containing at least seven introns (5' non-translated region, and amino acids 41/42, 121/122, 150, 204, 267 and 327/328: see *Table 21.4*). For example, the presence of introns at positions 121/122 and 204 in the genes of both sea urchin and rat (see *Table 21.4*) suggests that they have been retained from an early deuterostome actin gene. The finding that parts of these two introns in a sea urchin gene show considerable homology (>65%) to each other and to another sequence in the 5' coding-adjacent sequence[31] may indicate the operation of correction events or the more recent addition of a common sequence in the two introns.

The wide range of intron positions in protostomes and protists, none of which have been found in deuterostomes, and the much smaller number of introns found in these organisms (*Table 21.4*) might argue that the second model (intron addition) may be applicable to the earlier evolution of actin genes. It is difficult to imagine, however, how frequent intron addition could occur in actin genes without disrupting the highly conserved sequence of the actin protein. As proposed by Doolittle[42], protists may represent the more advanced evolutionary forms in that their smaller genomes may result from a 'streamlining' introduced during the much higher number of divisions that they have undergone. It is thus possible that the original actin gene contained all the 14 intron positions contained in the actin genes so far sequenced, plus, no doubt, many more to be discovered, and that these have largely been eliminated in protists.

The first evidence that this may be the case comes from recent results on one of the soybean actin genes[159]. The intron at amino acid 151 in this gene, which is also found in two other soybean actin genes and in a corn actin gene (D. M. Shah, personal communication) splits the same glycine codon, at the same nucleotide, as the intron at amino acid 150 in the rat skeletal muscle actin gene[137] and the chicken skeletal muscle actin gene[55]. A second intron in the corn and soybean genes, situated between amino acids 18 and 19 (using the numbering system for mammalian skeletal muscle actin, see footnote to *Table 21.4*) is in an identical position to an intron found in one of the nematode genes[52]. It would thus seem that there may have been introns at these positions in the primordial actin gene.

In a number of cases, structural and functional domains have been shown to be encoded by different exons[4, 82, 167]. The actin molecule has so far been shown to contain only two structural domains[171]. However, it has been postulated from the vast number of interactions of actin with other proteins[189] and its extraordinary evolutionary conservation (see below) that practically every amino acid in the actin molecule may be responsible for a specific interaction[93], i.e. that there is a large number of small functional domains. If the original actin gene contained many introns, it would be possible to invoke the idea of Gilbert[67] that 'exon shuffling' may have served to construct this gene from a series of small exons each having evolved to encode a polypeptide having a specific interaction. However, some of the exons in the primordial actin gene, if it contained all 14 intron positions shown in *Table 21.4*, would have encoded only a few amino acids (e.g. amino acids 4–13, 13–18). Unless introns are found at these positions (amino acids 4, 13) in other actin genes, it will remain equally plausible to argue that these introns represent later additions to the yeast and *Drosophila* actin genes.

Actin gene evolution

The actin protein sequence has been extraordinarily conserved during evolution. *Physarum* actin shows only 5% amino acid substitutions from mammalian cytoplasmic actins[177] and a sea urchin actin only 6% divergence from *Drosophila* actins[31]. The most diverged actin sequence so far detected is that of *Oxytricha fallax*[89], but this may encode an actin-related fibre protein (FFP-38), since the encoded polypeptide is smaller than other actins. Other actin genes exist in *Oxytricha*[88] which may encode actins more homologous to those found in higher organisms. Amongst other actins, the highest divergence is that between *Drosophila* and soybean actins (19% of their amino acids)[159].

Does the evolutionary conservation result from selective pressures at the nucleotide or at the protein level? The high silent nucleotide substitution between genes encoding similar or identical proteins[58, 118, 123] suggests that most of the selection must act on the protein. Comparison of actin genes widely separated in evolution e.g. sea urchin with *Drosophila* or yeast, also shows a large percentage of silent nucleotide substitutions (14–18%)[31]. However, similar or greater extents of nucleotide sequence divergence have been observed in other genes compared over a more restricted evolutionary period, e.g. rat and chicken tubulin genes[99], mammalian and avian globin genes[44]. It is thus possible that the entire actin gene sequence is evolving at a slower rate than the majority of genes, although this remains to be substantiated by more detailed comparisons over more limited evolutionary periods.

Within the vertebrates, different actin isoforms exhibit a limited number of amino acid substitutions which are tissue-specific rather than species-specific. For example, chicken and mammalian skeletal muscle actins are identical, as are chicken gizzard and mammalian γ smooth muscle actins and these differ from non-muscle actins at a small number of residues[175, 176, 178]. Those residues shared by skeletal and smooth muscle actins and different from non-muscle actins[178] might be involved in specific functions of muscle as opposed to non-muscle actins. This is clearly not the case for the muscle actins of *Drosophila* and nematodes which resemble the vertebrate non-muscle actins in their amino acid sequence[52, 58, 211]. If the amino acid differences shown by vertebrate muscle actins represent functional adaptations, these must thus have been imposed by the requirements of vertebrate muscle tissues. The most variable part of the actin sequence, the N-terminal acidic segment, has recently been implicated in the binding of actin to the myosin heavy chain[172].

The differences between different organisms in amino acid sequence of actin isoforms, in the organization of actin genes and in their intron positions indicate that actin multigene families have evolved several times during evolution. The creation of actin gene families may have been part of the process by which separate taxonomic groups arose[37]. However, we have observed the production of a sub-family of actin-like sequences in the mouse genome, resulting from a recent amplification event, which has occurred since the rat–mouse divergence[122]. These sequences are only distantly related to actin cDNA probes and seem likely to represent pseudogenes. It thus seems that the creation of multigene families may be a recurring event which may, at certain moments in evolution, confer a selective advantage leading to retention of the gene family.

A selective pressure for actin multigene families might result either from a requirement for isoforms fulfilling related but distinct functions or as a response to a requirement for greater regulatory flexibility[54]. The functional significance of the different isoforms remains to be established (see above p. 381). In *Dictyostelium*, multiple actin genes encoding identical actin isoforms have been shown to be differentially regulated[116]. Gene duplication and dispersion may therefore serve to place actin genes in different regulatory modules[37]. The possibility that a regulatory module corresponds to a genetic linkage group is discussed below.

Linkage of contractile protein genes

An important aspect of contractile protein gene organization which may have functional implications is the question of their linkage. This has been discussed for genomic sequences within the different contractile protein multigene families, and it is clear that some genes are linked in certain species, although the opposite is also true. If we now consider whether functional genes which are co-expressed are linked, the rather fragmentary data can be resumed as follows. Within the actin multigene families, *Drosophila* actin genes are probably expressed together in pairs[59], β- and γ-actins are present in the same cell[13] and the two striated muscle actin mRNAs accumulate together in fetal skeletal muscle and during the differentiation of a cloned mouse myoblast cell line[123]. In the corresponding genomes, the six actin genes of *Drosophila* are dispersed at different chromosomal loci[57], the β and γ non-muscle genes are found in different chromosomal fractions in the chick[25], and the two striated muscle actin genes do not co-segregate in recombinant inbred mouse lines and are therefore not closely linked[122]. Thus, non-linkage of these genes demonstrates that contractile protein genes which are expressed together are not necessarily linked.

Recent results in *Drosophila* show that the three linked tropomyosin genes (two muscle, one non-muscle) map in the same chromosomal region (88F) on the right arm of chromosome 3[6] as one of the actin genes (*DmA₁*), presumably a muscle actin, expressed in the thorax of *Drosophila*[59]. This is also a locus where a class of chemically-induced dominant flightless mutants map[126]. Mapping of the actin and tropomyosin genes in this region is based on *in situ* hybridization which localizes sequences to within 100 kb of each other. In fact, these genes are separated by 130 kb (E. Fyrberg, personal communication).

A preliminary report on a myosin heavy chain gene in *Drosophila* would suggest that it maps in a different region (36B on 2L) from any of the actins[9], one at which flight muscle mutations have also been reported[126].

In the nematode, *C. elegans*, the *unc* 54 myosin heavy chain gene belongs to linkage group I[186] and is thus not linked to at least three of the four actin genes of the nematode, which are clustered on linkage group V[52]. In vertebrates, a first report on the chromosomal localization of genes from different contractile protein families shows that the actin, myosin heavy chain and LC₂ myosin light chain genes expressed in adult skeletal muscle are on different mouse chromosomes[34].

Such evidence as there is would therefore tend to argue against the linkage of genes which are co-expressed in a given contractile protein phenotype. The simple model whereby such genes would be regulated by a single *cis*-acting regulatory sequence can be ignored, and a more complex version of this where a *trans*-acting signal regulates a set of dispersed genes, each containing a related *cis*-acting regulatory sequence, must be envisaged to explain the establishment of a phenotype. The proximity of tropomyosin and actin genes in *Drosophila* is intriguing. The non-muscle tropomyosin gene in this group is presumably co-expressed with the non-muscle actin genes localized elsewhere in the genome. It is possible that phenotypically-related genes were grouped together at some stage during evolution and that the processes of duplication, crossing-over and dispersion, which have led to the proliferation and distribution of sequences within a multigene family have also been operational on such phenotypically-related genes, with the resultant development of long-range regulatory systems.

Regulation of contractile protein gene expression

The important question is how differential regulation of genes within a multigene family, and coordinate regulation of genes between multigene families, are effected in order to establish a contractile protein phenotype. One approach to this problem is to study gene structure and organization, as discussed in the last sections.

There is not yet enough data to determine whether recognizable signals exist in the potential 5' regulatory regions adjacent to contractile protein genes, which may point to a common mechanism of activation or repression. Another approach is to look at gene expression *in vivo* in order to define the relative levels of protein and mRNA synthesis and accumulation for different contractile proteins within a phenotype, and when switching from one phenotype to another. One of the best model systems in which to study these questions is the terminal differentiation of skeletal muscle myoblasts into muscle fibres and the subsequent maturation of the fibres.

When does synthesis of contractile proteins begin?

The later stages of myogenesis can be reproduced in tissue culture[17]. Myoblasts isolated from developing skeletal muscles proliferate in culture and then enter a phase of cell fusion and formation of multinucleated fibres. A number of early studies showed that synthesis, accumulation and assembly of contractile proteins into myofibrils are activated during myoblast fusion[47, 141, 204]. Pulse labelling of myoblast cultures and analysis of newly synthesized proteins on two-dimensional gels has shown that the synthesis of a large number of contractile proteins begins around the time of cell fusion[38]. The kinetics of synthesis are more or less similar depending on the system examined[36, 38, 161]. It is difficult to determine the exact time for the onset of the synthesis of each contractile protein, since this depends on the level of sensitivity of detection for each protein, which need not necessarily be identical. It is clear that although the activation of contractile protein synthesis occurs coincidentally with cell fusion, or a few hours before fusion[161], it is not dependent on cell fusion, since it can be detected in cultures where fusion is blocked by lowering the Ca^{2+} concentration of the medium[47].

When does synthesis of the contractile protein mRNAs begin?

A number of indirect experiments suggested that storage of contractile protein mRNAs in myoblasts might represent an important regulatory control in muscle development[17]. In contrast, *in vitro* translation studies[36, 39, 203] and molecular hybridization experiments using cDNA probes to the abundant muscle mRNA class[1, 98, 140, 208] suggested that mRNA accumulation directly regulates contractile protein synthesis. The availability of cloned probes for individual contractile protein mRNAs now allows direct measurements of the synthesis and accumulation of these mRNAs. Experiments measuring steady state levels of contractile protein mRNAs during myoblast fusion have shown a close correlation between mRNA levels and synthesis of the corresponding protein: for myosin heavy chain mRNA in rat L_6E_9 cultures[8], for actin in primary chick cultures[158], and for actin, myosin light and heavy chain mRNAs, in rat L8 cultures[161], mouse T984 clone 10 cultures[20] and primary quail cultures[72]. These experiments indicate that large-scale storage of 'inactive' contractile protein mRNAs is not a major regulatory event in muscle cell differentiation. Any translational control (for review, see reference 75) acts in addition to controls regulating mRNA abundance.

When does activation of the contractile protein genes occur?

Experiments using a global cDNA probe to mRNAs accumulated during the differentiation of mouse myoblasts showed that the genes encoding these mRNAs were relatively resistant to DNase I digestion in the nuclei of dividing myoblasts[1]. This has been confirmed by experiments measuring the sensitivity of the skeletal muscle actin and myosin light chain 2 genes in myoblasts of L8 cultures to very limited DNase I digestion[21]. These results indicate that activation of the contractile protein genes occurs during the terminal stages of muscle differentiation when myoblasts cease dividing and start to fuse into multinucleated myotubes. That the change in conformation of contractile protein genes measured by DNase I digestion does indeed coincide with a transcriptional activation of these genes is indicated by measurements of *in vitro* elongation of transcripts initiated *in vivo* on the rat myosin heavy chain gene[131]. The proportion of myosin heavy chain transcripts in the nuclear 'run-off' transcripts increased greater than fifty-fold during the differentiation of rat L_6E_9 myoblasts.

How many molecules of contractile proteins and their mRNAs are synthesized?

The stoichiometry of the synthesis of several contractile proteins (myosin light chains 1 and 2, myosin heavy chain, α-actin, troponin C and α- and β-tropomyosins) has been examined in differentiating quail cultures[38]. These contractile proteins are synthesized at equimolar rates *in vivo* with the exception of α-actin, which is synthesized at a 2.5-fold higher rate. This stoichiometry seems to be mainly determined by the respective concentrations of the contractile protein mRNAs[39]. The number of mRNA molecules accumulated per myotube nucleus has been estimated to be about 40 000–50 000 mRNA molecules for actin mRNA in chick primary cultures[158] and approximately the same for myosin heavy chain mRNA in rat L_6E_9 cultures[8]. In both cases, the number of mRNAs reaches a maximum value and then declines in older cultures.

Is there amplification of contractile protein genes during muscle differentiation?

It can be calculated from the above figures for the accumulation of contractile protein mRNAs that for about 40 000 mRNAs to accumulate over a time period of 24–48 hours requires a transcription rate of 6–13 molecules/nucleus/minute, assuming all transcripts become functional mRNA[120] and that the mRNA is essentially stable (half life $\geqslant$ 50 hours). These values are similar to those obtained for ovalbumin mRNA synthesis in the chick oviduct (11–17 molecules/cell/minute)[139] and lower than the theoretical maximum rate of 32–48 gene transcripts/minute calculated by Kafatos[87] from measurements of RNA polymerase packing density and rates of polymerase elongation. If the mRNA is indeed stable, then it is not necessary, *a priori*, to invoke gene amplification to explain the accumulation of contractile proteins. One report[209] shows amplification of the skeletal muscle α-actin gene during the differentiation of embryonic chicken muscle, but to date there is no evidence that this represents a general mechanism regulating contractile protein gene expression during myogenesis[134].

Muscle development *in vivo*: how and why do developmental switches in muscle contractile protein expression occur?

Myogenesis of skeletal muscle represents a model system for investigating the idea that 'gene batteries' are regulated during cellular differentiation[37]. As discussed earlier, fetal to adult switches occur in the expression of muscle myosin heavy chain, light chain and actin genes. The modes of regulation in each case appear to be different. Developmental isoforms of myosin heavy chain appear sequentially[199], whereas the mRNAs for the two actin striated muscle isoforms are co-expressed in fetal muscle[123] and in a cloned mouse muscle cell line[20] but not in adult skeletal muscle. Similarly, both the fetal LC_{lemb} and adult LC_{IF} are synthesized in late fetal muscle[66, 195] and in early fusing primary cultures of rat and calf myoblasts[36, 196], although in cultures of the rat L6 cell line only the fetal isoform LC_{lemb} is found[195]. The apparent lack of correlation between the timing of the changes in the expression of the fetal and adult isoforms of different contractile protein gene families during skeletal muscle development[123] suggests that different regulatory programmes exist for each gene family.

Another situation in which two gene sets may be activated together and the expression of one gene set repressed later in development is that of the slow and fast isoforms. Cultures of cloned myoblasts from slow or fast avian skeletal muscles express both adult fast and slow isoforms of myosin light chains[91, 169], α-tropomyosin[127] and troponin C[72], whereas adult fast and slow muscles express preferentially their corresponding fast or slow isoforms.

Two extreme hypotheses can be conceived to explain the existence of developmental isoforms. Firstly, these changes may reflect a requirement for functionally different proteins. In the case of the actins, no firm conclusion can be reached at this stage. For myosin heavy chains, physiological differences between different isoforms have been noted[199]. The second hypothesis would be that this situation represents in some way a regulatory advantage for the organism, e.g. the possibility of expressing two genes to facilitate rapid protein accumulation or of evolving different regulatory signals for two genes in the same gene family. *In vitro* mutagenesis of cloned contractile protein genes followed by *in vivo* tests of the function of the encoded protein will determine whether naturally occurring amino acid differences between isoforms are functionally important.

Concluding remarks

The discussion of why there should be developmental isoforms of the contractile proteins can be extended to the general questions of why multigene families have evolved for those proteins and what is the significance of their tissue-specific expression. Apart from physiological considerations at the protein level, one can ask whether this reflects a requirement for tissue-specific contractile protein gene 'batteries' or 'regulatory modules'[37]. A simple model would predict that each type of muscle would express a specific battery of contractile protein genes. However, this appears not to be the case (see *Table 21.1*). The switches that occur during skeletal muscle maturation are a striking example of the fact that different multigene families have independently evolved different regulatory strategies. The generation of a contractile protein phenotype would thus seem to involve a highly complex system of regulatory interactions. The hope is that with these multigene families, as with others, the use of cloned genes in DNA-mediated gene transfer experiments in different muscle and non-muscle cell types *in vitro* should lead to some insight into their regulation. At present the structural data which are beginning to accumulate have not provided any further information of mechanistic significance beyond the consensus-type sequence data already described for other genes. Most of the observations of interest relate to the evolution of multigene families. In this respect, the high degree of sequence conservation of contractile proteins, particularly actins, makes comparative studies of these genes possible throughout eukaryotic evolution. Information on intron position in actin genes from many species provides new insight into the generation of intron dispersion and possible models for the assembly of an original actin gene from fragments of DNA. Similarly, data on the numbers, homologies and degree of dispersion of actin genes in different species contributes to understanding the kind of processes taking place during the generation of multigene families, and indeed of different species-specific strategies in this respect.

Acknowledgements

We are indebted to a number of colleagues, especially S. Alonso, G. Bugaisky, M. Fiszman, E. Kardami, D. Montarras, B. Robert and R. Whalen for helpful discussions and comments on the manuscript, and to all our colleagues and fellow workers who allowed us to see and cite their results prior to publication. The authors' laboratory is supported by grants from the Centre National de la Recherche Scientifique, the Délégation Générale à la Recherche Scientifique et Technique, the Institut National de la Santé et de la Recherche Médicale, the Commissariat à l'Energie Atomique, la Fondation pour la Recherche Médicale Française, the Ligue Nationale Française contre le Cancer, NATO, and the Muscular Dystrophy Association, USA.

References

1. Affara, N. A., Robert, B., Jacquet, M., Buckingham, M. E. and Gros, F. (1980). Changes in gene expression during myogenic differentiation. I. Regulation of mRNA sequences expressed during myotube formation. *Journal of Molecular Biology*, **140**, 441–458

2. Amphlett, G. W., Syska, H. and Perry, S. V. (1976). The polymorphic forms of tropomyosin and troponin I in developing rabbit skeletal muscle. *FEBS Letters*, **63**, 22–26

3. Arnold, H. H. and Siddiqui, M. A. Q. (1979). Cloning of synthetic DNA that codes for embryonic cardiac myosin light-chain polypeptide. *Biochemistry*, **18**, 5641–5647

4. Artymiuk, P. J., Blake, C. C. F. and Sippel, A. E. (1981). Genes pieced together–exons delineate homologous structures of diverged lysozymes. *Nature, London*, **290**, 287–288

5. Barker, W. C., Ketcham, L. K. and Dayhoff, M. O. (1978). Contractile system proteins. In *The Atlas of Protein Structure* (Dayhoff, M. O., Ed.). Silver Spring, Maryland, National Biomedical Research Foundation

6. Bautch, V. L., Storti, R. V., Mischke, D. and Pardue, M. L. (1982). Organization and expression of *Drosophila* tropomyosin genes. *Journal of Molecular Biology*, **162**, 231–251

7. Bender, W., Davidson, N., Kindle, K. L., Taylor, W. C., Silverman, N. and Firtel, R. A. (1978). The structure of M_6, a recombinant plasmid containing *Dictyostelium* DNA homologous to actin mRNA. *Cell*, **15**, 779–788

8. Benoff, S. and Nadal-Ginard, B. (1980). Transient induction of poly-A short myosin heavy chain mRNA during terminal differentiation of L_6E_9 myoblasts. *Journal of Molecular Biology*, **140**, 283–298

9. Bernstein, S. I., Mogami, K., Donady, J. J. and Emerson, C. P. (1983). *Drosophila* muscle myosin heavy chain is encoded by a single gene located in a cluster of muscle mutations. *Nature, London*, **302**, 393–397

10. Billeter, R., Heizmann, C. W., Howald, H. and Jenny, E. (1981). Analysis of myosin light and heavy chain types in single human skeletal muscle fibres. *European Journal of Biochemistry*, **116**, 389–395

11. Billeter, R., Heizmann, C. W., Reist, U., Howald, H. and Jenny, E. (1981). α- and β-tropomyosin in typed single fibres of human skeletal muscle. *FEBS Letters*, **132**, 133–136

12. Billeter, R., Heizmann, C. W., Reist, U., Howald, H. and Jenny, E. (1982). Two-dimensional peptide analysis of myosin heavy chains and actin from single-typed human skeletal muscle fibres. *FEBS Letters*, **139**, 45–48

13. Bravo, R., Fey, S. J., Small, J. V., Larsen, P. M. and Celis, J. E. (1981). Co-existence of three major isoactins in a single sarcoma 180 cell. *Cell*, **25**, 195–202

14. Breathnach, R. and Chambon, P. (1981). Organization and expression of eukaryotic split genes coding for proteins. *Annual Review of Biochemistry*, **50**, 349–383

15. Bremer, J. W., Busch, H. and Yeoman, L. C. (1981). Evidence for a species of nuclear actin distinct from cytoplasmic and muscle actins. *Biochemistry*, **20**, 2013–2017

16. Bronson, D. D. and Schachat, F. H. (1982). Heterogeneity of contractile proteins. Differences in tropomyosin in fast, mixed and slow skeletal muscles of the rabbit. *Journal of Biological Chemistry*, **257**, 3937–3944

17. Buckingham, M. E. (1977). Muscle protein synthesis and its control during the differentiation of skeletal muscle cells *in vitro*. In *International Review of Biochemistry, Volume II, Biochemistry of Cell Differentiation* (Paul, J., Ed.), pp. 269–333. Baltimore, University Park Press

18. Burridge, K. (1974). A comparison of fibroblast and smooth muscle myosins. *FEBS Letters*, **45**, 14–17

19. Burridge, K. and Bray, D. (1975). Purification and structural analysis of myosins from brain and other non-muscle tissues. *Journal of Molecular Biology*, **99**, 1–14

20. Caravatti, M., Minty, A., Robert, B., Montarras, D., Weydert, A., Cohen, A., Daubas, P. and Buckingham, M. E. (1982). Regulation of muscle gene expression: the accumulation of mRNAs coding for muscle specific proteins during myogenesis in a mouse cell line. *Journal of Molecular Biology*, **160**, 59–76

21. Carmon, Y., Czosnek, H., Nudel, U., Shani, M. and Yaffé, D. (1982). DNase sensitivity of genes expressed during myogenesis. *Nucleic Acids Research*, **10**, 3085–3098

22. Childs, G., Maxson, R. Cohn, R. H. and Kedes, L. (1981). Orphons: dispersed genetic elements derived from tandem repetitive genes of eukaryotes. *Cell*, **23**, 651–663

23. Chizzonite, R. A., Everett, A. and Zak, R. (1981). Possible identity of atrial myosin heavy chain to a ventricular form. *Journal of Cell Biology*, **91**, 346

24. Cleveland, D. W., Lopata, M. A., MacDonald, R. J., Cowan, N. J., Rutter, W. J. and Kirschner, M. W. (1980). Number and evolutionary conservation of α- and β-tubulin and cytoplasmic β- and γ-actin genes using specific cloned cDNA probes. *Cell*, **20**, 95–105

25. Cleveland, D. W., Hugues, S. H., Stubblefield, E., Kirschner, M. W. and Varmus, H. E. (1981). Multiple α and β tubulin genes represent unlinked and dispersed gene families. *Journal of Biological Chemistry*, **256**, 3130–3134

26. Cole, H. A. and Perry, S. V. (1975). The phosphorylation of troponin I from cardiac muscle. *Biochemical Journal*, **149**, 525–533

27. Collins, J. H. (1976). Homology of myosin DTNB light chain with alkali light chains, troponin C and parvalbumin. *Nature, London*, **259**, 699–700
28. Collins, J. H. and Elzinga, M. (1975). The primary structure of actin from rabbit skeletal muscle: completion and analysis of the amino acid sequence. *Journal of Biological Chemistry*, **250**, 5915–5920
29. Collins, J. H., Greaser, M. L., Potter, J. D. and Horn, M. J. (1977). Determination of the amino acid sequence of troponin C from rabbit skeletal muscle. *Journal of Biological Chemistry*, **252**, 6356–6362
30. Côté, G. P. and Smillie, L. B. (1981). Preparation and some properties of equine platelet tropomyosin. *Journal of Biological Chemistry*, **256**, 11004–11010
31. Cooper, A. D. and Crain, W. R. (1982). Complete nucleotide sequence of a sea urchin actin gene. *Nucleic Acids Research*, **10**, 4081–4092
32. Cummins, P. and Perry, S. V. (1974). Chemical and immunochemical characteristics of tropomyosin from striated and smooth muscle. *Biochemical Journal*, **141**, 43–49
33. Cummins, P., Price, K. N. and Littler, W. A. (1980). Foetal myosin light chain in human ventricle. *Journal of Muscle and Cell Motility*, **1**, 357–366
34. Czosnek, H., Nudel, U., Shani, M., Barker, P. E., Pravtcheva, D. D., Ruddle, F. H. and Yaffé, D. (1982). The genes coding for the muscle contractile proteins, myosin heavy chain, myosin light chain 2 and skeletal muscle actin are located on three different mouse chromosomes. *EMBO Journal*, **1**, 1299–1305
35. Dalla-Libera, L., Sartore, S. and Schiaffino, S. (1979). Comparative analysis of chicken atrial and ventricular myosins. *Biochimica et Biophysica Acta*, **581**, 283–294
36. Daubas, P., Caput, D., Buckingham, M. and Gros, F. (1981). A comparison between the synthesis of contractile proteins and the accumulation of their translatable mRNAs during calf myoblast differentiation. *Developmental Biology*, **84**, 133–143
37. Davidson, E. H., Thomas, T. L., Scheller, R. H. and Britten, R. J. (1982). The sea urchin actin genes and a speculation on the evolutionary significance of small gene families. In *Genome Evolution* (Dover, G. A. and Flavell, R. A., Eds.), pp. 177–192. New York, Academic Press
38. Devlin, R. B. and Emerson, C. P. (1978). Coordinate regulation of contractile protein synthesis during myoblast differentiation. *Cell*, **13**, 599–611
39. Devlin, R. B. and Emerson, C. P. (1979). Coordinate accumulation of contractile protein mRNAs during myoblast differentiation. *Developmental Biology*, **69**, 202–216
40. Dhoot, G. K. and Perry, S. V. (1979). Distribution of polymorphic forms of troponin components and tropomyosin in skeletal muscle. *Nature, London*, **278**, 714–718
41. Dodemont, H. J., Soriano, P., Quax, W. J. Ramaekers, T., Lenstra, J. A., Groenen, M. A. M., Bernardi, G. and Bloemendal, H. (1982). The genes coding for the cytoskeletal proteins actin and vimentin in warm-blooded vertebrates. *EMBO Journal*, **1**, 167–171
42. Doolitle, W. F. (1978). Genes in pieces: were they ever together? *Nature, London*, **272**, 581–582
43. Durica, D. S., Schloss, J. A. and Crain, W. R. (1980). Organization of actin gene sequences in the sea urchin: molecular cloning of an intron-containing DNA sequence coding for a cytoplasmic actin. *Proceedings of the National Academy of Sciences, USA*, **77**, 5683–5687
44. Efstradiatis, A., Posakony, J. W., Maniatis, T., Lawn, R. M. O'Connell, C., Spritz, R. A., Deriel, J. K., Forget, B. G., Weissman, S. M., Slightom, J. L., Blechl, A. E., Smithies, O., Baralle, F. E., Shoulders, C. C. and Proudfoot, N. J. (1980). The structure and evolution of the human β-globin family. *Cell*, **21**, 653–688
45. Elzinga, M. and Trus, B. (1980). In *Methods in Peptide and Protein Sequence Analysis* (Birr, C., Ed.), pp. 213–224. Amsterdam, Elsevier
46. Elzinga, M., Maron, B. J. and Adelstein, R. S. (1976). Human heart and platelet actins are products of different genes. *Science*, **191**, 94–95
47. Emerson, C. P. and Beckner, S. K. (1975). Activation of myosin synthesis in fusing and mononucleated myoblasts. *Journal of Molecular Biology*, **93**, 431–447
48. Engel, J. N., Gunning, P. W. and Kedes, L. M. (1981). Isolation and characterization of human actin genes. *Proceedings of the National Academy of Sciences, USA*, **78**, 4674–4678
49. Engel, J., Gunning, P. and Kedes, L. (1982). Human cytoplasmic actin proteins are encoded by a multigene family. *Molecular and Cellular Biology*, **2**, 674–684
50. Epstein, H. F., Berman, S. A. and Miller, D. M. (1982). Myosin synthesis and assembly in nematode body wall muscle. In *Muscle Development: Molecular and Cellular Control*, (Pearson, M. L. and Epstein, H. F., Eds.), pp. 419–429. New York, Cold Spring Harbor Press (in press)
51. Epstein, H. F., Miller, D. M., Gossett, L. A. and Hecht, R. M. (1982). Immunological studies of myosin isoforms in nematode development. In *Muscle Development: Molecular and Cellular Control* (Pearson, M. L. and Epstein, H. F., Eds.), pp. 7–15. New York, Cold Spring Harbor Press

52. Files, J. G., Carr, S. and Hirsh, D. (1982). The actin gene family of *C. elegans*. *Journal of Molecular Biology*, **164**, 355–376
53. Fine, R. E. and Blitz, A. L. (1975). A chemical comparison of tropomyosins from muscle and non-muscle tissue. *Journal of Molecular Biology*, **95**, 447–454
54. Firtel, R. A. (1981). Multigene families encoding actin and tubulin. *Cell*, **24**, 6–7
55. Fornwald, J. A., Kuncio, G., Peng, I. and Ordahl, C. P. (1982). The complete nucleotide sequence of the chick α-actin gene and its evolutionary relationship to the actin gene family. *Nucleic Acids Research*, **10**, 3861–3876
56. Frank, G. and Weeds, A. G. (1974). The amino acid sequence of the alkali light chains of rabbit skeletal muscle myosin. *European Journal of Biochemistry*, **44**, 317–334
57. Fyrberg, E. A., Kindle, K. L. and Davidson, N. (1980). The actin genes of *Drosophila*: a dispersed multigene family. *Cell*, **19**, 365–378
58. Fyrberg, E. A., Bond, B. J., Hershey, N. D., Mixter, K. S. and Davidson, N. (1981). The actin genes of *Drosophila*: protein coding regions are highly conserved but intron positions are not. *Cell*, **24**, 107–116
59. Fyrberg, E. A., Mahaffey, J. W., Bond, B. J. and Davidson, N. (1982). Transcripts of the six *Drosophila* actin genes accumulate in a stage and tissue specific manner. *Cell*, **33**, 115–123
60. Gallwitz, D. and Seidel, R. (1980). Molecular cloning of the actin gene from yeast *Saccharomyces cerevisiae*. *Nucleic Acids Research*, **8**, 1043–1059
61. Gallwitz, D. and Sures, I. (1980). Structure of a split yeast gene: complete nucleotide sequence of the actin gene in *Saccharomyces cerevisiae*. *Proceedings of the National Academy of Sciences, USA*, **77**, 2546–2550
62. Gallwitz, D., Perrin, F. and Seidel, R. (1981). The actin gene in yeast *Saccharomyces cerevisiae*: 5′ and 3′ end mapping, flanking and putative regulatory sequences. *Nucleic Acids Research*, **9**, 6339–6350
63. Garfinkel, L. I., Periasamy, M. and Nadal-Ginard (1982). Cloning and characterization of cDNA sequences corresponding to myosin light chains 1, 2 and 3, Troponin-C, Troponin-T, α-Tropomyosin, and α-Actin. *Journal of Biological Chemistry*, **257**, 11078–11086
64. Garrels, J. I. (1970). Changes in protein synthesis during myogenesis in a clonal cell line. *Developmental Biology*, **73**, 134–152
65. Garrels, J. I. and Gibson, W. (1976). Identification and characterization of multiple forms of actin. *Cell*, **9**, 793–805
66. Gauthier, G. F., Lowey, S., Benfield, P. A. and Hobbs, A. W. (1982). Distribution and properties of myosin isozymes in developing avian and mammalian skeletal muscle fibres. *Journal of Cell Biology*, **92**, 471–484
67. Gilbert, W. (1978). Why genes in pieces? *Nature, London*, **271**, 501
68. Gilliam, T. C., Gulick, J. and Robbins, J. (1983). Myosin heavy chain genes and their expression in normal and dystrophic chickens. *European Journal of Biochemistry* (in press)
69. Giometti, C. S. and Anderson, N. L. (1981). A variant of human non-muscle tropomyosin found in fibroblasts by using two-dimensional electrophoresis. *Journal of Biological Chemistry*, **256**, 11840–11846
70. Hamada, H., Leavitt, J. and Kakunaga, T. (1981). Mutated β-actin gene: co-expression with an unmutated allele in a chemically transformed human fibroblast cell line. *Proceedings of the National Academy of Sciences, USA*, **78**, 3634–3638
71. Hamada, H., Petrino, M. G. and Kakunaga, T. (1982). Molecular structure and evolutionary origin of human cardiac muscle actin gene. *Proceedings of the National Academy of Sciences, USA*, **79**, 5901–5905
72. Hastings, K. E. M. and Emerson, C. P. (1982). cDNA clone analysis of six co-regulated mRNAs encoding skeletal muscle contractile proteins. *Proceedings of the National Academy of Sciences, USA*, **79**, 1553–1557
73. Hentschel, C. C. and Birnstiel, M. L. (1981). The organization and expression of histone gene families. *Cell*, **25**, 301–313
74. Heilig, R., Perrin, F., Gannon, F., Mandel, J. L. and Chambon, P. (1980). The ovalbumin gene family: structure of the X gene and evolution of duplicated split genes. *Cell*, **20**, 625–637
75. Heywood, S. M., Thibault, M. C. and Siegel, S. (1981). Control of gene expression in muscle development. In *Cell and Muscle Motility, Volume III*. New York, Plenum Press
76. Hoh, J. F. Y., Yeoh, G. P. S., Thomas, M. A. W. and Higgenbottom, L. (1979). Structural differences in the heavy chains of rat ventricular myosin isoenzymes. *FEBS Letters*, **97**, 330–333
77. Hollis, G. F., Hieter, P. A., McBride, O. W., Swan, D. and Leder, P. (1982). Processed genes: a dispersed immunoglobulin gene bearing evidence of RNA-type processing. *Nature, London*, **296**, 321–325

78. Humphries, S. E., Whittall, R., Minty, A., Buckingham, M. E. and Williamson, R. (1981). There are approximately 20 actin genes in the human genome. *Nucleic Acids Research*, **9**, 4895–4908

79. Humphrey, J. E. and Cummins, P. (1982). Atrial and venticular tropomyosin and tripomin I in the developing and adult, bovine and human heart. Submitted

80. Huszar, G. (1972). Developmental changes in the primary structure and histidine methylation in rabbit skeletal muscle myosin. *Nature, London*, **240**, 260–264

81. Huszar, G. and Elzinga, M. (1972). Homologous methylated and non-methylated histidine peptides in skeletal and cardiac myosins. *Journal of Biological Chemistry*, **247**, 745–753

82. Hyldig-Nielsen, J. J., Jensen, E. O., Paludan, K., Wiborg, O., Garrett, R., Jorgensen, P. and Marcker, K. A. (1982). The primary structures of two leghemoglobin genes from soybean. *Nucleic Acids Research*, **10**, 689–701

83. Jacq, C., Miller, J. R. and Brownlee, G. G. (1977). A pseudogene structure in 5S DNA of *Xenopus laevis*. *Cell*, **12**, 109–120

84. Jakes, R., Northrop, F. and Kendrick-Jones, J. (1976). Calcium-binding regions of myosin 'regulatory' light chains. *FEBS Letters*, **70**, 229–234

85. Jeffreys, A. J. and Harris, S. (1982). Processes of gene duplication. *Nature, London*, **296**, 9–10

86. Jones, C. and Kafatos, F. C. (1980). Structure, organization and evolution of developmentally regulated chorion genes in a silkmoth. *Cell*, **22**, 855–867

87. Kafatos, F. C. (1972). The cocoonase zymogen cells of silk moths: a model of terminal cell differentiation for specific protein synthesis. In *Current Topics in Developmental Biology, Volume 7*, (Moscona, A. A. and Monroy, A., Eds.), pp. 125–192. New York, Academic Press

88. Kaine, B. P. and Spear, B. B. (1980). Putative actin genes in the macronucleus of *Oxytricha fallax*. *Proceedings of the National Academy of Sciences, USA*, **77**, 5336–5340

89. Kaine, B. P. and Spear, B. B. (1982). Nucleotide sequence of a macronuclear gene for actin in *Oxytricha fallax*. *Nature, London*, **295**, 430–432

90. Katcoff, D., Nudel, U., Zevin-Sonkin, D., Carmon, Y., Shani, M., Lehrach, H., Frischauf, A. M. and Yaffé, D. (1980). Construction of recombinant plasmids containing rat muscle actin and myosin light chain DNA sequences. *Proceedings of the National Academy of Sciences, USA*, **77**, 960–964

91. Keller, L. R. and Emerson, C. P. (1980). Synthesis of adult myosin light chains by embryonic muscle cultures. *Proceedings of the National Academy of Sciences, USA*, **77**, 1020–1024

92. Kindle, K. L. and Firtel, R. A. (1978). Identification and analysis of *Dictyostelium* actin genes, a family of moderately repeated genes. *Cell*, **15**, 763–778

93. Korn, E. D. (1978). Biochemistry of actomyosin-dependent cell motility (a review). *Proceedings of the National Academy of Sciences, USA*, **75**, 588–599

94. Leavitt, J. and Kakunaga, T. (1980). Expression of a variant form of actin and additional polypeptide changes following chemically induced *in vitro* neoplastic transformation of human fibroblasts. *Journal of Biological Chemistry*, **255**, 1650–1661

95. Leavitt, J., Bushar, G., Kakunaga, T., Hamada, H., Hirakawa, T., Goldman, O. and Merril, C. (1982). Variations in expression of mutant β-actin accompanying incremental increases in human fibroblast tumorigenicity. *Cell*, **28**, 259–268

96. Leder, A., Swan, D., Ruddle, F., d'Eustachio, P. and Leder, P. (1981). Dispersion of α-like globin genes of the mouse to three different chromosomes. *Nature, London*, **293**, 196–200

97. Leger, J. J. and Elzinga, M. (1977) Studies on cardiac myosin light chains: comparison of the sequences of cardiac and skeletal myosin LC-2. *Biochemical and Biophysical Research Communications*, **74**, 1390–1396

98. Leibovitch, M. P., Leibovitch, S. A., Harel, J. and Kruh, J. (1979). Changes in the frequency and diversity of mRNA populations in the course of myogenic differentiation. *European Journal of Biochemistry*, **97**, 321–326

99. Lemischka, I. R., Farmer, S., Racaniello, V. R. and Sharp, P. (1981). Nucleotide sequence and evolution of a mammalian α-tubulin mRNA. *Journal of Molecular Biology*, **151**, 101–120

100. Little, P. F. R. (1982). Globin pseudogenes. *Cell*, **28**, 683–684

101. Lomedico, P., Rosenthal, N. Efstradiatis, A., Gilbert, W., Kolodner, R. and Tizard, R. (1979). The structure and evolution of the two nonallelic rat preproinsulin genes. *Cell*, **18**, 545–558

102. Lompre, A. M., Mercardier, J. J., Wisnewsky, C., Bouveret, P., Pantaloni, C., d'Albis, A. and Schwartz, K. (1981). Species and age-dependent changes in the relative amounts of cardiac myosin isoenzymes in mammals. *Developmental Biology*, **84**, 286–290

103. Mackenzie, J. M. and Epstein, H. F. (1980). Paramyosin is necessary for determination of nematode thick filament length *in vivo*. *Cell*, **22**, 747–755

104. MacLeod, A. R. (1981). Construction of bacterial plasmids containing sequences complementary to chicken α-tropomyosin mRNA. *Nucleic Acids Research*, **9**, 2675–2689

105. MacLeod, A. R. (1982). Distinct α-tropomyosin mRNA sequences in chicken skeletal muscle. *European Journal of Biochemistry*, **126**, 293–297
106. MacLeod, A. R., Karn, J. and Brenner, S. (1981). Molecular analysis of the *unc* 54 myosin heavy-chain gene of *Caenorhabditis elegans*. *Nature, London*, **291**, 386–290
107. Mahdavi, V., Periasamy, M. and Nadal-Ginard, B. (1982). Molecular characterization of two myosin heavy chain genes expressed in the adult heart. *Nature, London*, **297**, 659–664
108. Maita, T., Umegane, T., Kato, Y. and Matsuda, G. (1980). Amino acid sequence of the L-1 light chain of chicken cardiac-muscle myosin. *European Journal of Biochemistry*, **107**, 565–575
109. Mak, A. S., Lewis, W. G. and Smillie, L. B. (1979). Amino acid sequences of rabbit skeletal β- and cardiac tropomyosins. *FEBS Letters*, **105**, 232–234
110. Mak, A. S., Smillie, L. B. and Stewart, G. R. (1980). A comparison of the amino acid sequences of rabbit skeletal muscle α- and β-tropomyosins. *Journal of Biological Chemistry*, **255**, 3647–3655
111. Marotta, C. A., Strocchi, P. and Gilbert, J. M. (1978). Microheterogeneity of brain cytoplasmic and synaptoplasmic actins. *Journal of Neurochemistry*, **30**, 1441–1451
112. Masaki, T., Bader, D. M., Reinach, F. C., Shimizu, T., Obinata, T., Shafig, S. A. and Fischman, D. A. (1982). Monoclonal antibody analysis of myosin heavy chain and C-protein isoforms during myogenesis. In *Muscle Development*: (Pearson, M. L. and Epstein, H. F., Eds.), pp. 405–419. New York, Cold Spring Harbor Press (in press)
113. Matsuda, G., Maita, T. and Unegane, T. (1981). The primary structure of L1 light chain of chicken fast skeletal muscle myosin and its genetic implication. *FEBS Letters*, **126**, 111–113
114. Matsuda, G., Maita, T., Kato, Y., Chen, J. and Unegane, T. (1981). Amino acid sequences of the cardiac L-2A and gizzard 17 000 Mr light chains of chicken muscle myosin. *FEBS Letters*, **135**, 232–236
115. Matsuda, R. and Tomino, S. (1981). Biosynthesis of tropomyosin and troponin during development of the chicken skeletal muscle. *Journal of Biochemistry*, **90**, 815–821
116. McKeown, M. and Firtel, R. A. (1981). Differential expression and 5' end mapping of actin genes in *Dictyostelium*. *Cell*, **24**, 799–807
117. McKeown, M. and Firtel, R. A. (1981). Evidence for sub-families of actin genes in *Dictyostelium* as determined by comparison of 3' end sequences. *Journal of Molecular Biology*, **151**, 593–606
118. McKeown, M., Hirth, K. P., Edwards, C. and Firtel, R. A. (1982). Examination of the regulation of the actin multigene family in *Dictyostelium discoideum*. In *Embryonic Development: Part A. Genetic Aspects* Burger, M. M. and Weber, R., Eds.), pp. 51–78. New York, A. R. Liss, Inc.
119. Medford, R. M., Wydro, R. M., Nguyen, H. T. and Nadal-Ginard, B. (1980). Cytoplasmic processing of myosin heavy chain mRNA: evidence provided by using a recombinant DNA *Proceedings of the National Academy of Sciences, USA*, **77**, 5749–5753
120. Minty, A. J. and Birnie, G. D. (1981). Messenger RNA populations in eukaryotic cells. Evidence from recent nucleic acid hybridization experiments bearing on the extent and control of differential gene expression. In *Biochemistry of Cellular Regulation, Volume III: Development and Differentiation*, (Buckingham, M., Ed.), pp. 43–82. Boca Raton (Florida), C. R. C. Press
121. Minty, A. J. and Newmark, P. (1980). Gene regulation: new, old and remote controls. *Nature, London*, **288**, 210–211
122. Minty, A. J., Alonso, S., Guenet, J. L. and Buckingham, M. E. (1983). Number and organization of actin related sequences in the mouse genome. *Journal of Molecular Biology*, **167**, 77–103
123. Minty, A. J., Alonso, S., Caravatti, M. and Buckingham, M. E. (1982). A fetal skeletal muscle actin mRNA in the mouse and its identity with cardiac actin mRNA. *Cell*, **30**, 185–192
124. Minty, A. J., Caravatti, M., Robert, B., Cohen, A., Daubas, P., Weydert, A., Gros, F. and Buckingham, M. E. (1981). Mouse actin mRNAS. Construction and characterization of a recombinant plasmid molecule containing a complementary DNA transcript of mouse α-actin mRNs. *Journal of Biological Chemistry*, **256**, 1008–1014
125. Moerman, D. G., Plurad, S., Waterston, R. H. and Baillie, D. L. (1982). Mutations in the *unc* 54 myosin heavy chain gene of *Caenorhabditis elegans* that alter contractility but not muscle structure. *Cell*, **29**, 773–781
126. Mogami, K. and Hotta, Y. (1981). Isolation of *Drosophila* flightless mutants which affect myofibrillar proteins of indirect flight muscle. *Molecular and General Genetics*, **185**, 409–417
127. Montarras, D. and Fiszman, M. Y. (1982). A new muscle phenotype is expressed by subcultured quail myoblasts from future fast and slow muscles. *Journal of Biological Chemistry*, **258**, 3883–3888
128. Montarras, D., Fiszman, M. Y. and Gros, F. (1981). Characterization of the tropomyosin present in various chick embryo muscle types and in muscle cells differentiated *in vitro*. *Journal of Biological Chemistry*, **256**, 4081–4086

129. Montarras, D., Fiszman, M. Y. and Gros, F. (1982). Changes in tropomyosin during development of chick embryonic skeletal muscles *in vivo* and during differentiation of chick muscle cells *in vitro*. *Journal of Biological Chemistry*, **257**, 545 548

130. Moos, M. and Gallwitz, D. (1982). Structure of a human β-actin-related pseudogene which lacks intervening sequences. *Nucleic Acids Research*, **10**, 7843–7849

131. Nadal-Ginard, B., Medford, R. M., Nguyen, H., Gubits, R. and Bekesi, E. (1982). Regulation of myosin heavy chain gene expression during *in vitro* L_6E_9 cell myogenesis. In *Muscle Development:* (Pearson, M. L. and Epstein, H. F., Eds.), pp. 143–169. New York, Cold Spring Harbor Press

132. Nellen, W. and Gallwitz, D. (1982). Actin genes and actin mRNA in *Acanthamoeba castellanii*. Nucleotide sequence of the split actin gene I. *Journal of Molecular Biology*, **159**, 1–18

133. Ng, R. and Abelson, J. (1980). Isolation and sequence of the gene in *Saccharomyces cerevisiae*. *Proceedings of the National Academy of Sciences, USA*, **77**, 3912–3916

134. Nguyen, H.T., Gubits, R. M., Wydro, R. M. and Nadal-Ginard, B. (1982). Sarcomeric myosin heavy chain is coded by a highly conserved multigene family. *Proceedings of the National Academy of Sciences, USA*, **79**, 5230–5234

135. Nudel, U., Katcoff, D., Carmon, Y., Zevin-Sonkin, D., Levi, Z., Shaul, Y., Shani, M. and Yaffé, D. (1980). Identification of recombinant phages containing sequences for different rat myosin heavy chain genes. *Nucleic Acids Research*, **8**, 2133–2146

136. Nudel, U., Shani, M., Zakut, R., Katcoff, D., Calvo, J., Carmon, Y., Finer, M. and Yaffé, D. (1981). Studies on the structure of genes coding for contractile proteins. In *Abstracts of Cold Spring Harbor Meeting on Molecular and Cellular Control of Muscle Development, Sept. 1981*, p. 14. New York, Cold Spring Harbor Laboratory

137. Nudel, U., Katcoff, D., Zakut, R., Shani, M., Carmon, Y., Finer, M., Czosnek, H., Ginzburg, I. and Yaffé, D. (1982). Isolation and characterization of rat skeletal muscle and cytoplasmic actin genes. *Proceedings of the National Academy of Sciences, USA*, **79**, 2763–2767

138. Overbeek, P. A., Merlino, G. T., Peters, N. K., Cohn, V. H., Moore, G. P. and Kleinsmith, L. J. (1981). Characterization of five members of the actin gene family in the sea urchin. *Biochemica et Biophysica Acta*, **656**, 195–205

139. Palmiter, R. D. (1973). Rate of ovalbumin mRNA synthesis in the oviduct of estrogen-primed chicks. *Journal of Biological Chemistry*, **248**, 8260–8270

140. Paterson, B. M. and Bishop, J. O. (1977). Changes in the mRNA population of chick myoblasts during myogenesis *in vitro*. *Cell*, **12**, 751–765

141. Paterson, B. and Strohman, R. C. (1972). Myosin synthesis in cultures of differentiating chicken embryo skeletal muscle. *Developmental Biology*, **29**, 113–138

142. Patient, R. K., Elkington, J. A., Kay, R. M. and Williams, J. G. (1980). Internal organization of the major adult α- and β-globin genes of *X. laevis*. *Cell*, **21**, 565–573

143. Potter, J. D. (1974). The content of troponin, tropomyosin, actin and myosin in rabbit skeletal muscle myofibrils. *Archives of Biochemistry and Biophysics*, **162**, 436–441

144. Price, K. M., Littler, W. A. and Cummins, P. (1980). *Biochemical Journal*, **191**, 571–580

145. Proudfoot, N. (1980). Pseudogenes. *Nature, London*, **286**, 840–841

146. Robbins, J., Freger, G. A., Chisholm, D. and Gillam, T. C. (1982). Isolation of multiple genomic sequences coding for chicken myosin heavy chain protein. *Journal of Biological Chemistry*, **257** 549–556

147. Robert, B., Weydert, A., Caravatti, M., Minty, A., Cohen, A., Daubas, P., Gros, F. and Buckingham, M. E. (1982). cDNA recombinant plasmid complementary to mRNAs for light chains 1 and 3 of mouse skeletal muscle myosin. *Proceedings of the National Academy of Sciences, USA*, **79**, 2437–2441

148. Rowlerson, A., Pope, B., Murray, J., Whalen, R. G. and Weeds, A. G. (1981). A novel myosin present in cat jaw-closing muscles. *Journal of Muscle Research and Cell Motility*, **2**, 415–438

149. Roy, R. K., Sreter, F. A. and Sarkar, S. (1979). Changes in tropomyosin subunits and myosin light chains during development of chicken and rabbit striated muscles. *Developmental Biology*, **69**, 15–30

150. Rozek, C. E. and Davidson, N. (1981). Isolation and preliminary characterization of a myosin heavy chain gene from *Drosophila melanogaster*. In *Abstracts of Cold Spring Harbor Meeting on Molecular and Cellular Control of Muscle Development, Sept. 1981*, p. 46. New York, Cold Spring Harbor Laboratory

151. Sanchez, F., Tobin, S. L., Rdest, U., Zulauf, E. and McCarthy, B. J. (1982). Two *Drosophila* actin genes in detail; gene structure, protein structure and transcription during development *Journal of Molecular Biology*, **163**, 533–551

152. Sartore, S., Dalla-Libera, L. and Schiaffino, S. (1979). Fractionation of rabbit ventricular myosins by affinity chromatography with insolubilized antimyosin antibodies. *FEBS Letters*, **106**, 197–201

153. Sartore, S., Pierobon-Bormioli, S. and Schiaffino, S. (1978). Immunohistochemical evidence for myosin polymorphism in the chicken heart. *Nature, London*, **274**, 82–83

154. Schel, T. and Dove, W. F. (1982). Mendelian analysis of the organisation of actin sequences in *Physarum polycephalum. Journal of Molecular Biology*, **160**, 41–57

155. Scheller, R. H., McAllister, L. B., Crain, W. R., Durica, D. S., Posakony, J. W., Thomas, T. L., Britten, R. J. and Davidson, E. H. (1981). Organization and expression of multiple actin genes in the sea urchin. *Molecular and Cellular Biology*, **1**, 609–628

156. Schuler, M. A. and Keller, E. B. (1981). The chromosomal arrangement of two linked actin genes in the sea urchin S. purpuratus. *Nucleic Acids Research*, **9**, 591–604

157. Schwartz, R. and Rothblum, K. (1980). Regulation of muscle differentiation: isolation and purification of chick actin messenger ribonucleic acid and quantitation with cDNA probes. *Biochemistry*, **19**, 2506–2514

158. Schwartz, R. J. and Rothblum, K. N. (1981). Gene switching in myogenesis: differential expression of the chicken actin multigene family. *Biochemistry*, **20**, 4122–4129

159. Shah, D. M., Hightower, R. C. and Meagher, R. B. (1982). Complete nucleotide sequence of a soybean actin gene. *Proceedings of the National Academy of Sciences, USA*, **79**, 1022–1026

160. Shani, M., Nudel, U., Zevin-Sonkin, D., Zakut, R., Givol, D., Katcoff, D., Carmon, Y., Reiter, J., Frischauf, A. M. and Yaffé, D. (1981). Skeletal muscle actin mRNA. Characterization of the 3′ untranslated region. *Nucleic Acids Research*, **9**, 579–589

161. Shani, M., Zevin-Sonkin, D., Saxel, O., Carmon, Y., Katcoff, D., Nudel, U.. and Yaffé, D. (1981). The correlation between the synthesis of skeletal muscle actin, myosin heavy chain and myosin light chain and the accumulation of corresponding mRNA sequences during myogenesis. *Developmental Biology*, **86**, 483–492

162. Sinha, A. M., Umeda, P. K., Rajamanickham, C., Kavinsky, C., Hsu, H. J., Jakovcic, S. and Rabinowitz, M. (1982). Molecular cloning of mRNA sequences for cardiac α and β form myosin heavy chains. Expression in ventricles of normal, hypothyroid and thyrotoxic rabbits. *Proceedings of the National Academy of Sciences, USA*, **79**, 5847–5851

163. Sodja, A., Arking, R. and Zafar, R. S. (1982). Actin gene expression during embryogenesis of *Drosophila melanogaster. Developmental Biology*, **90**, 363–368

164. Soriano, P., Szabo, P. and Bernardi, G. (1982). The scattered distribution of actin genes in the mouse and human genomes. *EMBO Journal*, **1**, 579–583

165. Sreter, F. A., Balint, M. and Gergely, J. (1975). Structural and functional changes of myosin during development. *Developmental Biology*, **46**, 317–325

166. Starr, R. and Offer, G. (1973). Polarity of the myosin molecule. *Journal of Molecular Biology*, **81**, 17–31

167. Stein, J. P., Catterall, J. F., Kristo, P., Means, A. R. and O'Malley, B. W. (1980). Ovomucoid intervening sequences specify functional domains and generate protein polymorphism. *Cell*, **21**, 681–687

168. Steinbach, J. H., Schubert, D. and Eldridge, I. (1980). α- and β-tropomyosin in typed single fibres of human skeletal muscle. *Experimental Neurology*, **67**, 655–669

169. Stockdale, F. E., Saden, H. and Raman, N. (1981). Slow muscle myoblasts differentiating *in vitro* synthesize both slow and fast myosin light chains. *Developmental Biology*, **82**, 168–171

170. Stone, D. and Smillie, L. B. (1978). The amino acid sequence of rabbit skeletal α-tropomyosin. *Journal of Biological Chemistry*, **253**, 1137–1148

171. Suck, D., Kabsch, W. and Mannherz, H. G. (1981). Three-dimensional structure of the complex of skeletal muscle actin and bovine pancreatic DNAse I at 6 Å resolution. *Proceedings of the National Academy of Sciences, USA*, **78**, 4319–4323

172. Sutoh, K. (1982). Identification of myosin-binding sites on the actin sequence. *Biochemistry*, **21**, 3657–3661

173. Tobin, S. L., Zulauf, E., Sanchez, F., Craig, E. A. and McCarthy, B. J. (1980). Multiple actin-related sequences in the *Drosophila melanogaster* genome. *Cell*, **19**, 121–131

174. Umeda, P. K., Kavinsky, C., Sinha, A. M., Hsu, H. J., Jakovcic, S. and Rabinowitz, M. (1982). Molecular cloning of myosin heavy chain cDNA: from skeletal and cardiac muscle. In *Muscle Development: Molecular and Cellular Control*, (Pearson, M. L. and Epstein, H. F., Eds.) pp. 169–177. New York, Cold Spring Harbor Press

175. Vandekerckhove, J. and Weber, K. (1978). The amino acid sequence of actin from chicken skeletal muscle actin and chick gizzard smooth muscle actin. *FEBS Letters*, **102**, 219–222

176. Vandekerckhove, J. and Weber, K. (1978). Actin amino acid sequences: Comparison of actins from calf thymus, bovine brain, and SV40-transformed mouse 3T3 cells with rabbit skeletal muscle actin. *European Journal of Biochemistry*, **90**, 451–462

177. Vandekerckhove, J. and Weber, K. (1978). The amino acid sequence of *Physarum* actin. *Nature, London*, **276**, 720–721
178. Vandekerckhove, J. and Weber, K. (1979). The complete amino acid sequence of actins from bovine aorta, bovine heart, bovine fast skeletal muscle and rabbit slow skeletal muscle. *Differentiation*, **14**, 123–133
179. Vandekerckhove, J. and Weber, K. (1980). Vegetative *Dictyostelium* cells containing 17 actin genes express a single major actin. *Nature, London*, **284**, 475–477
180. Vandekerckhove, J. and Weber, K. (1981). Actin typing on total cellular extracts: a highly sensitive protein-chemical procedure able to distinguish different actins. *European Journal of Biochemistry*, **113**, 595–603
181. Vandekerckhove, J., Franke, W. W. and Weber, K. (1981). Diversity of expression of non-muscle actin in amphibia. *Journal of Molecular Biology*, **152**, 413–426
182. Vandekerckhove, J., Leavitt, J., Kakunaga, T. and Weber, K. (1980). Co-expression of a mutant β-actin and the two normal β- and γ-actins in a stably transformed human cell line. *Cell*, **22**, 893–899
183. Van Eerd, J. P. and Takahashi, K. (1976). Determination of the complete amino acid sequence of bovine cardiac troponin C. *Biochemistry*, **15**, 1171–1180
184. Wahli, W., Dawid, I. B., Wyler, T., Weber, R. and Ryffel, G. U. (1980). Comparative analysis of the structural organization of two closely related vitellogenin genes. *Cell*, **20**, 107–117
185. Wasylyk, B. and Chambon, P. (1981). A T to A base substitution and small deletions in the conalbumin TATA box drastically decreases specific *in vitro* transcription. *Nucleic Acids Research*, **9**, 1813–1824
186. Waterston, R. H., Smith, K. C. Moerman, D. G. (1982). Genetic fine structure analysis of the myosin heavy chain gene *unc* 54 of *C. elegans*. *Journal of Molecular Biology*, **158**, 1–15
187. Watterson, D. M., Sharief, F. and Vanaman, T. C. (1980). The complete amino acid sequence of the Ca^{2+}-dependent modulator protein (calmodulin) of bovine brain. *Journal of Biological Chemistry*, **255**, 962–975
188. Weeds, A. G. (1976). Light chains from slow-twitch muscle myosin. *European Journal of Biochemistry*, **66**, 157–173
189. Weeds, A. G. (1982). Actin-binding proteins – regulators of cell architecture and motility. *Nature, London*, **296**, 811–816
190. Weeds, A. G. and Burridge, K. (1975). Myosin from cross-reinnervated cat muscles. Evidence for reciprocal transformation of heavy chains. *FEBS Letter*, **57**, 203–208
191. Westwood, S. A. and Perry, S. V. (1982). Two forms of the P light chain of myosin in rabbit and bovine hearts. *FEBS Letters*, **142**, 31–34
192. Weydert, A., Daubas, P., Caravatti, M., Minty, A., Bugaisky, G., Cohen, A., Robert, B. and Buckingham, M. (1983). Detection of a sequential accumulation of mRNAs encoding different myosin heavy chain isoforms during skeletal muscle development *in vivo*, using a recombinant plasmid coding for an adult fast myosin heavy chain from mouse skeletal muscle. *Journal of Biological Chemistry* (submitted)
193. Whalen, R. G. and Sell, S. M. (1980). Myosin from fetal hearts contains the skeletal muscle embryonic light chain. *Nature, London*, **286**, 731–733
194. Whalen, R. G., Butler-Browne, G. S. and Gros, F. (1976). Protein synthesis and actin heterogeneity in calf muscle cells in culture. *Proceedings of the National Academy of Sciences, USA*, **73**, 2018–2022
195. Whalen, R. G., Butler-Browne, G. S. and Gros, F. (1978). Identification of a novel form of myosin light chain present in embryonic muscle tissue and cultured muscle cells. *Journal of Molecular Biology*, **126**, 415–431
196. Whalen, R. G., Butler-Browne, G. S., Sell, S. and Gros, F. (1979). Transitions in contractile protein isozymes during muscle cell differentiation. *Biochimie*, **61**, 625–632
197. Whalen, R. G., Sell, S. M., Eriksson, A. and Thornell, L. E. (1982). Myosin subunit types in skeletal and cardiac tissues and their developmental distribution. *Developmental Biology*, **91**, 478–484
198. Whalen, R. G. Schwartz, K., Bouveret, P., Sell, S. M. and Gros, F. (1979). Contractile protein isozymes in muscle development: Identification of an embryonic form of myosin heavy chain. *Proceedings of the National Academy of Sciences, USA*, **76**, 5197–5201
199. Whalen, R. G., Sell, S. M., Butler-Browne, G. S., Schwartz, K., Bouveret P. and Pinset, I. (1981). Three myosin heavy chain isozymes appear sequentially in developing rat muscle. *Nature, London*, **292**, 805–809
200. Wikman-Coffelt, J. and Srivastava, S. (1979). Differences in atrial and ventricular myosin light chain LC_1. *FEBS Letters*, **106**, 207–212

201. Wilkinson, J. M. (1978). The components of troponin from chicken fast skeletal muscle. *Biochemical Journal*, **169**, 229–238
202. Wilkinson, J. M. and Grand, R. J. A. (1978). Comparison of amino acid sequence of troponin I from different striated muscles. *Nature, London*, **271**, 31–35
203. Yablonka, Z. and Yaffé, D. (1977). Synthesis of myosin light chains and accumulation of translatable mRNA coding for light-chain-like polypeptides in differentiating muscle cultures. *Differentiation*, **8**, 133–143
204. Yaffé, D. and Dym, H. (1972). Gene expression during differentiation of contractile muscle fibers. *Cold Spring Harbor Symposium on Quantitative Biology*, **37**, 543–547
205. Yamawaki-Kataoka, Y., Miyata, T. and Honjo, T. (1981). The complete nucleotide sequence of mouse immunoglobulin γ2a gene and evolution of heavy chain genes: further evidence for intervening sequence-mediated gene transfer. *Nucleic Acids Research*, **9**, 1365–1381
206. Zakut, R., Shani, M., Givol, D., Neuman, S., Yaffe, D. and Nudel, U. (1982). The nucleotide sequence of the rat skeletal muscle actin gene. *Nature, London*, **298**, 857–859
207. Zengel, J. M. & Epstein, H. F. (1980). Mutants altering the co-ordinate synthesis of specific myosins during nematode muscle development. *Proceedings of the National Academy of Sciences, USA*, **77**, 852–856
208. Zevin-Sonkin, D. and Yaffé, D. (1980). Accumulation of muscle-specific RNA sequences during myogenesis. *Developmental Biology*, **74**, 326–334
209. Zimmer, W. and Schwartz, R. J. (1982). Amplification of chicken actin genes during myogenesis. In *Gene Amplification*, (Schimke, R. T., Ed.). pp. 137–146. New York, Cold Spring Harbor Laboratory
210. Zimmer, E. A., Martin, S. L., Beverley, S. M., Kan, Y. W. and Wilson, A. C. (1980). Rapid duplication and loss of genes coding from the α-chains of hemoglobin. *Proceedings of the National Academy of Sciences, USA*, **77**, 2158–2162
211. Zulauf, E., Sanchez, F., Tobin, S. L., Rdest, U. and McCarthy, B. J. (1981). Developmental expression of a *Drosophila* actin gene encoding actin I. *Nature, London*, **292**, 556–558
212. Zweig, S. E. (1981). The muscle specificity and structure of two closely related fast twitch white muscle myosin heavy chain isozymes. *Journal of Biological Chemistry*, **256**, 11847–11853

Chapter 22

The silk fibroin gene

S. P. Gregory

Introduction

Fibroins are long protein filaments secreted by various members of the classes Arachnida and Insecta during the construction of their webs and cocoons[29, 30, 50]. The fibroin secreted by the mulberry silk moth, *Bombyx mori*, is of particular importance to the silk industry and has therefore received more attention than its counterparts in less economically attractive species. This chapter will review the current state of knowledge concerning the structure and expression of the *B. mori* silk fibroin gene.

B. mori is unique among insects in being entirely dependent on man for its survival. In more than 4000 years of sericulture this species has been subjected to artificial selection, not only for increased quality and quantity of silk production but also for ease of rearing and handling. The net result is that *B. mori* has become completely domesticated with individuals now unknown in the wild state.

It is only comparatively recently that *B. mori* has been adopted as a laboratory organism for studying gene expression during development[14]. In this respect it is well suited to such investigations. The docile flightless adults will breed throughout the year and large numbers of larvae (silkworms) can be raised on natural or artificial diets. Silk fibroin itself is produced predominantly, though not exclusively, at the end of larval life when it is used in cocoon formation. This large protein with a relatively simple amino acid sequence is synthesized in prodigious amounts by a single cell type. To cope with demand these cells possess a translational machinery adapted specifically for rapid fibroin synthesis. Regulation of expression at the transcriptional level has also been reported during the larval feeding and moulting stages[31, 55]. With the availability of genetic pedigrees and stocks of mutant strains deficient in various parameters of cocoon production, there is thus a considerable potential for identifying those factors responsible for correct expression of the silk fibroin gene.

Bombyx mori

The adult mulberry silk moth is of medium size and light in colouration. It is Chinese in origin and most closely related to the wild type *B. mandarina*. As in other Lepidoptera, the chromosomes are small and almost round in shape, making

397

identification difficult. The haploid chromosomal complement (n) for *B. mori* is 28, whilst for *B. mandarina* it is more usually 27.

Larval life encompasses a period of some 4–5 weeks and includes five instars. During each moulting phase there is a period of inactivity lasting 24 hours; larvae do not feed at this stage. The fifth instar larva takes eight days to mature and when fully grown attains a length of 5–8 cm. The cocoon (about 0.4 g in dry weight) is then woven from a single thread of silk approximately 300 m long. The principal constituents of silk are two proteins: fibroin (the fibrous component) and sericin (the adhesive component). Typical analysis of a cocoon would reveal 70–80% fibroin, 20–30% sericin and 2–3% other materials (waxes, carbohydrates, dyes etc.)[50].

This thread of silk originates from a single pair of modified salivary glands known as the silk glands.

The silk glands

Development

The cells of the silk gland are ectodermal in origin, arising from invaginations of the basal parts of the second maxillae. Rapid cell division gives rise to a morphologically fully-formed and functional pair of glands by the time the embryo hatches. However, all cell division in these glands is confined to the embryonic stage, so further growth during larval life occurs with concomitant polyploidization.

The early stages of silk gland development are in phase with the overall growth of the larva but at the onset of the fourth instar there is a marked acceleration in the rate of silk gland maturation[36]. This is particularly evident during the fifth instar when the silk gland weight increases twenty-fold relative to a six-fold increase in total larval weight. By the end of this stage the silk glands account for nearly half the body weight and the constituent cells have undergone a 1.6×10^5 increase in size[50].

Functional compartmentalization

Morphologically, each gland can be divided into three sections (see *Figure 22.1*). The posterior region comprises a long thin tube, 20 cm in length in the mature larva, which leads to a shorter but much wider section termed the middle region. The thin tube connecting the middle region to the spinneret, through which the silk is eventually secreted, forms the anterior region of the gland.

Fibroin is synthesized exclusively by the cells of the posterior section. An autoradiographic study utilizing ³H-glycine, the major component of fibroin, has recently revealed some details of this process and subsequent transportation events[44]. The initial site of synthesis is the rough endoplasmic reticulum which also undergoes a gross morphological change from a lamellar to a vesicular form during periods of peak fibroin production[57]. Pulse-chase experiments show the next stage to be a movement of label to the Golgi apparatus where fibroin condensation and storage occurs. The globules of fibroin so formed are transported via a microtubule organized system to the apical cytoplasm where secretion takes place. Treatment with colchicine or vinblastine inhibits this movement; treatment with cytochalasin B or D accelerates it.

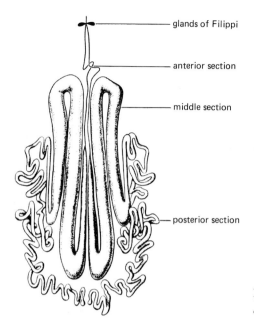

glands of Filippi

anterior section

middle section

posterior section

Figure 22.1. Schematic drawing of the silk glands of the *Bombyx mori* larva. Reproduced from Lucas *et al.*[30] with permission.

The fibroin is exported into the lumen of the gland as a concentrated (around 30% w/v) aqueous solution and this is then passed to the middle region by peristalsis. It is here that the bulk of the other constituents of silk, principally sericin, are added. In those strains producing coloured silks, a sharp boundary line can be seen between the pigment-producing cells of the middle section and the fibroin-producing cells of the posterior section. The viscous solutions of fibroin and sericin are then stored in the middle region as separate entities. On final extrusion from the gland, via the spinneret, coagulation of the mixture occurs with sericin acting as the 'glue protein' for fibroin filament formation. The final thread thus consists of two cores of fibroin, one from each half of the gland, surrounded by several layers of sericin. The insoluble nature of fibroin after spinning, and its coating with sericin have proved to be considerable barriers to protein chemists' attempts to elucidate the properties of fibroin.

The number of cells per silk gland is variable but an average value would be 300, 250 and 500 for the anterior, middle and posterior regions respectively[50]. In all productive strains, the largest cells, and those with the highest ploidy, occur in the middle section.

Silk biosynthesis

As we have seen, newly hatched larvae are capable of spinning small amounts of silk which are later used to hold the old skin to the substratum during moulting. Gene amplification, through increased ploidy, has not taken place to any great extent at this stage and is thus alone unlikely to account for the control of fibroin synthesis. Indeed, all larvae are capable of spinning silk, except during the moulting stages. Silk biosynthesis is subject then to a cyclical switching mechanism. During the feeding stage the components of silk are synthesized and stored in the lumen of the gland. During each moult the contents of the lumen are degraded and replacement synthesis ceases until the moult has finished.

Suzuki and Suzuki have shown that fibroin messenger RNA levels decrease during moulting[55]. The question of whether this is due to decreased messenger RNA (mRNA) stability, modified post-transcriptional processing and/or differential transcriptional activity has recently been addressed by Maekawa and Suzuki[31]. By hybridizing ³H-RNA (pulse-labelled for 10 or 30 minutes) to immobilized fibroin genomic sequences they have been able to estimate the relative transcription rate of the fibroin gene during the transition from late fourth instar to early fifth instar larva (see *Figure 22.2*). The observed decrease in transcriptional

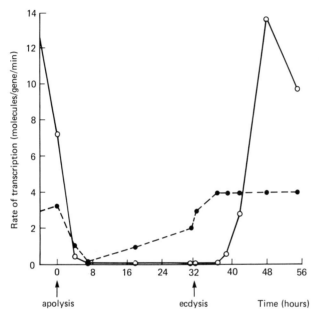

Figure 22.2. Relative rates of transcription of the fibroin and rRNA genes in the posterior silk gland during the transition from fourth instar to fifth instar larva; O—O, fibroin gene transcription; ●--●, rRNA gene transcription. Data from Maekawa and Suzuki[31]

activity during the moulting stage is not reflected in the rates of ribosomal RNA (rRNA) or heterogeneous nuclear RNA (hnRNA) synthesis, although a dramatic qualitative change in mRNA synthesis does take place. They conclude, therefore, that this control is primarily exerted at the level of fibroin gene transcription.

Fibroin mRNA constitutes approximately 2% of total RNA in posterior silk gland (PSG) cells of feeding third and fourth instar larvae (0.2 and 2 µg per pair of posterior glands, respectively), but it is during the fifth instar that fibroin biosynthesis reaches its peak of production (4.4% of total PSG cell RNA at day 7; 220 µg fibroin mRNA per pair of glands)[53,55]. The physiological activity of the posterior silk gland during this final instar can be divided into two phases (see *Figure 22.3*)[42,57]. The first four days represent a period of sustained growth during which the machinery for synthesis and secretion of fibroin is assembled. By the second day post-ecdysis fibroin message forms the predominant mRNA in the cell. Significant quantitative changes occur in the rest of the poly(A)+ population at this

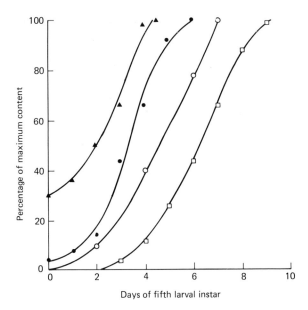

Figure 22.3. Accumulation of nucleic acids and fibroin in the posterior silk gland during the fifth larval instar. DNA (▲), total RNA (●), fibroin mRNA (○) and fibroin (□) contents are expressed as percentages of their maxima. From Prudhomme and Couble[42]

stage, but there are few, if any, qualitative changes amongst the 3000 species represented[5].

The transition to the synthetic phase after four days is associated with the cessation of DNA synthesis and the acceleration of fibroin protein synthesis. In the subsequent four days fibroin accounts for 75% of the total protein synthesized with a mean production rate of 0.03 fibroin molecules/mRNA/second[14, 50]. At the end of this phase each PSG cell will have produced on average 300 μg of silk fibroin[56].

The middle region of the silk gland undergoes a similar synthetic phase to the posterior region with sericin accounting for 50% of the total protein synthesized during the later stages of fifth instar life[14]. However, the details of sericin gene expression have yet to be characterized. The availability of cloned sericin sequences[40] will no doubt stimulate further studies and provide a means for determining how differential gene activation is achieved in different compartments of the silk gland[51].

Silk proteins

The silk proteins are characterized by their large size and unusual amino acid compositions. Fibroin is essentially composed of a repetitious Gly-Ala peptide interspersed with serine residues. Estimates of the exact amino acid composition[7] vary from 45% glycine, 29% alanine, 12% serine (see *Table 22.1*)[29] to 43% glycine, 35% alanine, 10% serine[48]. These determinations are generally in good agreement regardless of whether the fibroin is extracted from the cocoon or the silk gland. The

TABLE 22.1. Correlation between amino acid composition of fibroin and tRNA acylation in the posterior silk gland cells of late fifth instar larvae

	tRNA (pmoles)[a]		Amino acid composition(%)[b]	
Amino acid	4th day	6th and 8th days	Fibroin	Sericin
Glycine	14.0	120.0	44.5	14.7
Alanine	12.5	64.3	29.3	4.3
Serine	23.0	39.6	12.1	37.3
Tyrosine	14.5	27.7	5.2	2.6
Valine	-	-	2.2	3.6
Aspartic acid	9.0	10.6	1.3	14.8
Glutamic acid	5.0	8.6	1.02	3.4
Threonine	8.0	12.6	0.91	8.7
Iso-leucine	-	-	0.66	0.7
Phenylalanine	-	-	0.63	0.3
Leucine	-	-	0.53	1.4
Arginine	-	-	0.47	3.6
Proline	-	-	0.36	0.7
Lysine	-	-	0.32	2.4
Cysteine	-	-	0.15	0.5
Histidine	-	-	0.14	1.2
Tryptophan	-	-	0.11	0
Methionine	-	-	0.10	0

[a] Data from Garel et al.[15]. Acylations were performed with 20 μg tRNA for 20 mins at 30°C. Each incubation, in the presence of one labelled amino acid tracer, was carried out once with silk gland synthetases followed by two or three further incubations with rat liver synthetases.
[b] Data from Lucas and Rudall[29]; proteins solubilized from cocoons.

same is not true, however, for sericin. Estimates of the percentage amino acid composition vary considerably[7]. As we shall see, this may be due to the large number of polypeptide components which together make up the sericin family, and hence influence its overall amino acid content. The major residues appear to be serine, glycine and aspartic acid.

The size of these proteins has been determined by SDS–polyacrylamide gel electrophoresis and agarose–guanidine chromatography[48]. Since it is difficult to solubilize the cocoon proteins, the material used in these studies was extracted directly from the glands. Fibroin, in this case, was shown to consist of equimolar amounts of two large polypeptide chains (370 000 ± 20 000 d). It is likely that the presence of two components is due to the use of a hybrid strain which was expressing two distinct fibroin alleles. Certainly such alleles are known to exist and they can give rise to proteins of varying molecular weights[21, 48]. Indeed, it might be expected that genes which contain internally repetitious amino acid sequences, and impose no stringent demands on overall polypeptide chain dimensions, would give rise to sequences of differing lengths through a mechanism such as unequal crossing-over[34, 47].

A similar analysis of sericin components presented a much more complex picture[48]. At least 15 peptide chains (ranging in size from 20 000 to 220 000 d) were observed in the fluid from the anterior tip of the middle silk gland. The most abundant of these were the three largest polypeptides (130 kd, 210 kd and 220 kd) and so their amino acid compositions were determined. The similarities revealed (*Table 22.2*) are suggestive of a common origin, although these deviate appreciably

TABLE 22.2. Amino acid compositions of three polypeptides isolated from the anterior tip of the middle silk gland[(a)]

Amino acid (%)	220 kd	210 kd	130 kd
Aspartic acid	32	26	23
Glutamic acid	20	28	17
Serine	15	15	11
Glycine	7.8	8.7	8.7
Alanine	4.0	4.5	2.9

[(a)] Data from Sprague[48]

from the sericin composition derived from boiled cocoons (see *Table 22.1*). Gamo *et al.* consider sericin to be composed of five major polypeptides[13]. Moreover, these different components are synthesized in different sections of the middle silk gland. Like fibroin, genetic variants exist which encode sericin proteins of different lengths[12].

In the absence of complete amino acid sequence data, there has been considerable speculation by protein chemists about the possible configurations adopted by the giant fibroin molecule[14, 23, 27, 28, 37]. The bulk of the molecule appears to be organized into a crystalline β-pleated sheet comprising Gly-Ala repeats interspersed with serines and the occasional tyrosine. It is this region which imparts the characteristic mechanical properties to the silk thread[27]. The remainder of the molecule is rather amorphous, containing mannose and glucosamine residues and the potential to form disulphide bridges[14]. Glycine is again abundant within these regions but this time largely in association with less common amino acids. It is the amorphous regions then that are likely to be involved in solubilization and transport of the fibroin protein.

The advent of recombinant DNA technology now means that it is much easier to sequence a gene and derive its product theoretically than to sequence the protein product itself. The information derived to date from this approach is presented in *Figure 22.5*.

Fibroin messenger RNA

Structure

The observation that fibroin is composed largely of glycine and alanine residues allows us to make certain predictions concerning the primary structure of the messenger RNA. The codons for glycine and alanine are GGN and GCN, respectively, intimating that the corresponding message must contain at least 57% GC, of which 40% or more should be G residues[52]. This compares favourably with a GC content of 39% for total *B. mori* DNA and 48% for *B. mori* ribosomal DNA. Coupled with its high sedimentation constant (47s), it is thus a relatively simple task to purify fibroin mRNA by sucrose gradient centrifugation and identify it by partial sequence analysis[52]. Suzuki and Brown have digested this purified fibroin mRNA with RNase T_1 and pancreatic RNase and identified the sequences of the oligonucleotides produced[52]. Their results show the major codons utilized in the repetitive protein coding region of the mRNA to be GGU followed by GGA for glycine, GCU for alanine and UCA for serine.

Analysis of fibroin mRNA on denaturing (formaldehyde) agarose–acrylamide composite gels revealed a molecular weight of 6.0×10^6 d[25]. Estimates from electron microscopy place this value slightly lower at 5.6×10^6 d, so an average molecular weight of 5.8×10^6 daltons (1.6×10^4 bases) is usually assumed[25].

The proportion of the protein coding region within this message is difficult to ascertain since estimates of gene, mRNA and protein sizes have been carried out using different *B. mori* strains, and as we have seen different strains can carry fibroin genes of different lengths. However, from the genomic sequence, it appears that the 5′ untranslated region is short, spanning only 24 nucleotides (see *Figure 22.5*). The 3′ untranslated region is longer but to what extent has still to be determined.

Between 43% and 81% of fibroin mRNAs bind to oligo (dT)-cellulose through a 3′ poly(A) tract[25]. The RNase-resistant material retained by the column comprises approximately 0.6% of the molecule, suggesting an average of 100 residues for the poly(A) tail. Further details of the 3′ untranslated region are confined to a short sequence of 34 nucleotides immediately adjacent to this poly(A) tract. Hagenbüchle *et al.*[19] sequenced cDNA primed by oligo (dT) and discovered an unusually long oligo (T) stretch as well as the AATAAA motif postulated to act as a polyadenylation signal[6]. The equivalent mRNA sequence would read as follows:

$$5′\ \text{AUGUUUUUUUUUUCGUUAAUAAAUGAGAGCAUUUA}_{(n)}\ 3′$$

In common with other mRNAs[46], fibroin message has a blocked and methylated 5′ terminus, a so-called type II cap structure: $m^7G(5')_{ppp}\ (5')\ A^m{}_p U^m{}_p C_p{}^{64}$. Approximately 60% of fibroin mRNA molecules will have this structure; the remaining 40% possess the unmethylated cap structure $G_{ppp}A_p Py_p X_p X_p G_p{}^{61}$.

Synthesis during the fifth instar

As we have seen in *Figure 22.3*, fibroin mRNA accumulates at an almost constant rate (7–10 molecules/gene/minute) throughout the fifth larval instar[53]. However, since rRNA and hnRNA syntheses decrease over this same period, the relative rate of fibroin mRNA production is in effect continually increasing. Thus, in experiments in which RNA was pulse-labelled for 30 minutes *in vivo*, fibroin mRNA comprised 1% of total PSG cell RNA on the second day of the fifth instar, 3% on the fourth day and 8% on the seventh day[53]. At this stage an average fibroin gene exhibited a rate of transcription fifty-fold higher than a corresponding rRNA gene. This accentuated expression of the silk fibroin genes is also exhibited by PSG nuclei after they have been isolated from the cells and incubated in an *in vitro* transcribing medium[53]. Synthesis in this latter case was abolished by the inclusion in the incubation medium of 0.2 µg/ml α-amanitin, confirming that transcription of the fibroin gene was mediated by RNA polymerase II.

The posterior silk gland cells of the fifth instar larva are thus remarkably efficient at synthesizing fibroin. More than 99% of the fibroin mRNA in fifth instar larvae is synthesized *de novo* during this period at a synthetic rate approaching 20% of total RNA synthesis by the end of the instar[50]. The transcription and stabilization of the fibroin message occurs at rates exceeding those for ribosomal RNAs. It is not surprising then that the translational machinery is also adapted to rapid fibroin synthesis.

Translation

The phenomenon of functional transfer RNA (tRNA) adaptation has been recognized for a number of years, but nowhere has it been more clearly demonstrated than in the case of the posterior silk gland cells. In this cell type, as in others specializing in the production of luxury proteins, the pool of tRNAs reflects the codon content of the predominant message in the cell. Thus, the major tRNA species in the posterior silk gland cells from late fifth instar larvae are those recognizing glycine, alanine, serine and tyrosine codons (see *Table 22.1*)[15]. Together they represent more than 70% of the total PSG cell tRNAs[17]. The relative abundance of these tRNA species is substantially less at earlier stages of fifth instar development.

Moreover, this adaptation has been postulated to extend to the different iso-accepting tRNA species[16]. The relative concentrations of the different iso-acceptor tRNAs fluctuate during fifth instar development (see *Table 22.3*) and it has been proposed that they regulate fibroin biosynthesis by modulating the rate of translation. Thus, a specific $tRNA_{2a}^{Ala}$ is found in PSG cells but no other silkworm

TABLE 22.3. Relative amounts (in percentages) of the different iso-acceptor species of the major tRNAs present in posterior silk gland cells[a]

Fifth instar	Glycine			Alanine			Serine				
	I	II	III	I	II	III	I	II	III	IV	V
4th day	65	20	15	2	78	20	8	23	22	26	21
8th day	33	48	19	4	62	34	22	24	20	20	14

[a]Data from Garel et al.[16]

tissues[35]. A recent report suggests that in late fifth instar larvae $tRNA_1^{Ser}$ predominates in the middle silk gland whereas $tRNA_2^{Ser}$ is the major serine iso-acceptor species of the posterior silk gland[20]. This result is consistent with previous RNA sequence data[52] which suggested that the major serine codon in fibroin mRNA is the UCA triplet recognized by $tRNA_2^{Ser}$. Of the 18 serine codons so far identified at the 5' end of the genomic sequence, only one corresponds to the UCA triplet: 10 are in the form of AGU or AGC, both of which are recognized by $tRNA_1^{Ser}$ (see *Table 22.4*). This may explain the observed increase in the relative abundance of $tRNA_1^{Ser}$ in PSG cells as the fifth larva instar matures (see *Table 22.3*). The incidence in the partial genomic sequence of the various glycine and alanine codons (see *Table 22.4*) reflects their occurrence in the mRNA as a whole[52].

This adjustment of intracellular levels of (iso-acceptor) tRNA species to the codon frequency of the mRNA being decoded is presumably used as a way of optimizing translational efficiency. A qualitative and quantitative adaptation amongst the tRNA population might tend to increase both the rate of polypeptide chain elongation and the fidelity of translation. Chavancy and Garel have recently suggested that elongation is non-uniform with pauses occurring in the ribosome translation of the message[2]. Such pauses have previously been observed in the translation of fibroin message, both *in vivo* and *in vitro*[3, 26], and might occur if there

TABLE 22.4. Codon usage in fibroin mRNA as determined from the primary gene sequence[a]

Glycine	GGU	17	Serine	AGU	6
	GGA	10		AGC	4
	GGG	3		UCU	4
	GGC	1		UCC	3
				UCA	1
Alanine	GCU	15		UCG	0
	GCA	9			
	GCC	6	Tyrosine	UAU	8
	GCG	3		UAC	2

[a] Data from Tsujimoto and Suzuki[61,62]. Figures indicate the number of times
a particular codon occurs in the fibroin gene sequence so far determined
(*Figure 22.5*)

is mismatch between codon and anti-codon sequences. It is worth noting that the partial translation of fibroin mRNA in a heterologous cell-free system[18] was overcome by adding PSG tRNAs to the incubation mix[26].

The mechanisms(s) by which tRNA gene expression and fibroin gene expression are linked remains unknown. Selective amplification of tRNA genes during the fifth instar is not involved, though. Mazima and Shimura have reported that the genes for tRNAs[Gly] and tRNAs[Leu] comprise about 3.0–$3.5 \times 10^{-4}\%$ of the genome at both day 1 and day 7 of fifth instar development[50]. Preferential degradation of tRNAs has also not been observed[4], so it seems likely that there is an element of transcriptional control involved. The idea of selective activation of different clusters of tRNA genes, with characteristically different iso-acceptor species, is an attractive one[4].

The fibroin gene

The silk gland genome

The haploid DNA content of *B. mori* has been estimated to be 0.52 ± 0.01 pg (3.1×10^{11} d)[8,42]. During the differentiation of the silk gland, the cells undergo a number of cycles of replication without concomitant cell division. In the posterior region there are 17–18 rounds of DNA synthesis which result in a massive accumulation of DNA (0.2 µg/mature PSG cell) equivalent to $200\,000$ times more DNA than a normal diploid cell[9].

This polyploidization of the silk gland nucleus is not associated with the formation of giant polytene chromosomes akin to those seen in Dipteran salivary glands (see Chapter 7). Instead, the nucleus remains as if in interphase. Distinctive chromosomes are not visible at any stage in post-embryonic silk gland growth, but Feulgen staining reveals a multitude of small chromosome-like granules. The location of the fibroin and rRNA genes within these nuclei has been studied by *in situ* hybridization using ^{125}I-labelled fibroin mRNA and rRNA probes[58]. In all preparations examined (posterior and middle gland cells, feeding and moulting stages, second to fifth instar), both the ribosomal DNA and the fibroin genes were distributed randomly throughout the nuclei. Unlike the situation in polyteny, the daughter chromosomes generated by these individual endomitotic replications appear to separate and segregate at random.

Amplification of the genome occurs in a uniform manner. Repetitive sequence distribution remains the same[9] and there is no selective amplification of the fibroin[56] or ribosomal genes[9]. From reassociation kinetics, 55% of the *B. mori* genome would appear to contain single-copy sequences[8]. The rest can be divided more or less equally into sequences reiterated 500 times and 50 000 times. The ribosomal genes fall into the former category with an estimated 240 copies/haploid genome[9]. Determination of the multiplicity of the fibroin gene proved more difficult because of the extremely GC-rich nature of the gene and the internally repetitious simple sequences[56]. Using ^{125}I-labelled mRNA hybridization, under conditions of RNA excess, Gage and Manning were eventually able to demonstrate the presence of only one copy of the fibroin gene/haploid genome[10]. Although there is no selective gene amplification during silk gland differentiation[56], polyploidization will generate 4×10^5 copies of the fibroin gene/PSG cell.

It should be borne in mind that the fibroin gene is large and the *B. mori* genome relatively small. Thus, the fibroin gene accounts for 0.004% of the *B. mori* DNA, a percentage nearly two orders of magnitude higher than the equivalent occurrence of the globin gene in the mammalian genome.

Chromosomal location

Linkage analysis studies of fibroin allele segregation have shown that the gene always partitions with the *Nd* (naked pupa) mutation and is thus located on chromsome 23[22]. Two of the sericin polypeptide genes have been mapped to chromosome11[12].

The immediate chromosomal environment of the fibroin gene, like that for many other GC-rich genes, appears to have a high AT-base content (60–70%)[24]. Repetitive (hybridizing) sequences have been identified near to the 5' end of the gene, but none have been detected in a restriction fragment that includes the 600 bp upstream from the start site and 300 bp downstream[41, 61, 62]. The nature of these repetitive elements remains obscure in the absence of data concerning their primary structure and distribution throughout the rest of the genome.

The methylated state of the fibroin chromosomal locus has yet to be examined directly, but *B. mori* DNA in general is low in methyl groups. Less than 0.1% of the cytosines in silk gland DNA exist as 5-methyl-cytosine[9] (see Chapter 4). This is particularly important if restriction enzymes are to be used to analyse the state of the native fibroin gene.

Primary structure

Gage and Manning have carried out such a restriction enzyme analysis in order to derive a physical map of the natural fibroin gene[11, 33, 34]. The availability of a wide range of restriction enzymes with differing specificities now makes it possible to analyse gene structure in depth without resorting to DNA sequencing strategies. The pattern of restriction enzyme cleavages obtained can reveal characteristic features in the underlying sequence and predict certain primary structures. This is especially useful when analysing large genes such as those coding for silk fibroin and sericin.

The enzyme cleavages observed by Gage and Manning can be divided into three classes. Firstly, there are those enzymes that do not cut within the repetitive sequence 'core' of the gene (see *Figure 22.4*); surprisingly, this included Hae III and Tha I which might be expected to cut frequently in GC-rich regions. Secondly, there are enzymes such as Hpa II and Alu I which cleave throughout the core. Finally, there are a group of enzymes (Mbo II, Ava II and Hinf I) which cleave the core a limited number of times. Although the recognition sites for these enzymes differ, they are all clustered together in the same regions of the gene[11]. It has been

5 kb

Figure 22.4. Possible chromosomal organization of the fibroin gene. ▮▮, repetitive protein coding sequence; ▭, non-repetitive coding sequence; ▧, intron.

concluded that the core of the gene (15 kbp) is divided into at least 10 domains coding for the crystalline portions of fibroin, each interrupted by smaller domains coding for the amorphous regions. This has been incorporated into a recent model of fibroin structure[14].

Predictions can also be made about codon usage (see also reference 52). For instance, the absence of Sau 3A sites in the core shows that the glycine codon GGA is not used prior to serine codons, although it is frequently used prior to alanine codons. The information that can be gleaned from such restriction studies is limited, however, and not entirely unambiguous. Suzuki and his colleagues have gone further and cloned the *B. mori* fibroin gene and its flanking sequences[39, 54]. With such an abundant, easily purified message the cloning operation was relatively simple. DNA extracted from posterior silk glands was sheared and introduced into pMB9 by AT-tailing. On screening with fibroin ^{32}P-mRNA, 13 positive clones were identified[39].

Interest has centred on those clones containing the 5′ terminus of the gene and its flanking region for it is here that the sequences involved in the control of the initiation of transcription are likely to reside. The mRNA start site was localized by hybridizing 5′ end-labelled unmethylated mRNA to restriction fragments from this region[61]. S1 protection experiments[61] and cDNA sequencing[62] confirmed that the 5′ end of the mRNA was located at the second of two ATCAG repeats (the terminal nucleotide is labelled +1 in *Figure 22.5*). This is consistent with RNA sequencing data on the structure of the methylated cap (see p. 404)[61].

The polymerase II consensus promoter sequences identified for other protein coding genes[1] are also present in this gene. The Goldberg-Hogness or TATA box occurs at −30 to −27 and an upstream CAAT sequence is found at −88 to −85. The relevant 5′ flanking sequence is presented in *Figure 22.5*. Also shown is the position within this gene of the 970 bp intron. R-looping originally demonstrated the existence of a single major intervening sequence near the 5′ terminus of the fibroin gene[61] and its sequence later revealed that, in common with other introns, it conformed to the 5′ GT−AG 3′ rule governing intron/exon junctions[45].

This intron is present in cloned fibroin genes derived from both posterior silk gland DNA and total pupal DNA (the latter is a source of non-expressed genes). No differences have been observed in the restriction maps of genes from productive

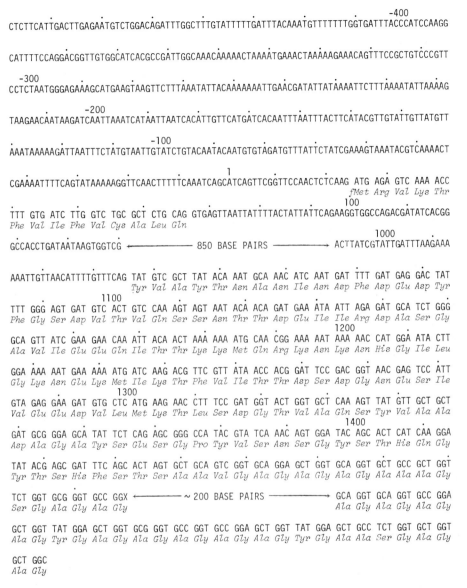

```
                                                              -400
CTCTTCATTGACTTGAGAATGTCTGGACAGATTTGGCTTTGTATTTTTGATTTACAAATGTTTTTTTGGTGATTTACCCATCCAAGG


CATTTTCCAGGACGGTTGTGGCATCACGCCGATTGGCAAACAAAAACTAAAATGAAACTAAAAAGAAACAGTTTCCGCTGTCCCGTT

    -300
CCTCTAATGGGAGAAAGCATGAAGTAAGTTCTTTAAATATTACAAAAAAATTGAACGATATTATAAAATTCTTTAAAATATTAAAAG

            -200
TAAGAACAATAAGATCAATTAAATCATAATTAATCACATTGTTCATGATCACAATTTAATTTACTTCATACGTTGTATTGTTATGTT

                        -100
AAATAAAAAGATTAATTTCTATGTAATTGTATCTGTACAATACAATGTGTAGATGTTTATTCTATCGAAAGTAAATACGTCAAAACT

                            1
CGAAAATTTTCAGTATAAAAAGGTTCAACTTTTTCAAATCAGCATCAGTTCGGTTCCAACTCTCAAG ATG AGA GTC AAA ACC
                                                                     fMet Arg Val Lys Thr
                                                                     100
TTT GTG ATC TTG GTC TGC GCT CTG CAG GTGAGTTAATTATTTTACTATTATTCAGAAGGTGGCCAGACGATATCACGG
Phe Val Ile Phe Val Cys Ala Leu Gln

                                                                   1000
GCCACCTGATAATAAGTGGTCG ←――――――― 850 BASE PAIRS ――――――→ ACTTATCGTATTGATTTAAGAAA


AAATTGTTAACATTTTGTTTCAG TAT GTC GCT TAT ACA AAT GCA AAC ATC AAT GAT TTT GAT GAG GAC TAT
                        Tyr Val Ala Tyr Thr Asn Ala Asn Ile Asn Asp Phe Asp Glu Asp Tyr
                    1100
TTT GGG AGT GAT GTC ACT GTC CAA AGT AGT AAT ACA ACA GAT GAA ATA ATT AGA GAT GCA TCT GGG
Phe Gly Ser Asp Val Thr Val Gln Ser Ser Asn Thr Thr Asp Glu Ile Ile Arg Asp Ala Ser Gly
                                                             1200
GCA GTT ATC GAA GAA CAA ATT ACA ACT AAA AAA ATG CAA CGG AAA AAT AAA AAC CAT GGA ATA CTT
Ala Val Ile Glu Glu Gln Ile Thr Thr Lys Lys Met Gln Arg Lys Asn Lys Asn His Gly Ile Leu

GGA AAA AAT GAA AAA ATG ATC AAG ACG TTC GTT ATA ACC ACG GAT TCC GAC GGT AAC GAG TCC ATT
Gly Lys Asn Glu Lys Met Ile Lys Thr Phe Val Ile Thr Thr Asp Ser Asp Gly Asn Glu Ser Ile
                    1300
GTA GAG GAA GAT GTG CTC ATG AAG AAC CTT TCC GAT GGT ACT GGT GCT CAA AGT TAT GTT GCT GCT
Val Glu Glu Asp Val Leu Met Lys Asn Leu Ser Asp Gly Thr Val Ala Gln Ser Tyr Val Ala Ala
                                                                    1400
GAT GCG GGA GCA TAT TCT CAG AGC GGG CCA TAC GTA TCA AAC AGT GGA TAC AGC ACT CAT CAA GGA
Asp Ala Gly Ala Tyr Ser Gln Ser Gly Pro Tyr Val Ser Asn Ser Gly Tyr Ser Thr His Gln Gly

TAT ACG AGC GAT TTC AGC ACT AGT GCT GCA GTC GGT GCA GGA GCT GGT GCA GGT GCT GCC GCT GGT
Tyr Thr Ser His Phe Ser Thr Ser Ala Ala Val Gly Ala Gly Ala Gly Ala Gly Ala Ala Ala Gly

TCT GGT GCG GGT GCC GGX ←―――― ~ 200 BASE PAIRS ―――――→ GCA GGT GCA GGT GCC GGA
Ser Gly Ala Gly Ala Gly                                 Ala Gly Ala Gly Ala Gly

GCT GGT TAT GGA GCT GGT GCG GGT GCC GGT GCC GGA GCT GGT TAT GGA GCT GCC TCT GGT GCT GGT
Ala Gly Tyr Gly Ala Gly Ala Gly Ala Gly Ala Gly Ala Gly Tyr Gly Ala Ala Ser Gly Ala Gly

GCT GGC
Ala Gly
```

Figure 22.5. Nucleotide sequence of the 5′ portion of the fibroin gene and its flanking region. 1 represents the putative initiation nucleotide of mRNA transcription. Amino acid residues are shown in italics where the protein coding sequence has been determined. Data from Tsujimoto and Suzuki[61, 62].

or non-productive cells, so gross DNA rearrangements cannot play a role in fibroin gene expression[61]. Moreover, the intron is transcribed to form part of a fibroin mRNA precursor[60]. The presence of introns in the RNA from posterior silk gland cells but not middle silk gland cells is further evidence for the control of fibroin biosynthesis at the transcriptional level[60].

Dissection of the promoter

Those sequences required for transcription of the fibroin gene *in vitro* have been studied by Tsujimoto *et al.*[63]. A whole cell extract derived from HeLa cells[32] was shown to be capable of supporting efficient and accurate transcription of the cloned fibroin gene[63]. By progressively deleting bases from the 5' or 3' directions, the boundaries of the 'promoter' sequence necessary for this transcription could be determined. Sequences downstream from −29 and upstream from +6 were essential. These results serve to emphasize the importance of the TATA box in *in vitro* expression systems. Deletion of the complete transcribable sequence of the gene leads to initiation at greatly reduced levels at a point approximately 30 bases downstream from the intact TATA box. This is consistent with the postulated role of the TATA box as a selector for transcription start sites[1]. Between 26 and 41 bp of non-specific sequence are also required upstream from the −29 deletion in order to get efficient, faithful transcription. The TATA box alone is thus not capable of directing initiation. Recently, 14 different single point mutants in the region of −30 to +1 of the fibroin DNA were obtained and tested for their transcriptional activities[20a]. The results indicated the importance of the TATA box and the −20 region.

Transcription efficiency in the HeLa extract was slightly less for the fibroin promoter than the Ad2 major late promoter. However, this situation was reversed when the fibroin gene was transcribed in a homologous cell-free system[59]. Extracts from posterior silk gland cells (active in fibroin synthesis) or middle silk gland cells (inactive) were both able to support faithful transcription of the cloned fibroin gene at levels greatly exceeding those displayed by the Ad2 major late or mouse β-globin promoters. Moreover, additional sequences upstream from the TATA box were required for this efficient expression of the fibroin gene. Recent studies have localized the regions responsible for this enhancement of transcription to the nucleotide positions −238 to −115 and −73 to −52 (M. Tsuda and Y. Suzuki, manuscript in preparation). Analyses of these upstream sequences may contribute to our understanding of the mechanisms by which the differential expression of eukaryotic promoters is brought about in this system.

Conclusions

In a period of 3–4 days, each fibroin gene in the posterior silk gland is transcribed 10000 times on average, with each mRNA going on to produce 100000 fibroin protein molecules[56]. Meanwhile, the fibroin genes of the middle silk gland are inactive; fibroin precursor or mature mRNA levels are below the limits of detection[31,60]. In contrast, the sericin genes are only active in middle silk gland cells.

How, then, is this differential expression accomplished in a single tissue? What factors facilitate the rapid transcription of the fibroin gene, the stabilization of the mRNA during the fifth instar and its efficient translation? Conversely, how is the fibroin gene switched off during the moulting stage and the message destabilized? The recent acquisition of cloned fibroin and sericin sequences will allow the unravelling of some of these sticky questions.

Acknowledgements

I am very grateful to Professor Y. Suzuki for his critical assessment of this manuscript and for supplying the unpublished data presented herein.

I should also like to thank Hilary and Caroline Gregory for their help in preparing the manuscript and June Webster for making an excellent job of the typing.

References

1. Breathnach, R. and Chambon, P. (1981). Organization and expression of eucaryotic split genes coding for protein. *Annual Review of Biochemistry*, **50**, 349–383
2. Chavancy, G. and Garel, J.-P. (1981). Does quantitative tRNA adaptation to codon content in mRNA optimize the ribosomal translation efficiency? Proposal for a translation system model. *Biochimie*, **63**, 187–195
3. Chavancy, G., Marbaix, G., Huez, G. and Cleuter, Y. (1981). Effect of tRNA pool balance on rate and uniformity of elongation during translation of fibroin mRNA in a reticulocyte cell-free system. *Biochimie*, **63**, 611–618
4. Chevallier, A. and Garel, J.-P. (1979). Studies on tRNA adaptation, tRNA turnover, precursor tRNA and tRNA gene distribution in *Bombyx mori* by using two-dimensional polyacrylamide gel electrophoresis. *Biochimie*, **61**, 245–262
5. Couble, P., Garel, A. and Prudhomme, J.-C. (1981). Complexity and diversity of polyadenylated mRNA in the silk gland of *Bombyx mori*: Changes related to fibroin production. *Developmental Biology*, **82**, 139–149
6. Fitzgerald, M. and Shenk, T. (1981). The sequence 5'-AAUAAA-3' forms part of the recognition site for polyadenylation of late SV40 mRNAs. *Cell*, **24**, 251–260
7. Fournier, A. (1979). Quantitative data on the *Bombyx mori* silkworm: A review. *Biochimie*, **61**, 283–320
8. Gage, L. P. (1974). The *Bombyx mori* genome: Analysis by DNA reassociation kinetics. *Chromosoma*, **45**, 27–42
9. Gage, L. P. (1974). Polyploidization of the silk gland of *Bombyx mori*. *Journal of Molecular Biology*, **86**, 97–108
10. Gage, L. P. and Manning, R. F. (1976). Determination of the multiplicity of the silk fibroin gene and detection of fibroin gene-related DNA in the genome of *Bombyx mori*. *Journal of Molecular Biology*, **107**, 327–348
11. Gage, L. P. and Manning, R. F. (1980). Internal structure of the silk fibroin gene of *Bombyx mori*, I. The fibroin gene consists of a homogeneous alternating array of repetitious crystalline and amorphous coding sequences. *Journal of Biological Chemistry*, **255**, 9444–9450
12. Gamo, T. (1982). Genetic variants of the *Bombyx mori* silkworm encoding sericin proteins of different lengths. *Biochemical Genetics*, **20**, 165–177
13. Gamo, T., Inokuchi, T. and Laufer, H. (1977). Polypeptides of fibroin and sericin secreted from the different sections of the silk gland in *Bombyx mori*. *Insect Biochemistry*, **7**, 285–295
14. Garel, J.-P. (1982). The silkworm, a model for molecular and cellular biologists. *Trends in Biochemical Sciences*, **7**, 105–108
15. Garel, J.-P., Mandel, P., Chavancy, G. and Daillie, J. (1970). Functional adaptation of tRNAs to fibroin biosynthesis in the silk gland of *Bombyx mori*. *FEBS Letters*, **7**, 327–329
16. Garel, J.-P., Mandel, P., Chavancy, G. and Daillie, J. (1971). Functional adaptation of tRNAs to protein biosynthesis in a highly differentiated cell system. III. Induction of isoacceptor tRNAs during the secretion of fibroin in the silk gland of *Bombyx mori*. *FEBS Letters*, **12**, 249–252
17. Garel, J.-P., Garber, R. L. and Siddiqui, M. A. Q. (1977). Transfer RNA in posterior silk gland of *Bombyx mori*: polyacrylamide gel mapping of mature transfer RNA, identification and partial structural characterization of major isoacceptor species. *Biochemistry*, **16**, 3618–3624
18. Greene, R. A., Morgan, M., Shatkin, A. J. and Gage, L. P. (1975). Translation of silk fibroin messenger RNA in an Ehrlich ascites cell-free extract. *Journal of Biological Chemistry*, **250**, 5114–5121
19. Hagenbüchle, O., Krikeles, M. S. and Sprague, K. U. (1979). The nucleotide sequence adjacent to poly(A) in silk fibroin messenger RNA. *Journal of Biological Chemistry*, **254**, 7157–7162

20. Hentzen, D., Chevallier, A. and Garel, J.-P. (1981). Differential usage of iso-accepting transfer RNA^{Ser} species in silk glands of *Bombyx mori*. *Nature, London*, **290**, 267–269

20a. Hirose, S., Takeuchi, K. and Suzuki, Y. (1982). *In vitro* characterization of the fibroin gene promoter by the use of single-base substitution mutants. *Proceedings of the National Academy of Sciences, USA*, **79**, 7258–7262

21. Hyodo, A. and Shimura, K. (1980). The occurrence of hereditary variants of fibroin in the silkworm *Bombyx mori*. *Japanese Journal of Genetics*, **55**, 203–209

22. Hyodo, A., Gamo, T. and Shimura, K. (1980). Linkage analysis of the fibroin gene in the silkworm *Bombyx mori*. *Japanese Journal of Genetics*, **55**, 297–300

23. Lim, V. I. and Steinberg, S. V. (1981). A novel structural model for silk fibroin: $\alpha_L \alpha_R$ β-structure. *FEBS Letters*, **131**, 203–207

24. Lizardi, P. M. and Brown, D. D. (1975). The length of the fibroin gene in the *Bombyx mori* genome. *Cell*, **4**, 207–215

25. Lizardi, P. M., Williamson, R. and Brown, D. D. (1975). The size of fibroin messenger RNA and its polyadenylic acid content. *Cell*, **4**, 199–205

26. Lizardi, P M., Madhvai, V., Shields, D. and Candelas, G. (1979). Discontinuous translation of silk fibroin in a reticulocyte cell-free system and in intact silk gland cells. *Proceedings of the National Academy of Sciences, USA*, **76**, 6211–6215

27. Lotz, B. and Colonna Cesari, F. (1979). The chemical structure and the crystalline structures of *Bombyx mori* silk fibroin. *Biochimie*, **61**, 205–214

28. Lotz, B., Gonthier-Vassel, A., Brack, A. and Magoshi, J. (1982). Twisted single crystals of *Bombyx mori* fibroin and related model polypeptides with β structure. A correlation with the twist of the β sheets in globular proteins. *Journal of Molecular Biology*, **156**, 345–357

29. Lucas, F. and Rudall, K. M. (1968). Extracellular fibrous proteins: the silks. In *Comprehensive Biochemistry*, Volume 26, Part B (Florkin, M. and Stotz, E. H., Eds.), pp. 475–558. Amsterdam, London, New York, Elsevier

30. Lucas, F., Shaw, J. T. B. and Smith, S. G. (1958). The silk fibroins. *Advances in Protein Chemistry*, **13**, 107–242

31. Maekawa, H. and Suzuki, Y. (1980). Repeated turn-off and turn-on of fibroin gene transcription during silk gland development of *Bombyx mori*. *Developmental Biology*, **78**, 394–406

32. Manley, J. L., Fire, A., Cano, A., Sharp, P. A. and Gefter, M. L. (1980). DNA-dependent transcription of adenovirus genes in a soluble whole-cell extract. *Proceedings of the National Academy of Sciences, USA*, **77**, 3855–3859

33. Manning, R. F. and Gage, L. P. (1978). Physical map of the *Bombyx mori* DNA containing the gene for silk fibroin. *Journal of Biological Chemistry*, **253**, 2044–2052

34. Manning, R. F. and Gage, L. P. (1980). Internal structure of the silk fibroin gene of *Bombyx mori*. II. Remarkable polymorphism of the organization of crystalline and amorphous coding sequences. *Journal of Biological Chemistry*, **255**, 9451–9457

35. Meza, L., Araya, A., Leon, G., Krauskopf, M., Siddiqui, M. A. Q. and Garel, J.-P. (1977). Specific alanine-tRNA species associated with fibroin bisoynthesis in the posterior silk gland of *Bombyx mori*. *FEBS Letters*, **77**, 255–260

36. Morimoto, T., Matuura, S., Nagata, S. and Tashiro, Y. (1968). Studies on the posterior silk gland of the silkworm, *Bombyx mori*. III. Ultrastructural changes of posterior silk gland cells in the fourth larval instar. *Journal of Cell Biology*, **38**, 604–614

37. Nakajima, B., Hirata, K., Nishi, N. and Noguchi, J. (1982). Silk fibroin model, Poly(L-alanylglycyl–L-alanylglycyl–L-alanylglycyl–L-serylglycine). *International Journal of Biological Macromolecules*, **3**, 46–52

38. Nunome, J. (1937). Development of the silk gland in *Bombyx mori*. *Bulletin of Applied Zoology, Japan*, **9**, 68–92

39. Ohshima, Y. and Suzuki, Y. (1977). Cloning of the silk fibroin gene and its flanking sequences. *Proceedings of the National Academy of Sciences, USA*, **74**, 5363–5367

40. Okamoto, H., Ishikawa, E. and Suzuki, Y. (1982). Structural analysis of sericin genes: homologies with fibroin gene in the 5′ flanking nucleotide sequences. *Journal of Biological Chemistry*, **257**, 15192–15199

41. Pearson, W. R., Mukai, T. and Morrow, J. F. (1981). Repeated DNA sequences near the 5′-end of the silk fibroin gene. *Journal of Biological Chemistry*, **256**, 4033–4041

42. Prudhomme, J. C. and Couble, P. (1979). The adaptation of the silk gland cell to the production of fibroin in *Bombyx mori*. *Biochimie*, **61**, 215–227

43. Rasch, E. M. (1974). The DNA content of sperm and hemocyte nuclei of the silkworm, *Bombyx mori* L. *Chromosoma*, **45**, 1–26

44. Sasaki, S., Nakajima, E., Fujii-Kuriyama, Y. and Tashiro, Y. (1981). Intracellular transport and secretion of fibroin in the posterior silk gland of the silkworm *Bombyx mori*. *Journal of Cell Science*, **50**, 19–44

45. Sharp, P. A. (1981). Speculations on RNA splicing. *Cell*, **23**, 643–646

46. Shatkin, A. J. (1976). Capping of eukaryotic mRNAs. *Cell*, **9**, 645–654

47. Smith, G. P. (1976). Evolution of repeated DNA sequences by unequal crossover. *Science*, **191**, 528–535

48. Sprague, K. U. (1975). The *Bombyx mori* silk proteins: characterization of large polypeptides. *Biochemistry*, **14**, 925–931

49. Sprague, K. U., Roth, M. B., Manning, R. F. and Gage, L. P. (1979). Alleles of the fibroin gene coding for proteins of different lengths. *Cell*, **17**, 407–413

50. Suzuki, Y. (1977). Differentiation of the silk gland. In *Results and Problems in Cell Differentiation*, Volume 8, (Beerman, W., Ed), pp. 1–44. Berlin and Heidelberg, Springer-Verlag

51. Suzuki, Y. (1982). Studies of fibroin gene transcription by *in vitro* genetics. In *Embryonic Development*: Proceedings of the IX Congress of the International Society of Developmental Biologists, (Burger, M. and Weber, R., Eds.), pp. 305–325. New York, Alan R. Liss Inc.

52. Suzuki, Y. and Brown, D. D. (1972). Isolation and identification of the messenger RNA for silk fibroin from *Bombyx mori Journal of Molecular Biology*, **63**, 409–430

53. Suzuki, Y. and Giza, P. E. (1976). Accentuated expression of silk fibroin genes *in vivo* and *in vitro*. *Journal of Molecular Biology*, **107**, 183–206

54. Suzuki, Y. and Ohshima, Y. (1977). Isolation and characterization of silk fibroin gene with its flanking sequences. *Cold Spring Harbor Symposium on Quantitative Biology*, **42**, 947–957

55. Suzuki, Y. and Suzuki, E. (1974). Quantitative measurements of fibroin messenger RNA synthesis in the posterior silk gland of normal and mutant *Bombyx mori*. *Journal of Molecular Biology*, **88**, 393–407

56. Suzuki, Y., Gage, L. P. and Brown, D. D. (1972). The genes for silk fibroin in *Bombyx mori*. *Journal of Molecular Biology*, **70**, 637–649

57. Tashiro, Y., Morimoto, T., Matsuura, S. and Nagata, S. (1968). Studies on the posterior silk gland of the silkworm, *Bombyx mori*. I. Growth of posterior silk gland cells and biosynthesis of fibroin during the fifth larval instar. *Journal of Cell Biology*, **38**, 574–588

58. Thomas, C. and Brown, D. D. (1976). Localization of the genes for silk fibroin in silk gland cells of *Bombyx mori*. *Developmental Biology*, **49**, 89–100

59. Tsuda, M. and Suzuki, Y. (1981). Faithful transcription initiation of fibroin gene in a homologous cell-free system reveals an enhancing effect of 5′ flanking sequence far upstream. *Cell*, **27**, 175–182

60. Tsuda, M., Ohshima, Y. and Suzuki, Y. (1979). Assumed initiation site of fibroin gene transcription. *Proceedings of the National Academy of Sciences, USA*, **76**, 4872–4876

61. Tsujimoto, Y. and Suzuki, Y. (1979). Structural analysis of the fibroin gene at the 5′ end and its surrounding regions. *Cell*, **16**, 425–436

62. Tsujimoto, Y. and Suzuki, Y. (1979). The DNA sequence of *Bombyx mori* fibroin gene including the 5′ flanking mRNA coding, entire intervening and fibroin protein coding regions. *Cell*, **18**, 591–600

63. Tsujimoto, Y., Hirose, S., Tsuda, M. and Suzuki, Y. (1981). Promoter sequence of fibroin gene assigned by *in vitro* transcription system. *Proceedings of the National Academy of Sciences, USA*, **78**, 4838–4842

64. Yang, N. S., Manning, R. F. and Gage, L. P. (1976). The blocked and methylated 5′ terminal sequence of a specific cellular messenger: the mRNA for silk fibroin of *Bombyx mori*. *Cell*, **7**, 339–347

Chapter 23

Keratin genes

G. E. Rogers

Introduction

The keratins are intracellular structural proteins and constitute the main products of the activities of vertebrate cells in the epidermis and its appendages such as feather, hair, wool and horn[5].

Avian and mammalian keratins have been studied the most. The application of protein chemical methods involving breakage of the disulphide bonds in the presence of denaturants, separation of individual polypeptide chains and, where possible, amino acid sequencing, has for many years revealed a high degree of polypeptide variability explicable mainly on the basis of an underlying genetic complexity and the existence of families of genes[15]. The feather keratin family consists of proteins all closely related (homologous) in primary structure. In the more complex example of wool and hair (mammalian keratins) the protein families consist of sub-families, each sub-family being a group of homologous polypeptide chains. The human epidermal keratins also appear to be a family of sequence-related polypeptide chains although the degree of relatedness is not yet clear (see below).

The methods of protein chemistry are limited when applied to keratin protein families so that the details of their genetic relatedness and evolution will only become clear when the keratin gene-containing regions in the DNA of different animal species are selected by recombinant DNA methods and submitted to detailed study by DNA sequencing and allied mapping procedures. Practically no information on these aspects of keratins have yet appeared in the literature so that what is presented below was unpublished at the time of writing.

Avian keratins

There are two main families of avian keratins, the group which constitute feathers and that found in the cells of leg and foot scales. Feather keratins have polypeptide chains of molecular weight $10\,500$[13] whereas those of the scales are $14\,500$[32]. Both families form approximately 3 nm diameter filaments[4, 29]. The complete amino acid sequences of two species of bird feather keratin have been published[21] as well as partial sequences for chick feather keratin[33]. The amino acid sequences of chick feather and scale keratin are distinct (*Figure 23.1*) and include an unusual tripeptide repeat found in the scale protein which essentially accounts for the molecular size difference between the scale and feather families[32]. Both feather keratins[15] and scale keratins (S. Wilton and G. Rogers, unpublished observations) rapidly appear after 13 days of embryonic chick development and there is a massive commitment to their continued synthesis (*Figure 23.2*) until around 19 days. The evidence[16] indicates that the homologous chains are all coordinately expressed.

Feather	N-acetyl	*Ser*	*Cys*	*Phe*	*Asp*	*Leu*	*Cys*	*Arg*	*Pro-*	*-Cys*	*Gly*	*Pro-*		*-Thr*	*Pro*
Scale	N-acetyl	*Ser*	*Cys*	*Ser*	*Ser*	*Leu*	*Cys*	*Ala*	*Pro*	*Ala*	*Cys*	*Gly*	*Val*	*Ala*	*Thr*	*Pro*

Feather	*Leu*	*Ala*	*Asn-*	*-Cys*	*Asx*	*Glx*	*Pro*	*Cys*	*Val*	*Arg*	*Gln*	*Cys*	*Gln*	*Asn*	*Ser*	*Arg*
Scale	*Leu*	*Ala*	*Asx*	*Ser*	*Cys*	*Asx*	*Glx*	*Pro*	*Cys*	*Val*	*Arg*[26]	––––––––––––––––––––

Feather	*Val* *Val* *Ile* *Glx* *Pro* *Ser*[35–84,] ––––––––––––––––––––––––––––––––––––––
Scale	–––––––––––––––––––– ––––––––––––––––––––––––––––––––––––

Feather	––––––– [85]*LeuSer* *Gly* *Arg* *Phe-* *-Cys* *Gly* *Arg* *Arg* *Cys* *Leu* *Pro* *Cys-* OH
Scale	–––––––––––––––––– *Arg* *Tyr* *Leu* *Ser* *Gly-* - *-Cys* *Gly* *Pro* *Cys-* OH
Scale clone (pCSK12)	*Arg* *Tyr* *Arg* *Arg* *Gly* *Ser-* *-Cys* *Gly* *Pro* *Cys* OH

Figure 23.1. Partial N and C terminal amino acid sequences of chick feather and chick scale keratins described by Walker and Rogers[33] and Walker and Bridgen[32] arranged to demonstrate their similarities. Large gaps of sequence, approximately 50 residues for feather and 100 residues for scale (including the Gly-Gly-X repeat), are being completed via DNA sequencing (see *Figure 23.3, 23.6* and *23.8*). The C terminal sequence derived from scale clone pCSK12 (text, p. 421 and *Figure 23.8*) and included for comparison, differs sufficiently for it to have come from a low abundance mRNA species and not detected by direct protein sequencing.

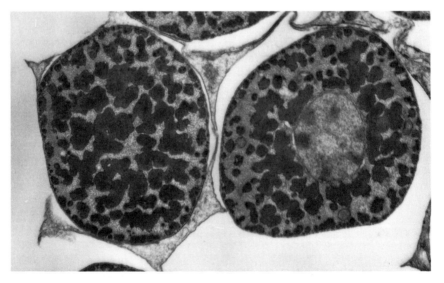

Figure 23.2. Electron micrograph of a cross section of two embryonic feather barb cells at 15 days. The ribosomal cytoplasm of both cells (one with central nucleus) is packed with aggregates of keratin which are long rod-like structures in cross section. The magnification is too low to visualize the 3 nm diameter keratin filaments. Parts of flattened fibroblasts are seen around the barb cells. Magnification ×6000.

The 12S mRNA for feather keratin was isolated[22] and molecular hybridization studies with this mRNA (molecular weight 240 000; 800 bases long) and its cDNA copy[14] suggested that the complexity of the mRNA was equivalent to around 35 species and was expressed in embryonic feather cells from perhaps a total complement of 100–200 genes in the genome. Although it was postulatd that the keratin genes for feather might be clustered as a result of gene duplication[18], this origin of multiple feather (and scale) keratin proteins is only now becoming apparent from molecular genetic studies.

Feather keratins

A cDNA library has been prepared from 14 day embryonic feather keratin mRNA and two of the clones selected using total embryonic feather keratin cDNA as a probe have been totally sequenced and identified as feather keratin (R. Saint, J. Crowe, C. Morris and G. Rogers, in preparation). These clones contain only part of the coding region and the sequence of one of them, pCFK23, is shown in *Figure 23.3*. The first indication that feather keratin genes are clustered was obtained when two genomic clones, λCFK1 and λCFK2, were selected from a chick genomic

```
TCC CGT GTG GTG ATC CAG CCC TCT CCC GTG GTG GTG ACC CTG CCT GGT CCC ATC CTC AGC
ser arg val val ile gln pro ser pro val val val thr leu pro gly pro ile leu ser

TCC TTC CCC CAG AAC ACA GCT GTG GGA TCC AGC ACC TCT GCT GCT GTT GGC AGC ATC CTG
ser phe pro gln asn thr ala val gly ser ser thr ser ala ala val gly ser ile leu

AGC CAA GAG GGA GTA CCC ATC TCT TCT GGA GGC TTT GGC TTC TCT GGC CTA GGT GGC CGG
ser gln glu gly val pro ile ser ser gly gly phe gly phe ser gly leu gly gly arg

TTC TCT GGC AGG AGG TGC CTG CCA TGC TAA AGCCAAGGTGAACGTCCTCTGAGCGCATCCCAGTGATGC
phe ser gly arg arg cys leu pro cys STOP

TCGAGCCAGCACCAGACTGAGGATGTAGCTGCTGGCCGGGGTTTCCGGATGGGCTGACCACCCTCCTGCCCTCCTGCAA


AGCAGAGAGGGAAGCCAGGGTGCCAGCCTGTGCTGTCTGGAAACACAGCCAGCAAACATCTTCTTCTCCTGCTTTCTTC


TCATCATCACAAGTCTTCGTTGTCTCGTTGCTGTCCTGTGCCATGGGTTCATCCTGAAGCAAGCTGAGAGGGCCCTACTT


CTTCCTCTCGCCACATGAGGGAGGAAGACTCGCACATCCTATTATGCAGTTGCAGCTGATGCCAATCTGTTTAACAGCT


GCCTTGTAGCAGCTTTAAACTATGCGCTGCTTTGATTCTTCTTTAGATTCAATAAAATTTATGCTGCATTGTAATCTCA21
```

Figure 23.3. Nucleotide sequence of the cDNA clone pCFK23 consisting of 207 bases of coding for a chick feather keratin and the 3' non-coding region of about 440 bases. This cDNA copy was incomplete so that the derived protein sequence is missing about 29 amino acid residues at the N terminus (see *Figure 23.1*). The polyadenylation sequence in the 3' non-coding region is underlined and the polyadenylate tail at the 3' end of the original mRNA is noted.

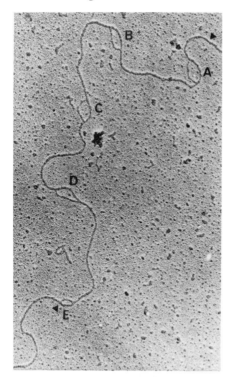

Figure 23.4. R loops produced by hybridizing the chick genomic clone λCFK1 (insert length 15.5 kb) with feather keratin mRNA. Five genes, A–E, are present each about 700 bp except for E which is smaller. It is now known that gene E is represented in the cloned fragment as only the 3' end of the gene (see *Figure 23.5*), The five feather keratin genes are regularly spaced by an intergenic distance of 2.6 kb but it is not known whether this feature is conserved in the total cluster, the extent of which is also unknown. From Molloy *et al.*[20], with permission.

library (prepared by Dr J. Dodgson and Dr J. Engel and donated by Dr T. Maniatis) with total cDNA. When λCFK1 was R-looped with embryonic feather keratin mRNA an array of five genes, each about 0.7 kb, was visualized with intergenic distances consistently of 2.6 kb (*Figure 23.4*). Subsequent mapping by Southern blot hybridization with the cDNA probe was performed under conditions allowing cross-hybridization to occur and the five genes, A–E, have been precisely located (*Figure 23.5*). The only gene to be totally sequenced so far is gene C[20] but some data has also been obtained for the others. The sequence for gene C from the cap to the 3' terminus is given in *Figure 23.6* and for comparison the partial coding region for an expressed gene contained in pCFK23 (*Figure 23.3*) is included. It can

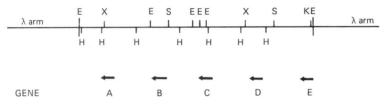

Figure 23.5. The positions of the feather keratin genes A–E (*Figure 23.4*) in relation to restriction sites in the 15.5 kb insert of λCFK1 genomic clone, determined by Southern blot hybridization. The direction of transcription for genes B–E runs consistently E → B as shown. The restriction sites are denoted E, Eco RI; X, Xba I; S, Sal I; K, Kpn I; H, Hind III. From Molloy *et al.*[20], with permission.

be seen from these examples that there is a high degree of sequence conservation with changes mainly limited to the third base. There are no introns (intervening sequences) within the coding sequence of gene C. Further, the conceptual amino acid sequence is directly equivalent to the known sequences[21] (see *Figure 23.1*) so that in accord with earlier findings from the translation of feather keratin mRNA in heterologous systems, there is no possibility in this instance of there being a keratin precursor, a pre-keratin (or pro-keratin).

```
                                                                                        125
ATCCACTTCTCTTGCCTTCTCCTCCTTGGTGAACAAGGTGAGCTGCTGTGGCTTTCTTCTCACTC
```

```
                                                                                        250
TTGCTCTGCTTCTCCTCTTGCTCTTTGTGCCTCCATTAAGTCTTTGCCTTTGTGCAGACCTTGGTATGAGCCTTGCTCCTTGTTGTGCTCCCACCAGCTCTGTTCTGCTTGCCAGTGTGAGGTGT
```

```
                                                                                        375
GGTGGGAGAAGGCCTTTGGCTGACTCTACTGGTGATTCTGGGGATCTGGGCTCGCCAAGCCTCTGTCCCTGCTCCGCCTGTCCCTGATCCCCGGCCTGTGCCTCCTGTCTTGGCAATCAGGCCAC
```

```
                                                                                        484
CAGGCTACTTCTCACGCAGCCTGTGCTTTTCTTGTCCTCTCTCCAGGTCTACTCCCATCCTACAGCC ATG TCC TGC TTC GAT CTG TGC CGT CCC TGT GGC CCG ACC CCA
                                                                        met ser cys phe asp leu cys arg pro cys gly pro thr pro
```

```
                                                                                                            577
CTG GCC AAC AGC TGC AAC GAG CCC TGT GTG CGC CAG TGC CAG GAC TCC CGG GTG GTG ATC CAG CCC TCT CCC GTG GTG GTG ACC CTG CCT GGA
                                                                                            ACC CTG CCT GGT
leu ala asn ser cys asn glu pro cys val arg gln cys gln asp ser arg val val ile gln pro ser pro val val val thr leu pro gly
```

```
                                                                                                            670
CCC ATC CTC AGC TCC TTC CCC CAG AAC ACC GCT GCG GGC TCC AGC ACC TCT GCT GCT GTT GGC AGT ATC CTG AGT GAG GAG GGA GTG CCC ATC
CCC ATC CTC AGC TCC TTC CCC CAG AAC ACA GCT GTG GGA TCC AGC AGC ATC CTG AGC CAA GAG GGA GTA CCC ATC
pro ile leu ser ser phe pro gln asn thr ala ala gly ser ser thr ala ala val gly ser ile leu ser glu glu gly val pro ile
                                          val                                                                  gln
```

```
                                                                                                            772
TCC TCT GGT GGC TTT GGC ATC TCT GGC CTG GGT AGC CGC TTC TCT GGC AGG AGG TGT CTG CCC TGC TAA GGACGAGGTGTTCATCCCATGGATGCATCCTCA
TCT TCT GGA GGC TTT GGC TTC TCT GGC CTA GGT GGC CGG TTC TCT GGC AGG AGG TGC CTG CCA TGC TAA
ser ser gly gly phe gly ile ser gly leu gly ser arg phe ser gly arg arg cys leu pro cys STOP
                      phe                    gly
```

```
                                                                                        897
GGAAACCCAAAGCTTGGTGCTGGACTGCTGACTGAGCTTCTGAGCAGGATCCACTGAGCACCCTCCTGCTCTCCTGCAAAGCAAAGAGGGAATTCAAGTTGCCAGCCTGTGCTGCCTGTAGACAC
```

```
                                                                                        1022
AGACAGCAGCTGTCTTCTTCTTTTCCTTCTTTCTTTCATCATTAGGGGTTCTTGTTGTGTGTCCTTTGTCCTGTGCCCTGGGTTCATCCTGAAGCAAGTTGAGATGGCCCTGCTTCTTCTTCCACTT
```

```
                                                                                        1146
GTCTTGTGATGGGGAAGACATGCATCCCATCTTCCTGTAGTTTCCCTCCTTATGGCCAATATGTTTGCCAGCTGTATTTGTAGCAGCCTTTAACTTTCACTTTTTTGATTCCTTCTTTGAGCTC
```

```
            1200
AATAAAATTTATGCTGCATTGTAATCTCAGTCTCCTCATGTTTCTTGCCTCATA
```

Figure 23.6. Nucleotide sequence of chick feather keratin gene C of the genomic clone λCFK1 beginning at the 'cap' site. The 5′ non-coding region (underlined, bold type) is divided by the intron (lighter type). The coding sequence has the conceptual amino acid sequence (97 amino acids) beneath (*cf. Figure 23.1* and reference 21). For comparison, the nucleotide sequence of the coding region from the cDNA clone pCFK23 (*Figure 23.3*) is included. Base changes are underlined and amino acid changes are given below the gene C sequence. The 3′ non-coding region, approximately 440 bases, runs down to the putatitive 3′ terminus of the mRNA transcript (end of bold type). The polyadenylation of sequence near the terminus is underlined. From Molloy *et al.*[20], with permission.

Of special note is the presence of a 320 bp intron in the 5′ non-coding region. Preliminary data indicate that there are introns not only similarly located in all five genes, but that they are closely related in sequence. There are no other examples so far in the literature in which introns are present solely in the 5′ non-coding region. The significance of their presence and their high degree of sequence conservation are of great interest as is the finding that the 3′ non-coding sequences of different feather keratin genes have zones which are highly conserved (*Figure 23.7a*). The intergenic regions of λCFK1 have been partly sequenced and TATAA sequences and other putative control regions have been found upstream from the cap site suggesting that each structural gene has its own complement of promoter-type signals (data not shown). Other signals such as the polyadenylation site AATAAA and transcription stops are present on the 3′ side.

From all the genomic and cDNA sequence data an overall picture of feather keratin mRNA structure can be derived (*Figure 23.7b*). However, major questions remain, including how many genes there are altogether and whether they are one cluster or several. Certainly the gene comparisons possible so far have substantiated the view that gene duplication has occurred.

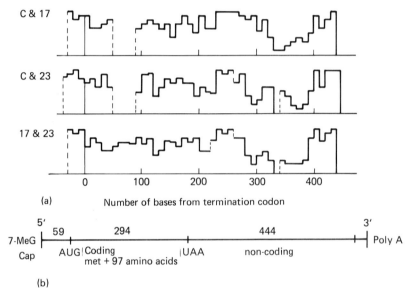

(a)   Number of bases from termination codon

(b)

Figure 23.7(a). Representation of the conservation of zones in the 3′ non-coding region of feather keratin mRNAs. Three cloned sequences, of gene C and of the cDNA clones pCFK23 and pCFK17 (data not detailed) are compared in groups of 10 nucleotides. Maxima in the histogram indicating complete identity, occur at the 3′ terminus and at around 150 nucleotides from it. Intermediate levels of homology are to be seen elsewhere. (b) Structure of feather keratin mRNA based mainly on sequence data from genomic and cDNA clones. The 5′-capped and 3′-polyadenylated mRNA has a 59 base 5′ non-coding region and a coding sequence significantly shorter than the 3′ non-coding region. The total length of the mRNA would be 847 bases, including 50 bases in the polyadenylate tail. This compares favourably with estimate by gel electrophoresis (see text).

Scale keratins

The examination of the genes for this family is not as advanced as for feather. An RNA fraction enriched in mRNA approximately 1000 bases long, that translates *in vitro* to yield only scale keratin, has been copied, the cDNA cloned into pBR322 and several clones, selected using total cDNA, have been sequenced (data not shown, S. Wilton and G. Rogers, in preparation). Preliminary results on genomic clones selected by cloned scale keratin cDNA sequences indicate that introns are not present in the coding region and, if any are present in the 5' region of scale genes they are not homologous with the 5' intron of feather genes. The location of scale genes in the genome in relation to those of feather is not known.

Sequencing of the coding region of one of the scale genes (*Figure 23.8*) has yielded the complete amino acid sequence of a scale keratin for the first time.

ATGTCCTGCTATGACCTGTGCCCACCAACATCGTGCATCAGCCGCCCCCAGCCCATCGCTGAC

| MET | SER | CYS | TYR | ASP | LEU | CYS | PRO | PRO | THR | SER | CYS | ILEU | SER | ARG | PRO | GLN | PRO | ILEU | ALA | ASP |

AGTGGGAATGAGCCATGCGTCCGACAGTGCCCTGACTCCACAACTGTGATCCAGCCACCTCCT

| SER | GLY | ASN | GLU | PRO | CYS | VAL | ARG | GLN | CYS | PRO | ASP | SER | THR | THR | VAL | ILEU | GLN | PRO | PRO | PRO |

GTTGCCGTCACCTTCCCTGGACCCATCCTCAGCTCCTTCCCCCAGGATTCAGTTGTGGGATCC

| VAL | ALA | VAL | THR | PHE | PRO | GLY | PRO | ILEU | LEU | SER | SER | PHE | PRO | GLN | ASP | SER | VAL | VAL | GLY | SER |

TCTGGAGCACCCATCTTT(GGNGGCTCCTCCCTGGGCNNNNGGGGCCTGTATGGCTAT)5GGT

| SER | GLY | ALA | PRO | ILEU | PHE | GLY | GLY | SER | SER | LEU | GLY | X | X | GLY | LEU | TYR | GLY | TYR | GLY |

AGATCCTATGGTTCTGGCTACTGCAGCCCTTACTCCTACCGG TACAACAGGTACCGCCGTGGC

| ARG | SER | TYR | GLY | SER | GLY | TYR | CYS | SER | PRO | TYR | SER | TYR | ARG | TYR | ASN | ARG | TYR | ARG | ARG | GLY |

AGCTGCGGGCCCTGCTAA

| SER | CYS | GLY | PRO | CYS |

Figure 23.8. The complete nucleotide sequence of the coding region of an avian scale gene isolated from the chick genome. The conceptual amino acid sequence is also shown. Normally the initiating Met residue is removed and the N terminal Ser residue is acetylated. After nucleotide 198 there is the 39-nucleotide sequence written as a consensus sequence that is contiguously repeated five times. The amino acid sequence of this region is rich in glycine residues and is presumed to be the Gly-Gly-X repeat region described in scale keratin by Walker and Bridgen[32]. The short nucleotide sequences denoted by --- are inverted repeats within each of the 39-base repeats. These 12 bp segments could form large loops of secondary structure by annealling with equivalent segments in adjacent 39-base repeats (K. Gregg, S. Wilton and G. Rogers, submitted).

Notable in the nucleotide sequence is the presence of a region of five contiguous repeats each of 39 bases. These give rise to repeats of 13 amino acids in the scale keratin molecule, three of which are *identical* and two others are very similar. This appears to be the region identified by Walker and Bridgen (1976) as the Gly-Gly-X rich region. Comparison of gene sequences of scale and feather (*Figures 23.6* and *23.8*) reveals a striking homology. The repeating sequence constitutes the major difference. The findings support the notion that both avian scales and feathers evolved from a primordial scale structure (K. Gregg, S. Wilton and G. Rogers, submitted).

Mammalian keratins

Hair and wool keratins

The keratins of hair and wool consist essentially of three families of proteins, each family having distinct molecular properties of molecular size and amino acid composition (*Table 23.1*). Thus, the low-sulphur family consists of proteins with molecular weights 38 000–58 000 and a much lower cystine content than the high-sulphur group with molecular weights of 11 000 to about 20 000. The third family of high-glycine/tyrosine proteins enriched in these two amino acids have molecular weights varying from 7000–10 000. Sub-families exist and some complete amino acid sequences are now known[1,2] including one for the low-sulphur proteins (W. G. Crewther, personal communication).

TABLE 23.1. The classes of proteins constituting wool and hair keratin[a]

Protein class name	Protein families identified within the class[b]	Designation[c]	Molecular weight (daltons)	Approx. half-cystine (moles %)	Sedimentation value of equivalent mRNA
Low-sulphur	Component 5 (1)	SCMKA-5	56 000	4	
	Component 7 (3)	SCMKA-7a, 7b, 7c	58 000	6	20
	Component 8 (4)	SCMKA-8a, 8b		6	
		8c-1, 8c-2	38–43 000	6	
High-sulphur	Group 1	SCMKB-IIIB	11 000	16	9
	Group 2 (>20)	SCMKB-IIIA	16 000	24	12
	Group 3	SCMKB-2A, 2B, 2C	19 000	21	13
High-Gly/Tyr	Type I (>10)		6–9 000	6	7
	Type II (?)		6–9 000	10	

[a] Data mostly from Crewther[1] and Swart et al.[30].
[b] The proteins of each of these families bear homologies between the families as well as within a class. The numbers in parentheses are the numbers of individual chains in each family. They are approximate and minimal values, but indicate the complexity of these proteins.
[c] The prefix SCMK indicates that the proteins are S-carboxymethylated derivatives of the sulphydryl (reduced) form of the protein.

The proteins are synthesized in the keratinizing zone of the hair follicle. The low-sulphur proteins have a tertiary structure which is highly α-helical. They assemble into 8 nm diameter filaments which package together with the inclusion of the proteins of the other two families to form the large fibrous aggregates (fibrils) characteristic of hair and wool cortical cells[5]. Changes in the regulatory processes controlling the transcription of the gene families and the translational rates of their mRNAs must account for the variations in the relative amounts of individual keratin proteins found in different species and in altered states produced genetically, by the environment or disease. For example, the content of high-glycine/tyrosine proteins is markedly lowered in the 'Naked' mouse mutation[31]. Likewise, between two strains of sheep they can differ by more than ten-fold and in man, hair and nail are polymorphic with respect to the low-sulphur and high-sulphur proteins[11]. Nutritional states can also affect the regulatory mechanisms, one well-documented instance being the lowering of the cystine content of hair in the protein deficiency disease Kwashiorkor[6]. Virtually nothing is known of the molecular basis of these processes, nor is it known for certain how many of the proteins that make up each family or sub-family are subject to fluctuations. The focussing of recombinant DNA methods on keratinization is the only way to gain some understanding of these problems.

Earlier studies of protein biosynthesis in guinea pig[28] and wool follicles[36] were not rapidly followed by the successful isolation of mRNAs because of extensive degradation of RNA by the ribonucleases of adult skin. The problem was partially overcome by use of inhibitors[26] or by the isolation of intact polysomes[35] and recently, dramatically improved by the use of liquid nitrogen temperatures followed by guanidine denaturants (M. Frenkel, E. Kuczek and G. Rogers, unpublished observations). A pure cDNA representing one of the high-sulphur protein sequences has been isolated from a library of cDNA sequences derived from the mRNA of wool follicle polysomes (M. Sleigh and K. Ward, unpublished observations) and a start has been made in the investigation of the different wool keratin families in the sheep genome. Such studies will increase as representative probes become available, but at the time of writing the unpublished data is limited.

The high-sulphur protein family

A genomic clone, λSWK-1, has been selected from a sheep genomic library containing genes coding for proteins of the high-sulphur group, SCMKB-2 (*Table 23.1*). The formation of R loops indicated the presence of two genes (*Figure 23.9a*) separated by an intergenic distance of about 2 kb and an adjoining length of about

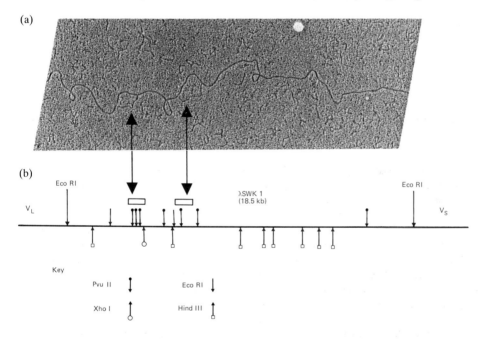

Figure 23.9. (a) R loops formed by hybridizing the sheep genomic clone λSWK1 (insert size 18.5 kb) with a preparation of poly(A)$^+$ RNA from wool follicles. The two genes were presumed to be for high-sulphur proteins because the clone had been selected with a cDNA probe (pSWK10, Dr K. Ward) containing the equivalent sequence. The R loops are about 800 nucleotides long. Their difference in length, although unexplained, may be due to sequence mismatch through cross-hybridization. The genes are separated by a distance of about 2 kb. About 10 kb of DNA, apparently devoid of wool keratin genes is present in the insert. (b) The two genes for high-sulphur wool proteins (identified by DNA sequencing, *Figure 23.10*) are shown in relation to restriction sites mapped in the 18.5 kb clone λSWK1. Their location relative to that in (a) is shown by arrows. It should be noted that (b) is presented at a smaller scale than (a).

GATCCTTAATTGAATGACCACTTAGGAAGGGAATGTTCAGTTTAGACAAACTCTGCCTAAAGGCTATCAAAGGACATTTATGACAAAGCAAGCCAGGGCAACGGTTAGCCAAAGTCTAAT 120

CTATCACACAACTTTCATGATGAGTCTTGGCTTGACCCAAAGGCAAAGAAGAAACATAATGAGATTTAATTGAACAATGACAAAAGCAGGCTATTCAAAATGCTATTTATAACCTTCTTA 240

ACAACTCAAATATT[AGCCAATCA]GAGACTGAATTTCAATGGTATCCAGGAAAAATGCTAAATGTTAGGGAAATAACATAGCTCTACAGGATG[TATAAAA]GGACAGATGCAGAAGGTGGAG 360

CCAAA[A]CTCAAAAACTTCTCTTAACAACCCTCCTCTCAACCCAACTCCTGACACCATGGCCTGCTGTTCCACCAGCTTCTGTGGATTTCCCATCTGTTCCACTGGTGGGACCTGTGGCTC 480
      MetAlaCysCysSerThrSerPheCysGlyPheProIleCysSerThrGlyGlyThrCysGlySe

CAGTCCCTGCCAGCCGACCTGCTGCCAGACCAGCTGCTGCCAGCCAACCTCCATCCAGACCAGCTGCTGCCAACCGATCTCCATCCAGACCAGCTGCTGCCAGCCAACCTCCATCCAGAC 600
rSerProCysCysGlnProThrCysCysGlnThrCysCysGlnProThrSerIleGlnThrSerCysCysGlnProIleSerIleGlnThrSerCysCysGlnProThrSerIleGlnTh

CAGCTGCTGCCAGCCAACCTGCCTCCAGACCAGTGGCTGTGGCATTGGTGGCATTGGCTATGGCCAGGTGGGTAGCAGCGGAGCTGTGAGCAGCGGAGCACCAGGTG 720
rSerCysCysGlnProThrCysLeuGlnThrSerGlyCysGluThrGlyIleGlyGlyIleGlyLeuGlySerIleGlyTyrGlyGlnValGlySerSerGlyAlaValSerSerArgThrArgTr

GTGCCGCCCTGACTGCCGCGTGGAGGGCACCAGCCTGCCTCCCTGCTGTGTGGTGAGCTGCACACCCCCGTCCTGCTGCCAGCTGTACTATGCCCAGGCCTCCTGCTGCCGCCCATCCTA 840
pCysArgProAspCysArgValGluGlyThrSerLeuProProCysCysValValSerCysThrProProCysCysGlnLeuTyrTyrAlaGlnAlaSerCysCysArgProSerTy

CTGTGGACAGTCCTGCTGCCGCCCAGCCTGCTGCTGCCAGCCCACCTGCATTGAGCCCATCTGTGAGCCCAGCTGCTGTGAGCCCACCTGCTGAAAGCAAGGCTGCTGATTGCTTAAGAA 960
*rCysGlyGlnSerCysCysProAlaCysCysCysGlnProThrCysIleGluProIleCysGluProSerCysCysGluProThrCys****

TGAAGGGGGAAGCAAAACAAACCTGCGCTGATCTCTGAAGAACCATTCAATTTTTCTTGGATGCTAAGTATCATGAAGTTTCTAAGCTGCTCAGAAATGAAAGTCTTAGGACACTTTTGA 1080

ATGTTCTGAAATAGCACGCTGAAAACAGGCCTCAGCCACGGTGTCTGGATTTCCTGCCTTGCTTGAGCAGAGTACCAGACTCGAGTAGCTCTTCATTTCAGTGACATCTCTGATCTCTTT 1200

CCTCTCTCTTGTTCCCATATATTTTCAGAC[CTGCTTT]TCTCCTTCAATATC[CTGTTAATTT]CA[AATAAA]ATATTTAAACTTGCAAAGCAAACTGAGTTATTTTCTTCCTTCTACCTTCAAG 1320

AGTGAATTTGACAGTAATGCTAAGGCATGAACATAAAGTCTAAGCCAGAGAGATTAAATACAATTACATTATGCTTAAATCTATTGCTTAGCAAAATGCTCCATGCATTAAATAATCTTT 1440

GTAGTTGTTCAAATACTTACCTGATCTAATCTGTTGTGAAAAGAAAGAAAGTGAAGTCATTTAGTCGTGTCCAACTCTTTGCAACCCTATGGACTGTAGCCCACCAGAGTCCTCTATCCA 1560

TGGGGATTCTCCAGGCAAGAATACTGGAGTGGGTTGCCATTTCCTTCTCCAGGGGATCTTGCCAATCCAGAGATCAAATCCAGGTCTCCCACACTGCAGGCAGACTCTTTACCCTCTGAG 1680

CCACCAGGCTCTTTTTTACCTAACATTTTCAGATTGATCAAAGCTACAGATGTAATCCTATCATATTTTAATCTCCTTCTGCTTAATAGAGTAGATGTTCCTCCTCATGCAGGCCCATATC 1800

TAAAGAAATTGTATTACCTATTAACTAGATTTCAAAATATCAACCAGATACTATCAGATATCTATCTATTTTCCAAGATCATTTATGTTCAGTTCCTATACCTTAGCCTATTGTATTCTTC 1920

CCGCTGATTCCCAAATTGCTCTCTAAGTTTCAAACCCACACTAACTGCCATTAACTGTATTTTTATGGTGTATCTATGGTCTCAATGAGAATAGAACAAATAGCTGCAGAGATAGCCAG 2040

TGAGTTCTACCATCCCCATTCTAATGGATGTGTCAGACACATCCATTAGAAATCTGAAACAACTATGTATTTTATACTTTGCAACCCAATCTGAAAAATGTTCCCAATAAATGCACACAC 2160

ATCTGGAACAAAAAATACCTCTTAAAATTTAATTTTTTCAGTTCTTAAGTATGTGAGCAACAACATACAAAACACAATAGGAGAAAACGTAGAACACATTGAGAACAATATTTTTGAGACA 2280

ATTACAAGCTCCCGGCACTCACCTCTATCATTCAGCTGGTGGCAGTAGGTCACACACAAGATTACTTATTTCAGAGAAGAATAAAGTAGATACAGTGCACCCTTCAAAATGTCAGAG 2400

CTTATCGATGCTTTTTCTTCAGACTTAGTTCAACTATTAAATACAAAAGATGTGTCAGCTTCTCCTGTTCCCTTTTGACAATCATAAGGACTGCTTCATGAGAGGCACTACAGCTTAAAAG 2520

TTCTATTTAATCATTTATAAGTGGTATAGACACTCATGAATCATGCCTTTTACTCCACCCATTGAAAATCAGATTCAATAAACATTTATTAAGGTTCTACCAGAAGCGAAGCACTGTGGT 2640

GGTGATCTTTCTGTATACTATATCATTTGGTTCACAAGGAAATATTTCTGGCCGTGGAACACAAGTAATATTTACACATTTCAAATTGGTTGAAATCAAAGGAAAAATGATGGTGATTTA 2760

TTTGATGTTGGAATAAATGAGTTGGAAGCTTTTAATTATGTGTGACCAAAGGTACCTTTCAATTCAGTGGATGTCTTGATGTTCTAAATTCAAGAGAATTCAGTCTATATATTCAAGAGG 2880

GGGAAAAAAATAGCTAAAAACAGCACAATTACTCTTAGCCTTAGAGACCTAGACAGAGAATCACATGGAATAACTACCTCTACAAAATGAGGGTGTTGACATGGACAAAAATGGGAGGGC 3000

TCTTTGAACATGGAAAACCAACAAACCTATTCCAAAGAGCAAAGGACCAATATGAGGAGTCCAAATGAAACTATAAAACGAAGTAATTTCTAAACTGCTGCTCATAATGTATCATAGGTAT 3120

TTCAATGTA[AGCCAATCA]GAGGACCACTGAAAAACAAATTTATACAAAAACAAGGAAGATATGCCAAGAAGTC[TATAAAA]TAGCAGCTAGCGCGATA[TATAAAA]GGCCTGACACAGAAGTC 3240

ACC[A]CCCAAATCCTCCAAGCATTACAATTCTCAGCCCAACTCCTGACACCATGGCCTGTTGCTCCACCAGCTTCTGTGGATTTCCCACTTGCTCCACTGGTGGGACCTGTGGTTCCAAC 3360
      MetAlaCysCysSerThrSerPheCysGlyPheProThrCysSerThrGlyGlyThrCysGlySerAsn

TTTTGCCAGCCAACCTGCTGCCAGACCAGCTGCTGCCAGCCAACCTCCATTCAGACCAGCTGCTGCCAGCCGACCTCTATCCAGACCAGCTGCTGCCAGCCAACTTCCATCCAAACCAGC 3480
PheCysGlnProThrCysCysGlnThrSerCysCysGlnProThrSerIleGlnThrSerCysCysGlnProThrSerIleGlnThrSerCysCysGlnProThrSerIleGlnThrSer

TGCTGCCAACCGATCTCCATCCAGACCAGCTGCTGCCAGCCAACCTGCCTCCAGACCAGTGGCTGTGGCAGGGCTGTGGCATTGGTGGCAGCATTGGCTATGGCAAGTGGGTAGCAGC 3600
CysCysGlnProIleSerIleGlnThrSerCysCysGlnProThrCysLeuGlnThrSerGlyCysGluThrGlyLeuGlyGlySerIleGlyTyrGlyLysValGlySerSer

GGAGCTGTGAGCAGCCGCACCAAGTGGTGCCGCCCTGACTGCCGCGTGGAGGGCACCAGCCTGCCTCCCTGCTGTGTGGTGAGCTGCACATCCCCGTCCTGCTGCCAGCTGTACTATGCC 3720
GlyAlaValSerSerArgThrLysTrpCysArgProAspCysArgValGluGlyThrSerLeuProProCysCysValValSerCysThrSerProSerCysCysGlnLeuTyrTyrAla

CAAGCCTCCTGCTGCCGCCCATCCTACTGTGGACAGTCCTGCTGCCGCCCAGCCTGCTGCTGCCAGCCCACCTGCATTGAGCCCGTCTGTGAGCCCACCTGCTGTGAGCCCACCTGCTGA 3840
*GlnAlaSerCysCysArgProSerTyrCysGlyGlnSerCysCysArgProAlaCysCysCysGlnProThrCysIleGluProValCysGluProThrCysCysGluProThrCys****

AAGCAAGG[TTGCTCATT]TAAAATTGCCCAAGACACAGTATCTCTGAATAATTTATCGCCTCAACCACCCATGGACAGCTAACAAGCTCTTAGCTCCCATTTGGGTTTTTGTTATGGGCGC 3960

TACAGAGTATATGAGTTCTATCTGATTTCATTCTACAATGAATATCTGTACTTTCCACGGCAGATGCTGCGTACCACCTGTCATCATCAAATTGCTTTGGCTATACTATTTCTGATTTTG 4080

ATGCAAGGTTGAATATTGCTGACATATTGTGGAATTTATCCTTTTGAACTGCGCACAAGCCTATTTCTCCTGCTTTCTGATCTATTTTTAGCTTCTGTTTTGTCCCCAAATTTTTACAACG 4200

TCAAAGGCACCTGAGTATAGATAGTATATCTTCAGATGTCACCAATGAGAATAGAAGCTCTTCACCCAATGTTCAGCTTCTAAGAAGGAGACTAGACTTTTCCATATTTCAACATCTGAT 4320

TCCATCCTCCTCTTTGCCTGTTGGCTATTTCAATTATATCTGTGATTCAGTGTCTTCTGTAATTTTTAATAAATGTTACATATGGGGCAAAGAAATCACAGCCTTTGTGTCCACTTACA 4440

CACAAAATGGGTTAATTTCTTCTGATTAAAAGATGCCTGTTTTAAAGAAATACATTTGTTTCTTTCTTAAAGGGACTGAAGTCATATACAGTGTAATAATTTAGTCATTTTTCCAGTAAA 4560

AATTTATTGACTACCACTTCTAAAGCAAGAACTAAGAAGATCAGAAAAGATGAAAAATTGTTTCTTATCCTTGAGGACTTTACTATCCAACAAGAAAAAGTATATGTGAAAGTAATTACAT 4680

ACAATATGGCAAGATAAATTGTATTCCCTTAGCAGCTG 4800

10 kb apparently devoid of any other keratin genes. The restriction map locating these coding sequences is shown in *Figure 23.9(b)* and the sequence of both genes is given in *Figure 23.10*. The first coding sequence is for an amino acid sequence that corresponds completely with that of the high-sulphur protein SCMKB-2A (*Table 23.1*). The other gene has a coding sequence that conceptually translates to a variant of SCMKB-2A. This suggests that SCMKB-2A represents a sub-family of genes. The two gene sequences show no evidence of introns (B. Powell, M. Sleigh, K. Ward and G. Rogers, in preparation). A further salient finding is that their 3' non-coding regions show no regions of homology, quite the reverse of that pertaining to feather keratin genes (see p. 417).

Whether the features described are extensive and apply to other families of the hair and wool keratins will become known as more DNA sequences are obtained. Little more can be said at this time except that since the amino acid sequences equivalent to the DNA sequences match the known protein sequences it is unlikely that any of the high sulphur keratin family are expressed as pre (or pro) keratins.

Epidermal keratins

Amphibian epidermis

Several keratin-like proteins in the molecular weight range 20 000–69 000 are specifically synthesized in the epidermis of *Xenopus laevis*[23]. Peptide mapping, although not definitive, suggested that some chains synthesized from the keratin mRNA were slightly larger than adult proteins and might be processed to a smaller size.

Of particular note for keratin gene studies is that this system is subject to hormonal control by 3,3',5-triiodo-L-thyroxine, administration of which causes metamorphosis in the larva and precocious keratinization in the larval epidermis. It is not yet certain whether the activation of keratin gene expression represents a major switch mechanism although it is clear that keratin mRNA levels increase with consequent increases in the keratin-like proteins[24]. A modulation of gene expression by vitamin A is readily demonstrable in cultured amphibian epidermal cells. The vitamin depresses the rate of keratinization although it apparently does not alter the nature of the keratin products of differentiation[25].

Figure 23.10. A genomic sequence of 4718 nucleotides embracing two genes for the high-sulphur keratin protein of the B2A type (see *Table 23.1*). Flanking sequences and about 2 kb of intergenic DNA are included. The CAAT and TATA sequences which are known to be involved in promoter function are shown as boxes 5' to each of the genes, the second gene, interestingly has two TATA boxes. The proposed 5' start for each of the mRNAs is denoted by $\boxed{A}$. The 3' termini are not shown but would be 15–30 bases downstream from the AATAAA polyadenylation signals. Other features of note are the underlined sequences marked I, II, III and IV which are faithfully repeated in the 3' non-coding regions of each gene, two of them being adjacent to the polyadenylation signals. Their significance is not known. In the intergenic region there is an inverted repeat of 14 nucleotides that would form a stable stem-loop structure. The total number of high-sulphur keratin protein genes and the extent of conservation of these organizational features has yet to be elucidated. From B. Powell *et al.* (in preparation).

Human epidermis

The keratin proteins expressed in human epidermal cells are arbitrarily referred to as small keratins, molecular weight 46 000–58 000 and large keratins, molecular weight 63 000–67 000. At least ten different proteins have been distinguished by sizing on acrylamide gels but nevertheless have been shown to be related by immunological crossreaction and very similar amino acid composition[7]. Presumably they all participate in forming intermediate filaments of the cytoskeleton[17]. It was earlier concluded that some smaller keratins might be products of larger precursors. This suggestion became largely unacceptable when peptide maps revealed amino acid sequences in the smaller keratins not present in the larger keratins and furthermore, different mRNAs could be resolved each translating *in vitro* into a protein product indistinguishable in molecular weight from the *in vivo* protein[8]. The immunological and compositional relatedness of the different keratin chains could still mean that repeating or nearly repeating sequences are shared between them. The present view is that they constitute a family of closely related (homologous) proteins and are probably the product of separate genes, different gene sets being activated and controlled during the outgrowth of the cells[9]. The small keratins are made in the basal layers whereas the large keratins, molecular weight 63 000–67 000, increase in relative proportions in the outer layers and dominate the protein composition of the outermost epidermal layer, the stratum

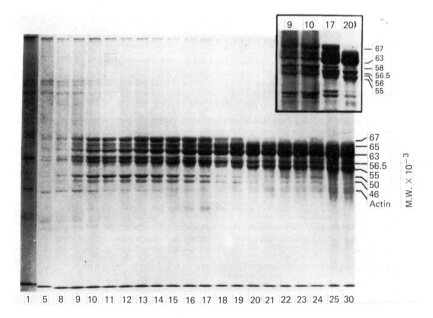

Figure 23.11. Gel electrophoretic demonstration of the changing proportion of different molecular weight polypeptide species during differentiation and maturation of human epidermal cells. Basal cells are tracks 8–16, granular layer cells, tracks 17–19 and stratum corneum in tracks 20–30. Selected representative tracks are given in the insert. It is seen that not only do the higher molecular weight chains (63 000–67 000) increase in relative proportion to the smaller molecular weight polypeptides (55 000–58 000) but the latter group consists of major chains of even lower molecular weight than in the more basal layers. From Fuchs and Green[9], with permission.

corneum (*Figure 23.11*). As judged by molecular weight, additional small keratin proteins appear and the large keratins disappear when the epidermal cells are cultured, demonstrating the pronounced influence of the cell environment on the expression of keratin genes.

The mRNAs for all of the proteins may be expressed from separate genes in the basal cells as these cells migrate outwards or they might arise from changes in the pattern of processing of nuclear RNA transcribed from a smaller set of genes or even a single gene. This vital question, together with the other possibility that some of the large keratin chains themselves may be processed after translation, can only be answered by preparing pure mRNA sequences via recombinant DNA techniques and subjecting them to DNA sequencing in parallel with a detailed study of the equivalent genes selected from human genomic libraries. Such studies are in progress (E. Fuchs, private communication).

Mouse epidermis

Mouse epidermis is being investigated in a similar fashion to that described above for human epidermis. The keratin proteins produced are in the 40 000–70 000 molecular weight range, the main ones having molecular weights of 55 000, 59 000 and 67 000 d. Clones carrying coding sequences for them have been isolated from a cDNA library (D. Roop, P. Hawley-Nelson and S. Yuspa, private communication). Partial sequences of these three clones have yielded protein sequence data which show large stretches of amino acids compatible with the formation of α-helices, a major structural feature of epidermal proteins (P. Steinert, D. Roop, W. Idler and R. Rice, private communication). The availability of the specific keratin clones will enable investigation of the changes that occur in keratin gene expression in primary cell cultures where it is known that some of the proteins are no longer expressed, a similar phenomenon to that observed for human epidermal cells.

General conclusions

Except for the immunoglobulins, keratins are the most heterogeneous of proteins. The molecular basis of this heterogeneity is of considerable interest in relation to mechanisms of differentiation as well as specifically in understanding the development of keratinizing tissues. The expression of a multiplicity of genes is probably the sole origin of heterogeneity although for the present, changes in RNA or protein processing cannot be excluded. The evolution of multiple genes and their divergence to produce large families remaining under a coordinated control is an intriguing question.

Gene multiplicity is a form of amplification possibly required by the cells to provide the requisite levels of mRNA for the rapid synthesis of keratin. Application of recombinant DNA methods to the study of keratin genes in different species will undoubtedly bring about dramatic changes in our understanding as they already have in other eukaryotic systems described in other chapters of this book. In the future it will be possible to define the number of genes for a keratin family and to determine how many of these are, in fact, expressed during growth and development. Initial progress in research findings has already revealed some important and unexpected facts. For example, there are the introns in feather keratin genes confined to the 5' non-coding part, as yet a unique finding, and the high degree of nucleotide sequence conservation in specific locations along the 3' non-coding regions. These features presumably have some importance for function. The 5' intron means that pre-mRNA should be found in embryonic feather RNA. So far as the coding regions are concerned however, there is no evidence that there are any extensions at, say, the 5' ends which would give rise to pre-keratin proteins. Although the spacing between feather genes is approximately 2.5 kb and about 2 kb for the two mammalian high-sulphur wool keratin genes, it is too early to say whether dimensions of this order will be consistently found within and between species.

Now that DNA fragments carrying defined keratin coding sequences and their associated putative control regions are becoming available, the possibility exists to undertake experiments concerning gene functions; that is, to study the transcriptional events specifying the mRNAs for particular sets of keratin proteins in the correct temporal and regional framework of keratinization. Such studies will *inter alia* involve microinjection into oocytes or mammalian cells. It is even conceivable that signals relating to the inductive influences of mesenchymal tissues[3, 19, 27] or the effects of thyroxine[24], steroid hormones[5, 34], certain depilatory compounds[12] and vitamin A[10] on biosynthetic activities of the epidermis and its derivatives will be definable at the nuclear level. Finally, the many nucleotide sequences for keratin genes that will come forth in the relatively near future will enhance the opportunity to compare amino acid sequences. It should become possible, for example, to define the limits to the variation of the primary structure of keratins within which the fibrous framework itself can assemble. Comparisons between the low-sulphur proteins of hair and wool and the equivalent proteins found in human and mouse epidermis (keratin intermediate filament proteins) will be especially interesting in this regard. The comparison of nucleotide sequences should also throw considerable light on evolutionary events such as the relationship between keratins of reptilian, avian and mammalian species including the putative evolution of feathers from reptilian scales[5]. It is already apparent from the presence of a pentapeptide repeat that the high-sulphur proteins of wool keratin evolved from an ancestral protein based on this sequence[30].

Acknowledgements

The author wishes to thank several colleagues of his laboratory for the use of unpublished data and also is grateful to Drs Kevin Ward and Merilyn Sleigh, CSIRO, Sydney, Dr Dennis Roop, National Cancer Institute, NIH, Bethesda and Dr Elaine Fuchs, Department of Biochemistry, University of Chicago, for information about their latest work.

Some of the data used (avian and mammalian keratins) came from studies supported by grants to the author from the Australian Research Grants Committee and for the Wool Research Trust Fund on the recommendation of the Australian Wool Corporation.

References

1. Crewther, W. G. (1975). Primary structure and chemical properties of wool. *5th International Wool Textile Research Conference, Aachen*, Volume I, 1–102
2. Crewther, W. G., Dowling, L. M. and Inglis, A. S. (1980). Amino acid sequence data from a microfibrillar protein of alpha keratin. *6th International Wool Textile Research Conference, Pretoria*, Volume II, 79–91
3. Dhouailly, D., Rogers, G. E. and Sengel, P. (1978). The specification of feather and scale protein synthesis in epidermal–dermal recombination. *Developmental Biology*, **65**, 58–68
4. Filshie, B. K. and Rogers, G. E. (1963). An electron microscope study of the fine structure of feather keratin. *Journal of Cell Biology*, **13**, 1–12
5. Fraser, R. D. B., MacRae, T. P. and Rogers, G. E. (1972). *Keratins, Their Composition, Structure and Biosynthesis*. Illinois, Charles C. Thomas
6. Friedman, M. and Orraca-Tetteh, R. (1978). Hair as an index of protein malnutrition in nutritional improvement of food and feed proteins. (Friedman, M., Ed.), pp. 131–154. New York, Plenum Publishing Corporation
7. Fuchs, E. and Green, H. (1978). The expression of keratin genes in epidermis and cultured epidermal cells. *Cell*, **15**, 887–897
8. Fuchs, E. and Green, H. (1979). Multiple keratins of cultured human epidermal cells are translated from different mRNA molecules. *Cell*, **17**, 573–582
9. Fuchs, E. and Green, H. (1980). Changes in keratin gene expression during terminal differentiation of the keratinocyte. *Cell*, **19**, 1033–1042
10. Fuchs, E. and Green, H. (1981). Regulation of terminal differentiation of cultured human keratinocytes by vitamin A. *Cell*, **25**, 617–625
11. Gillespie, J. M. and Marshall, R. C. (1980). Variability in the proteins of wool and hair. *6th International Wool Textile Research Conference, Pretoria*, Volume II, 67–77
12. Gillespie, J. M., Frenkel, M. J. and Reis, P. J. (1980). Changes in the matrix proteins of wool and mouse hair following the administration of depilatory compounds. *Australian Journal of Biological Sciences*, **33**, 125–136
13. Harrap, B. S. and Woods, E. F. (1964). Soluble derivatives of feather keratin. I. Isolation, fractionation and amino acid composition. *Biochemical Journal*, **92**, 8–18
14. Kemp, D. J. (1975). Unique and repetitive sequences in multiple genes for feather keratin. *Nature, London*, **254**, 573–577
15. Kemp, D. J. and Rogers, G. E. (1972). Differentiation of avian keratinocytes. Characterisation and relationships of the keratin proteins of adult and embryonic feathers and scales. *Biochemistry*, **11**, 969–975
16. Kemp, D. J., Dyer, P. Y. and Rogers, G. E. (1974). Keratin synthesis during development of the embryonic chick feather. *Journal of Cell Biology*, **62**, 114–131
17. Lazarides, E. (1980). Intermediate filaments as mechanical integrators of cellular space. *Nature, London*, **283**, 249–256
18. Lockett, T. J., Kemp, D. J. and Rogers, G. E. (1979). Organization of the unique and repetitive sequences in feather keratin messenger ribonucleic acid. *Biochemistry*, **18**, 5654–5663
19. McAleese, S. R. and Sawyer, R. H. (1981). Correcting the phenotype of the epidermis from chick embryos homozygous for the gene *scaleless* (*sc/sc*). *Science*, **214**, 1033–1034
20. Molloy, P. L., Powell, B. C., Gregg, K., Barone, E. and Rogers, G. E. (1982). Organisation of feather keratin genes in the chick genome. *Nucleic Acids Research*, **10**, 6007–6021
21. O'Donnell, I. J. and Inglis, A. S. (1974). Amino acid sequence of a feather keratin from silver gull (*Larus novae-hollandiae*) and comparison with one from emu (*Dromaius novae-hollandiae*). *Australian Journal of Biological Sciences*, **27**, 369–382
22. Partington, G. A., Kemp, D. J. and Rogers, G. E. (1973). Isolation of feather keratin mRNA and its translation in a rabbit reticulocyte cell-free system. *Nature, New Biology*, **246**, 33–36
23. Reeves, O. R. (1975). Adult amphibian epidermal proteins: Biochemical characterization and developmental appearance. *Journal of Embryology and Experimental Morphology*, **34**, 55–73
24. Reeves, R. (1977). Hormonal regulation of epidermis-specific protein and messenger RNA synthesis in amphibian metamorphosis. *Developmental Biology*, **60**, 163–179
25. Reeves, O. R. and Laskey, R. A. (1975). *In vitro* differentiation of a homogeneous cell population – the epidermis of *Xenopus laevis*. *Journal of Embryology and Experimental Morphology*, **34**, 75–92
26. Rogers, G. E., Frenkel, M. J. and Lock, R. A. (1981). Ribonucleic acids coding for the keratin complex of hair. *1st International Congress of Hair Research, Hamburg, Germany*. In *Hair Research* (Orfanos, C. W., Montagna, W. and Stuttgen, G., Eds.), pp. 85–94. Springer Verlag

27. Sengel, P. (1976). *Morphogenesis of Skin*. London, Cambridge University Press
28. Steinert, P. M. and Rogers, G. E. (1973). *In vitro* studies on the synthesis of guinea pig hair keratin proteins. *Biochimica et Biophysica Acta*, **312**, 403–412
29. Stewart, M. (1977). The structure of chicken scale keratin. *Journal of Ultrastructure Research*, **60**, 27–33
30. Swart, L. S., Joubert, F. J. and Parris, D. (1976). Homology in the amino acid sequences of the high sulphur proteins from keratins. *5th International Wool Textile Research Conference, Aachen*, Volume II, 254–264
31. Tenenhouse, H. S. and Gold, R. J. M. (1976). Loss of a homologous group of proteins in a dominantly inherited ectodermal malformation. *Biochemical Journal*, **159**, 149–160
32. Walker, I. D. and Bridgen, J. (1976). The keratin chains of avian scale tissue. Sequence heterogeneity and the number of scale keratin genes. *European Journal of Biochemistry*, **67**, 283–293
33. Walker, I. D. and Rogers, G. E. (1976). The structural basis for the heterogeneity of chick down feather keratin. *European Journal Biochemistry*, **69**, 341–350
34. Wallace, A. L. C. (1979). The effect of hormones on wool growth. In *Physiological and Environmental Limitations to Wool Growth*, (Black, J. L. and Reis, P. J., Eds.), pp. 257–268. University of New England Publishing Unit, N. S. W. Austrialia
35. Ward, K. A. and Kasmarik, S. E. (1980). The isolation of wool keratin messenger RNA from sheep. *Journal of Investigative Dermatology*, **75**, 244–248
36. Wilkinson, B. R. (1971). Cell-free biosynthesis of wool keratin proteins. *Biochemical Journal*, **125**, 371–373

Plant genes

Introduction

This section is devoted to the genes of plants and fungi. The striking difference in the number of chapters in this and in the previous section reflects, at least in part, a genuine differential in the extent of present knowledge of animal and plant systems. But some of the earlier chapters in Section IV, and especially Chapter 15, do include information gleaned from plant systems. Only occasionally do plants display the extremes of differentiation and dedication to a particular restricted repertoire of gene expression which characterizes many animal tissues. And since it is these particular situations which provide rich sources of particular mRNA molecules, there is some difficulty in getting a handle on an individual plant gene. The situation is, however, changing rapidly and if this book were being compiled a decade hence the situation would almost certainly be very different.

The two chapters included in this section are inevitably more generalized than most of the others in this book, and the task for the authors was therefore the more difficult. Chapter 24, on fungal genes, covers the now extensive work done with such organisms as the budding yeast *Saccharomyces*, and the filamentous fungus *Neurospora*. It is also particularly timely to devote reasonable page space to the genes of *Saccharomyces* since it now competes with *E. coli* as a favoured cloning organism for both prokaryotic and eukaryotic genes.

Chapter 24 indicates that some fungal genes possess intron sequences, as do many other eukaryotic structural genes, and also that fungal mRNA is both capped and tailed. Certain aspects of gene regulation in fungi are also particularly noteworthy. One is the presence of wide-domain structural genes where, as in the *are*A gene of *Aspergillus nidulans*, numerous structural gene sequences appear to be repressed by a common molecule. Another is the prevalence of postively controlled sequences in fungi. Many prokaryotic sequences, like the *lac* operon of *E. coli*, are under negative regulatory control, but many, though not all, fungal genes are positively regulated, involving effector molecules which are necessary to ensure efficient promotion or transcription. This chapter also emphasizes the existence of both 'cluster-genes' (genes coding for polyfunctional polypeptides) and 'gene clusters' (closely linked genes whose products are functionally related) in fungi, although the easy confusion of the two terms is unfortunate.

An aspect of fungal genetics which has received considerable attention of late is the interesting phenomenon of mating-type genes in yeast. Some discussion of this topic will be found in Chapter 15 of this volume. Although the conversion phenomenon occurring in the mating type genes may not be a true transposition, it has been thought best to discuss the topic under the heading of transposition since the situation is often cited as one involving transposition of genetic elements.

Chapter 25 undertakes a broad review of what is currently known about plant genomes and plant genes. It stresses that in terms of the range of genome size, the unique sequence/repetitious sequence interspersion patterns, and the presence of TATA, CAAT and other consensus regulatory sequences, plant genomes are essentially similar to those of animals. The prevalence of single sequence structural genes is also common to both plants and animals. About a dozen different plant genes have been cloned as genomic or cDNA copies, and of these one, the soybean storage protein gene, is discussed in some detail both in terms of structure and expression. An additional aspect of particular interest in this chapter is the information about messenger and hnRNA complexity in different tissues of tobacco. This account provides an interesting insight into gene regulation and post-transcriptional control during differentiation in a plant species, data which is not widely available for animal species although broadly similar information has been obtained for some species such as *Drosophila melanogaster*.

There is no doubt that the cloning and consequent sequencing of plant genes will develop very rapidly in the next few years, if for no other reason than that the insertion of novel advantageous genes into plants of agricultural importance is likely to be an important area of the exploitation of artificial genetic transformation in eukaryotes.

Chapter 24

Fungal systems

H. N. Arst, Jr.

Introduction

Since genetic and biochemical analysis of fungi is relatively easy, genetic characterization of these organisms is proceeding at an astonishing pace. Although eukaryote in organization, a number of fungi offer an ease of manipulation rivalling that of prokaryotes, making them increasingly popular for a variety of genetic studies. Readers should note the availability of a comprehensive monograph entitled *Fungal Genetics*[40] and the recent publication of a multi-authored treatise, *The Molecular Biology of the Yeast Saccharomyces*[101, 102]. Certain aspects of fungal systems are covered elsewhere in this volume, including extranuclear genes, histones, histone genes, transfer RNA genes, ribosomal RNA genes and mobile genetic elements (including mating type regulation in yeasts).

A particularly attractive feature of the fungi is small genome size. The budding yeast *Saccharomyces cerevisiae* and the fission yeast *Schizosaccharomyces pombe* have haploid genomes no more than four times as large as *Escherichia coli*, whilst for the filamentous fungi *Aspergillus nidulans* and *Neurospora crassa* the corresponding factor is about seven. A feature of ever increasing importance is the availability of transformation systems in three of these organisms. Transformation in *N. crassa*, first reported by Mishra and Tatum[72] was confirmed with physical evidence by Case *et al.*[19]. In *S. cerevisiae* it was reported by Hinnen *et al.*[46] and hybrid plasmids capable of replication in both *S. cerevisiae* and *E. coli* were first made by Beggs[10] and are capable of transforming *S. pombe*[9].

Gene structure

Some of the earliest studies of gene structure were performed in fungi[40]. The best characterized fungal gene is *CYC1*, encoding iso-l-cytochrome *c*, the major cytochrome *c* in *S. cerevisiae*[92]. Beyond contributing to its own characterization, studies using *CYC1* have advanced understanding in such areas as mutagen specificity, genetic recombination, protein structure and the translational apparatus. A recent example is the demonstration of a low frequency of recombination between *CYC1* and *CYC7*, an unlinked gene encoding iso-2-cytochrome *c*, the minor form of cytochrome *c* whose amino acid sequence is ~80% homologous with iso-1-cytochrome *c*[37].

A series of strains having a mutated AUG initiation codon (plus where necessary other mutations) in *CYC1* were reverted with the object of obtaining mutations resulting in initiation of translation in other codons (in the previously obtained knowledge that the N-terminal ten amino acids are dispensable) (reference 91 and earlier references therein). Translation initiation mutations were obtained in a region surrounding the normal initiation codon spanning 37 nucleotides. Sherman *et al.* conclude the absence of an absolute requirement for a particular sequence (e.g. ribosome binding site) 5′ to the initiation codon so that translation begins at the most 5′ proximal AUG codon of the mRNA (see also reference 62 and references therein). These studies further indicated that whereas GUG can also act as initiation codon in prokaryotes, only AUG can initiate translation of *CYC1*[90]. Stiles *et al.*[98] have studied the nucleotide changes of a mutation mapping in the 5′ leader region of *CYC1* mRNA and conclude that a base change in front of the normal initiation codon creates a proximal AUG codon out of phase with the normal reading frame which probably usurps the initiation function.

Another well characterized fungal gene is *am* of *N. crassa*, encoding NADP-linked glutamate dehydrogenase[40]. *am* has played a similar, if less extensive, role to *CYC1* in the development of molecular genetics of this organism.

Gene expression

The promoter of the *S. cerevisiae HIS3* gene (encoding imidazoleglycerolphosphate dehydratase) has been studied by Struhl[103], by selecting deletions arising by 'illegitimate' recombination in a hybrid bacteriophage λ containing yeast DNA including *HIS3*. *HIS3* expression was then monitored in both *S. cerevisiae* and *E. coli*. Expression in *S. cerevisiae* required retention of a much larger DNA sequence adjacent to the 5′ end of the coding region than expression in *E. coli* and mere retention of the sequence up to and including the Goldberg–Hogness T-A-T-A-A-A-T-A box in front of *HIS3* ensures only partial expression in *S. cerevisiae*. Struhl noted that this apparently larger promoter size in *S. cerevisiae* might result from required nucleosome phasing, compact or higher order DNA structure bringing distant sequences closer to the transcriptional initiation site, and/or interaction of specific non-histone proteins with RNA polymerase II. In a related study, Struhl[104] has shown that DNA sequences over 300 base pairs from the coding region can influence gene expression.

HIS3 is but one of a number of cloned *S. cerevisiae* genes encoding enzymes of primary metabolism which are expressed in *E. coli* (see reference 82 for list). Such genes from *N. crassa*[110], *S. pombe*[118], *A. nidulans*[35,61] and *Kluyveromyces lactis*[34] have also been cloned by expression in *E. coli*. In some cases (e.g. reference 43), expression is enhanced by mutations on the plasmid or *E. coli* chromosome which might, for example, enable transcription from a prokaryotic promoter[30,113] or stabilize mRNA[44]. In other cases cloned fungal genes specifying enzymes of primary metabolism are not expressed in *E. coli*, even though they are expressed in homologous transformation experiments[47,88]. The *S. cerevisiae* 3-phosphoglycerokinase gene is not expressed in *E. coli* because, as detected immunologically, a non-functional enzyme 5 kd smaller than the normal enzyme is produced[47]. In still other cases such as *S. cerevisiae* genes involved in the cell cycle[24] or gene regulation[63] or specifying the arginine permease[15] cloning in *E. coli* would be unfeasible but cloning in *S. cerevisiae* was successful. Several centromeres have been cloned by homologous transformation in *S. cerevisiae* owing to the mitotic stability they confer on hybrid plasmids carrying a selectable gene[52].

Just as a number of fungal genes are expressed in *E. coli*, a number of heterologous genes are expressed in *S. cerevisiae*, including bacterial plasmid genes for β-lactamase[22, 30, 84], aminoglycoside phosphotransferase-3' (I)[58] and chloramphenicol acetyltransferase[25] and the *E. coli lacZ* gene for β-galactosidase[78]. In the case of β-lactamase *S. cerevisiae* is able to process correctly an inactive preprotein by cleavage of a signal peptide[84]. On the other hand, the *E. coli lysA* gene is apparently not expressed in *S. cerevisiae*[21]. The greatest heterologous transformation feat yet described in *S. cerevisiae* is the chromosomal integration of the entire nitrogen fixation cluster (at least 15 genes) from *Klebsiella pneumoniae*[119] although there is no evidence that any of the genes is expressed.

There are also a few reports of heterologous eukaryotic gene expression in fungi. The *S. cerevisiae LEU2* gene is expressed in *S. pombe*[9]. Cloned *Drosophila melanogaster* DNA is able to complement an *ade8* mutation in *S. cerevisiae*[45]. A cloned gene for mature human leukocyte interferon D is expressed in *S. cerevisiae*[48]. Choice of this gene was based partly on its lack of introns and its similarity of codon usage to that of abundant yeast proteins, and Hitzeman *et al.* replaced its 5'-promoter-leader region with that of a yeast gene. A cloned rabbit chromosomal β-globin gene is transcribed in yeast, but the transcript is abnormally processed and no mRNA splicing occurs[11]. (In contrast, a cloned yeast tRNA$_3^{Leu}$ gene is transcribed and correctly spliced in a HeLa cell extract[95].)

DNA sequences are available for a few *S. cerevisiae* genes. Intervening sequences or introns are present in some, not in others. Colby *et al.*[26] have found that mutation at a splice junction in a tRNA gene affects the efficiency but not the fidelity of processing. Hopper *et al.*[50] have selected several allelic mutations which lead to loss of ability to excise intervening sequences from at least ten tRNAs (including all eight species of tRNA^{Tyr} chain termination suppressors) but probably have no effect on tRNA or mRNA synthesis. The authors attribute the viability of these mutants (as required by the selection procedure) even under conditions of tRNA precursor accumulation to leakiness of the mutations. Bromley *et al.*[16] have presented evidence for a role of the *RNA2* gene product in the processing of at least some intron-containing mRNAs.

The *am* gene of *N. crassa* contains a 67 bp intron in codon fifteen (J. R. S. Fincham, personal communication). The presence or absence of introns in various *S. cerevisiae* tRNA genes is summarized by Broach *et al*[14]. Wallace *et al.*[112] have specifically deleted the intron from a *S. cerevisiae* tRNA^{Tyr} ochre suppressor gene. The intron-deleted gene retains suppressor activity, indicating that the intron is dispensable for gene expression. An intron of over 300 bp occurs in the fourth codon of the *S. cerevisiae* actin gene[42, 74] but there are no introns in the *CYC1*[94], *CYC7*[73] or several other structural genes of this organism[49]. In the actin gene the intron is A + T-rich (68–70%) and splice junction sequences are similar to those in higher eukaryotes and at least the 5' junction sequence is essential for splicing[41]. In comparing 5' flanking sequences for several yeast genes, Montgomery *et al.*[73] speculate that the length of DNA between the initiation codon and a Goldberg–Hogness box structure might be positively correlated with the level of gene expression.

Fungal mRNA, like that of higher eukaryotes, contains a 5' terminal cap structure (reviewed in reference 8). A 3' polyadenylate tail intermediate in length between that of prokaryotes and higher eukaryotes is also present on a variety of fungal mRNAs (reviewed in reference 17). A deletion mutation in *CYC*1 mapping beyond the 3′ end of the translated region interferes with transcription termination[120].

'Cluster-genes'

The fungi seem to have more and larger 'cluster-genes' encoding polyfunctional polypeptides than prokaryotes. Some of the more spectacular examples include *arom* of *N. crassa* (aromatic amino acid biosynthesis)[71], *FAS*1 and *FAS*2 of *S. cerevisiae* (fatty acid synthesis)[87] and *HIS*4 of *S. cerevisiae*[60]. The kinetic, thermodynamic and metabolic (i.e. compartmentation of metabolites) advantages of polyfunctional enzymes are discussed by Metzenberg[71] and Welch and Gaertner[114]. To explain why 'cluster-genes' are more prevalent in eukaryotes than in prokaryotes, Metzenberg[71] has suggested that reducing the number of osmotically active particles might be an important priority for larger cells. Another possible reason is the scarcity of polycistronic mRNA in eukaryotes (see below p. 441). 'Cluster-genes' are an alternative means of maintaining parity between different enzyme activities.

Regulatory genes

Work in numerous laboratories has indicated that most fungal gene regulation occurs at the level of transcription[27, 79, 99]. In *S. cerevisiae* two mRNAs transcribed from a single structural gene for cytoplasmic and secreted invertases differ in length at the 5′ end so that promoter selection and/or differential processing provide a possible means of modulating gene expression[18, 80]. A number of *trans* acting (i.e. specifying a diffusible product) fungal regulatory genes have been identified. In some, mutant alleles of opposing phenotypes (reduced vs. increased structural gene expression) have been selected. Characterization of regulatory genes involves assessment of pleiotropy, dominance relationships, mutational frequencies for each class of mutation and determination of the phenotype by a loss of function allele (ideally a large but intragenic deletion). The regulatory gene can then be classified as positive or negative acting and, according to the structural genes it controls, as wide-domain, pathway-specific or integrator. In *A. nidulans* and *N. crassa* (but not in *S. cerevisiae*) there is a clear preponderance of positive acting regulatory genes. Metzenberg[71] has argued that positive control requires fewer molecules of diffusible regulator and thus less osmotic burden, which would be advantageous for larger cells.

Arguably the best characterized wide-domain regulatory gene is *are*A of *A. nidulans* (and its probable *N. crassa* equivalent *nit*-2[2, 4, 67]). The syntheses of a large number of enzymes and permeases involved in nitrogen nutrition are repressed by the preferred nitrogen source ammonium. The positive acting *are*A gene product is necessary for the expression of structural genes whose products are subject to nitrogen metabolite (i.e. ammonium) repression. Loss of function mutations, designated *are*A[r], result in inability to utilize nitrogen sources other than ammonium and low or undetectable levels of *are*A-controlled enzymes and permeases. Other, much rarer alleles, designated *are*A[d], lead to de-repression of one or more ammonium-repressible activities. An important characteristic of *are*A which provides compelling evidence for its direct regulatory role is non-hierarchical heterogeneity of mutant phenotypes[3]. For example, a single mutant allele might have an $areA^r$ phenotype with respect to one actvity, an $areA^d$ phenotype for a second and an $areA^+$ (wild type) phenotype for a third. Or consider two different $areA^r$ alleles. The first might be thermosensitive for expression of activity A but null at any temperature for expression of activity B whereas the second is thermosensitive for B and null for A. There is genetic evidence for the existence of an $areA$-related pseudogene[109]. The probable *N. crassa nit*-2 product has been isolated as a non-histone nuclear protein which binds to DNA-cellulose but is specifically eluted by glutamine (reviewed in reference 67). This is consistent with other evidence[67] that ammonium must be converted to glutamine to repress. Glutamine probably reduces the affinity of the *nit*-2 and $areA$ products for receptor sites adjacent to structural genes under their control.

Nitrogen metabolite repression is analogous to carbon catabolite repression in that it ensures preferential utilization of the most favoured carbon sources. Because of the involvement of cyclic AMP in this regulatory mechanism in *E. coli*, there has been no shortage of speculative extrapolation to the fungi. Experimental results have generally indicated that such speculation is unwarranted[3, 68, 77]. A physiological role for cyclic AMP in the activation of a protein kinase in *S. cerevisiae* has been established[68]. A bewildering variety of mutations affecting carbon catabolite repression in *S. cerevisiae* has been reported, but little attempt has been made to relate these to each other and no coherent comprehensive model is presently possible, although there is evidence that chromatin conformational changes accompany transcriptional activation of a carbon catabolite-repressed gene[93]. A negative acting regulatory gene designated *creA* mediates carbon catabolite repression in *A. nidulans*[2,3].

Other wide-domain regulatory genes are involved in phosphorus and sulphur nutrition[71] and a general control (probably involving positive regulation in *S. cerevisiae*[105]) of amino acid biosynthesis[31, 70, 75].

Amongst the best characterized pathway-specific fungal regulatory genes are *uaY* and *nirA* of *A. nidulans* and *GAL4* of *S. cerevisiae*. *uaY* regulates expression of at least eight unlinked structural genes involved in purine uptake and catabolism, mediating induction by uric acid[86]. The probable *uaY* product has been isolated as a protein able to bind phosphocellulose, DNA-cellulose, uric acid and the gratuitous inducer 2-thiouric acid[83]. The *uaY* product shows an apparent nuclear limitation *in vivo*, resulting in differences of complementation behaviour when heterokaryons are compared with diploids[85, 86]. Apparent nuclear limitation could be a trivial consequence of non-random nuclear distribution in a heterokaryon if a gene product is limiting. The interesting and trivial interpretations can be rigorously distinguished in *Coprinus lagopus* where dikaryons (with two nuclei per cell) exist[20]. Where this has been done, dikaryons follow the diploid pattern, indicating the apparent nuclear limitation to be an artefact[40] but the nature of the genes tested has not been firmly established.

nirA controls the genes involved in nitrate and nitrite assimilation in a positive manner, mediating induction by nitrate and nitrite[29]. In addition to $nirA^-$ (loss of function) mutations leading to non-inducibility and inability to utilize nitrate or nitrite, two kinds of mutations increasing structural gene expression are known[108]. $nirA^c$ (constitutive) mutations bypass the need for induction whereas genetically separable $nirA^d$ mutations alleviate nitrogen metabolite repression and partially bypass the requirement for the $areA$ product. Doubly mutant alleles (designated *nir*A$^{c/d}$) have an additive phenotype. These findings suggest that the *nir*A gene product contains two separate domains, a co-inducer binding region (defined by *nir*A^c mutations) and a region interacting with the *are*A gene product or with initiator sites adjacent to structural genes under *are*A and *nir*A control (defined by *nir*A^d mutations). A strong parallel can be drawn between *nir*A and the pathway-specific positive (but also negative) acting regulatory gene *ara*C of *E. coli*[108]. It is tempting to draw an analogy between *are*A and the positive acting wide-domain *E. coli* regulatory gene *crp*, whose product and its effector cyclic AMP are required for the expression of structural genes subject to carbon catabolite repression (including those under *ara*C control[3]).

The *GAL*4 product is required for syntheses of α-galactosidase and activities necessary for galactose utilization in *S. cerevisiae*[51,76]. Constitutivity mutations map in the coding region and thus alter the structure of the *GAL*4 product rather than increase its amount[76], but increasing its amount (by transformation with a multiple copy plasmid containing *GAL*4) also results in constitutivity (S. Johnston and J. E. Hopper, unpublished results). In addition to *GAL*4 there is a negative acting regulatory gene *GAL*80. Both groups of workers favour a dosage titration model in which the *GAL*80 product in the absence of co-inducer complexes the *GAL*4 product to render it inactive. *GAL*4 constitutivity mutations would then occur at the *GAL*80 binding site[76]. Johnston and Hopper also note that the data are consistent with competition between the *GAL*4 and *GAL*80 proteins for overlapping DNA receptor sites. In an overlapping site model, *GAL*4 constitutive mutations would presumably define a DNA binding site as would *GAL*80 super-repressed mutations.

The existence of both positive and negative acting regulatory genes in a number of other fungal systems has favoured extrapolation of the dosage titration model. Oshima and co-workers[76] have proposed a similar model for the regulation of phosphatases and phosphate transport systems in *S. cerevisiae*. An additional gene product ('mediator') is apparently necessary because the effector (inorganic phosphate) acts as a co-repressor rather than as a co-inducer. Cooper[28] has also put forward a dosage titration model for the regulation of five structural genes involved in allantoin degradation and urea active transport in *S. cerevisiae* (but the model requires modification to accommodate a second positive regulatory gene identified by Cooper's group and independently by Jacobs *et al*[56]. Another possible application of the model is for regulation of alcohol dehydrogenase II in *S. cerevisiae* where identification of a negative control gene (*ADR*4) has accompanied characterization of a positive control gene (*ADR*1)[33]. Metzenberg and his colleagues[71] have independently developed similar models to account for the regulation of phosphorus-providing activities in *N. crassa*.

Negative acting regulatory genes are involved in arginine biosynthesis and catabolism in *S. cerevisiae*[28, 36, 115]. The *carg*R (*CAR*80) product represses catabolic structural genes whilst a presumed heteropolymer of the *arg*RI (*ARG*80), *arg*RII (*ARG*81) and *arg*RIII (*ARG*82) products represses anabolic structural genes. However, this heteropolymeric repressor also plays a formally positive role, leading Wiame[115] to propose the term 'ambivalent repressor'. In the presence of an effector (e.g. arginine), ambivalent *arg*R repressor both prevents expression of the anabolic genes and elicits expression of the catabolic genes, presumably by titration of the *carg*R repressor, (ensuring that anabolism and catabolism are mutually exclusive processes). Thus loss of function mutations in the *arg*R genes lead to non-inducibility of the catabolic enzymes[115] and a mutation designated *arg*RII^d

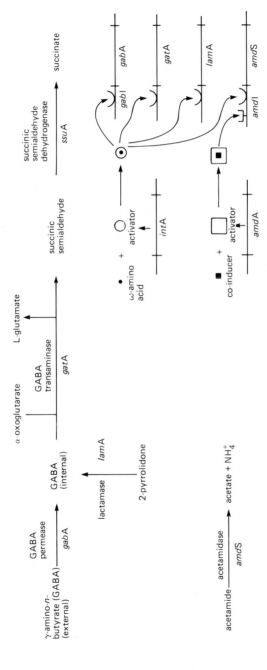

Figure 24.1. Pathways of γ-amino-*n*-butyrate and acetamide catabolism in *A. nidulans* with reference to the role of the integrator gene *int*A. Makins *et al.*[66] have shown that *lam*A probably specifies a lactamase. *Cis*-acting regulatory regions have been identified by mutation only for *gab*A and *amd*S[2]. The order of receptor sites adjacent to *amd*S is based on the fine structure map in *Figure 24.2*. *amd*A is a positive acting regulatory gene acting in parallel with *int*A to control *amd*S expression. Both acetyl-CoA and benzoate independently induce acetamidase[53] but recent evidence suggests that neither of these forms of induction is mediated by *amd*A[55]. *ssu*A is not under *int*A control[1].

increases catabolic gene expression and would seem to define a site of interaction with the *carg*R repressor[36].

Britten and Davidson[13] proposed a model for gene regulation which allows induction of the synthesis of any given gene product through the action of any one out of two or more positive acting regulatory genes. The genome therefore need only contain a single copy of a structural gene rather than a copy for each context in which it is expressed. As each regulatory gene can control a unique (but not necessarily mutually exclusive) set of structural genes, a given regulatory gene can integrate the expression of a set of structural genes in a particular context. Thus an integrator gene is a positive acting regulatory gene which mediates, through induction, the concomitant expression of two or more independently transcribed structural genes of which at least one is subject to independent regulation mediated by another positive regulatory gene (of either the integrator or pathway-specific type)[1,2]. The *int*A gene of *A. nidulans* mediates induction, by ω-amino acids, of acetamidase, γ-amino-*n*-butyrate (GABA) permease, GABA transaminase and a lactamase[1,2] (*Figure 24.1*). In the presence of ω-amino acids, *int*A integrates *amd*S expression with that of *gab*A, *lam*A and *gat*A whereas in other contexts (such as the presence of acetate, benzoate or an effector of the *amd*A product) *amd*S can be expressed independently.

Receptor sites

An important prediction of these models involving diffusible regulatory products acting at transcription is the existence of *cis*-acting receptor sites adjacent to controlled structural genes. Tight linkage and *cis*-acting behaviour alone are insufficient to establish that a regulatory mutation resides in a receptor site because these are also characteristic of rearrangements which place structural genes under the control of different promoters. A number of *cis*-acting mutations associated with increased gene expression in *S. cerevisiae* result from insertion of a repetitive transposable element[38, 116] which probably contains a promoter[23]. The increased expression phenotype is strongly influenced by genes controlling conjugation, providing a characteristic feature for recognition, perhaps indicating that these elements normally occur adjacent to genes involved in conjugation[38]. Transposition of extended segments (including several genes) of the *S. cerevisiae* genome is also possible[97]. Deletion mutations which greatly increase expression of the *CYC7* gene of *S. cerevisiae*, presumably by fusing it to a new regulatory region have been selected[65]. A reciprocal translocation having a similar result is described by Sherman *et al.*[92]. Arst *et al.*[7] have described an insertional translocation which apparently fuses the *A. nidulans* structural gene for nitrite reductase to a new promoter.

Nevertheless, apparently genuine receptor site mutations are known. These include operator constitutive mutations adjacent to structural genes involved in arginine biosynthesis[69, 106] and catabolism[115] in *S. cerevisiae* and a super-repressed mutation adjacent to the negatively controlled sulphate permease structural gene of *A. nidulans*[64]. An initiator constitutive mutation with an 'up-promoter' effect for the xanthine–uric acid permease of *A. nidulans* was selected by its ability to accommodate a mutant form of the *are*A product, implying functional overlap between the promoter and two initiator sites[5]. This phenotype is dependent on a functional *ua*Y allele, suggesting that the mutation increases receptor site affinity for a particular conformer of the *ua*Y product[85].

An extremely impressive collection of *cis*-acting regulatory mutations for acetamidase in *A. nidulans* has been assembled by Hynes[55] (references therein and personal communication). These include:

1. An 'up-promoter' mutation *amd*I18;
2. Three extreme 'down-promoter' mutations *amd*-406 and *amd*-407 (short deletions) and *amd*-205;
3. *amd*I9 leading to increased induction by sources of acetyl-CoA;
4. *amd*I93 preventing *int*A-mediated induction (see *Figure 24.1*) with a partial 'down-promoter' effect; and
5. *amd*I66 increasing acetamidase levels in strains carrying a constitutive *amd*A allele (see *Figure 24.1*).

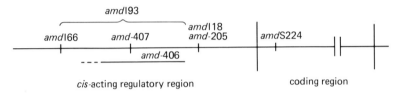

Figure 24.2. Fine structure map (not to scale) of the *cis*-acting regulatory region adjacent to *amd*S, the structural gene for acetamidase in *A. nidulans*[54,55]. The position of *amd*I93 within the bracket is unknown. *amd*I9 has not been mapped relative to other regulatory mutations. Characteristics of the regulatory mutations are described in the text.

Hynes' current map is *Figure 24.2*. These mutations are 'upstream' from the 5' end of the coding region and the two examined act at transcription (M. J. Hynes and C. M. Corrick, personal communication).

Gene clusters

Functionally related genes are seldom clustered in fungi. Most mRNA in *S. cerevisiae* is monocistronic[81], and efforts to select reinitiation mutations 'downstream' from a chain termination codon (giving a dicistronic mRNA of sorts) have either failed[89] or have to date yielded only transposable element insertions and complex rearrangements[39] (Lacroute in discussion in reference 92; M.-L. Bach, personal communication). Current investigation into the organization of several fungal gene clusters is outlined below.

The *qa* gene cluster of *N. crassa* is involved in the first three steps of quinic acid catabolism[42a, 71, 79, 88] (and references in these). There are six genes including structural genes for the three enzymes, a positive acting regulatory gene (*qa*-1) whose expression is constitutive but subject to further autogenous transcriptional regulation and two genes of unknown function (*qa*-x and *qa*-y) (*Figure 24.3*). The *qa*-1 gene product mediates quinate induction of the syntheses of mRNAs for the other five genes. All transcripts are monocistronic and the direction of transcription is available only for *qa*-3.

*arg*B and *arg*C, specifying apparently separate but sequential enzymes of arginine biosynthesis in *S. cerevisiae* are tightly linked[57] (and references therein). An operator constitutive mutation, *cis*-dominant with respect to both genes, maps in the order: operator – *arg*B – *arg*C. Jacobs *et al.*[57] propose three possible models:

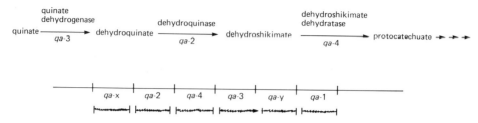

Figure 24.3. The pathway of quinic acid catabolism in *N. crassa*. The map order and transcript size and orientation is based on Giles[42a], Schweizer *et al.*[88] and Patel *et al.*[79] and references therein. Transcripts are indicated by wavy lines.

1. Both genes are translated as a single polypeptide which is then cleaved *in vivo* to yield two proteins;
2. The genes are organized as a bacterial operon with separate translation from a dicistronic mRNA; and
3. The genes are transcribed as a dicistronic mRNA which is processed into two monocistronic mRNAs for separate translation.

A possible argument in favour of model (1) is that polypeptide chain termination mutations in *arg*B are polar with respect to expression of *arg*C. Such polarity would not be expected for eukaryotic polycistronic mRNA containing separate translation initiation sites (e.g. models (2) and (3)) because transcription and translation take place on opposite sides of the nuclear membrane.

The *prn* cluster involved in proline catabolism in *A. nidulans* is shown in *Figure 24.4*[6,59]. It comprises structural genes for a permease and the two enzymes and a positive acting regulatory gene *prn*A. *Cis*-acting regulatory mutations affecting expression of *prn*B map in the central regulatory region[6] (and references therein). Deletions beginning in *prn*A or *prn*D and ending in *prn*B reduce but do not abolish expression of *prn*C in *cis*[6] (and references therein; K. K. Sharma and Arst, unpublished). This suggests that a *prn*B *prn*C dicistronic mRNA is initiated in the central regulatory region but that *prn*C is also expressed via an overlapping mono- (as shown) and/or tetracistronic transcript(s). This interpretation requires the direction of transcription shown in *Figure 24.4*.

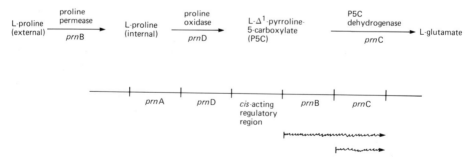

Figure 24.4. The pathway of L-proline catabolism in *A. nidulans*. *prn*A is a positive regulatory gene controlling expression of *prn*D, *prn*C and, to a lesser extent, *prn*B[59] (and references therein). Evidence for size and orientation of transcripts (wavy lines) is given in text. A tetracistronic transcript (for *prn*C expression) initiated to the left of *prn*A is possible in place of the monocistronic transcript shown or in addition to it.

The nitrate assimilation gene cluster of *A. nidulans* is shown in *Figure 24.5*[29, 109a] (A. G. Brownlee and Arst, unpublished). Insight into its organization is provided by an insertional translocation fusing *nii*A to a promoter from linkage group II and inserting a considerable portion of linkage group II between *nii*A and *nia*D (in linkage group VIII)[7]. Expression of *nia*D and *crn*A is unaffected. *nii*A expression

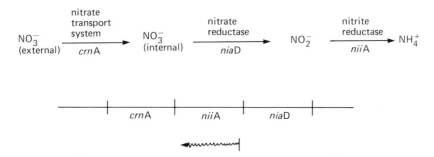

Figure 24.5. The nitrate assimilation pathway in *A. nidulans*. The reasoning behind placement of initiation site for transcript (wavy line) shown is given in the text. It cannot be excluded that this transcript also includes *crn*A. Overlapping transcripts are also possible.

remains partially responsive to its normal regulatory signals (induction by nitrate, repression by ammonium) indicating that the normal *nii*A promoter/initiator is present in tandem to the translocated promoter, implying the point of initiation and direction of transcription shown. This demonstrates unequivocally that *nii*A expression cannot occur solely via a di- or tricistronic transcript initiated on the *nii*A-distal side of *nia*D. The lack of effect on *crn*A expression (Arst, unpublished) leads to a similar conclusion. The translocation gives no evidence concerning the direction(s) of transcription of *nia*D and *crn*A. It does reduce maximal *nii*A expression, a finding open to several interpretations including the possibility that it prevents synthesis of an overlapping *nia*D *nii*A di- or tricistronic transcript.

Figure 24.6 shows the gene cluster specifying the first three enzymes of galactose metabolism in *S. cerevisiae*[51, 76]. St. John and Davis[99] and St. John et al.[100] have demonstrated monocistronic transcripts for each of the three genes and shown that *GAL1* is transcribed in the opposite direction to *GAL10* and *GAL7*. An overlapping *GAL10 GAL7* transcript was also detected. A deletion removing *GAL1* and ending within *GAL10* (leaving *GAL7* intact) does not prevent *GAL7* expression. This demonstrates that physiological expression of *GAL7* occurs via the monocistronic transcript. The data presented do not, however, exclude the possibility that some fraction of *GAL7* (and *GAL10*) expression occurs through translation of the dicistronic mRNA.

A developmentally regulated cluster of six poly(A)$^+$ RNA-specifying regions expressed specifically in conidiospores of *A. nidulans* has recently been reported[107]. These regions (presumably representing at least six genes but whose roles have not been established) specify RNAs ranging from (approximately) 590–2500 nucleotides in length and within the six transcript cluster there are two changes in direction of transcription.

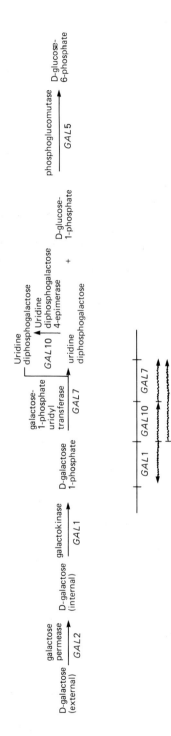

Figure 24.6. The Leloir pathway or pathway of galactose utilization in *S. cerevisiae*. *GAL2* and *GAL5* are unlinked to the gene cluster shown. *GAL2* is regulated by the same *trans*-acting regulatory genes which regulate expression of the genes of the cluster but *GAL5* is expressed constitutively[51,76] (and references therein). Size and orientation of transcripts is based on St. John and Davis[99] and St. John *et al.*[100]

Other aspects of genetic regulation

Another level at which gene expression can be influenced is the degree of stability of gene product; Cooper[28] discusses studies of mRNA stability in *S. cerevisiae* and Wolf[117] reviews the regulatory role played by proteases. The location of gene products is similarly crucial to regulation. Analysis of the steps involved in transport of extracellular and vascuolar enzymes by yeast is progressing[96] (and references therein). Various aspects of the role of compartmentation in regulating fungal metabolism and gene expression are discussed by Davis[31], Davis *et al.*[32], Cooper[27] and Brandriss and Magasanik[12]. An interesting and possibly unique mechanism in this regard has been studied by Wiame and his colleagues[111] (and references therein). The catabolic enzyme arginase binds to and prevents activity of the anabolic enzyme ornithine carbamoyltransferase thereby preventing futile metabolite cycling and waste of considerable ATP. Vissers *et al.*[111] have made an extensive study of the occurrence of this phenomenon (termed 'epiarginasic regulation') in 32 yeast species representing 11 genera. It was absent in all species where ornithine carbamoyltransferase is mitochondrial (arginase being cytosolic) and among those in which it is cytosolic, its presence or absence correlated to some extent with energy metabolism so as to suggest that selective pressure towards development of epiarginasic regulation might be greater the less efficient the organism is in synthesizing ATP.

References

1. Arst, H. N. Jr. (1976). Integrator gene in *Aspergillus nidulans*. *Nature, London*, **262**, 231–234
2. Arst, H. N. Jr. (1981). Aspects of the control of gene expression in fungi. In *Genetics as a Tool in Microbiology*, Society for General Microbiology Symposium 31, (Glover, S. W. and Hopwood, D. A., Eds.), pp. 131–160. Cambridge, Cambridge University Press
3. Arst, H. N. Jr. and Bailey, C. R. (1977). The regulation of carbon metabolism in *Aspergillus nidulans*. In *Genetics and Physiology of Aspergillus*, (Smith, J. E. and Pateman, J. A., Eds.), pp. 131–146. London, Academic Press
4. Arst, H. N. Jr. and Cove, D. J. (1973). Nitrogen metabolite repression in *Aspergillus nidulans*. *Molecular and General Genetics*, **126**, 111–141
5. Arst, H. N. Jr. and Scazzocchio, C. (1975). Initiator constitutive mutation with an 'up-promoter' effect in *Aspergillus nidulans*. *Nature, London*, **254**, 31–34
6. Arst, H. N. Jr., Jones, S. A. and Bailey, C. R. (1981). A method for the selection of deletion mutations in the L-proline catabolism gene cluster of *Aspergillus nidulans*. *Genetical Research*, **38**, 171–195
7. Arst, H. N. Jr., Rand, K. N. and Bailey, C. R. (1979). Do the tightly linked structural genes for nitrate and nitrite reductases in *Aspergillus nidulans* form an operon? Evidence from an insertional translocation which separates them. *Molecular and General Genetics*, **174**, 89–100
8. Banerjee, A. K. (1980). 5'-terminal cap structure in eucaryotic messenger ribonucleic acids. *Microbiological Reviews*, **44**, 175–205
9. Beach, D. and Nurse, P. (1981). High-frequency transformation of the fission yeast *Schizosaccharomyces pombe*. *Nature, London*, **290**, 140–142
10. Beggs, J. D. (1978). Transformation of yeast by a replicating hybrid plasmid. *Nature, London*, **275**, 104–109
11. Beggs, J. D., van den Berg, J., van Ooyen, A. and Weissmann, C. (1980). Abnormal expression of chromosomal rabbit β-globin gene in *Saccharomyces cerevisiae*. *Nature, London*, **283**, 835–840
12. Brandriss, M. C. and Magasanik, B. (1981). Subcellular compartmentation in control of converging pathways for proline and arginine metabolism in *Saccharomyces cerevisiae*. *Journal of Bacteriology*, **145**, 1359–1364

13. Britten, R. J. and Davidson, E. H. (1969). Gene regulation for higher cells: a theory. *Science*, **165**, 349–357

14. Broach, J. R., Friedman, L. and Sherman, F. (1981). Correspondence of yeast UAA suppressors to cloned tRNA$^{Ser}_{UCA}$ genes. *Journal of Molecular Biology*, **150**, 375–387

15. Broach, J. R., Strathern, J. N. and Hicks, J. B. (1979). Transformation in yeast: development of a hybrid cloning vector and isolation of the *CAN*1 gene. *Gene*, **8**, 121–133

16. Bromley, S., Hereford, L. and Rosbash, M. (1982). Further evidence that the *rna2* mutation of *Saccharomyces cerevisiae* affects mRNA processing. *Molecular and Cellular Biology* , **2**, 1205–1211

17. Carlin, R. K. (1978). The poly(A) segment of mRNA: (1) Evolution and function and (2) The evolution of viruses. *Journal of Theoretical Biology*, **71**, 323–338

18. Carlson, M. and Botstein, D. (1982). Two differentially regulated mRNAs with different 5′ ends encode secreted and intracellular forms of yeast invertase. *Cell*, **28**, 145–154

19. Case, M. E., Schweizer, M., Kushner, S. R. and Giles, N. H. (1979). Efficient transformation of *Neurospora crassa* by utilizing hybrid plasmid DNA. *Proceedings of the National Academy of Sciences, USA*, **76**, 5259–5263

20. Casselton, L. A. and Lewis, D. (1967). Dilution of gene products in the cytoplasm of heterokaryons in *Coprinus lagopus*. *Genetical Research*, **9**, 63–71

21. Chenais, J., Richaud, C., Ronceray, J., Cherest, H., Surdin-Kerjan, Y. and Patte, J.-C. (1981). Construction of hybrid plasmids containing the *lys*A gene of *Escherichia coli*: studies of expression in *Escherichia coli* and *Saccharomyces cerevisiae*. *Molecular and General Genetics*, **182**, 456–461

22. Chevallier, M. R. and Aigle, M. (1979). Qualitative detection of penicillinase produced by yeast strains carrying chimeric yeast-coli plasmids. *Federation of European Biochemical Societies Letters*, **108**, 179–180

23. Ciriacy, M. and Williamson, V. M. (1981). Analysis of mutations affecting *Ty*-mediated gene expression in *Saccharomyces cerevisiae*. *Molecular and General Genetics*, **182**, 159–163

24. Clarke, L. and Carbon, J. (1980). Isolation of the centromere-linked *CDC*10 gene by complementation in yeast. *Proceedings of the National Academy of Sciences, USA*, **77**, 2173–2177

25. Cohen, J. D., Eccleshall, T. R., Needleman, R. B., Federoff, H., Buchferer, B. A. and Marmur, J. (1980). Functional expression in yeast of the *Escherichia coli* plasmid gene coding for chloramphenicol acetyltransferase. *Proceedings of the National Academy of Sciences, USA*, **77**, 1078–1082

26. Colby, D., Leboy, P. S. and Guthrie, C. (1981). Yeast tRNA precursor mutated at a splice junction is correctly processed *in vivo*. *Proceedings of the National Academy of Sciences, USA*, **78**, 415–419

27. Cooper, T. G. (1980). Selective gene expression and intracellular compartmentation: two means of regulating nitrogen metabolism in yeast. *Trends in Biochemical Sciences*, **5**, 332–334

28. Cooper, T. G. (1982). The regulation of yeast gene expression by multiple control elements. In *Genetic Engineering of Microorganisms for Chemicals*, (Hollaender, A., De Moss, R. D., Kaplan, S., Konisky, J., Savage, D. and Wolfe, R. S., Eds.), pp. 143–161. New York, Plenum Press

29. Cove, D. J. (1979). Genetic studies of nitrate assimilation in *Aspergillus nidulans*. *Biological Reviews*, **54**, 291–327

30. Crabeel, M., Messenguy, F., Lacroute, F. and Glansdorff, N. (1981). Cloning *arg*3, the gene for ornithine carbamoyltransferase from *Saccharomyces cerevisiae*: expression in *Escherichia coli* requires secondary mutations; production of plasmid β-lactamase in yeast. *Proceedings of the National Academy of Sciences, USA*, **78**, 5026–5030

31. Davis, R. H. (1975). Compartmentation and regulation of fungal metabolism: genetic approaches. *Annual Review of Genetics*, **9**, 39–65

32. Davis, R. H., Weiss, R. L. and Bowman, B. J. (1978). Intracellular metabolite distribution as a factor in regulation in *Neurospora*. In *Microenvironments and Metabolic Compartmentation*, (Srere, P. A. and Estabrook, R. W., Eds.), pp. 197–210. New York, Academic Press

33. Denis, C. L., Ciriacy, M. and Young, E. T. (1981). A positive regulatory gene is required for accumulation of the functional messenger RNA for the glucose-repressible alcohol dehydrogenase from *Saccharomyces cerevisiae*. *Journal of Molecular Biology*, **148**, 355–368

34. Dickson, R. C. and Markin, J. S. (1978). Molecular cloning and expression in *E. coli* of a yeast gene coding for β-galactosidase. *Cell*, **15**, 123–130

35. Dmochowska, A. T., Bal, J., Bartnik, E. and Weglenski, P. (1980). Cloning of *Aspergillus nidulans* DNA. I. Selection and analysis of recombinant plasmids capable of complementing *pyr*F, *arg*IF and *pro*AB mutations in *Escherichia coli*. *Acta Microbiologica Polonica*, **29**, 213–225

36. Dubois, E., Hiernaux, D., Grenson, M. and Wiame, J.-M. (1978). Specific induction of catabolism and its relation to repression of biosynthesis in arginine metabolism of *Saccharomyces cerevisiae*. *Journal of Molecular Biology*, **122**, 383–406

37. Ernst, J. F., Stewart, J. W. and Sherman, F. (1982). Formation of composite iso-cytochromes c by recombination between non-allelic genes of yeast. *Journal of Molecular Biology*, **161**, 373–394

38. Errede, B., Cardillo, T. S., Wever, G. and Sherman, F. (1981). Studies on transposable elements in yeast. I. ROAM mutations causing increased expression of yeast genes: their activation by signals directed toward conjugation functions and their formation by insertion of *Ty1* repetitive elements. *Cold Spring Harbor Symposia on Quantitative Biology*, **45**, 593–607

39. Exinger, F. and Lacroute, F. (1979). Genetic evidence for the creation of a reinitiation site by mutation inside the yeast *ura 2* gene. *Molecular and General Genetics*, **173**, 109–113

40. Fincham, J. R. S., Day, P. R. and Radford, A. (1979). *Fungal Genetics*, 4th Edition. Oxford, Blackwell Scientific Publications

41. Gallwitz, D. (1982). Construction of a yeast actin gene intron deletion mutant that is defective in splicing and leads to the accumulation of precursor RNA in transformed yeast cells. *Proceedings of the National Academy of Sciences, USA*, **79**, 3493–3497

42. Gallwitz, D. and Sures, I. (1980). Structure of a split yeast gene: complete nucleotide sequence of the actin gene in *Saccharomyces cerevisiae*. *Proceedings of the National Academy of Sciences, USA*, **77**, 2546–2550

42a. Giles, N. H. (1978). The organization, function and evolution of gene clusters in eucaryotes. *The American Naturalist*, **112**, 641–657

43. Harashima, S., Sidhu, R. A., Toh-E., A. and Oshima, Y. (1981). Cloning of the *HIS5* gene of *Saccharomyces cerevisiae* by yeast transformation. *Gene*, **16**, 335–341

44. Hautala, J. A., Bassett, C. L., Giles, N. H. and Kushner, S. R. (1979). Increased expression of a eukaryote gene in *Escherichia coli* through stabilization of its messenger RNA. *Proceedings of the National Academy of Sciences, USA*, **76**, 5774–5778

45. Henikoff, S., Tatchell, K., Hall, B. D. and Nasmyth, K. A. (1981). Isolation of a gene from *Drosophila* by complementation in yeast. *Nature, London*, **289**, 33–37

46. Hinnen, A., Hicks, J. B. and Fink, G. R. (1978). Transformation of yeast. *Proceedings of the National Academy of Sciences, USA*, **75**, 1929–1933

47. Hitzeman, R. A., Clarke, L. and Carbon, J. (1980). Isolation and characterization of the yeast 3-phosphoglycerokinase gene (PGK) by an immunological screening technique. *Journal of Biological Chemistry*, **255**, 12073–12080

48. Hitzeman, R. A., Hagie, F. E., Levine, H. L., Goeddel, D. V., Ammerer, G. and Hall, B. D. (1981). Expression of a human gene for interferon in yeast. *Nature, London*, **293**, 717–722

49. Holland, M. J., Holland, J. P., Thill, G. P. and Jackson, K. A. (1981). The primary structure of two yeast enolase genes. Homology between the 5′ noncoding flanking regions of yeast enolase and glyceraldehyde-3-phosphate dehydrogenase genes. *Journal of Biological Chemistry*, **256**, 1385–1395

50. Hopper, A. K., Schultz, L. D. and Shapiro, R. A. (1980). Processing of intervening sequences: a new yeast mutant which fails to excise intervening sequences from precursor tRNAs. *Cell*, **19**, 741–751

51. Hopper, J. E. (1981). Regulation of gene expression in the galactose-melibiose regulon: review of a new model invoking protein-protein interaction between constitutively synthesized positive and negative regulators. In *Molecular Genetics in Yeast*, Alfred Benzon Symposium 16, (von Wettstein, D., Friis, J., Kielland-Brandt, M. and Stenderup, A., Eds.), pp. 199–205. Copenhagen, Munksgaard

52. Hsiao, C. L. and Carbon, J. (1981). Direct selection procedure for the isolation of functional centromeric DNA. *Proceedings of the National Academy of Sciences, USA*, **78**, 3760–3764

53. Hynes, M. J. (1978). Multiple independent control mechanisms affecting the acetamidase of *Aspergillus nidulans*. *Molecular and General Genetics*, **161**, 59–65

54. Hynes, M. J. (1979). Fine-structure mapping of the acetamidase structural gene and its controlling region in *Aspergillus nidulans*. *Genetics*, **91**, 381–392

55. Hynes, M. J. (1982). A *cis*-dominant mutation in *Aspergillus nidulans* affecting the expression of the *amdS* gene in the presence of mutations in the unlinked gene, *amdA*. *Genetics*, **102**, 139–147

56. Jacobs, E., Dubois, E., Hennaut, C. and Wiame, J.-M. (1981). Positive regulatory elements involved in urea amidolyase and urea uptake induction in *Saccharomyces cerevisiae*. *Current Genetics*, **4**, 13–18

57. Jacobs, P., Jauniaux, J.-C. and Grenson, M. (1980). A *cis*-dominant regulatory mutation linked to the *argB–argC* gene cluster in *Saccharomyces cerevisiae*. *Journal of Molecular Biology*, **139**, 691–704

58. Jiminez, A. and Davies, J. (1980). Expression of a transposable antibiotic resistance element in *Saccharomyces. Nature, London*, **287**, 869–871

59. Jones, S. A., Arst, H. N. Jr. and McDonald, D. W. (1981). Gene roles in the *prn* cluster of *Aspergillus nidulans. Current Genetics*, **3**, 49–56

60. Keesey, J. K., Bigelis, R. and Fink, G. R. (1979). The product of the *his4* gene cluster in *Saccharomyces cerevisiae*. A trifunctional polypeptide. *Journal of Biological Chemistry*, **254**, 7427–7433

61. Kinghorn, J. R. and Hawkins, A. R. (1982). Cloning and expression in *Escherichia coli* K-12 of the biosynthetic dehydroquinase function of the *arom* cluster gene from the eucaryote, *Aspergillus nidulans. Molecular and General Genetics*, **186**, 145–152

62. Kozak, M. (1981). Possible role of flanking nucleotides in recognition of the AUG initiator codon by eukaryotic ribosomes. *Nucleic Acids Research*, **9**, 5233–5352

63. Losson, R. and Lacroute, F. (1982). Cloning of a eukaryotic regulatory gene. *Molecular and General Genetics*, **184**, 394–399

64. Lukaszkiewicz, Z. and Paszewski, A. (1976). Hyperrepressible operator-type mutant in sulphate permease gene of *Aspergillus nidulans. Nature, London*, **259**, 337–338

65. McKnight, G. L., Cardillo, T. S. and Sherman, F. (1981). An extensive deletion causing overproduction of yeast iso-2-cytochrome *c. Cell*, **25**, 409–419

66. Makins, J. F., Holt, G. and Macdonald, K. D. (1981). *lam A* mutants of *Aspergillus nidulans* show decreased β-lactamase activity but do not influence penicillin production. *Aspergillus News Letter*, **15**, 30–33

67. Marzluf, G. A. (1981). Regulation of nitrogen metabolism and gene expression in fungi. *Microbiological Reviews*, **45**, 437–461

68. Matsumoto, K., Uno, I., Oshima, Y. and Ishikawa, T. (1982). Isolation and characterization of yeast mutants deficient in adenylate cyclase and cAMP-dependent protein kinase. *Proceedings of the National Academy of Sciences, USA*, **79**, 2355–2359

69. Messenguy, F. (1976). Regulation of arginine biosynthesis in *Saccharomyces cerevisiae*: isolation of a *cis*-dominant, constitutive mutant for ornithine carbamoyltransferase synthesis. *Journal of Bacteriology*, **128**, 49–55

70. Messenguy, F. (1979). Concerted repression of the synthesis of the arginine biosynthetic enzymes by aminoacids: a comparison between the regulatory mechanisms controlling aminoacid biosyntheses in bacteria and in yeast. *Molecular and General Genetics*, **169**, 85–95

71. Metzenberg, R. L. (1979). Implications of some genetic control mechanisms in *Neurospora. Microbiological Reviews*, **43**, 361–383

72. Mishra, N. C. and Tatum, E. L. (1973). Non-Mendelian inheritance of DNA-induced inositol independence in *Neurospora. Proceedings of the National Academy of Sciences, USA*, **70**, 3875–3879

73. Montgomery, D. L., Leung, D. W., Smith, M., Shalit, P., Faye, G. and Hall, B. D. (1980). Isolation and sequence of the gene for iso-2-cytochrome *c* in *Saccharomyces cerevisiae. Proceedings of the National Academy of Sciences, USA*, **77**, 541–545

74. Ng, R. and Abelson, J. (1980). Isolation and sequence of the gene for actin in *Saccharomyces cerevisiae. Proceedings of the National Academy of Sciences, USA*, **77**, 3912–3916

75. Niederberger, P., Miozzari, G. and Hütter, R. (1981). Biological role of the general control of amino acid biosynthesis in *Saccharomyces cerevisiae. Molecular and Cellular Biology*, **1**, 584–593

76. Oshima, Y. (1982). Regulatory circuits for gene expression: the metabolism galactose and phosphate. In *The Molecular Biology of the Yeast Saccharomyces, Metabolism and Gene Expression*, Cold Spring Harbor Monograph 11B, (Strathern, J. N., Jones, E. W. and Broach, J. R., Eds.), pp. 159–180. Cold Spring Harbor

77. Pall, M. L. (1981). Adenosine 3′,5′-phosphate in fungi. *Microbiological Reviews*, **45**, 462–480

78. Panthier, J.-J., Fournier, P., Heslot, H. and Rambach, A. (1980). Cloned β-galactosidase gene of *Escherichia coli* is expressed in the yeast *Saccharomyces cerevisiae. Current Genetics*, **2**, 109–113

79. Patel, V. B., Schweizer, M., Dykstra, C. C., Kushner, S. R. and Giles, N. H. (1981). Genetic organization and transcriptional regulation in the *qa* gene cluster of *Neurospora crass. Proceedings of the National Academy of Sciences, USA*, **78**, 5783–5787

80. Perlman, D., Halvorson, H. O. and Cannon, L. E. (1982). Presecretory and cytoplasmic invertase polypeptides encoded by distinct mRNAs derived from the same structural gene differ by a signal sequence. *Proceedings of the National Academy of Sciences, USA*, **79**, 781–785

81. Petersen, N. S. and McLaughlin, C. S. (1973). Monocistronic messenger RNA in yeast. *Journal of Molecular Biology*, **81**, 33–45

82. Petes, T. D. (1980). Molecular genetics of yeast. *Annual Review of Biochemistry*, **49**, 845–876

83. Philippides, D. and Scazzocchio, C. (1981). Positive regulation in a eukaryote, a study of the *ua*Y gene of *Aspergillus nidulans*. II. Identification of the effector binding protein. *Molecular and General Genetics*, **181**, 107–115

84. Roggenkamp, R., Kustermann-Kuhn, B. and Hollenberg, C. P. (1981). Expression and processing of bacterial β-lactamase in the yeast *Saccharomyces cerevisiae*. *Proceedings of the National Academy of Sciences, USA*, **78**, 4466–4470

85. Scazzocchio, C. and Arst, H. N. Jr. (1978). The nature of an initiator constitutive mutation in *Aspergillus nidulans*. *Nature, London*, **274**, 177–179

86. Scazzocchio, C., Sdrin, N. and Ong, G. (1982). Positive regulation in a eukaryote, a study of the *ua*Y gene of *Aspergillus nidulans*: I. Characterisation of alleles, dominance and complementation studies, and a fine structure map of the *ua*Y–*oxp*A cluster. *Genetics*, **100**, 185–208

87. Schweizer, E., Werkmeister, K. and Jain, M. K. (1978). Fatty acid biosynthesis in yeast. *Molecular and Cellular Biochemistry*, **21**, 95–107

88. Schweizer, M., Case, M. E., Dykstra, C. C., Giles, N. H. and Kushner, S. R. (1981). Identification and characterization of recombinant plasmids carrying the complete *qa* gene cluster from *Neurospora crassa* including the *qa*-1$^+$ regulatory gene. *Proceedings of the National Academy of Sciences, USA*, **78**, 5086–5090

89. Sherman, F. and Stewart, J. R. (1975). The use of iso-1-cytochrome *c* mutants of yeast for elucidating the sequences that govern initiation of translation. In *Organization and Expression of the Eukaryotic Genome; Biochemical Mechanisms of Differentiation in Prokaryotes and Eukaryotes*, Proceedings of the 10th FEBS Meeting 38, (Bernardi, G. and Gross, F., Eds.), pp. 175–191. Amsterdam, North Holland

90. Sherman, F., McKnight, G. and Stewart, J. W. (1980). AUG is the only initiation codon in eukaryotes. *Biochimica et Biophysica Acta*, **609**, 343–346

91. Sherman, F., Stewart, J. W. and Schweingruber, A. M. (1980). Mutants of yeast initiating translation of iso-1-cytochrome *c* within a region spanning 37 nucleotides. *Cell*, **20**, 215–222

92. Sherman, F., Cardillo, T. S., Errede, B., Friedman, L., McKnight, G. and Stiles, J. I. (1981). Yeast mutants overproducing iso-cytochromes *c*. In *Molecular Genetics In Yeast*, Alfred Benzon Symposium 16, (Von Wettstein, D., Friis, J., Kielland-Brandt, M. and Stenderup, A., Eds.), pp. 154–174. Copenhagen, Munksgaard

93. Sledziewski, A. and Young, E. T. (1982). Chromatin conformational changes accompany transcriptional activation of a glucose-repressed gene in *Saccharomyces cerevisiae*. *Proceedings of the National Academy of Sciences, USA*, **79**, 253–256

94. Smith, M., Leung, D. W., Gillam, S., Astell, C. R., Montgomery, D. L. and Hall, B. D. (1979). Sequence of the gene for iso-1-cytochrome *c* in *Saccharomyces cerevisiae*. *Cell*, **16**, 753–761

95. Standring, D. N., Venegas, A. and Rutter, W. J. (1981). Yeast tRNA$_3^{Leu}$ gene transcribed and spliced in a HeLa cell extract. *Proceedings of the National Academy of Sciences, USA*, **78**, 5963–5967

96. Stevens, T., Esmon, B. and Schekman, R. (1982). Early stages in the yeast secretory pathway are required for transport of carboxypeptidase Y to the vacuole. *Cell*, **30**, 439–448

97. Stiles, J. I., Friedman, L. R., Helms, C., Consaul, S. and Sherman, F. (1981). Transposition of the gene cluster *CYC1–OSM1–RAD7* in yeast. *Journal of Molecular Biology*, **148**, 331–346

98. Stiles, J. I., Szostak, J. W., Young, A. T., Wu, R., Consaul, S. and Sherman, F. (1981). DNA sequence of a mutation in the leader region of the yeast iso-1-cytochrome *c* mRNA. *Cell*, **25**, 277–284

99. St. John, T. P. and Davis, R. W. (1981). The organization and transcription of the galactose gene cluster of *Saccharomyces*. *Journal of Molecular Biology*, **152**, 285–315

100. St. John, T. P., Scherer, S., McDonell, M. W. and Davis, R. W. (1981). Deletion analysis of the *Saccharomyces* GAL gene cluster. Transcription from three promoters. *Journal of Molecular Biology*, **152**, 317–334

101. Strathern, J. N., Jones, E. W. and Broach, J. R. (Eds.) (1981). *The Molecular Biology of the Yeast Saccharomyces, Life Cycle and Inheritance*. Cold Spring Harbor Monograph 11A. Cold Spring Harbor

102. Strathern, J. N., Jones, E. W. and Broach, J. R. (Eds.) (1982). *The Molecular Biology of the Yeast Saccharomyces, Metabolism and Gene Expression*, Cold Spring Harbor Monograph 11B. Cold Spring Harbor

103. Struhl, K. (1981). Deletion mapping a eukaryotic promoter. *Proceedings of the National Academy of Sciences, USA*, **78**, 4461–4465

104. Struhl, K. (1981). Position effects in *Saccharomyces cerevisiae*. *Journal of Molecular Biology*, **152**, 569–575

105. Struhl, K. (1982). Regulatory sites for *his3* gene expression in yeast. *Nature, London*, **300**, 284–287

106. Thuriaux, P., Ramos, F., Piérard, A., Grenson, M. and Wiame, J. M. (1972). Regulation of the carbamoylphosphate synthetase belonging to the arginine biosynthetic pathway of *Saccharomyces cerevisiae*. *Journal of Molecular Biology*, **67**, 277–287

107. Timberlake, W. E. and Barnard, E. C. (1981). Organization of a gene cluster expressed specifically in the asexual spores of *A. nidulans*. *Cell*, **26**, 29–37

108. Tollervey, D. W. and Arst, H. N. Jr. (1981). Mutations to constitutivity and derepression are separate and separable in a regulatory gene of *Aspergillus nidulans*. *Current Genetics*, **4**, 63–68

109. Tollervey, D. W. and Arst, H. N. Jr. (1982). Domain-wide, locus-specific suppression of nitrogen metabolite repressed mutations in *Aspergillus nidulans*. *Current Genetics*, **6**, 79–85

109a. Tomsett, A. B. and Cove, D. J. (1979). Deletion mapping of the *nii*A and *nia*D gene region of *Aspergillus nidulans*. *Genetical Research*, **34**, 19–32

110. Vapnek, D., Hautala, J. A., Jacobson, J. W., Giles, N. H. and Kushner, S. R. (1977). Expression in *Escherichia coli* K-12 of the structural gene for catabolic dehydroquinase of *Neurospora crassa*. *Proceedings of the National Academy of Sciences, USA*, **74**, 3508–3512

111. Vissers, S., Urrestarazu, L. A., Jauniaux, J.-C. and Wiame, J.-M. (1982). Inhibition of ornithine carbamoyltransferase by arginase among yeasts: correlation with energy production, subcellular localization and enzyme synthesis. *Journal of General Microbiology*, **128**, 1235–1247

112. Wallace, R. B., Johnson, P. F., Tanaka, S., Schöld, M., Itakura, K., Abelson, J. (1980). Directed deletion of a yeast transfer RNA intervening sequence. *Science*, **209**, 1396–1400

113. Walz, A., Ratzkin, B. and Carbon, J. (1978). Control of expression of a cloned yeast (*Saccharomyces cerevisiae*) gene (*trp5*) by a bacterial insertion element (IS2). *Proceedings of the National Academy of Sciences, USA*, **75**, 6172–6176

114. Welch, G. R. and Gaertner, F. H. (1980). Enzyme organization in the polyaromatic-biosynthetic pathway: the *arom* conjugate and other multienzyme systems. *Current Topics in Cellular Regulation*, **16**, 113–162

115. Wiame, J. M. (1971). The regulation of arginine metabolism in *Saccharomyces cerevisiae*: exclusion mechanisms. *Current Topics in Cellular Regulation*, **4**, 1–38

116. Williamson, V. M., Young, E. T. and Ciriacy, M. (1981). Transposable elements associated with constitutive expression of yeast alcohol dehydrogenase II. *Cell*, **23**, 605–614

117. Wolf, D. H. (1980). Control of metabolism of yeast and other lower eukaryotes through the action of proteinases. *Advances in Microbial Physiology*, **21**, 267–338

118. Yamamoto, M., Asakura, Y. and Yanagida, M. (1981). Cloning of a gene from the fission yeast *S. pombe* which complements *E. coli pyr*B, the gene for aspartate transcarbamylase. *Molecular and General Genetics*, **182**, 426–429

119. Zamir, A., Maina, C. B., Fink, G. R. and Szalay, A. A. (1981). Stable chromosomal integration of the entire nitrogen fixation gene cluster from *Klebsiella pneumoniae* in yeast. *Proceedings of the National Academy of Sciences, USA*, **78**, 3496–3500

120. Zaret, K. S. and Sherman, F. (1982). DNA sequence required for efficient transcription termination in yeast. *Cell*, **28**, 563–573

Chapter 25

Structure and function of plant genes

R. B. Goldberg

Introduction

During the past few years there has been an explosion of interest in plant molecular biology. This largely stems from the realization that molecular genetic engineering procedures have a tremendous potential for generating improved plant varieties. Plant cells are ideally suited for this purpose since in many cases entire, fertile plants can be obtained from single cells grown in culture[13]. It is not unreasonable to expect that in the not too distant future it will be possible to introduce any gene into a plant cell, and then generate an improved crop from the genetically modified cells. Clearly, any rational program of improving plants by molecular genetic engineering procedures must be based on a knowledge of how genes are organized in plant chromosomes, and how they are expressed during development. Although information on plant gene organization and expression still lags behind that which exists for animal cells, plant molecular biologists are rapidly closing the gap. In this chapter, the limited information which is available on the structure and function of plant genes is reviewed.

Plant genomes

Plant nuclei contain approximately 2×10^5 to 1×10^8 kilobases (kb) of DNA depending upon the species[48, 51]. This range of genome sizes is similar to that observed for animals, and indicates that plant genomes are very complex, and possess an enormous genetic potential. Like other eukaryotes, plant chromosomes contain both repeated and single copy DNA sequences[57]. Although the ratio of these sequences varies from plant to plant, many plant genomes contain a high proportion of repeated sequences[18]. For example, 60% of soybean DNA sequences are reiterated[22, 27a] and almost 80% of those in wheat[19]. Repeated DNA sequences in plant chromosomes vary in complexity and copy number. While it has not been possible to determine the precise number of different repeat families, estimates range from 1000–40 000 depending upon the plant[57].

A large number of experiments, in several laboratories, have shown that repeated and single copy DNA sequences are arranged in precise patterns in plant chromosomes, although these patterns may vary considerably[18, 53, 57]. Most plants

451

studied to date have DNA sequences organized in a short period interspersion pattern similar to that found in many animal genomes[1, 14]. In plants such as tobacco[61] and soybean[22, 27a, 46], short, 0.3 kb repetitive sequences are dispersed at 1–2 kb intervals throughout most of the single copy DNA. Generally, the dispersed repeats constitute 25–50% of the total amount of repetitive DNA, the remainder being organized into long tandem or clustered blocks which exceed 2 kb in length[46]. In contrast to this form of organization, plants such as the mung bean[43a] and pearl millet[59] have genomes organized in a long period interspersion pattern analogous to that found in *Drosophila*[1, 41]. In these genomes the repetitive elements are dispersed at much larger intervals (>5 kb) throughout the single copy DNA.

The exact functional significance, if any, of these organizational patterns is unknown. It may be that interspersion patterns simply reflect a genome which is fluid, and is in the process of change[8]. This view is in accord with studies carried out by Flavell and his colleagues who showed that homologous repetitive sequences in related cereals vary in copy number and chromosomal location[3, 18, 49]. Their findings suggest that repeated sequences in plant chromosomes undergo a cyclical process of amplification and transposition on an evolutionary scale.

Gene expression in plant cells

Irrespective of the exact organizational pattern of plant genomes, DNA excess hybridization studies with cDNAs synthesized from mRNA templates have shown that most diverse plant structural genes are present approximately once per haploid genome[21, 23, 24, 37]. For example, tobacco leaf cDNA hybridizes to excess nuclear DNA with kinetics expected for single copy DNA sequences[24]. That is not to say that all single copy DNA sequences are structural genes, or that all structural genes are single copy. Certainly, many important plant proteins are encoded by repeated gene sequences which are organized into discrete, multigene families[30]. The gene families which encode prevalent proteins of the soybean seed will be discussed below. In terms of the total number of different structural genes expressed in a plant cell, however, repeated structural genes are in the minority.

Estimates of the number of diverse structural genes expressed in plant cells have been obtained from the results of DNA/RNA hybridization experiments with both cDNA and single copy DNA[2, 21, 24, 25, 35, 37]. These studies indicate that approximately 15 000–25 000 structural genes are expressed at the mRNA level, or about 5% of the single copy DNA. These values are considerably higher than those observed for simple eukaryotic organisms such as *Aspergillus*[54] and yeast[29], but are comparable to gene number estimates obtained with typical animal cells[40, 42]. Not all of the genes expressed in plant cells, however, encode mRNAs which are present in the same cellular concentrations. Indeed, plant mRNAs vary considerably in prevalence, and range from as many as 10^5 molecules to as few as 1 molecule/cell[21, 24, 25]. The vast majority of diverse structural genes, however, code for mRNAs present only a few times per cell[24, 35]. These are the rare or complex class mRNAs.

Recent DNA/RNA hybridization experiments which compared the mRNA sequence sets of all organ systems of the tobacco plant have demonstrated that qualitative changes in gene expression programs are required to both maintain and establish distinct differentiated states in higher plants[35]. This point is illustrated by the experiments summarized in *Figure 25.1*. It can be seen that each organ system contains a set of approximately 25 000 diverse mRNAs; however, the sequence composition of each set differs significantly. Most importantly, each organ system contains a subset of mRNAs which are not detectable at the cytoplasmic level in the cells of other organ systems. For example, anther cells contain approximately 11 000 unique mRNAs, ovary 10 000, leaf 7000, stem 6000, and root 7000. Petal is the sole exception, since there is virtually complete identity between its message set and that of the leaf. This is not unexpected, however, since leaf and petal are indistinguishable from a morphogenetic point of view[16]. Only about one-third of each mRNA set is shared by all organ systems. This common subset represents about 8000 diverse mRNAs, or 1.5% of the single copy DNA. Presumably these common mRNAs fulfill 'housekeeping' functions which are required by all cells. Clearly though, each unique morphogenetic state in a higher plant is associated with a developmentally-specific subset of mRNAs.

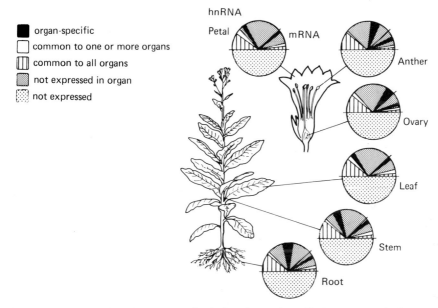

Figure 25.1. Regulation of gene expression in the tobacco plant. Circles represent single copy DNA sequences in the tobacco genome. Each comprises 40% of the genome, or 6.4 $\times 10^5$ kb of diverse sequences[61]. hnRNA represents the fraction of single copy DNA expressed as nuclear RNA, and mRNA represents the single copy DNA fraction expressed as mRNA[34, 35, 36]. Colors depicting each single copy DNA sequence subset were taken from the results of DNA/RNA hybridization experiments which compared the nuclear RNA and mRNA sequence sets of each organ system[34, 35, 36]. The mRNA sector, or single copy DNA coding regions, represents 12% of the single copy DNA or 5% of the genome. The hnRNA + mRNA sectors comprise 50% of the single copy DNA or 20% of the genome.

From the studies summarized in *Figure 25.1* it is possible to make a limit estimate of the total number of structural genes which are expressed in the entire plant. This estimate suggests that about 60 000 different structural genes, or 12% of the single copy DNA, is required to program and maintain the vegetative and reproductive states of a higher plant. This is certainly a minimum estimate, since genes encoding very rare mRNAs, or mRNAs present in rare cell types of complex organ systems, would have gone undetected by the hybridization procedures used. However, it is clear that a large amount of genetic information is required to program plant differentiation, and that plant gene expression is strikingly regulated.

Regulation of plant gene expression

Although about 5% of the single copy DNA in a plant cell is represented in the mRNA population, a larger fraction is represented in the nuclear RNA. A series of hybridization experiments between single copy DNA and nuclear RNA from different tobacco organ systems are summarized in *Figure 25.1*[34, 36]. These studies show that about 20% of the single copy DNA is represented in the nuclear RNA of each organ system, or about four times that which is represented in the mRNA. Thus, only one-fourth of the diverse RNA sequences in the nucleus of a plant cell are exported and become associated with polysomes. This indicates that post-transcriptional selection events must occur in order to generate each unique mRNA set.

Are post-transcriptional selection processes sufficient, however, to generate the developmentally-specific mRNA populations found in different plant cells? Two extreme models can be envisioned to explain the striking regulation of plant gene expression. The post-transcriptional model states that all structural genes of the plant are constitutively expressed at the nuclear RNA level, and that post-transcriptional selection events operate to give rise to each organ-specific mRNA population. On the other hand, the transcriptional model states that genes which are uniquely expressed in a given organ system (e.g. anther-specific genes) are repressed in other organ systems. Each model leads to testable predictions. The post-transcriptional model predicts that the nucleus of each plant cell, irrespective of differentiated state, will contain the entire mRNA collection of the plant (i.e. 60 000 diverse messages). This is not unreasonable since 12% of the single copy DNA in the tobacco plant consists of coding sequences expressed during the entire life cycle, and 20% of the single copy DNA is represented in the nuclear RNA of each organ system[34, 36]. The transcriptional model, however, predicts that organ-specific mRNAs will be found only in the nuclear RNA population of that organ system, and no other. In fact, both models are correct.

Figure 25.1 shows that the tobacco stem contains approximately 7000 mRNAs which are not detectable in the cytoplasm of other organ systems. Recently, Kamalay and Goldberg[35] showed that virtually all of these stem-specific mRNAs are present in the leaf nuclear RNA population, supporting the proposition that post-transcriptional processes play an important role in the regulation of plant gene expression. Experiments summarized in *Figure 25.1*, however, indicate that transcriptional control processes are important as well. Comparisons of the nuclear RNA sequence sets of all tobacco organ systems have demonstrated that each organ system has a nuclear RNA subset which is developmentally regulated, and is detectable only in the nuclear RNA of that organ system[34, 36]. This finding is contrary to what is expected if post-transcriptional processes are solely responsible for controlling the composition of each mRNA population (unless, of course, the organ-specific nuclear RNAs are not structural gene transcripts). What are the organ-specific nuclear RNAs? Kamalay and Goldberg[36] determined that these RNA species consist primarily of structural gene transcripts. For example, the 11 000 anther-specific mRNAs are found only in the anther nuclear RNA population, and are undetectable in the nuclear RNA of other organ systems. This is the expected result for structural genes which are under transcriptional control. Thus, both transcriptional *and* post-transcriptional processes play important roles in the regulation of plant gene expression.

The hybridization studies described above, and summarized in *Figure 25.1*, permit a limit estimate to be made of the total fraction of single copy DNA which is expressed in the nuclear RNA of the entire tobacco plant. This estimate indicates that at least 50% of the single copy DNA, or 3×10^5 kb of diverse sequences, is transcribed. While this is certainly a minimum estimate, it indicates that an enormous amount of genetic information is expressed in a higher plant, and that complex regulatory processes must operate to enable this information to be expressed in different cell types. Since 12% of the tobacco single copy DNA is expressed at the mRNA level, and 50% is expressed in the nuclear RNA, what is the additional nuclear RNA sequence complexity? Most likely it consists of intervening sequence transcripts (see below), transcripts which extend downstream from the 3′ end of structural genes, and transcripts whose function and relationship to structural gene sequences are not yet understood.

Structure and function of specific plant genes

In order to unravel the complex processes which control plant gene expression, it is necessary to study specific plant genes. During the past few years recombinant DNA procedures have been used by several laboratories to investigate the organization and expression of specific plant genes. *Table 25.1* presents a non-exhaustive list of plant genes for which genomic and/or cDNA clones now exist. All of these genes were chosen for investigation because they encode prevalent products which can be isolated and identified, and because they are involved in important cellular processes; for example, light control of gene expression, hormone induction of gene activity, and regulation of gene expression in development. Recently, structural genes which code for rare plant mRNAs have been identified on cloned genomic segments[17]. It is not yet known, however, what

TABLE 25.1. Molecular characteristics of cloned plant genes

Gene	Gene length (kb)	mRNA length (kb)	Number of introns	Gene/mRNA length ratio[a]	Reference
SOYBEAN					
Glycinin	2.8	2.1	1	1.3	17
β-Conglycinin (α, α′-type)	3.1	2.5	5	1.2	Beachy, R. N., unpublished results
Lectin	1.1	1.1	0	1.0	Goldberg, R. B. and Vodkin, L. O., unpublished results
Kunitz trypsin inhibitor	0.9	0.9	0	1.0	Jofuku, D. O. and Goldberg, R. B., unpublished results
15 kd protein	0.85	0.85	0	1.0	17
Actin	1.5	1.2	3	1.3	50
Leghemoglobin *c*	1.3	0.7	3	1.9	7, 31
Leghemoglobin *a*	1.6	0.7	3	2.3	31
Ribulose bisphosphate carboxylase	1.4	0.95	2	1.5	5
FRENCH BEAN					
Phaseolin (β-type)	2.0	1.5	5	1.3	Slightom, J. and Hall, T. C., unpublished results
CORN					
Zein	0.75	0.75	0	1.0	45

[a] The average gene/mRNA length ratio for these genes is 1.35. This ratio compares to 5.1 for that of fifteen randomly selected animal genes including: globin, ovalbumin, ovomucoid, lysozyme, vitellogenin, serum albumin, dihydrofolate reductase, collagen, insulin, amylase, conalbumin and interferon[40].

proteins these genes encode, or to which cellular processes they contribute. In the section below, one example will be presented to show how recombinant DNA procedures have been used to study the expression of specific genes in plant development.

Soybean seed protein gene expression

The development of soybean embryos is shown in *Figure 25.2*. Four periods can be identified in which distinct morphogenetic and physiological events occur[56]. During the first period cellular differentiation and morphogenesis occur resulting in the formation of all cell, tissue, and organ types of the embryo. The second period is

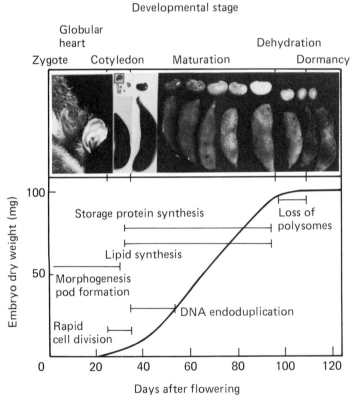

Figure 25.2. Development of soybean embryos. Adapted from Goldberg *et al*[25]. and Dure[15].

one of intense cell division, and results in the formation of ~99% of the 3×10^6 cells found in the mature embryo. The third period, called maturation, marks a dramatic change in the events which occur during embryogenesis. In this period, cell division has stopped, the embryo expands to its maximum size, and massive amounts of seed proteins accumulate. These proteins include storage globulins, lectins, and trypsin inhibitors. The storage globulins, which comprise ~50% of the seed protein mass, are used as a 'food source' by the plant at seed germination. Thus, during the maturation period a small number of genes direct the synthesis of a massive amount of gene product which will be used at a later stage of development. During the final period of embryogenesis dehydration occurs, cellular processes become quiescent, and the embryo enters a dormant period as a mature, dry seed.

cDNAs for a number of soybean seed protein mRNAs have been inserted into bacterial plasmids[23]. These cDNA clones represent mRNAs for the major storage protein glycinin, the minor storage protein β-conglycinin, the Kunitz trypsin inhibitor, lectin, and an unidentified, 15 500 dalton polypeptide. A number of questions regarding seed protein gene expression can be answered using these cDNA clones. First, what is the temporal program of seed protein gene expression during development? *Figure 25.3* summarizes the results of titration studies in which the cloned seed protein cDNAs were used to quantitate the levels of each seed protein mRNA during development[23]. Clearly, seed protein gene expression

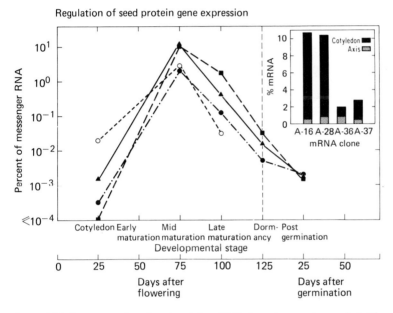

Figure 25.3. Representation of seed protein mRNAs in soybean embryos. ▲ A-16, β-conglycinin; ■ A-28, glycinin; ● A-36, 15 kd protein; ○ A-37 Kunitz trypsin inhibitor. Adapted from Goldberg *et al.*[23]

is strikingly regulated. Seed protein mRNAs accumulate during early maturation and then decay during late maturation when the dormancy program sets in. The magnitude of these prevalence changes varies from 10^2 to 10^5 depending upon the message. In addition, the accumulation and diminution of seed protein mRNAs parallel the accumulation of seed proteins, indicating that the relative seed protein synthesis rates are controlled primarily by the availability of seed protein mRNAs.

Is seed protein gene expression regulated with respect to embryonic cell type as well? *Figure 25.2* shows that soybean embryos contain two organ systems, the axis and the cotyledons. The axis gives rise to the root, stem, and leaves of the plant after germination. The cotyledons are responsible primarily for the synthesis and storage of seed proteins during embryogeny, and senesce after seed germination.

Thus, these organ systems have very different developmental fates. The inset to *Figure 25.3* summarizes the results of experiments which compared seed protein mRNA prevalences in the axis and cotyledons[23]. As seen in *Figure 25.3* the concentration of each seed protein mRNA is quite different in the cotyledon and axis. While the four mRNAs studied comprised 25% of the cotyledon mRNA mass, they represented only 3% of the axis mRNA mass. Thus, not only is seed protein gene expression regulated temporally in development, but it is regulated with respect to embryonic cell type as well.

Is seed protein gene expression regulated primarily at the transcriptional or post-transcriptional level? This question was answered by investigating seed protein gene expression in the leaves of the mature plant[23]. Titration studies similar to those summarized in *Figure 25.3* demonstrated that the leaf cytoplasm does not contain seed protein mRNAs. This result indicates that seed protein gene expression is restricted to a specific time of the soybean life cycle, the period of embryo development. If seed protein gene expression is regulated post-transcriptionally, as outlined above for a large number of tobacco stem-specific genes[35,36], then seed protein gene transcripts should be found in leaf nuclear RNA. Alternatively, if seed protein gene expression is regulated at the transcriptional level, then seed protein gene transcripts should be undetectable in leaf nuclear RNA. Hybridization studies between leaf nuclear RNA and cloned seed protein cDNA probes failed to detect seed protein gene transcripts in leaf nuclear RNA, indicating that seed protein gene expression is regulated primarily at the transcriptional level[23].

Soybean seed protein gene structure

Clearly, from the above discussion it is evident that seed protein genes are under striking developmental control. How are seed protein genes organized in soybean chromosomes, and how do seed protein gene structures compare with those of analogous animal genes characterized to date? Hybridization studies with labeled seed protein cDNA probes and soybean chromosomal DNA indicated that seed protein genes are repeated approximately two (e.g. lectin genes) to five times (e.g. β-conglycinin genes) per haploid genome[23,25] (Goldberg *et al.*, unpublished results). These results indicate that seed protein genes are organized into small, discrete multigene families, and that they differ from the vast majority of structural genes expressed in a plant which are single copy.

Cloned genomic fragments have been used to investigate the structure and arrangement of seed protein genes in soybean chromosomes[17] (Goldberg *et al.*, unpublished results). Electron micrographs of a number of soybean seed protein genes are shown in *Figure 25.4*. These structures represent R loops formed between seed protein genes and embryo mRNA[33]. Two important conclusions can be obtained from the electron microscope data. First, an intervening sequence is present in the glycinin gene indicating that plant genes, like those of other eukaryotes, possess introns. Secondly, several seed protein genes do not possess introns which are visible in the electron microscope (e.g lectin and Kunitz trypsin inhibitor genes). S1 nuclease protection studies[4] have shown that there are no introns in the lectin or 15 kd protein genes. Although it is possible that the other genes contain introns smaller than those detectable in the electron microscope (<0.10 kb), these findings indicate that seed protein genes have relatively simple intragenic structures.

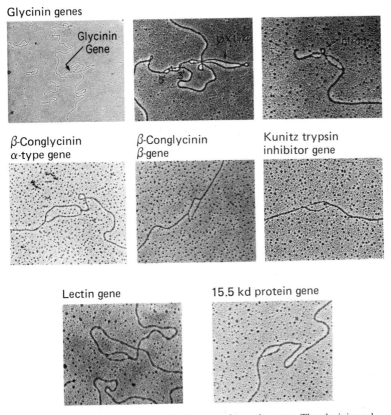

Figure 25.4. Intragenic structures of soybean seed protein genes. The glycinin and 15 kd protein R loops were taken from Fischer and Goldberg[17]. The lectin, Kunitz trypsin inhibitor, and β-conglycinin R loops are unpublished data of Goldberg *et al*.

How do soybean seed protein gene structures compare with those of other plant genes studied to date? *Table 25.1* shows that many plant genes have either no introns (e.g. lectin and zein), or a small number of short introns. Indeed, the largest total gene to coding length ratio observed thus far is 2.3 for the leghemaglobin *a* gene[31]. This result differs dramatically from what is observed for animal genes. The average total gene to coding length ratio of 15 randomly selected animal genes is about five, indicating that on the average <20% of an animal gene is represented in its mRNA (*Table 25.1*). A specific example illustrates this point. The major egg storage protein genes of the toad and chicken, vitellogenin and ovalbumin, possess thirty-three[55] and seven[28] introns each, respectively. In addition, these introns are large, and greatly exceed the size of coding regions. On the other hand, the major seed storage protein gene of soybean, glycinin, contains only one, moderate-sized intron[17] and that of the french bean *Phaseolus* only five, small introns[52]. Clearly, the plant genes investigated thus far have simpler structures than most animal genes characterized to date.

Do all plant genes have simple intragenic structures? From the limited data presented in *Table 25.1* no generalizations can be made. However, since plant nuclear RNA contains about four times the mRNA sequence complexity[24, 34], it is not unreasonable to expect that most plant genes will have more complex structures than the genes described in *Table 25.1* and shown in *Figure 25.4*, particularly those which code for rare plant mRNAs which are responsible for most of the mRNA sequence complexity[24]. This expectation has been partially realized by the structure of a soybean gene which codes for a leaf mRNA which comprises only $8 \times 10^{-5}\%$ of the leaf mRNA mass, or about 1 molecule/cell[17]. This gene possesses three introns and four exons. More importantly, however, the total gene to coding ratio is about three, indicating that most of the gene represents non-coding, intervening sequences. Whether other plant genes have complex structures, remains to be established.

Putative controlling sequences within and contiguous to plant genes

Analysis of the primary structures of a large number of animal genes has revealed several highly conserved DNA sequences which have been postulated to function in transcription, and in post-transcriptional processing and modification of primary transcripts. These sequences, and their putative functions, are portrayed in *Figure 25.5*. Do plant genes possess analogous sequences? During the past year a number

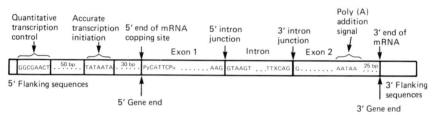

Figure 25.5. Highly conserved DNA sequences within and surrounding plant and animal genes. Adapted from Lewin[40]. The intron/exon junction sequences were compiled by Mount[43], the poly(A) addition signal was hypothesized by Proudfoot and Brownlee[47], and the functional significance of the quantitative control and accurate initiation 'boxes' was established by Grosveld *et al.*[27]

of plant genes have been sequenced in their entirety. These include: the leghemoglobin[7,31], actin[50] and ribulose biphosphate carboxylase[5] genes of soybean; the zein genes of corn[45]; and the phaseolin gene of the french bean[52]. Each of these genes has been found to contain conserved regions analogous to those present in animals. For example, leghemoglobin genes possess a TATAAA sequence 30 bp upstream from the 5′ end of the gene, a CCAAT sequence 90 bp upstream from the 5′ end of the gene, an AATAAA sequence 30 bp upstream from the 3′ gene end, and introns which obey the GT/AG rule shown in *Figure 25.5*[7,31]. Similarly, zein genes possess a TATA sequence 33 bp 5′ of the cap site, a CCAT sequence 117 by 5′ of the cap site, and an AATAAA sequence 24 bp after the stop codon terminating the polypeptide[45]. These data indicate that sequences required for initiation and modulation of transcription[27], splicing of primary transcripts[9], and perhaps poly(A) addition to primary transcripts[47] have been highly conserved within plant and animal genomes.

Future prospects

The central question is, of course, what are the sequences responsible for the cell-specific expression of plant genes? Certainly, they cannot be those described in *Figure 25.5* since these sequences are common to all eukaryotic genes. Clues to the types of sequences involved in the developmental regulation of eukaryotic gene expression have been obtained recently from transformation studies with several animal genes[6, 12, 32, 38, 39]. In each case a putative regulatory sequence responsible for the induction of gene expression was found to reside on the transformed genomic insert, and in some cases was shown to be within 300 bp of the 5' gene end (Corces *et al.*, unpublished data). In addition, these studies suggested that the regulatory sequences are short (<15 bp), and are in the proximity of the promoter region (i.e. the CCATT and ATA boxes).

At the present time (March 1982) analogous studies have not been carried out with isolated plant genes. Indeed, a generalized transformation procedure for introducing cloned genes into plant cells has not yet been established. Although several approaches are being tested (e.g. microinjection[10]; liposomes[20]; chromosome isolation[26]), the most promising utilize the Ti and Ri plasmids of *Agrobacterium* bacteria[11, 44, 58, 60]. These plasmids incite tumors on many plant species, and in the process transfer a portion of their DNA to plant chromosomes. The transferred DNA, or T-DNA, is then passed on from one generation to the next[44]. Clearly, it will be essential in the near future to utilize transformation experiments in order to identify those sequences responsible for the developmental-specific expression of plant genes. A knowledge of the regulatory circuitry controlling plant gene expression, coupled with the ability to introduce genes into plant cells and generate genetically modified plants, should provide exciting prospects for engineering new plant species in the not too distant future.

References

1. Angerer, R. C. and Hough-Evans, B. R. (1977). Sequence organization of eukaryotic DNA. In *Receptors and Hormone Action*, Volume 1, (O'Malley, B. W. and Birnbaumer, L., Eds.), pp. 1–30. New York, Academic Press
2. Auger, S., Baulcombe, D. and Verma, D. P. S. (1979). Sequence complexities of the poly(A)-containing mRNA in the uninfected root and the nodule tissue developed due to the infection by *Rhizobium*. *Biochimica et Biophysica Acta*, **363**, 496–507
3. Bedbrook, J. R., Jones, J., O'Dell, M., Thompson, R. D. and Flavell, R. B. (1980). A molecular description of telomeric heterochromatin in *Secale* species. *Cell*, **19**, 545–560
4. Berk, A. J. and Sharp, P. A. (1977). Sizing and mapping of early adenovirus mRNAs by gel electrophoresis of S1 endonuclease digested hybrids. *Cell*, **12**, 721–732
5. Berry-Lowe, S. L., Shah, D. M. and Meagher, R. B. (1982). The nucleotide sequence, expression, and evolution of one member of a multigene family encoding the small subunit of ribulose-1,5-biphosphate carboxylase in soybean. *Journal of Molecular and Applied Genetics*, **1**, 483–498
6. Birnster, R. L., Chen, H. Y., Thrumbauer, M., Senear, A. W., Warren, R. and Palmiter, R. D. (1981). Somatic expression of herpes thymidine kinase in mice following injection of a fusion gene into eggs. *Cell*, **27**, 223–231
7. Brisson, N. and Verma, D. P. S. (1982). Soybean leghemaglobin gene family: normal, pseudo, and truncated genes. *Proceedings of the National Academy of Sciences, USA*, **79**, 3193–3197
8. Britten, R. J. and Kohne, D. E. (1968). Repeated sequences in DNA. *Science*, **161**, 529–540
9. Busslinger, M., Moschonas, N. and Flavell, R. A. (1981). β+ Thallassemia: aberrant splicing results from a single point mutation in an intron. *Cell*, **27**, 289–298
10. Capecchi, M. R. (1980). High efficiency transformation by direct microinjection of DNA into cultured mammalian cells. *Cell*, **22**, 479–488

11. Chilton, M. D., Tepfer, D. A., Petit, A., David, C., Casse-Delbart, F. and J. Tempe (1982). *Agrobacterium rhizogenes* inserts T-DNA into the genomes of the host plant root cells. *Nature, London*, **295**, 432–434
12. Corces, V., Pellicer, A., Axel, R. and Meselson, M. (1981). Integration, transcription and control of a *Drosophila* heat shock gene in mouse cells. *Proceedings of the National Academy of Sciences, USA*, **78**, 7038–7042
13. Cove, D. J. (1979). The uses of isolated protoplasts in plant genetics. *Heredity*, **43**, 295–314
14. Davidson, E. H., Galau, G. A., Angerer, R. C. and Britten, R. J. (1975). Comparative aspects of DNA organization in metazoa. *Chromosoma*, **51**, 253–259
15. Dure, L. S. (1975). Seed formation. *Annual Review of Plant Physiology*, **26**, 259–278
16. Esau, K. (1977). *Anatomy of Seed Plants*. New York, John Wiley and Sons
17. Fischer, R. L. and Goldberg, R. B. (1982). Structure and flanking regions of soybean seed protein genes. *Cell*, **29**, 651–660
18. Flavell, R. (1980). The molecular characterization and organization of plant chromosomal DNA sequences. *Annual Review of Plant Physiology*, **31**, 569–596
19. Flavell, R. B. and Smith, D. B. (1976). Nucleotide sequence organization in the wheat genome. *Heredity*, **37**, 231–254
20. Fraley, R. T., Dellaporta, S. L. and Papahadjopoulos, D. (1982). Liposome-mediated delivery of tobacco mosaic virus RNA into tobacco protoplasts: A sensitive assay for monitoring liposome–protoplast interactions. *Proceedings of the National Academy of Sciences, USA*, **79**, 1859–1863
21. Galau, G. A. and Dure, L. (1981). Developmental biochemistry of cotton embryogenesis and germination: changing messenger RNA populations as shown by reciprocal heterologous cDNA–mRNA hybridization. *Biochemistry*, **20**, 4169–4178
22. Goldberg, R. B. (1978). DNA sequence organization in the soybean plant. *Biochemical Genetics*, **16**, 45–68
23. Goldberg, R. B., Hoschek, G., Ditta, G. S. and Breidenbach, R. W. (1981). Developmental regulation of cloned superabundant embryo mRNAs in soybean. *Developmental Biology*, **83**, 218–231
24. Goldberg, R. B., Hoschek, G., Kamalay, J. C. and Timberlake, W. E. (1978). Sequence complexity of nuclear and polysomal RNA in leaves of the tobacco plant. *Cell*, **14**, 123–131
25. Goldberg, R. B., Hoschek, G., Tam, S. H., Ditta, G. S. and Breidenbach, R. W. (1981). Abundance, diversity and regulation of mRNA sequence sets in soybean embryogenesis. *Developmental Biology*, **82**, 201–217
26. Griesbach, R. J., Malmberg, R. L. and Carlson, P. S. (1982). An improved technique for the isolation of higher plant chromosomes. *Plant Science Letters*, **24**, 55–60
27. Grosveld, G. C., de Boer, E., Shewmaker, C. K. and Flavell, R. A. (1982). DNA sequences necessary for transcription of the rabbit β-globin genes *in vivo*. *Nature, London*, **295**, 120–126
27a. Gurley, W. B., Hepburn A. G. and Key J. L. (1979). Sequence organization and the soybean genome. *Biochimica et Biophysica Acta*, **561**, 167–183
28. Helig, R., Perrin, F., Ganon, F., Mandel, J. L. and Chambon, P. (1980). The ovalbumin family: structure of the X gene and evolution of duplicated split genes. *Cell*, **20**, 625–637
29. Hereford, L. M. and Rosbash, M. (1977). Number and distribution of poly(A) RNA sequences in yeast. *Cell*, **10**, 453–462
30. Hood, L., Campbell, J. H. and Elgin, S. C. R. (1975). The organization, expression, evolution of antibody genes and other multigene families. *Annual Review of Genetics*, **9**, 305–354
31. Hyldig-Nielsen, J. J., Jensen, E. O., Wiborg, O., Garrett, R., Jorgensen, P. and Marcker, K. A. (1982). The primary structures of two leghemoglobin genes. *Nucleic Acids Research*, **10**, 689–701
32. Hynes, N. E., Kennedy, N., Rahmsdorf, U. and Groner, B. (1981). Hormone-responsive expression of an endogenous proviral gene of mouse mammary tumor virus after molecular cloning and gene transfer into cultured cells. *Proceedings of the National Academy of Sciences, USA*, **78**, 2038–2042
33. Kaback, D. B., Angerer, L. M. and Davidson, N. (1979). Improved methods for the formation and stabilization of R-loops. *Nucleic Acids Research*, **6**, 2499–2517
34. Kamalay, J. C. (1981). Regulation of structural gene expression in tobacco. PhD Thesis, University of California, Los Angeles
35. Kamalay, J. C. and Goldberg, R. B. (1980). Regulation of structural gene expression in tobacco. *Cell*, **19**, 935–946
36. Kamalay, J. C. and Goldberg, R. B. (1983). Regulation of nuclear RNA sequence sets in the tobacco plant. Submitted for publication
37. Kiper, M., Bartels, D., Herzfeld, F. and Richter, G. (1979). The expression of a plant genome in hnRNA and mRNA. *Nucleic Acids Research*, **6**, 1961–1978

38. Kurtz, D. (1981). Hormonal inducibility of rat α-2u globulin genes in transfected mouse cells. *Nature, London*, **291**, 629–631

39. Lee, F., Mulligan, R., Berg, P. and Ringold, G. (1981). Glucosteroids regulate expression of dihydrofolate reductase cDNA in mouse mammary tumour virus chimaeric plasmids. *Nature, London*, **284**, 228–232

40. Lewin, B. (1980). *Gene Expression Volume 2, Eukaryotic Chromosomes*. New York, John Wiley and Sons

41. Manning, J. E., Schmid, C. W. and Davidson, N. (1975). Interspersion of repetitive and non-repetitive DNA sequences in the *Drosophila melanogaster* genome. *Cell*, **4**, 141–155

42. Minty, A. J. and Birnie, G. D. (1981). Messenger RNA populations in eukaryotic cells: evidence from recent nucleic acid hybridization experiments bearing on the extent and control of differential gene expression. In *Biochemistry of Cellular Regulation, Volume 3*, (Clemens, M. J., Ed.), pp. 43–82. Boca Raton, CRC Press

43. Mount, S. M. (1982). A catalog of splice junction sequencs. *Nucleic Acids Research*, **10**, 459–472

43a. Murray, M. G., Palmer, J. D., Cuellar, R. E. and Thompson, W. F. (1979). DNA sequence organization in the many bean genome. *Biochemistry*, **18**, 5259–5266

44. Otten, L., DeGreve, H., Hernalsteens, J. P., Van Montagu, M., Schieder, O., Straub, J. and Schell, J. (1981). Mendelian transmission of genes introduced into plants by the Ti plasmids of *Agrobacterium tumefaciens*. *Molecular and General Genetics*, **183**, 209–211

45. Pederson, K., Devereux, J., Wilson, D. R., Sheldon, E. and Larkins, B. A. (1982). Cloning and sequence analysis reveals structural variation among related zein genes in maize. *Cell*, 29, 1015–1026

46. Pellegrini, M. and Goldberg, R. B. (1979). DNA sequence organization in soybean investigated by electron microscopy. *Chromosoma*, **75**, 309–326

47. Proudfoot, N. J. and Brownlee, G. G. (1976). 3′ noncoding region sequences in eukaryotic mRNA. *Nature, London*, **263**, 211–214

48. Rees, H. and Jones, R. N. (1972). The origin of wide species variation in nuclear DNA content. *International Review of Cytology*, **32**, 53–92

49. Rimpau, J., Smith, D. and Flavell, R. (1978). Sequence organization analysis of the wheat and rye genomes by interspecies DNA/DNA hybridzation. *Journal of Molecular Biology*, **129**, 327–359

50. Shah, D. M., Hightower, R. C. and Meagher, R. B. (1982). Complete nucleotide sequence of a soybean actin gene. *Proceedings of the National Academy of Sciences, USA*, **79**, 1022–1026

51. Sparrow, A. H. and Nauman, A. F. (1976). Evolution of genome size by DNA doublings. *Science*, **192**, 524–529

52. Sun, S. M., Slightom, J. L. and Hall, T. C. (1980). Intervening sequences in a plant gene: comparison of the partial sequence of cDNA and genomic DNA of french bean *Phaseolus*. *Nature, London*, **289**, 37–41

53. Thompson, W. F. and Murray, M. G. (1981). The nuclear genome: structure and function. In *The Biochemistry of Plants*, Volume 6, (Marcus, A., Ed.), pp. 1–81. New York, Academic Press

54. Timberlake, W. E. (1980). Developmental gene regulation in *Aspergillus nidulans*. *Developmental Biology*, **78**, 497–510

55. Wahli, W., David, I. B., Ryffel, G. U. and Weber, R. (1981). Vitellogenesis and the vitellogenin gene family. *Science*, **212**, 298–304

56. Walbot, V. (1978). Control mechanisms in plant embryogeny. In *Dormancy and Developmental Arrest* (Clutter, M., Ed.), pp. 113–166. New York, Academic Press

57. Walbot, V. and Goldberg, R. B. (1979). Plant genome organization and its relationship to classical plant genetics. In *Nucleic Acids in Plants*, Volume 1, (Hall, T. C. and Davies, J. W., Eds.), pp. 3–40. Boca Raton, CRC Press

58. White, F. F., Ghidossi, G., Gordon, M. P. and Nester E. W. (1982). Tumor induction by *Agrobacterium rhizogenes* involves the transfer of plasmid DNA to the plant genome. *Proceedings of the National Academy of Sciences, USA*, **79**, 3193–3197

59. Wimpee, C. F. and Rawson, J. R. Y. (1979). Characterization of the nuclear genome of pearl millet. *Biochimica et Biophysica Acta*, **562**, 192–206

60. Zambryski, P., Holster, M., Kruger, K., Depicker, A., Schell, J., Von Montagu, M. and Goodman, H. M. (1980). Tumor DNA structure in plant cells transformed by *A. tumefaciens*. *Science*, **209**, 1385–1391

61. Zimmerman, J. L. and Goldberg, R. B. (1977). DNA sequence organization in the genome of *Nicotiana tabacum*. *Chromosoma*, **59**, 227–252

Index

465